D0932458

Communications and Control Engineering Series

Editors: A. Fettweis · J. L. Massey · M. Thoma

Jürgen Ackermann

Sampled-Data Control Systems

Analysis and Synthesis, Robust System Design

With 152 Figures

Springer-Verlag
Berlin Heidelberg New York Tokyo

Dr. J. ACKERMANN
DFVLR – Deutsche Forschungs- und Versuchsanstalt
für Luft- und Raumfahrt e.V.
Oberpfaffenhofen
8031 Weßling

ISBN 3-540-15610-0 Springer-Verlag Berlin Heidelberg New York Tokyo
ISBN 0-387-15610-0 Springer-Verlag New York Heidelberg Berlin Tokyo

Library of Congress Cataloging in Publication Data.
Ackermann, Jürgen:
Sampled data control systems:
analysis and synthesis, robust system design/J. Ackermann.
Berlin; Heidelberg; New York; Tokyo: Springer, 1985.
(Communications and control engineering series)

Printed in Germany

Offsetprinting: Mercedes-Druck, Berlin
Bookbinding: Lüderitz & Bauer, Berlin
2161/3020 5 4 3 2 1 0

Preface

The first German edition of this book appeared in 1972, and in Polish translation in 1976. It covered the analysis and synthesis of sampled-data systems. The second German edition of 1983 extended the scope to design, in particular design for robustness of control system properties with respect to uncertainty of plant parameters. This book is a revised translation of the second German edition. The revisions concern primarily a new treatment of the finite effect sequences and the use of nice numerical properties of Hessenberg forms.

The introduction describes examples of sampled-data systems, in particular digital controllers, and analyzes the sampler and hold; also some design aspects are introduced. Chapter 2 reviews the modelling and analysis of continuous systems. Pole shifting is formulated as an affine mapping, here some new material on fixing some eigenvalues or some gains in a design step is included.

Chapter 3 treats the analysis of sampled-data systems by state-space and z-transform methods. This includes sections on inter-sampling behavior, time-delay systems, absolute stability and non-synchronous sampling. Chapter 4 treats controllability and reachability of discrete-time systems, controllability regions for constrained inputs and the choice of the sampling interval primarily under controllability aspects. Chapter 5 deals with observability and constructability both from the discrete and continuous plant output. Full and reduced order observers are treated as well as disturbance observers.

The ingredients of chapters 3 to 5 are put together in chapter 6 on control-loop synthesis. Controller structures and the separation theorem are treated first from a state-space point of view, then they are translated into a synthesis procedure by polynomial equations. Further topics are cancellations, prefilters and disturbance compensation.

The remaining chapters of the book provide some design tools and illustrate their use. It is certainly not possible to cover all aspects of control systems design with many - possibly conflicting - design goals and debatable formulations and bounds of design specifications. A method is particularly helpful if it gives the designer insight, which of his specifications are conflicting, such that he learns about the possible tradeoffs during the design process and does not have to decide in advance about his preferences. Two such methods are parameter-space design and optimization of vector performance indices. Both methods are treated with particular emphasis on robustness of desirable system properties with respect to uncertain physical parameters in a known plant model structure. The selection of the material is of course subjective, following the author's own line of research. Therefore the character of the book changes more to a monograph. Nevertheless the style is still that of a textbook and I hope that also these chapters will be used in graduate courses and research at universities.

Chapter 7 gives the theoretical foundations for pole region assignment, i.e. a description of the set of all gains in an assumed feedback structure which place the eigenvalues into a prescribed region in the complex plane. In chapter 8 the robustness problem for large plant parameter uncertainty is formulated as a multi-model problem. It is solved by simultaneous pole region assignment for a family of plant models. The possible tradeoffs in selecting a controller from the admissible set are formalized. This approach is illustrated by a detailed design study for the robust stabilization of an aircraft. The Pareto optimization of a vector of performance indices is used for introducing additional design specifications in a systematic tradeoff procedure.

In chapter 9 on multivariable systems only a few topics are selected and presented in a new form. The theory of finite-effect sequences is first developed from an open-loop point of view.

This exhibits the controllability structure, the input normalization and the effects of closing the loop in a nicely parameterized form. A section on quadratic optimal control concludes the chapter.

Appendix A compiles results on canonical forms from the literature. The nice numerical properties of Hessenberg forms are exploited and further results from matrix theory are collected. Appendix B treats the z-transform. Appendix C reviews stability criteria for linear discrete-time systems. Appendix D describes two application examples: aircraft stabilization and track-guided bus.

The literature is referenced by years of publication, e.g. [49.2] is the second reference in the section 1949 of the list. I have tried to give reference to original work, the reader may excuse that I have succeeded only partially in this respect. For results from other fields I have given reference to easily accessible and well readable publications.

The remarks supplement the main text but are not prerequisites for understanding the main text. The beginner better skips them.

Already the first edition of this book evolved from an annual short course in continuing education offered by the Carl-Cranz-Gesellschaft. The discussions in these courses, in particular discussions with practising engineers, have contributed much to the selection, arrangement and style of presentation of the material.

The book may be used in university courses at different levels. Examples are:

i) Introductory course on z-transform and sampled-data systems: Appendix B, chapters 1 and 3, selected material from chapters 2, 4 and 5.
ii) Analysis and synthesis of sampled-data systems: chapters 1 to 6.
iii) Design of sampled-data systems: Selected topics from chapters 3 to 5, section 2.6, chapters 7 and 8, appendix D.
iv) Advanced course on robust control: Section 2.6, chapters 7 to 9, sections A.4 and A.5.

Acknowledgements

I like to thank my teachers, co-workers and reviewers of this book. W. Oppelt has motivated me to study sampled-data control systems. As a student of E. Jury I have learned the classical theory of sampled-data systems. G. Knausenberger paved the way for my studies in Berkeley and initiated the continuing education courses mentioned above. He also encouraged my research on control system synthesis. I particularly enjoyed the joint research with R.S. Bucy and B.D.O. Anderson. New ideas on the parameter-space method ripened during my sabbatical year at Urbana, Illinois, in fruitful discussions with J. Cruz, P. Kokotovic, D. Looze, W. Perkins. Essential contributions to the further development of the parameter-space method and its application to flight control and track-guiding of a bus were made by my co-workers N. Franklin, D. Kaesbauer, P. Sondergeld and S. Türk.

I am particularly indebted to E. Jury, W. Perkins and M. Smith for their careful review of the manuscript. Also G. Knausenberger, W. Kortüm and H. Selbuz have given many helpful corrections and suggestions. A first draft of the English translation was made by M. Walker.

I extend my thanks to Mrs. G. Kieselbach for typing the manuscript and to Mrs. C. Bell for making the illustrations. Thanks also to A. von Hagen and his co-workers at the Springer-Verlag.

I gratefully acknowledge the support of the German Aerospace Research Establishment (DFVLR = Deutsche Forschungs- und Versuchsanstalt für Luft- und Raumfahrt). In particular W. Hasenclever, F. Thomas and G. Hirzinger have helped me to pursue my research interests and to spend a sabbatical year in spite of my duties in managing a research laboratory at DFVLR. I also acknowledge the support of the US Air Force during my sabbatical and the support of the Volkswagen-Foundation for my co-worker P. Sondergeld.

Oberpfaffenhofen February 1985 — Jürgen Ackermann

Contents

1 Introduction

1.1 Sampling, Sampled-Data Controllers

Dear reader, you are a sampled-data controller!

If you are driving a car and want to know your speed, you do not have to watch the speedometer continuously. It is sufficient that you monitor the continuously indicated velocity by an occasional glance, and that is "Sampling". Likewise the continuously changing voter opinion of political parties is sampled every few years by an election and the composition of the congress is then fixed for the following term of office. Stock market quotations are set every working day, and the temperature of a sick person is checked several times per day.

Such sampling also occurs in various technical control systems because of the applied measuring procedures. A gas chromatograph is a chemical analysis tool that requires a certain amount of time to examine a sample before it can handle the next one. Rotating radar antennas measure a target only once per revolution and this is also the case for radial measurements which are made from a spinning satellite. Radar-range measurement is again a sampling problem. Star sensors measure the position of satellites by taking the angle between the optical axis and the target star. A part of the sensor rotates in the optical system and the angle is measured only once per revolution. In the register control of a multicolor printing process the register error can only be determined at the arrival of the register marks.

Sampling also occurs when the actuator signal in a control system can be changed only at certain time instants. In alternating current rectifiers the ignition angle can be determined only once per period. This is a special form of pulse width modulation. The position of a spin stabilized satellite can be controlled in the two axes perpendicular to the spin axis by a fixed gas jet. Therefore it is possible to produce a moment in a desired inertially fixed direction only once per revolution.

Sampled-data controllers can also be applied to systems with continuous measurement and actuation. Probably the oldest sampled-data controller is the Gouy controller for the temperature control of an oven [1897]. A metal bar dips periodically into the mercury of a thermometer. This contact drives a relay which opens and closes the circuit for the heating. If the temperature of the oven produces a higher mercury level, then the heating is turned off for a longer time interval. This measurement device produces a pulse-width modulated signal. A further historical example is the chopper bar galvanometer [1944]. Here a sensitive voltage meter can adjust in a short time interval. A periodically moving chopper bar stops the indicator needle in a second time interval and switches the energy by the operation of a contact for positive, negative or zero voltage. This is held until the bar releases the indicator needle again. The chopper bar galvanometer produces a pulse-amplitude modulated signal with a three-point nonlinearity that can be greatly amplified by a relay switch.

Sampling is also necessary whenever one wants to use an expensive device which must perform different tasks in sequence. This time-sharing is for example applied if a digital computer is used as a controller for several control loops, or if a data transmission channel is used for the transmission of many signals in a time-multiplex-operation (e.g. a telemetry connection between a satellite and earth).

The main reason for the recent wide spread applications of sampled-data control is the availability of cheap microprocessors. In a digital signal processor the continuous input signal e(t) is first sampled, quantized, and coded, i.e. it is transformed into a sequence of binary numbers which the digital computer can digest. The computer produces another sequence of binary numbers. A continuous signal u(t) is generated from the output sequence by decoding and holding. Figure 1.1 illustrates these operations in the analog to digital converter (ADC), digital computer and digital to analog converter (DAC).

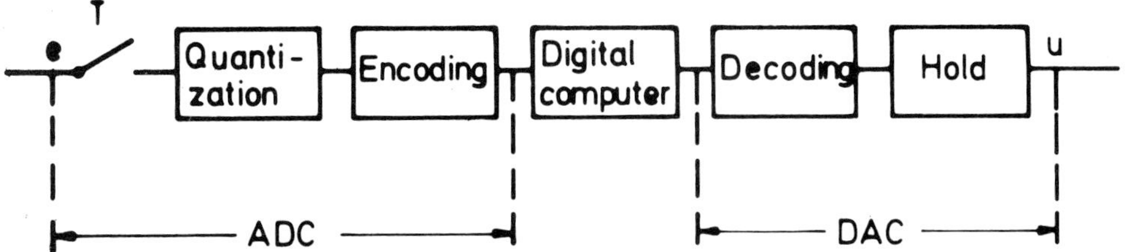

Figure 1.1 Digital computer with AD- and DA-converters

For the dynamic processes the operations "encoding" and "decoding" are significant only as far a certain time is required, which can be included in the computing time T_c, for the computation of a new value of the output sequence. This yields the block diagram of figure 1.2 in which the output-roundoff is represented by a second quantization.

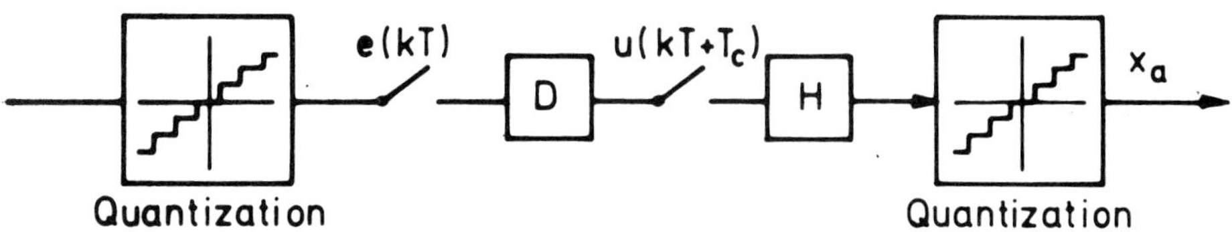

Figure 1.2 Block diagram for the arrangement of diagram 1.1

The digital computer program is symbolized by a $\underline{D}$. It must be causal, that is, for the computation of a new value $u(kT+T_c)$ of the output only the actual value of the input $e(kT)$ and previously stored values of input and output can be used. The general causal program is therefore

$$u(kT+T_c) = f[e(kT),\ e(kT-T)\ldots,\ u(kT-T+T_c),\ u(kT-2T+T_c)\ldots] \tag{1.1.1}$$

For $k = 0,1,2\ldots$ the output sequence is produced through recursive computation. The calculated output values appear again in quantized form, i.e. the amplitudes $u(kT+T_c)$ are rounded to the next integer multiple of the quantization step.

Now the following four simplifying assumptions will be made

- The quantization nonlinearity is neglected,
- The computation time T_c is small compared to the sampling interval T and is approximately set to zero,
- The computation program processes only the last m input and output values,
- The control algorithm is linear.

Then eq. (1.1.1) takes the form

$$u(kT) = d_m e(kT)+d_{m-1}e(kT-T)+\ldots+d_o e(kT-mT)-$$
$$-c_{m-1}u(kT-T)-\ldots-c_o u(kT-mT) \qquad (1.1.2)$$

Only the multiplication $d_m \times e(kT)$ and addition of this term is performed at time $t = kT$. All other terms are precalculated during the interval $kT-T < t < kT$. Thus also for higher m the computation time T_c is only the time for one multiplication and one addition, and it can be justifiably neglected. In later figures we do not include the second synchronous sampler at the output of D.

An important mathematical tool for the description of such input-output relations is the z-transform

$$f_z(z) = \mathcal{Z}\{f_k\} := \sum_{k=0}^{\infty} f_k z^{-k} \qquad (1.1.3)$$

It transforms a sequence f_k, $k = 0,1,2\ldots$, $f_k = 0$ for $k < 0$, into a function of the complex variable z. It is assumed in this book that the reader is familiar with the z-transform. If this is not the case, it is recommended that the reader works through appendix B, since the z-transform will be used frequently from chapter three onwards. The historical roots of the z-transform and related methods can be found in [47.1], [49.1], [51.1], [52.1], [52.2].

From the right shifting theorem of the z-transform, eq. (B.3.1),

$$\mathcal{Z}\{f_{k-i}\} = z^{-i}\mathcal{Z}\{f_k\}\ , \quad i \geq 0 \tag{1.1.4}$$

applied to eq. (1.1.2)

$$u_z = d_m e_z + d_{m-1} z^{-1} e_z + \ldots + d_o z^{-m} e_z - c_{m-1} z^{-1} u_z - \ldots - c_o z^{-m} u_z$$

Then the z-transfer function of the controller is

$$d_z(z) = \frac{u_z(z)}{e_z(z)} = \frac{d_m z^m + d_{m-1} z^{m-1} + \ldots + d_o}{z^m + c_{m-1} z^{m-1} + \ldots + c_o} \tag{1.1.5}$$

The causality of the transfer behavior according to eq. (1.1.2) is reflected in the fact that the degree of the denominator of the z-transfer function is at least equal to the degree of its numerator. In other words: $d_z(\infty)$ is finite. Eqs. (1.1.2) and (1.1.5) show that a pole excess (= numerator degree minus denominator degree) of p, i.e. $d_m = d_{m-1} = \ldots d_{m-p+1} = 0$ indicates that a reaction at the output to any input is delayed by p sampling intervals.

1.2 Sampled-Data Systems

In the first section examples of control systems were given in which both continuous time functions f(t) and sequences f_k, $k = 0,1,2\ldots$, occur. A single number f_k of the sequence characterizes a signal in the time interval $t_k < t < t_{k+1}$. In most cases the time instants t_k are fixed, typically equally spaced with interval T, that is $t_k = kT$.

Remark 1.1:

> In "signal-dependent sampling" t_k depends on signals in the control system. This difficult nonlinear case will not be treated in this book. An example is an anticipated election called before the end of a term of office.

A number sequence can influence a continuous system, for example as an actuator input, if a continuous signal is formed by an impulse forming element. Figure 1.3 shows two examples of impulse forming

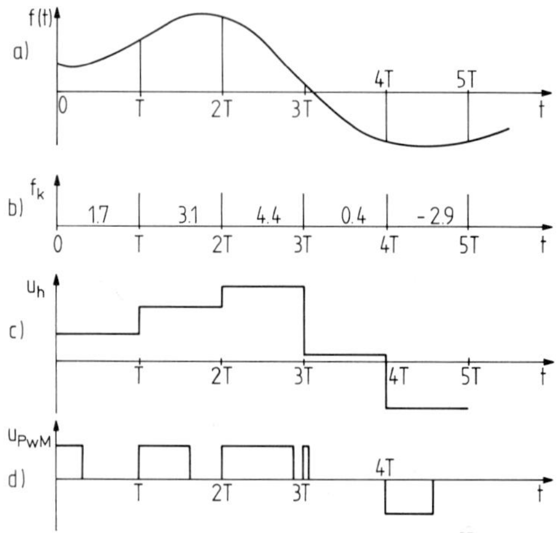

Figure 1.3 a) continuous signal
b) sampled sequence
c) pulse-amplitude modulated sampled signal
d) pulse-width modulated sampled signal

In figure 1.3 a the continuous signal f(t) is given. The "sampled sequence", figure 1.3 b is generated by sampling the amplitude values $f_k = f(kT)$. It should be noted here that the number sequence f_k can also originate from other sources, for example through the algorithm (1.1.1). In figure 1.3 c extrapolation by a hold element is shown, where

$$u_h(t) = f_k \text{ for } kT < t < kT+T \tag{1.2.1}$$

In figure 1.3 d the impulses are formed by a pulse-width modulator

$$u_{PWM}(t) = \begin{cases} \operatorname{sgn} f_k & kT<t<c|f_k| \\ 0 & c|f_k|<t<kT+T \end{cases} \tag{1.2.2}$$

(Saturation is avoided by choosing $c \leq T/|f_k|\max$).

There exist other impulse forming elements, for example pulse-frequency modulators or extrapolation elements of higher order. In practice, the hold element is most common, that is, the zero order extrapolation element of eq. (1.2.1).

The output of the impulse forming element is called "sampled signal". Every dynamic system in which at least one sampled signal and one continuous signal occurs is called a sampled-data system. Systems in which sampled sequences exclusively arise are "discrete-time systems". This notion is also used if all continuous signals f(t) are described only by their discrete values $f_k = f(kT)$.

Corresponding to the two forms of signals

. continuous signal f(t)

. sampled sequence (discrete signal) f_k

a sampled-data system can contain four types of elements:

1. Input and output continuous:

   for example the plant of a control system, described by a transfer function or state-space model.

2. Input continuous, output discrete:

   for example the sampler which picks the amplitude values $f_k = f(kT)$.

3. Input discrete, output discrete:

   for example the digital algorithm of eq. (1.1.1).

4. Input discrete, output continuous:

   for example the hold element or the pulse-width modulator.

The hold element of 4. can also be directly connected to the sampler of 2. This especially important combination "sampler and hold" will now be examined in the frequency domain.

A sampler and hold is a transfer element which takes the samples $f(kT)$ from its input signal $f(t)$ at the time instants $t = kT$ and holds them constant until the next sampling, see figure 1.4.

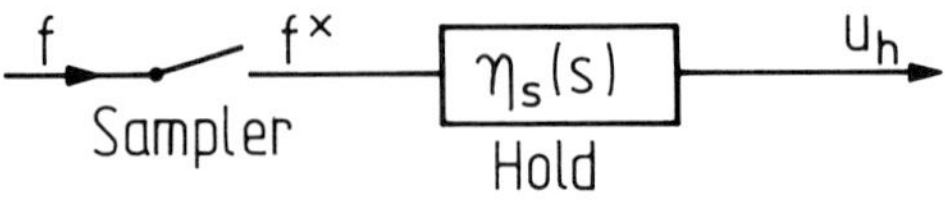

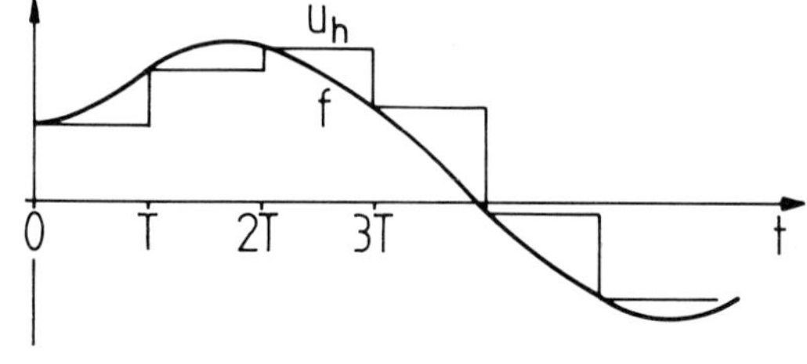

Figure 1.4 Sampler and hold

The staircase output signal is therefore

$$u_h(t) = f(kT) \text{ for } kT < t < kT+T, \; k = 0,\pm1,\pm2 \ldots \tag{1.2.3}$$

Remark 1.2:

If the input signal $f(t)$ has a discontinuity at the instant $t = kT$, then by convention it is assumed that the right-sided boundary value $f(kT+0)$ is sampled. In some cases this does not correspond with the technical implementation of the sampler and hold, for example the chopper bar galvanometer samples $f(kT-0)$. This convention is made, however, to conform with the sampled-data literature. For example in the sampling of a step function

$$1(t) := \begin{cases} 0 \text{ for } t < 0 \\ 1 \text{ for } t > 0 \end{cases} \tag{1.2.4}$$

for $t = 0$ the output variable $u_h(0) = 1$ is assumed. This is in agreement with the behavior of the technical devices, if the step occurs in the time interval $-T < t < 0$. If it is desired to examine the exact simultaneous occurance of step and sampling, then one must analyze the dynamics of the sampler and hold implementation exactly without idealizations. This case, however, has little practical importance. For simplification, the following notation will be used

$$f(kT+0) = f(kT) \tag{1.2.5}$$

Note, that the inverse Laplace-transform $\mathcal{L}^{-1}\{1/s\} = 1(t)$ represents the value $[1(+0)+1(-0)]/2 = 1/2$ at the time $t = 0$.

The output signal $u_h(t)$ of the sampler and hold can be decomposed into the "windows" $f(kT)\times[1(t-kT)-1(t-kT-T)]$ where $1(t)$ is defined in eq. (1.2.4). The above expression has the value $f(kT)$ in the kth interval, and is zero otherwise. Assume $f(kT) = 0$ for $k < 0$, and the entire output signal is

$$u_h(t) = \sum_{k=0}^{\infty} f(kT)[1(t-kT)-1(t-kT-T)] \tag{1.2.6}$$

The sum is convergent, because for every t exactly one element of the sum is different from zero. Its Laplace transform is

$$u_s(s) = \mathcal{L}\{u_h(t)\} = \sum_{k=0}^{\infty} f(kT)[\tfrac{1}{s}e^{-kTs} - \tfrac{1}{s}e^{-(k+1)Ts}]$$

$$= \frac{1-e^{-Ts}}{s} \times \sum_{k=0}^{\infty} f(kT)e^{-kTs} \tag{1.2.7}$$

In this representation, by sampling, a signal $f^*(t)$ is produced with the Laplace transform

$$f_s^*(s) := \sum_{k=0}^{\infty} f(kT)e^{-kTs} \tag{1.2.8}$$

This signal is the input of the hold element in figure 1.4 with the transfer function

$$\eta_s(s) := \frac{1-e^{-Ts}}{s} \tag{1.2.9}$$

Note that in this representation $f^*(t) = \mathcal{L}^{-1}\{f_s^*(s)\}$ is a signal uniquely described by the sequence $f(kT)$.

Remark 1.3:

With the substitution $z = e^{Ts}$ into eq. (1.2.8) the z-transform is obtained

$$f_s^*(\frac{1}{T}\ln z) = \sum_{k=0}^{\infty} f(kT)z^{-k} = f_z(z) \qquad (1.2.10)$$

In the s-domain this correspondence is

$$f_z(e^{Ts}) = f_s^*(s) \qquad (1.2.11)$$

The sampled signal $f^*(t) = \mathcal{L}^{-1}\{f_s^*(s)\}$ can be calculated by inverse Laplace transform from eq. (1.2.8)

$$f^*(t) = \sum_{k=0}^{\infty} f(kT)\,\mathcal{L}^{-1}\{e^{-kTs}\} = \sum_{k=0}^{\infty} f(kT)\delta(t-kT) \qquad (1.2.12)$$

Eq. (1.2.12) can also be interpreted as pulse-amplitude modulation. From a theorem of distribution theory [60.1], [63.1]

$$f(kT)\delta(t-kT) = f(t)\delta(t-kT) \qquad (1.2.13)$$

and with $f(t) = 0$ for $t < 0$

$$f^*(t) = f(t)\sum_{k=0}^{\infty} \delta(t-kT) = f(t)\sum_{k=-\infty}^{\infty} \delta(t-kT) \qquad (1.2.14)$$

The sum represents a periodic pulse of δ-functions. It is modulated by f(t) in the amplitude.

Remark 1.4:

The interpretation of sampling as pulse-amplitude modulation also shows a correspondence between the spectrum $f_s(j\omega)$ of the continuous signal and the spectrum $f_s^*(j\omega)$ of the sampled signal $(s=\sigma+j\omega, j=\sqrt{-1})$. For this purpose the periodic pulse of δ-functions is expressed by its Fourier series [60.1]

$$f^*(t) = f(t)\times\frac{1}{T}\sum_{m=-\infty}^{\infty} e^{-jm\omega_A t}\ , \ \omega_A = 2\pi/T \qquad (1.2.15)$$

The corresponding Fourier transform is

$$f_s^*(j\omega) = \int_{-\infty}^{\infty} f^*(t)e^{-j\omega t}dt = \frac{1}{T}\sum_{m=-\infty}^{\infty}\int_{-\infty}^{\infty} f(t)e^{-j(\omega+m\omega_A)t}dt$$

$$f_s^*(j\omega) = \frac{1}{T}\sum_{m=-\infty}^{\infty} f_s(j\omega+jm\omega_A) \qquad (1.2.16)$$

where $f_s(j\omega)$ is the Fourier transform of the continuous signal $f(t)$. Eq. (1.2.16) shows how pulse-amplitude modulation generates side-band frequencies of integer multiples of the sampling frequency ω_A. Figure 1.5 gives an example of the magnitudes of $f_s(j\omega)$ and $f_s^*(j\omega)$. The shown parts of the spectrum in $f_s^*(j\omega)$ can be added in consideration of the phase angle.

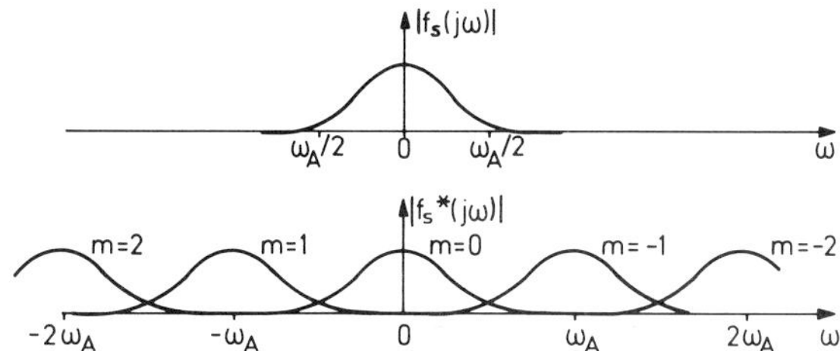

Figure 1.5 Spectrum of $f(t)$ and its sampled signal $f^*(t)$

In stable, continuous, time-invariant, linear transfer elements, which are periodically stimulated, only the same frequencies occuring in the input arise in the output after the transients have decayed. A basic difference in the case of the sampler is that in the output additional sideband frequencies arise. A similar effect is observed for non-linear elements where harmonic frequencies, that is, integer multiples of the stimulation frequency, occur at the output. The sampler is a linear but time-varying transfer element.

The hold element causes a phase delay, which is determined from its frequency response. From eq. (1.2.9) with $s = j\omega$:

$$\eta_s(j\omega) = \frac{1-e^{-j\omega T}}{j\omega}$$

$$= \frac{e^{j\omega T/2}-e^{-j\omega T/2}}{j\omega}\, e^{-j\omega T/2}$$

$$= \frac{2\sin\omega T/2}{\omega}\, e^{-j\omega T/2} \qquad (1.2.17)$$

The phase delay amounts to $\omega T/2$, i.e. it corresponds to a time-delay of one half sampling period.

A sampler and hold can be approximately implemented by the circuit in figure 1.6. It is also used in the simulation of sampled-data systems on the analog computer.

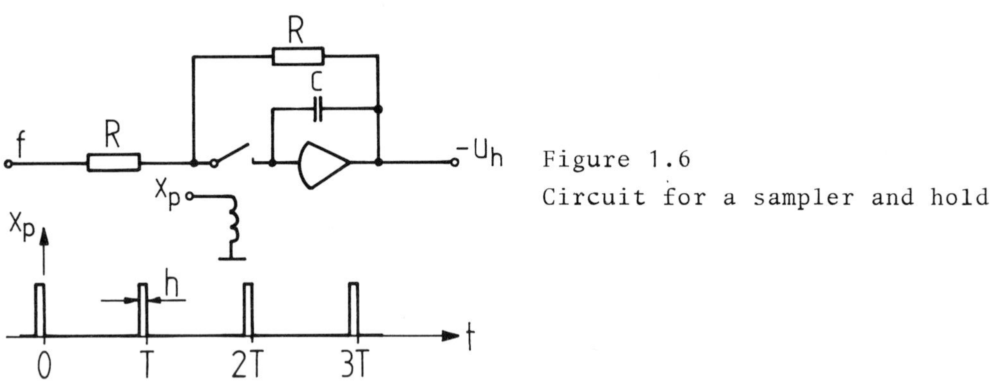

Figure 1.6
Circuit for a sampler and hold

During the short duration h of closing the switch, the capacitor C is loaded with the time constant RC to the value f(kT). When the switch is open the amplifier operates as an integrator, whose output remains constant. R is chosen as small as the maximum load of the amplifier allows. With the dimensioning of h and C a compromise must be made so that on the one hand $RC \ll h \ll T$, but on the other hand C is not noticably discharged in the hold interval T.

Remark 1.5:

If one knows certain properties of the signal, the extrapolation can be improved. For example knowing the spectrum of the signal, one can extrapolate optimally in the sense of least squares approximation. It is more sensible, however, to optimize the entire system, and not merely the extrapolation element [68.1].

The assumption is frequently made that the signal can be approximated by a polynomial $s_h(t) = c_o+c_1t+...+c_mt^m$. The coefficients c_i are determined so that the polynomial curve passes through the sampled values f(kT), f(kT-T)...f(kT-mT). m determines the order of the extrapolation elements. The extrapolation element of zero order is the hold element. For m=1 the extrapolation is linear:

$$u_h(t) = f(kT)+[f(kT)-f(kT-T)](t-kT)/T, \quad kT<t<kT+T \qquad (1.2.18)$$

The polynomial assumption is, however, disadvantageous for many signals which arise in control systems, for example step functions or stochastic signals. For a sinusoidal signal the phase shift increases with the order of the extrapolation element [58.2], [59.1]. This is disadvantageous for the stability of a control loop. Moreover, the technical implementation becomes more complicated. Therefore in practice the hold element is used most frequently.

1.3 Design Problems for Sampled-Data Loops

A simple sampled-data control loop is represented in figure 1.7

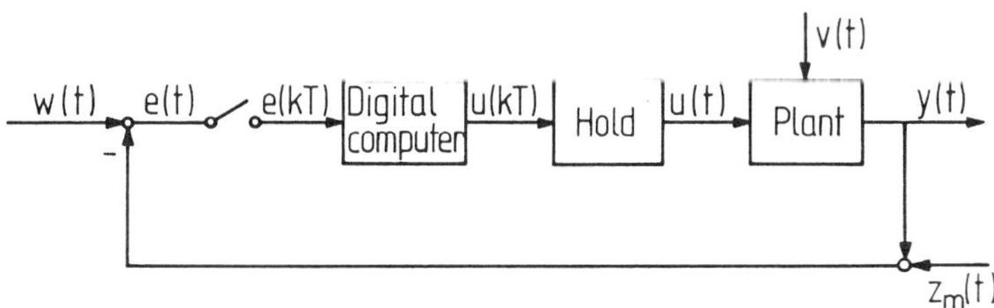

Figure 1.7 A typical sampled control loop

The controlled variable y(t) at the output of the plant is compared to the reference input w(t). From the continuous error signal e(t) = w(t)-y(t) only the sampled sequence e(kT) is supplied to a digital computer, which calculates a second sequence u(kT) from the first one. A hold element produces a staircase function

$$u(t) = u(kT) \ , \ kT < t < kT+T \ , \tag{1.3.1}$$

which is given as actuator input to the plant. The purpose of this control loop is to keep the error e(t) small in a sense to be specified, such that the controlled variable y(t) follows the reference variable w(t) as closely as possible. In comparison with an open loop control (in which y is not fed back) the closed loop control has three advantages:

1. e(t) can be kept small for all times also for unstable plants, if the feedback stabilizes the control loop (stabilization).

2. The influence of disturbance inputs v(t) at the plant can be reduced via disturbance compensation and filtering.

3. e(t) can be kept small for all times also for uncertain mathematical models of the plant (robust control).

A disadvantage of the feedback structure is that new uncertainties can be introduced by non-ideal sensors. This is indicated by measurement noise $z_m(t)$ in figure 1.7.

The control loop structure of figure 1.7 is the only choice, if the sensor measures e = w-y, e.g. a relative position with respect to a target. Other control loop structures (for example observer and state vector feedback) can be used if y(t) and w(t) are available separately. In all cases the measured variable y is fed back. The design methods for continuous-time controllers can be applied with the necessary modifications to the design of sampled-data controllers. In state space methods the controller order results from the method used and increases with the order of the plant. In graphic frequency domain methods (root locus, Bode, Nyquist, Nichols diagrams) one can try to find satisfactory solutions with controllers of lower order. For example in figure 1.7 a low order controller can be assumed according to eq. (1.1.5). The cost of a controller used to depend upon its order, particularly in the case of analog controllers. However, the costs of the implementation of a control algorithm by eq. (1.1.2) with today's microelectronics is small compared to other expenses, such as reliable sensors and design effort. The possible lower controller order is therefore no longer a significant advantage of graphical frequency domain design methods.

State space considerations lead to a better understanding of pole-zero cancellations, which are especially critical for sampled-data systems. Observers and state feedback provide a controller structure. The reference-response from w to y and the disturbance-response from v to y can be designed to some extent independently of each other in this structure. Moreover the generalization to multi-variable systems is straightforward.

On the other hand the frequency domain analysis also offers insights (e.g. integral controllers, admissible modeling inaccuracy of the plant at high frequencies, anti-aliasing filter, the allowable sector of nonlinearity of the actuator) and computational advantages (base-free calculation with polynomials and polynomial matrices). In the following chapters therefore both kinds of representation will be used together in order to combine their advantages.

The design process furthermore depends on the answers to the following questions:

a) What is known about the type of reference input w(t) and the disturbances v(t) (e.g. step, impulse, periodic or stochastic signal)? Which disturbances can be measured?

b) Which constraints exist for actuator input u(t) and the state variables?

c) Is it essentially a regulator problem (w(t) constant over a long period of time, compensation of disturbances v(t), good stationary accuracy) or more a servo problem (fast changes of the reference input w(t), fast transients, good tracking $y(t) \approx w(t)$) or a compromise between the two?

d) What is the form of the given plant model (state equations from physical laws, frequency response measurements, step response, realization from deterministic input-output observations, identification from stochastic input-output observations, dependency of the model coefficients on changing physical parameters, family of system models for different operating conditions)?

e) How is the performance of the controlled system measured (step response, quadratic-cost function, position of eigenvalues, time optimal solution, satisfaction of inequality constraints, analysis of the singular values of the return difference, disturbance and reference responses, final value control or combinations of some of them)?

No existing book can answer all of these questions. In addition to the standard design approaches we will express the demands on the control system primarily by an admissible region for the closed-loop eigenvalues. An important design goal is to achieve robustness of such eigenvalue requirements with respect to large parameter uncertainty of the plant. In addition, specific sampling problems will be considered: selection of the sampling period, anti-aliasing filters, multirate sampling, design for finite transients, and behavior between the sampling instants.

Readers interested in the historical development of sampled-data systems should read the review by Jury [84.8]. The early use of computers in control systems was reviewed by Åström and Wittenmark [83.4], see also [53.1].

The hardware and software selection and implementation of a control algorithm in microprocessors and AD and DA converters is a matter of rapidly changing technology. It is not discussed in this book. The reader is referred to the literature (e.g. [81.17], [83.5], [84.10]).

1.4 Excercises

1.1 Determine the z-transfer function of a discrete integrator
 a) with rectangular approximation
 b) with trapezoidal approximation.

1.2 A signal $f(t) = \cos \frac{\pi}{3}t + \cos 2\pi t$ is sampled at $t = kT$, $T = 1$.
 a) Plot the sampled signal and its spectrum.
 b) Use the anti-aliasing filter with transfer function

$$g_S(s) = \frac{\pi^2}{s^2+\sqrt{2}\pi s+\pi^2}$$

before the sampler. Plot the stationary response, its samples and its spectrum.

1.3 Plot the frequency response of the hold element for $T = 1$.

2 Continuous Systems

There are several reasons to include this chapter on continuous systems into a book devoted to sampled-data systems.

. Most sampled-data control systems contain a continuous-time plant. In this case with a reasonable choice of the sampling interval the analysis of controllable and observable eigenvalues and canonical decompositions may be performed in continuous-time. This makes it easier to keep track of the influence of uncertain physical parameters as will be illustrated in modelling and analysis of a loading bridge.

. The calculation of the transition matrix is an essential step in discretizing the plant.

. Pole assignment, also partial assignment of some poles and some gains leads to the same mathematical problem in continuous and discrete time. It is solved in this chapter for both cases.

. Specifications for control systems are usually given in continuous time. Some typical ones are introduced here and transformed later to the discrete-time case.

. For the example of the loading bridge introduced in this chapter many further exercise problems are given in the following chapters. These exercises will lead finally to the design of a robust digital controller with dynamic output feedback. It is strongly recommended that the reader works through this series of exercise problems in order to acquire a working familiarity with the theoretical material presented.

For the above reasons the analysis of continuous systems will be recapitulated briefly using the example of the loading bridge.

The essential relations and notations will be compiled for later reference. The reader will adjust his reading speed to his personal background: The beginner may have to go back to the literature on linear systems, eg. [55.1], [63.1], [69.1], [70.3], [71.1], [71.8], [72.3], [80.2], the reader, who had studied continuous systems some time ago, may recapitulate the selected material. A reader, familiar with the analysis of continuous single-input systems, should examine section 2.6 on the interpretation of pole assignment as an affine mapping, after which he can proceed to chapter 3.

2.1 Modelling, Linearization

The mathematical model of the plant can be determined in two ways:

a) from measurements of the input and output variables, where one considers the plant as a "black box" having an unknown internal structure,

b) from the relations for the dynamics of single components and their interaction in a known structure.

Frequently the two possibilities must be combined. While b) provides the structure of the mathematical model (for example for the dynamics of an airplane), the numerical values which appear as parameters can be determined by measurements (for example wind tunnel simulation or flight test). An advantage of b) is that variables with an intuitive, practical meaning can be introduced as state variables, such as position and velocity of a mass point, current in an inductor, voltage at a capacitor, temperature, pressure, inventory of a storehouse, etc.

We now give an elementary derivation of the nonlinear differential equation describing the loading bridge of figure 2.1. (Approach b) requires a detailed analysis of the respective plant. The loading bridge can therefore serve only as an example.)

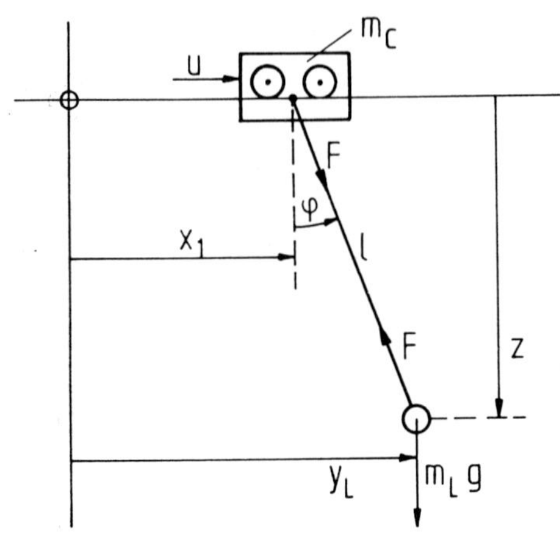

Figure 2.1 Loading bridge

For simplicity assume that

- no friction exists between the trolley and the bridge,
- the cable length ℓ is constant during the motion,
- the cable has no mass and no elasticity, and
- the dynamics of the motor and gear generating the input force u, which accelerates the trolley, are neglected.

Output y_L is the position of the load. The physical parameters of the loading bridge are

m_C = mass of the trolley (crab), augmented by the motor's moment of inertia transmitted via the gear ratio
m_L = load mass
ℓ = cable length
g = normal acceleration.

Using the variables x_1 (trolley-position), φ (cable angle), z (vertical distance from trolley to load), and F (cable force), the differential equations (force balances) are:

a) trolley, horizontal

$$m_C\ddot{x}_1 = u + F\sin\varphi \tag{2.1.1}$$

b) load, horizontal

$$m_L\ddot{y}_L = -F\sin\varphi \tag{2.1.2}$$

c) load, vertical

$$m_L\ddot{z} = -F\cos\varphi + m_L g \tag{2.1.3}$$

By elimination of the cable force F, for $m_L \neq 0$

$$m_C\ddot{x}_1 + m_L\ddot{y}_L = u \tag{2.1.4}$$

$$\ddot{y}_L\cos\varphi - \ddot{z}\sin\varphi = -g\sin\varphi \tag{2.1.5}$$

For constant cable length ℓ

$$z = \ell\cos\varphi \qquad y_L = x_1 + \ell\sin\varphi$$

$$\dot{z} = -\ell\dot{\varphi}\sin\varphi \qquad \dot{y}_L = \dot{x}_1 + \ell\dot{\varphi}\cos\varphi$$

$$\ddot{z} = -\ell\ddot{\varphi}\sin\varphi - \ell\dot{\varphi}^2\cos\varphi \qquad \ddot{y}_L = \ddot{x}_1 + \ell\ddot{\varphi}\cos\varphi - \ell\dot{\varphi}^2\sin\varphi$$

Substituting into eqs. (2.1.4) and (2.1.5) yields

$$(m_L+m_C)\ddot{x}_1 + m_L\ell(\ddot{\varphi}\cos\varphi - \dot{\varphi}^2\sin\varphi) = u \tag{2.1.6}$$

$$\ddot{x}_1\cos\varphi + \ell\ddot{\varphi} = -g\sin\varphi \tag{2.1.7}$$

This is a linear system of equations in $\ddot{x}_1$ and $\ddot{\varphi}$ whose coefficients are dependent upon φ, $\dot{\varphi}$ and u. Solving the equations gives

$$\ddot{x}_1 = \frac{u + (g\cos\varphi + \ell\dot{\varphi}^2)m_L\sin\varphi}{m_C + m_L\sin^2\varphi} \tag{2.1.8}$$

$$\ddot{\varphi} = -\frac{u\cos\varphi + (g + \ell\dot{\varphi}^2\cos\varphi)m_L\sin\varphi + gm_C\sin\varphi}{\ell(m_C + m_L\sin^2\varphi)} \tag{2.1.9}$$

The "state" $\underline{x}$ of a system is the part of the previous history of the system which determines its future behavior. Stated more exactly: $\underline{x}(t_1)$ is uniquely determined by $\underline{x}(t_0)$ and $u(t)$, $t_0 \le t \le t_1$.

For the system of eq. (2.1.8) and (2.1.9) the variables x_1, $\dot{x}_1$, φ and $\dot{\varphi}$ are suitable to describe the state of the system. Introducing the state vector

$$\underline{x} = \begin{bmatrix} x_1 \\ x_2 \\ x_3 \\ x_4 \end{bmatrix} = \begin{bmatrix} x_1 \\ \dot{x}_1 \\ \varphi \\ \dot{\varphi} \end{bmatrix} \tag{2.1.10}$$

eqs. (2.1.8) and (2.1.9) may be written as

$$\underline{\dot{x}} = \begin{bmatrix} x_2 \\ \dfrac{u+(g\cos x_3+\ell x_4^2)m_L \sin x_3}{m_C+m_L \sin^2 x_3} \\ x_4 \\ -\dfrac{u\cos x_3+(g+\ell x_4^2\cos x_3)m_L \sin x_3+gm_C \sin x_3}{\ell(m_C+m_L \sin^2 x_3)} \end{bmatrix} \tag{2.1.11}$$

In general the state variables must be chosen such that no derivatives appear on the right hand side. Eq. (2.1.11) is a nonlinear vector differential equation of the form

$$\underline{\dot{x}} = \underline{f}(\underline{x},u) \tag{2.1.12}$$

This plant model may be used for simulation and computer-aided trial and error methods for controller design. Most analysis and controller synthesis methods require a linear plant model. This can be obtained by linearizing eq. (2.1.12) for small deviations from an equilibrium point ($\underline{x}_o$, u_o) or reference trajectory ($\underline{x}_o(t)$, $u_o(t)$).

$$\underline{\dot{x}} = \underline{F}\underline{x}+\underline{g}u \text{ with } \underline{F} = \left.\frac{\partial \underline{f}}{\partial \underline{x}}\right|_{\underline{x}_o,u_o}, \quad \underline{g} = \left.\frac{\partial \underline{f}}{\partial u}\right|_{\underline{x}_o,u_o} \tag{2.1.13}$$

The controller is then designed for the linear system and is usually tested or refined in a simulation with the nonlinear system (2.1.12).

For the loading bridge the equilibrium point is characterized by the cable angle $\varphi = x_3 = 0$ and the cable angular velocity $\dot{\varphi} = x_4 = 0$. In the neighborhood of this equilibrium

$$\begin{aligned} \cos x_3 &\approx 1 , \quad \sin x_3 \approx x_3 \\ \sin^2 x_3 &\approx 0 \\ x_4^{\,2} \sin x_3 &\approx 0 \end{aligned} \tag{2.1.14}$$

Substituting into eq. (2.1.11)

$$\underline{\dot{x}} = \begin{bmatrix} 0 & 1 & 0 & 0 \\ 0 & 0 & f_{23} & 0 \\ 0 & 0 & 0 & 1 \\ 0 & 0 & f_{43} & 0 \end{bmatrix} \underline{x} + \begin{bmatrix} 0 \\ g_2 \\ 0 \\ g_4 \end{bmatrix} u \tag{2.1.15}$$

$$f_{23} = \frac{m_L}{m_C} g \qquad\qquad g_2 = \frac{1}{m_C}$$

$$f_{43} = -\frac{m_L + m_C}{m_C} \times \frac{g}{\ell} \qquad\qquad g_4 = \frac{-1}{m_C \ell}$$

Taking the position y_L of the load as the controlled variable the output equation is

$$y_L = x_1 + \ell \sin x_3 \tag{2.1.16}$$

and in linearized form

$$y_L = [1 \quad 0 \quad \ell \quad 0]\underline{x} = \underline{c}_L'\underline{x} \tag{2.1.17}$$

The row vector $\underline{c}_L'$ is written as the transpose (') of a column vector $\underline{c}_L$. The measured output may be different. If for example the trolley position x_1 and the cable angle x_3 are measured, the measurement equation is

$$\underline{y} = \begin{bmatrix} 1 & 0 & 0 & 0 \\ 0 & 0 & 1 & 0 \end{bmatrix} \underline{x} = \underline{Cx} \qquad (2.1.18)$$

(Note that in the usual notation the matrix $\underline{C}$ is not written as transposed.)

2.2 Basis of the State Space

With the derivation of the state equations from physical laws, a basis of physically meaningful coordinates is determined for the state space. For the loading bridge example the selection of the state variables in eq. (2.1.10) yields the desired coordinates. This leads to a system description

$$\begin{aligned} \dot{\underline{x}} &= \underline{Fx} + \underline{g}u \\ y &= \underline{c}'\underline{x} + du \end{aligned} \qquad (2.2.1)$$

Here a scalar measurement y is assumed, which may be directly influenced by the input via the term du (feedthrough).

Let n be the dimension of the state space, in which the state vector $\underline{x}$ lives. Mathematically we can use any set of n linearly independent vectors as a basis for this state space. They do not necessarily have a physical meaning, e.g. a linear combination of a position and a velocity could be used. Generally any nonsingular transformation

$$\underline{x}^* = \underline{Tx} \ , \ \det \underline{T} \neq 0 \qquad (2.2.2)$$

may be applied to change the system description to

$$\begin{aligned} \dot{\underline{x}}^* &= \underline{F}^*\underline{x}^* + \underline{g}^*u \ , \ \underline{F}^* = \underline{TFT}^{-1} \ , \ \underline{g}^* = \underline{Tg} \\ y &= \underline{c}^{*\prime}\underline{x}^* + du \ , \ \underline{c}^{*\prime} = \underline{c}'\underline{T}^{-1} \end{aligned} \qquad (2.2.3)$$

The n^2 coefficients of $\underline{T}$ can be chosen so that n^2 elements of $\underline{F}^*$, $\underline{g}^*$, $\underline{c}^{*\prime}$ are fixed to zero or one in a canonical form. The most important canonical forms and the corresponding transformation matrices are given in appendix A. Figure 2.2 illustrates the linear transformation by a block diagram. If one considers this as an analog computer circuit diagram, then it is obvious that through the linear transformation instead of $\dot{\underline{x}}$, n other variables $\dot{\underline{x}}^* = \underline{T}\dot{\underline{x}}$ are integrated and transformed back to $\underline{x} = \underline{T}^{-1}\underline{x}^*$.

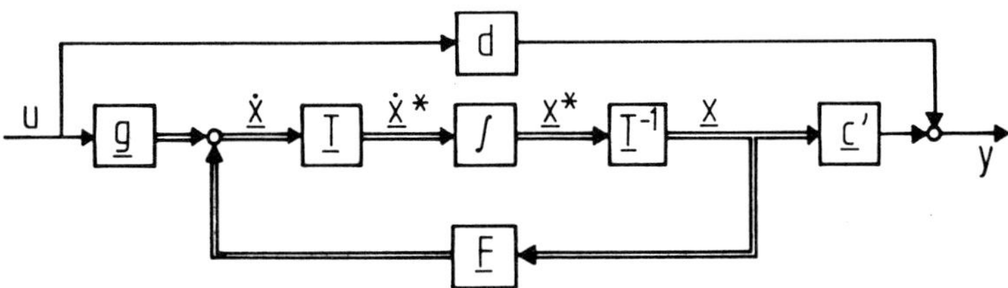

Figure 2.2 Transformation of the state representation

In canonical forms some calculations can be simplified, also some structural properties of systems can be visualized. For interpretation, the results are transformed back to physically meaningful variables. In some cases it is also possible to choose physical variables such that the state space model is simplified. See exercise 2.1.

2.3 System Properties

This section will treat several variables and properties of a system which are independent of the chosen basis of the state space. These are the characteristic polynomial, eigenvalues and stability, the transfer function, controllability and observability.

2.3.1 Eigenvalues and Stability

The characteristic polynomial of $\underline{F}^*$ is

$$\begin{aligned} Q(s) = \det(s\underline{I}-\underline{F}^*) &= \det[\underline{T}(s\underline{I}-\underline{F})\underline{T}^{-1}] \\ &= \det\underline{T}\times\det(s\underline{I}-\underline{F})\times\det\underline{T}^{-1} \qquad (2.3.1) \\ &= \det(s\underline{I}-\underline{F}) \end{aligned}$$

The eigenvalues s_i are the zeros of $Q(s)$. The system $\dot{\underline{x}} = \underline{F}\underline{x}$ is "asymptotically stable", if its solution $\underline{x}(t)$ for any initial condition $\underline{x}(t_0)$ is bounded and approaches $\underline{x} = \underline{0}$ asymptotically. This property is given if and only if the real parts of all eigenvalues of $\underline{F}$ are negative.

Example: Loading bridge

$$Q(s) = s^2(s^2 - f_{43}) = s^2(s^2 + \omega_L^2), \quad \omega_L^2 = \frac{m_L + m_C}{m_C} \times \frac{g}{\ell} \tag{2.3.2}$$

Eigenvalues $s_{1,2} = 0$, $s_{3,4} = \pm j\omega_L$, not asymptotically stable.

In complicated numerical examples eigenvalues can be computed accurately by the QR algorithm [65.1], [76.2].

Necessary and sufficient conditions for a real polynomial to have only roots with negative real parts are given in appendix C.

2.3.2 Transfer Function

Applying the Laplace transform $\underline{x}_s(s) = \mathcal{L}\{\underline{x}(t)\}$ to eq. (2.2.1) yields

$$\mathcal{L}\{\dot{\underline{x}}(t)\} = \mathcal{L}\{\underline{F}\underline{x}(t) + \underline{g}u(t)\}$$

$$s\underline{x}_s(s) - \underline{x}(0) = \underline{F}\underline{x}_s(s) + \underline{g}u_s(s)$$

$$\underline{x}_s(s) = (s\underline{I} - \underline{F})^{-1}\underline{x}(0) + (s\underline{I} - \underline{F})^{-1}\underline{g}u_s(s) \tag{2.3.3}$$

$$\begin{aligned} y_s(s) &= \underline{c}'\underline{x}_s(s) + du_s(s) \\ &= \underline{c}'(s\underline{I} - \underline{F})^{-1}\underline{x}(0) + [\underline{c}'(s\underline{I} - \underline{F})^{-1}\underline{g} + d]u_s(s) \end{aligned} \tag{2.3.4}$$

For the initial state $\underline{x}(0) = \underline{0}$ it follows that $y_s(s) = g_s(s) \times u_s(s)$, where

$$g_s(s) = \frac{y_s(s)}{u_s(s)} = \underline{c}'(s\underline{I} - \underline{F})^{-1}\underline{g} + d \tag{2.3.5}$$

is the transfer function of the system. It is invariant under a basis transformation in state space, eq. (2.2.2):

$$g_s(s) = \underline{c}^{*\prime}(s\underline{I}-\underline{F}^{*})^{-1}\underline{g}^{*}+d = \underline{c}'(s\underline{I}-\underline{F})^{-1}\underline{g}+d \qquad (2.3.6)$$

The resolvent $(s\underline{I}-\underline{F})^{-1}$ can be computed via Leverrier's algorithm [59.4] as follows:

$$(s\underline{I}-\underline{F})^{-1} = \frac{\underline{D}_{n-1}s^{n-1}+\underline{D}_{n-2}s^{n-2}+\ldots+\underline{D}_o}{s^n+q_{n-1}s^{n-1}+\ldots+q_o} = \frac{\underline{D}(s)}{Q(s)} \qquad (2.3.7)$$

The matrices $\underline{D}_i$ and the characteristic polynomial coefficients q_i follow from eq. (A.7.35) with the substitution

$$\underline{A} = \underline{F}\ ,\quad a_i = q_i\ ,\quad \lambda = s \qquad (2.3.8)$$

Substituting eq. (2.3.7) into the transfer function (2.3.5)

$$g_s(s) = \frac{\underline{c}'\underline{D}(s)\underline{g}}{Q(s)} + d \qquad (2.3.9)$$

Example: Loading bridge of eq. (2.1.15)

$$\underline{D}_3 = \underline{I}$$

$$q_3 = 0 \qquad \underline{D}_2 = \begin{bmatrix} 0 & 1 & 0 & 0 \\ 0 & 0 & f_{23} & 0 \\ 0 & 0 & 0 & 1 \\ 0 & 0 & f_{43} & 0 \end{bmatrix}$$

$$q_2 = -f_{43} \qquad \underline{D}_1 = \begin{bmatrix} -f_{43} & 0 & f_{23} & 0 \\ 0 & -f_{43} & 0 & f_{23} \\ 0 & 0 & 0 & 0 \\ 0 & 0 & 0 & 0 \end{bmatrix}$$

$$q_1 = 0 \qquad \underline{D}_o = \begin{bmatrix} 0 & -f_{43} & 0 & f_{23} \\ 0 & 0 & 0 & 0 \\ 0 & 0 & 0 & 0 \\ 0 & 0 & 0 & 0 \end{bmatrix}$$

$$q_o = 0 \qquad "\underline{D}_{-1}" = \underline{0}$$

$$(s\underline{I}-\underline{F})^{-1} = \frac{\underline{D}(s)}{Q(s)} = \frac{1}{s^2(s^2-f_{43})}\begin{bmatrix} s^3-f_{43}s & s^2-f_{43} & f_{23}s & f_{23} \\ 0 & s^3-f_{43}s & f_{23}s^2 & f_{23}s \\ 0 & 0 & s^3 & s^2 \\ 0 & 0 & f_{43}s & s^3 \end{bmatrix}$$

$$\underline{D}(s)\underline{g} = \begin{bmatrix} (s^2-f_{43})g_2+f_{23}g_4 \\ (s^3-f_{43}s)g_2+f_{23}sg_4 \\ s^2g_4 \\ s^3g_4 \end{bmatrix} = \begin{bmatrix} (s^2+g/\ell)/m_C \\ s(s^2+g/\ell)/m_C \\ -s^2/m_C\ell \\ -s^3/m_C\ell \end{bmatrix} \tag{2.3.10}$$

The transfer function depends on the chosen output:

a) output = load position

$$y_L = \underline{c}_L'\underline{x}\ , \quad \underline{c}_L' = [1 \quad 0 \quad \ell \quad 0]$$

$$g_s = \frac{y_{Ls}}{u_s} = \frac{\underline{c}_L'\underline{D}(s)\underline{g}}{Q(s)} = \frac{g/\ell m_C}{s^2(s^2+\omega_L^2)} \tag{2.3.11}$$

b) output = trolley position

$$x_1 = \underline{c}_1'\underline{x}\ , \quad \underline{c}_1' = [1 \quad 0 \quad 0 \quad 0]$$

$$g_s = \frac{x_{1s}}{u_s} = \frac{\underline{c}_1'\underline{D}(s)g}{Q(s)} = \frac{(s^2+g/\ell)m_C}{s^2(s^2+\omega_L{}^2)} \tag{2.3.12}$$

Figure 2.3 shows the poles and zeros.

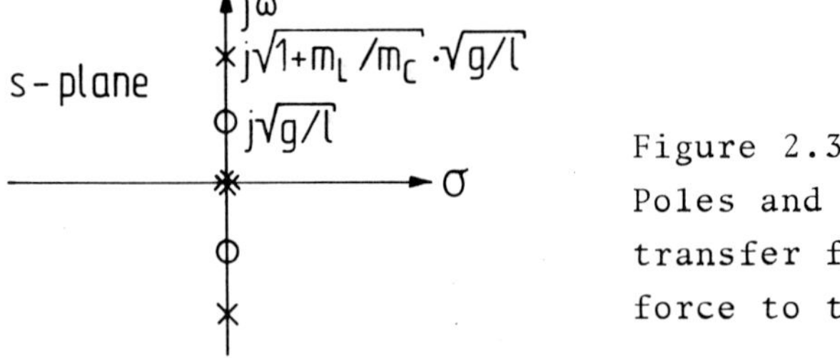

Figure 2.3
Poles and zeros of the transfer function from the force to the trolley position

c) output = cable angle

$$x_3 = \underline{c}_3'\underline{x}\ ,\ \underline{c}_3' = [0 \quad 0 \quad 1 \quad 0]$$

$$g_s = \frac{x_{3s}}{u_s} = \frac{\underline{c}_3'\underline{D}(s)g}{Q(s)} = \frac{-s^2/m_c\ell}{s^2(s^2+\omega_L^2)}$$

If s^2 is cancelled, there remains the second order transfer function

$$g_s = -\frac{1}{m_c\ell} \times \frac{1}{s^2+\omega_L{}^2} \qquad (2.3.13)$$

Leverrier's algorithm is particularly suited for symbolic calculation of $[s\underline{I}-\underline{F}(\underline{\theta})]^{-1}$ where $\underline{\theta}$ is a vector of physical plant parameters, as the above example showed. For numerical calculations in higher order systems Leverrier's algorithm is inaccurate. For this case it is recommended that $\underline{F}$ is transformed to Hessenberg form [78.7]. See appendix A.5.

The transfer function $g_s(s)$ of eq. (2.3.5)

$$g_s(s) = \underline{c}'(s\underline{I}-\underline{F})^{-1}\underline{g}+d =: \frac{B(s)}{A(s)}$$

can also be evaluated by computing the eigenvalues of two matrices. Note first, that the poles of the uncancelled transfer function are the eigenvalues of $\underline{F}$: $A(s) = \det(s\underline{I}-\underline{F})$. For simplicity let $d = 0$. Now the numerator polynomial $B(s)$ can be computed by comparing closed-loop eigenvalues for an arbitrary constant feedback gain α. Figure 2.4 shows the two equivalent representations.

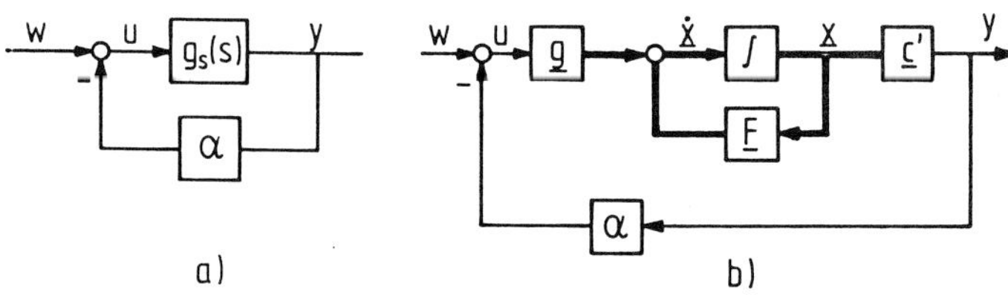

Figure 2.4 Scalar feedback around system
a) in transfer function representation
b) in state space representation

In representation a) the characteristic polynomial is

$$P(s) = A(s) + \alpha B(s) \tag{2.3.14}$$

and in representation b)

$$P(s) = \det(s\underline{I}-\underline{F}+\alpha\underline{gc}') \tag{2.3.15}$$

Thus

$$B(s) = \frac{1}{\alpha}\left[\det(s\underline{I}-\underline{F}+\alpha\underline{gc}')-\det(s\underline{I}-\underline{F})\right] \quad \text{for all } \alpha$$

or

$$g_s(s) = \frac{1}{\alpha}\left[\frac{\det(s\underline{I}-\underline{F}+\alpha\underline{gc}')}{\det(s\underline{I}-\underline{F})} - 1\right] \quad \text{for all } \alpha \tag{2.3.16}$$

For $\alpha = 1$ this relation was derived by Brockett [65.3], [70.4]. α is the gain parameter of a root locus, see section 2.6.4. The eigenvalues of $\underline{F}-\alpha\underline{gc}'$ approach the poles for $\alpha \to 0$ and they approach the zeros at infinity and in B(s) for $\alpha \to \infty$. Thus for good accuracy a medium value for α may be chosen and $g_s(s)$ can be evaluated by computing the eigenvalues of $\underline{F}$ and $\underline{F}-\alpha\underline{gc}'$.

2.3.3 Controllability

A time-invariant, continuous system $\dot{\underline{x}} = \underline{Fx}+\underline{gu}$ is controllable if there exists a finite time $t_1 > t_0$ such that every initial state $\underline{x}(t_0)$ can be transferred to the zero state $\underline{x}(t_1) = \underline{0}$ by a suitable input signal $u(t)$, $t_0 \le t \le t_1$. A system is controllable if and only if

$$\det \underline{\mathcal{C}} \neq 0 \tag{2.3.17}$$

where $\underline{\mathcal{C}} := [\underline{g}, \underline{Fg}\ldots\underline{F}^{n-1}\underline{g}]$ is the controllability matrix. (See Kalman [60.4], [60.6].)

Example: Loading bridge

The controllability matrix is

$$[\underline{g}, \underline{F}\underline{g}, \underline{F}^2\underline{g}, \underline{F}^3\underline{g}] = \begin{bmatrix} 0 & g_2 & 0 & g_4 f_{23} \\ g_2 & 0 & g_4 f_{23} & 0 \\ 0 & g_4 & 0 & g_4 f_{43} \\ g_4 & 0 & g_4 f_{43} & 0 \end{bmatrix} \qquad (2.3.18)$$

Because of the structure of this matrix, only the first and third columns or the second and fourth columns can be linearly dependent. Both cases arise simultaneously for

$$g_4(g_2 f_{43} - g_4 f_{23}) = g/\ell^2 m_C^2 = 0.$$

Thus the system is uncontrollable only under zero gravity. Note, however, that for the load $m_L = 0$ the cancellation of m_L in the derivation of eq. (2.1.5) is not allowed. In this case the system is reduced to the second order trolley system alone and the massless pendulum is not controllable.

For the transformed system (2.2.3) the controllability matrix is

$$\begin{aligned} \underline{\mathcal{C}}^* &= [\underline{g}^*, \underline{F}^*\underline{g}^* \ldots \underline{F}^{*n-1}\underline{g}^*] \\ &= \underline{T}[\underline{g}, \underline{F}\underline{g} \ldots \underline{F}^{n-1}\underline{g}] \\ &= \underline{T}\underline{\mathcal{C}} \end{aligned} \qquad (2.3.19)$$

$\underline{T}$ is nonsingular and therefore

$$\text{rank } \underline{\mathcal{C}}^* = \text{rank } \underline{\mathcal{C}} \qquad (2.3.20)$$

In other words the property of controllability is not dependent on the chosen basis in the state space. This fact may be used for a numerically favorable controllability test by transforming the system to Hessenberg form (see appendix A.5). For a controllable system, eq.(2.3.19) may be solved for $\underline{T} = \underline{\mathcal{C}}^* \underline{\mathcal{C}}^{-1}$, i.e. for the transformation between two representations of the same system. The rank of the controllability matrix also remains unchanged under state feedback $u = -\underline{k}'\underline{x}+w$. For the closed-loop system

$\dot{\underline{x}} = \underline{H}\underline{x}+\underline{g}w \; , \quad \underline{H}:= \underline{F}-\underline{g}\underline{k}'$

the controllability matrix is related to the open-loop controllability matrix by

$$[\underline{g},\ \underline{F}\underline{g}\ldots\underline{F}^{n-1}\underline{g}] = [\underline{g},\ \underline{H}\underline{g}\ldots\underline{H}^{n-1}\underline{g}]\begin{bmatrix} 1 & \underline{k}'\underline{g} & \underline{k}'\underline{F}\underline{g}\ldots\underline{k}\underline{F}^{n-2}\underline{g} \\ 0 & 1 & \underline{k}'\underline{g} & & \\ 0 & 0 & 1 & \ddots & \\ & & & \ddots & \underline{k}'\underline{g} \\ 0 & & 0 & & 1 \end{bmatrix} \tag{2.3.21}$$

The last matrix is nonsingular (i.e. its determinant is nonzero), therefore rank $[\underline{g},\ \underline{F}\underline{g}\ldots\underline{F}^{n-1}\underline{g}]$ = rank $[\underline{g},\ \underline{H}\underline{g}\ldots\underline{H}^{n-1}\underline{g}]$. If there are numerical difficulties with testing controllability, the situation may be improved by a feedback $\underline{k}'$.

2.3.4 Controllable Eigenvalues

The rank of the controllability matrix is equal to the order of the controllable subsystem. This is obvious if $\underline{F}$ has diagonal or Jordan form (see [68.5] and appendix A.2). In diagonal form

$$\dot{\underline{x}} = \begin{bmatrix} \lambda_1 & 0 & \ldots & 0 \\ 0 & \lambda_2 & & \\ \vdots & & \ddots & \\ 0 & & & \lambda_n \end{bmatrix}\underline{x} + \begin{bmatrix} g_1 \\ \vdots \\ g_n \end{bmatrix} u$$

The controllability matrix is

$$\mathcal{C} = \begin{bmatrix} g_1 & 0 & \ldots & 0 \\ 0 & g_2 & & \\ \vdots & & \ddots & \\ 0 & & & g_n \end{bmatrix}\begin{bmatrix} 1 & \lambda_1 & \ldots & \lambda_1^{n-1} \\ 1 & \lambda_2 & \ldots & \lambda_2^{n-1} \\ & & & \\ 1 & \lambda_n & \ldots & \lambda_n^{n-1} \end{bmatrix} \tag{2.3.22}$$

The rank of the controllability matrix can become less than n in two ways:

a) $g_i = 0$, i.e. the input u is not connected to the subsystem with eigenvalue λ_i and this mode is uncontrollable.

b) $\lambda_i = \lambda_j$, i.e. two identical parallel subsystems. They can be transformed into a controllable subsystem with the summed input g_i+g_j and into an uncontrollable subsystem with zero input.

For the general case an eigenvalue λ of the diagonal form is now replaced by a Jordan block and the respective input vector $\underline{b}_i$ is introduced.

$$\underline{J}_i = \begin{bmatrix} \lambda & 1 & & & 0 \\ 0 & \lambda & 1 & & \\ & & \ddots & \ddots & \\ & & & \ddots & 1 \\ 0 & & & & \lambda \end{bmatrix} \qquad \underline{b}_i = \begin{bmatrix} b_1 \\ b_2 \\ \vdots \\ b_r \end{bmatrix}$$

The controllability matrix for this subsystem is

$$\underline{\mathscr{C}}_i = [\underline{b}_i, \underline{J}_i\underline{b}_i \ldots \underline{J}_i^{r-1}\underline{b}_i] = \begin{bmatrix} b_1 & b_2 & & b_r \\ b_2 & & \cdot & 0 \\ & \cdot & & \\ & b_r & & \\ b_r & 0 & & 0 \end{bmatrix} \begin{bmatrix} 1 & \lambda & \lambda^2 & \ldots & \lambda^{r-1} \\ 0 & 1 & 2\lambda & & \vdots \\ & & \ddots & & (r-1)\lambda \\ 0 & & & 0 & 1 \end{bmatrix} \tag{2.3.23}$$

It is nonsingular if and only if $b_r \neq 0$. In the case $b_r = 0$, $b_{r-1} \neq 0$ the rank is r-1 and the last component x_{ir} of the respective state vector is not controllable. This case is illustrated in the block diagram 2.5.

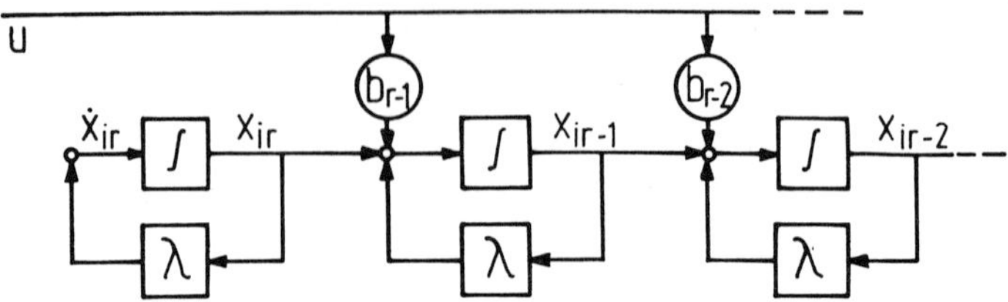

Figure 2.5 Block diagram of a Jordan block with noncontrollable state variable x_{ir}

Thus rank $\underline{\mathscr{C}}_i$ is the order of the controllable part of the Jordan block with eigenvalue λ. The composition of two Jordan blocks is illustrated by the following example

$$\underline{F} = \left[\begin{array}{ccc|cc} \lambda_1 & 1 & & & \\ & \lambda_1 & 1 & & \\ & & \lambda_1 & & \\ \hline & & & \lambda_2 & 1 \\ & & & & \lambda_2 \end{array}\right] \qquad \underline{g} = \left[\begin{array}{c} b_1 \\ b_2 \\ b_3 \\ \hline c_1 \\ c_2 \end{array}\right]$$

$$\underline{\mathscr{C}} = \left[\begin{array}{ccccc} b_1 & b_2 & b_3 & 0 & 0 \\ b_2 & b_3 & 0 & 0 & 0 \\ b_3 & 0 & 0 & 0 & 0 \\ \hline 0 & 0 & 0 & c_1 & c_2 \\ 0 & 0 & 0 & c_2 & 0 \end{array}\right] \left[\begin{array}{ccccc} 1 & \lambda_1 & \lambda_1^2 & \lambda_1^3 & \lambda_1^4 \\ 0 & 1 & 2\lambda_1 & 3\lambda_1^2 & 4\lambda_1^3 \\ 0 & 0 & 1 & 3\lambda_1 & 6\lambda_1^2 \\ \hline 1 & \lambda_2 & \lambda_2^2 & \lambda_2^3 & \lambda_2^4 \\ 0 & 1 & 2\lambda_2 & 3\lambda_2^2 & 4\lambda_2^3 \end{array}\right] \qquad (2.3.24)$$

The system is controllable for $\lambda_1 \neq \lambda_2$, $b_3 \neq 0$, $c_2 \neq 0$. In the case $\lambda_1 = \lambda_2$, $b_3 \neq 0$ only the third order subsystem is controllable. A matrix is called "cyclic" if the Jordan blocks are associated with distinct eigenvalues. (See also section A.7.) In terms of the above controllability results: a matrix $\underline{F}$ is cyclic, if there exists a vector $\underline{g}$ such that $(\underline{F}, \underline{g})$ is controllable.

The numerical calculation of Jordan forms and its use for testing system properties like controllability are studied in [81.15]. The

Jordan form directly exhibits the eigenvalues of controllable or uncontrollable subsystems. They are called controllable or uncontrollable eigenvalues. The controllability of an eigenvalue can also be examined without transformation to Jordan form by the Hautus test [69.9], [72.2].

An eigenvalue λ_i of $\underline{F}$ is controllable
if rank $[\underline{F}-\lambda_i\underline{I},\underline{G}] = n$ (2.3.25)
In the single-input case $\underline{G} = \underline{g}$.

An essential effect of controllability is, that all controllable eigenvalues can be shifted arbitrarily by a state vector feedback.

$$u = -\underline{k}'\underline{x} \tag{2.3.26}$$

This property was shown for the single-input case in [60.5], [62.1] and for the multi-variable case in [64.2], [64.11], [67.1], see also section 2.6. This property can be used as a controllability test: Produce a vector $\underline{k}'$ by a random number generator and compare the eigenvalues of $\underline{F}$ and $\underline{F}-\underline{g}\underline{k}'$. The eigenvalues which are common to the two matrices are almost certainly uncontrollable [78.4], [81.4].

A pair $(\underline{F}, \underline{g})$ is "stabilizable" if all unstable eigenvalues of $\underline{F}$ are controllable. Then they can be shifted to the left half s-plane by a state-vector feedback (2.3.26).

2.3.5 Linear Dependencies in the Controllability Matrix

Let rank $\mathscr{C} = r < n$. Then there is a basis change in the state space transforming the system to the following form

$$\begin{bmatrix} \dot{\underline{z}}_1 \\ \dot{\underline{z}}_2 \end{bmatrix} = \begin{bmatrix} \underline{F}_{11} & \underline{F}_{12} \\ 0 & \underline{F}_{22} \end{bmatrix} \begin{bmatrix} \underline{z}_1 \\ \underline{z}_2 \end{bmatrix} + \begin{bmatrix} \underline{g}_1 \\ 0 \end{bmatrix} u \text{ with } (\underline{F}_{11}, \underline{g}_1) \text{ controllable} \tag{2.3.27}$$

where $\underline{z}_1$ is a vector in an r-dimensional controllable subspace and $\underline{z}_2$ a vector in the complementary uncontrollable subspace of dimension n-r [63.2], [63.3]. The form (2.3.27) is not unique, because for each of the subspaces the basis can still be chosen. The form (2.3.27) can be achieved in a numerically efficient way by making use of the Hessenberg form of a matrix, see eq. (A.5.1).

The controllable subsystem is

$$\dot{\underline{z}}_1 = \underline{F}_{11}\underline{z}_1+\underline{g}_1 u+\underline{F}_{12}\underline{z}_2 \tag{2.3.28}$$

and the uncontrollable one

$$\dot{\underline{z}}_2 = \underline{F}_{22}\underline{z}_2 \tag{2.3.29}$$

The controllability matrices of (2.3.27) and the original system representation ($\underline{F}$, $\underline{g}$) are related by eq. (2.3.19)

$$\underline{\mathscr{C}}^* = \begin{bmatrix} \underline{g}_1 & \underline{F}_{11}\underline{g}_1 \ldots \underline{F}_{11}^{n-1}\underline{g}_1 \\ \underline{0} & \underline{0} \qquad \underline{0} \end{bmatrix} = \underline{T}[\underline{g},\ \underline{F}\underline{g}\ldots\underline{F}^{n-1}\underline{g}] \tag{2.3.30}$$

There must arise the same linear dependencies between the columns of the controllability matrices on both sides of eq. (2.3.30), i.e. $\underline{\mathscr{C}}^*\underline{v} = \underline{T}\,\underline{\mathscr{C}}\,\underline{v}$. Such linear relations result from the Cayley-Hamilton theorem, [59.4], eq. (A.7.30):

> Every square matrix satisfies its own characteristic equation, i.e. if
>
> $$\det(s\underline{I}-\underline{F}) = q_o s^o+q_1 s^1+\ldots+q_{n-1}s^{n-1}+s^n = 0 \tag{2.3.31}$$
>
> is the characteristic equation of $\underline{F}$, then
>
> $$q_o\underline{I}+q_1\underline{F}+\ldots+q_{n-1}\underline{F}^{n-1}+\underline{F}^n = \underline{0} \tag{2.3.32}$$

Applying this theorem to $\underline{F}_{11}$:

$$\det(s\underline{I}-\underline{F}_{11}) = p_o+p_1 s+\ldots+p_{r-1}s^{r-1}+s^r \tag{2.3.33}$$

implies

$$p_o\underline{I}+p_1\underline{F}_{11}+\dots p_{r-1}\underline{F}_{11}^{r-1}+\underline{F}_{11}^{r} = \underline{0} \quad (2.3.34)$$

Therefore in eq. (2.3.30)

$$\begin{bmatrix} \underline{g}_1 & \underline{F}_{11}\underline{g}_1 \dots \underline{F}_{11}^r\underline{g}_1 \\ \underline{0} & \underline{0} \quad\quad \underline{0} \end{bmatrix} \begin{bmatrix} p_o \\ p_1 \\ \vdots \\ p_{r-1} \\ 1 \end{bmatrix} = \underline{T}[\underline{g},\ \underline{Fg}\dots\underline{F}^r\underline{g}] \begin{bmatrix} p_o \\ p_1 \\ \vdots \\ p_{r-1} \\ 1 \end{bmatrix} = \underline{0} \quad (2.3.35)$$

Thus the characteristic polynomial (2.3.33) of the controllable subsystem of a pair ($\underline{F}$, $\underline{g}$) can be determined as follows: Check the columns $\underline{g}$, $\underline{Fg}$, $\underline{F}^2\underline{g}$... of the controllability matrix for their linear dependence. The first linear dependent one may be expressed as

$$\underline{F}^r\underline{g} = -p_o\underline{g} - p_1\underline{Fg} - \dots - p_{r-1}\underline{F}^{r-1}\underline{g} \quad (2.3.36)$$

The p_i are the characteristic polynomial coefficients of the controllable subsystem. The polynomial with coefficients p_i is also called minimal annihilating polynomial of the vector $\underline{g}$ with respect to the matrix $\underline{F}$ [59.4].

2.3.6 Observability

A time-invariant continuous system

$$\begin{aligned} \dot{\underline{x}} &= \underline{Fx}+\underline{g}u \\ y &= \underline{c}'\underline{x} \end{aligned} \quad (2.3.37)$$

is observable iff the state $\underline{x}(t_o)$ can be determined from u(t) and y(t) in a finite interval $t_o \le t \le t_1$. A system is observable if and only if

$$\det \underline{\mathcal{O}} \neq 0 \quad (2.3.38)$$

where

$$\underline{\mathcal{O}} := \begin{bmatrix} \underline{c}' \\ \underline{c}'\underline{F} \\ \vdots \\ \underline{c}'\underline{F}^{n-1} \end{bmatrix} \tag{2.3.39}$$

is the observability matrix. Algebraically this is the same type of property as controllability:

$$\det \underline{\mathcal{O}} = \det \underline{\mathcal{O}}' = \det[\underline{c}, \underline{F}'\underline{c}, \ldots \underline{F}'^{n-1}\underline{c}] \tag{2.3.40}$$

This means $(\underline{c}', \underline{F})$ is observable if and only if $(\underline{F}', \underline{c})$ is controllable. Because of this duality, all results of the previous sections can be immediately applied to observability.

Example 1: Loading bridge with the measured variable trolley position

The rows of the observability matrix are

$$\begin{aligned} \underline{c}' &= [1 \quad 0 \quad 0 \quad 0\;] \\ \underline{c}'\underline{F} &= [0 \quad 1 \quad 0 \quad 0\;] \\ \underline{c}'\underline{F}^2 &= [0 \quad 0 \quad f_{23} \quad 0\;] \\ \underline{c}'\underline{F}^3 &= [0 \quad 0 \quad 0 \quad f_{23}] \end{aligned} \tag{2.3.41}$$

The plant is observable. However, for small load mass m_L, also $f_{23} = m_L g/m_C$ is small and the system is "weakly observable".

Example 2: Loading bridge with the measured variable cable angle.

The rows of the observability matrix are

$$\begin{aligned} \underline{c}' &= [0 \quad 0 \quad 1 \quad 0\;] \\ \underline{c}'\underline{F} &= [0 \quad 0 \quad 0 \quad 1\;] \\ \underline{c}'\underline{F}^2 &= [0 \quad 0 \quad f_{43} \quad 0\;] \\ \underline{c}'\underline{F}^3 &= [0 \quad 0 \quad 0 \quad f_{43}] \end{aligned} \tag{2.3.42}$$

Since $\det \underline{\mathcal{O}} = 0$, the system is unobservable. A subsystem of second order is observable, because rank $\underline{\mathcal{O}} = 2$. Here equation (2.3.36) has the form $\underline{c}'\underline{F}^2 = -p_0\underline{c}' - p_1\underline{c}'\underline{F} = f_{43}\underline{c}'$, therefore $p_0 = -f_{43}$, $p_1 = 0$. The characteristic polynomial of the observable subsystem is $s^2 - f_{43}$.

The dual form of eq. (2.3.27) is

$$\begin{bmatrix} \dot{\underline{z}}_1 \\ \dot{\underline{z}}_2 \end{bmatrix} = \begin{bmatrix} \underline{F}_{11} & \underline{0} \\ \underline{F}_{21} & \underline{F}_{22} \end{bmatrix} \begin{bmatrix} \underline{z}_1 \\ \underline{z}_2 \end{bmatrix} + \begin{bmatrix} \underline{g}_1 \\ \underline{g}_2 \end{bmatrix} u \tag{2.3.43}$$

$$y = [\underline{c}_1' \quad \underline{0}] \begin{bmatrix} \underline{z}_1 \\ \underline{z}_2 \end{bmatrix}, \quad (\underline{c}_1', \underline{F}_{11}) \text{ observable}$$

the observable subsystem is

$$\begin{aligned} \dot{\underline{z}}_1 &= \underline{F}_{11}\underline{z}_1 + \underline{g}_1 u \\ y &= \underline{c}_1' \underline{z}_1 \end{aligned} \tag{2.3.44}$$

Example 2 (continued):

After rearrangement of the state variables eqs. (2.1.15) and (2.3.42) are already in the form (2.3.43). The observable subsystem is the pendulum

$$\begin{bmatrix} \dot{x}_3 \\ \dot{x}_4 \end{bmatrix} = \begin{bmatrix} 0 & 1 \\ f_{43} & 0 \end{bmatrix} \begin{bmatrix} x_3 \\ x_4 \end{bmatrix} + \begin{bmatrix} 0 \\ g_4 \end{bmatrix} u$$

$$y = [1 \quad 0] \begin{bmatrix} x_3 \\ x_4 \end{bmatrix}$$

The unobservable subsystem is the trolley.

2.3.7 Canonical Decomposition, Pole Zero Cancellations

As Kalman [63.2] and Gilbert [63.3] have shown, every constant, finite-dimensional, linear system

$$\begin{aligned} \dot{\underline{x}} &= \underline{F}\underline{x}+\underline{G}\underline{u} \\ \underline{y} &= \underline{H}\underline{x} \end{aligned} \tag{2.3.45}$$

can be decomposed into four subsystems with the following properties:

A: controllable, unobservable

B: controllable, observable

C: uncontrollable, observable

D: uncontrollable, unobservable

This means that a transformation

$$\begin{bmatrix} \underline{x}_A \\ \underline{x}_B \\ \underline{x}_C \\ \underline{x}_D \end{bmatrix} = \underline{T}\underline{x} \ , \ \det \underline{T} \neq 0 \tag{2.3.46}$$

exists such that

$$\begin{aligned} \begin{bmatrix} \dot{\underline{x}}_A \\ \dot{\underline{x}}_B \\ \dot{\underline{x}}_C \\ \dot{\underline{x}}_D \end{bmatrix} &= \begin{bmatrix} \underline{F}_{11} & \underline{F}_{12} & \underline{F}_{13} & \underline{F}_{14} \\ \underline{0} & \underline{F}_{22} & \underline{F}_{23} & \underline{0} \\ \underline{0} & \underline{0} & \underline{F}_{33} & \underline{0} \\ \underline{0} & \underline{0} & \underline{F}_{43} & \underline{F}_{44} \end{bmatrix} \begin{bmatrix} \underline{x}_A \\ \underline{x}_B \\ \underline{x}_C \\ \underline{x}_D \end{bmatrix} + \begin{bmatrix} \underline{G}_1 \\ \underline{G}_2 \\ \underline{0} \\ \underline{0} \end{bmatrix} \\ \underline{y} &= [\underline{0} \quad \underline{H}_2 \quad \underline{H}_3 \quad \underline{0}] \begin{bmatrix} \underline{x}_A \\ \underline{x}_B \\ \underline{x}_C \\ \underline{x}_D \end{bmatrix} \end{aligned} \tag{2.3.47}$$

where the subsystems $\underline{F}_{11}$ and $\underline{F}_{22}$ are controllable and the subsystems $\underline{F}_{22}$ and $\underline{F}_{33}$ are observable. Eqs. (2.3.27) and (2.3.43) are contained here as special cases. Figure 2.6 displays eq. (2.3.47).

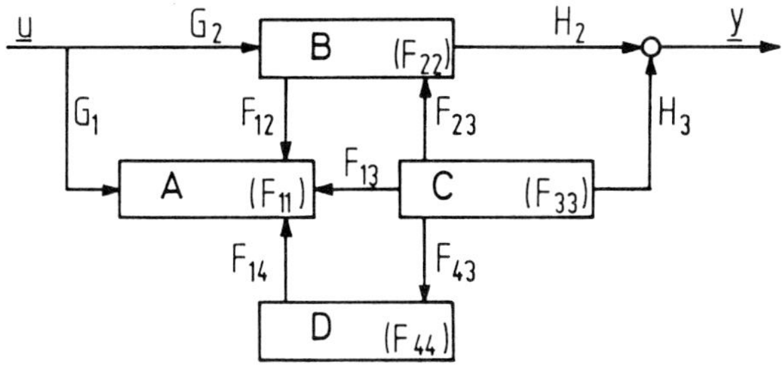

Figure 2.6 Canonical decomposition.

Only the controllable and observable part B enters into the transfer function from $\underline{u}$ to $\underline{y}$, that is

$$\underline{y}_s = \underline{G}_s \underline{u}_s$$
$$\underline{G}_s = \underline{H}(s\underline{I}-\underline{F})^{-1}\underline{G} = \underline{H}_2(s\underline{I}-\underline{F}_{22})^{-1}\underline{G}_2 \qquad (2.3.48)$$

Example: Loading bridge with output cable angle.

> In the transfer function, eq. (2.3.13), the double pole at $s = 0$ is cancelled. It belongs to the unobservable trolley subsystem.

Only the eigenvalues of subsystem B can be changed by feedback from $\underline{y}$ to $\underline{u}$. For the design of a controller, one is therefore interested in first removing the subsystems A, C, and D from the description of the plant. In this way only a minimal realization B of the transfer behavior from $\underline{u}$ to $\underline{y}$ becomes the basis of the design. This is acceptable only, if C is stable and if A cannot be put into a dangerous state by internal instability. The practical execution of the canonical decomposition can be performed by assigning to B those eigenvalues which have been altered by an arbitrary feedback $\underline{u} = -\underline{K}\underline{y}$ [78.4]. There are, however, numerically critical cases in this procedure [81.4]. For numerical aspects see also [81.14]. In the design problems treated in this book, it is assumed that, if necessary, this separation of part B has previously taken place. The controllability and observability of the plant is therefore assumed. In terms of the transfer function this means that numerator and denominator do not have common zeros. If an eigenvalue is either uncontrollable or unobservable, then it cancels in eq. (2.3.16) and there only remains part B.

Connecting subsystems can result in the cancellation of poles and zeros, this in turn leading to a reduction in the order of the transfer function compared to the sum of the orders of the subsystems. In this case a subsystem of type A, C, or D arises whose eigenvalues are equal to the cancelled poles. Subsystem B is described by the transfer function with all cancellations performed. Figure 2.7 shows three typical cases for systems with one input and one output.

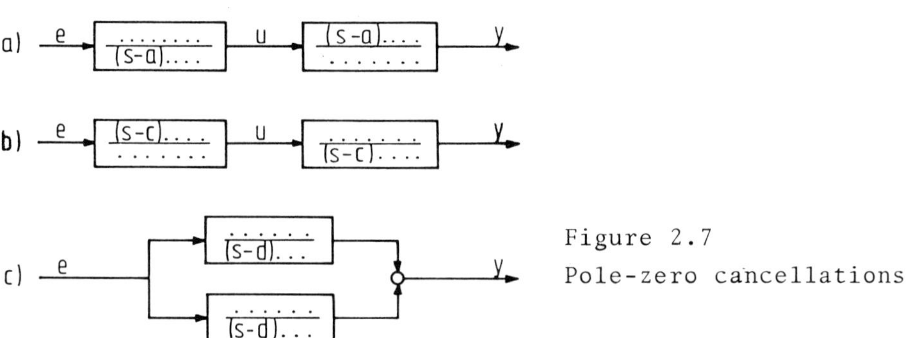

Figure 2.7
Pole-zero cancellations

In case a) the eigenvalue $s = a$ is controllable; because of the cancellation it is not observable from y (type A). In case b) the eigenvalue $s = c$ is observable; it is not controllable from e (type C). In case c) two subsystems of first order with the eigenvalue $s = d$ can be produced by partial fraction decomposition. The difference of their state variables is neither controllable nor observable (type D).

In frequency domain design methods the cancellation compensation is sometimes used for simplification. Then in figures 2.7 a and b the first subsystem is the controller, the second one is the plant, and the loop is closed by $e = w-y$. For such cancellation compensation the following points should be noted.

1. If a plant pole is cancelled by a compensator zero then the combined system will have a corresponding observable, uncontrollable mode. This pole cannot be shifted by feedback from

y to e. Thus also the closed-loop system will have this mode which can be excited by plant disturbances or initial conditions. This is not admissible for unstable poles. For slow and unsufficiently damped poles it results in an unfavorable disturbance response, although the cancelled reference response may be excellent.

2. If a plant zero is cancelled by a compensator pole, then the combined system will have a corresponding controllable, unobservable mode. This mode will again be present in the closed-loop system. It can be excited by the reference input or by measurement noise in y, and it is observable from u. This is not admissible for unstable poles. For fast and unsufficiently damped poles this results in fast actuator activity excited by high-frequency noise entering with the measurement y or with the reference input w.

2.4 Solutions of the Differential Equation

The differential equation

$$\dot{\underline{x}}(t) = \underline{F}\underline{x}(t) + \underline{g}u(t) \tag{2.4.1}$$

with initial state $\underline{x}(t_o)$ and given input $u(t)$, $t \geq t_o$ has the solution (see for example [55.1], [63.1]):

$$\underline{x}(t) = e^{\underline{F}(t-t_o)}\underline{x}(t_o) + \int_{t_o}^{t} e^{\underline{F}(t-\tau)}\underline{g}u(\tau)d\tau \tag{2.4.2}$$

$e^{\underline{F}t}$ is called "transition matrix".

2.4.1 Calculation of the Transition Matrix

The exponential function of a matrix is defined by the following power series (see eq. (A.7.54))

$$e^{\underline{F}t} := \sum_{i=0}^{\infty} \frac{\underline{F}^i t^i}{i!} \tag{2.4.3}$$

This series with a suitable truncation criterion can be used for the numerical computation of the transition matrix. The calculation for a fixed numerical value of t will be discussed in section 3.1. Methods for the numerical computation of $e^{\underline{F}t}$, also for higher order systems can be found in the literature, e.g. [78.3].

If the eigenvalues of $\underline{F}$ are known, then the transition matrix can be calculated as a function of t. Two possibilities are mentioned here:

a) by Laplace transform.

In eq. (2.4.2) let $t_o = 0$, $u(t) \equiv 0$

$$\underline{x}(t) = e^{\underline{F}t}\underline{x}(0) \tag{2.4.4}$$

on the other hand from eq. (2.3.3) with $u_s = 0$

$$\underline{x}_s(s) = (s\underline{I}-\underline{F})^{-1}\underline{x}(0) \tag{2.4.5}$$

Inverse Laplace transform yields

$$\underline{x}(t) = \mathcal{L}^{-1}\left\{(s\underline{I}-\underline{F})^{-1}\right\}\underline{x}(0) \tag{2.4.6}$$

Since eqs. (2.4.4) and (2.4.6) hold for every $\underline{x}(0)$,

$$e^{\underline{F}t} = \mathcal{L}^{-1}\left\{(s\underline{I}-\underline{F})^{-1}\right\} \tag{2.4.7}$$

Here $(s\underline{I}-\underline{F})^{-1}$ can be determined by the Leverrier algorithm, eq. (2.3.7). For the general inversion of the Laplace transform, for example via partial fractions decomposition, the denominator must be factored, i.e. the eigenvalues of $\underline{F}$ must be known.

b) By eq. (A.7.55), there exist coefficients c_i such that

$$e^{\underline{F}t} = \sum_{i=0}^{n-1} c_i(t)\underline{F}^i \tag{2.4.8}$$

The c_i are determined from the condition that eq. (2.4.8) is satisfied not only by the matrix $\underline{F}$ but also by its eigenvalues s_j. For simple eigenvalues one obtains the n equations.

$$e^{s_j t} = \sum_{i=0}^{n-1} c_i s_j^{\,i} \;, \quad j = 1,2\ldots n \tag{2.4.9}$$

If an eigenvalue s_j of multiplicity p arises, eq. (2.4.9) is (p-1) times differentiated with respect to s_j:

$$te^{s_j t} = \sum_{i=1}^{n-1} c_i i s_j^{\,i-1}$$

$$\vdots$$

$$t^{p-1} e^{s_j t} = \sum_{i=p-1}^{n-1} c_i i(i-1)\ldots(i-p+2) s_j^{\,i-p+1} \tag{2.4.10}$$

Another approach is suitable if the eigenvectors of $\underline{F}$ have been determined anyway. Then $\underline{F}$ can be transformed to its Jordan form

$$\underline{F} = \begin{bmatrix} \underline{J}_1 & & & \\ & \underline{J}_2 & & \\ & & \ddots & \\ & & & \underline{J}_m \end{bmatrix} , \quad \underline{J}_i \begin{bmatrix} s_i & 1 & 0 & 0 \\ 0 & s_i & 1 & \\ \vdots & & \ddots & 1 \\ 0 & & & s_i \end{bmatrix} \tag{2.4.11}$$

$$e^{\underline{F}t} = \begin{bmatrix} e^{\underline{J}_1 t} & \underline{0} & \ldots & \underline{0} \\ \underline{0} & e^{\underline{J}_2 t} & & \\ \vdots & & \ddots & \\ 0 & & & e^{\underline{J}_m t} \end{bmatrix} \tag{2.4.12}$$

and for an r×r Jordan block

$$e^{\underline{J}_i t} = e^{s_i t}\begin{bmatrix} 1 & t & t^2/2 \ldots t^{r-1}/(r-1)! \\ 0 & 1 & t \\ \vdots & & \ddots & \ddots \\ \vdots & & & \ddots & t \\ 0 & & & & 1 \end{bmatrix} \qquad (2.4.13)$$

The terms $e^{s_i t}$ are called "modes" of the system.

2.4.2 Impulse and Step Responses

In eq. (2.4.2) let $\underline{x}(t_o) = 0$, $t_o = 0$. Also an output relation $y = \underline{c}'\underline{x}+du$ is assumed. The response $y(t)$ is examined for typical inputs $u(t)$. For a δ-function input $u(t) = \delta(t)$ the impulse response is obtained as

$$g(t) = \underline{c}'e^{\underline{F}t}\underline{g} + d\delta(t) \qquad (2.4.14)$$

$g(t)$ is also designated as "weighting function". With $g(t)$ the response to any $u(t)$ can be expressed according to eq. (2.4.2) by the convolution integral

$$y(t) = \int_0^t g(t-\tau)u(\tau)d\tau \qquad (2.4.15)$$

By this representation also the response of linear nonrational transfer function systems can be described which cannot be represented by an ordinary differential equation. Besides rational systems, only time-delay systems with $g(t) = 0$ for $0 < t < T_t$, will be treated in this book.

The Laplace transform of the weighting function is the transfer function. This follows from eqs. (2.4.14) and (2.4.7)

$$g_s(s) = \mathcal{L}\{g(t)\} = \underline{c}'\mathcal{L}\{e^{\underline{F}t}\}\underline{g} + d = \underline{c}'(s\underline{I}-\underline{F})^{-1}\underline{g} + d \qquad (2.4.16)$$

which is the transfer function from eq. (2.3.5). More general, i.e. also for nonrational systems, the application of the convolution theorem of the Laplace transform to eq. (2.4.15) yields

$$y_s(s) = g_s(s) \times u_s(s) \tag{2.4.17}$$

For a time-delay system with $y(t) = u(t-T_t)$ and via shifting theorem one obtains

$$y_s(s) = u_s(s) \times e^{-sT_t} \tag{2.4.18}$$

The step response is the response to a unit step

$$1(t) = \begin{cases} 0 & t < 0 \\ 1 & t > 0 \end{cases} \tag{2.4.19}$$

From eq. (2.4.15) the step response is

$$y_{step}(t) = \int_0^t g(t-\tau)d\tau = \int_0^t g(v)dv$$

$$= \underline{c}'\int_0^t e^{\underline{F}v}dv\underline{g} + d \tag{2.4.20}$$

and for det $\underline{F} \neq 0$

$$y_{step}(t) = \underline{c}'(e^{\underline{F}t}-\underline{I})\underline{F}^{-1}\underline{g} + d \tag{2.4.21}$$

2.5 Specifications

For the assessment of the performance of a control system, the closed-loop step response is of special interest. Specifications are formulated both for the stationary response and for the dynamic transition from the zero state. The differential equation of the closed-loop with reference input r is

$$\dot{\underline{x}} = \underline{H}\underline{x}+\underline{g}r, \quad \det \underline{H} \neq 0 \text{ (because of the stable design there is no eigenvalue at } s = 0) \tag{2.5.1}$$

$$y = \underline{c}'\underline{x}$$

The step response from eq. (2.4.21) is

$$y_{step}(t) = \underline{c}'\int_0^t e^{\underline{H}v}dv\underline{g} = \underline{c}'(e^{\underline{H}t}-\underline{I})\underline{H}^{-1}\underline{g} \qquad (2.5.2)$$

2.5.1 Steady-State Response and Integral Controller

The steady-state response results for $t \to \infty$. Because the control loop is designed to be stable, $\dot{\underline{x}}(\infty) = \underline{0}$, and from eq. (2.5.1)

$$\begin{aligned} \underline{x}(\infty) &= -\underline{H}^{-1}\underline{g}r(\infty) \\ y(\infty) &= -\underline{c}'\underline{H}^{-1}\underline{g}r(\infty) \end{aligned} \qquad (2.5.3)$$

In order to achieve $y(\infty) = w(\infty) = 1$ in the stationary response to a reference step $w(t) = 1(t)$, $r(t)$ can be formed by a factor $V = -1/\underline{c}'\underline{H}^{-1}\underline{g}$ from $w(t)$

$$r(t) = V \times w(t) = -\frac{1}{\underline{c}'\underline{H}^{-1}\underline{g}}\, w(t) \qquad (2.5.4)$$

This solution approach requires that $\underline{c}'\underline{H}^{-1}\underline{g}$ is known exactly. The steady-state accuracy is therefore not robust with respect to parameter uncertainties of the plant. Moreover steady-state accuracy is only reached for the reference variable w. This accuracy is not preserved under disturbance inputs with nonzero stationary value.

It is more advantageous to use a controller structure in which $e(t) = w(t)-y(t)$ is integrated in the controller. Let the plant be

$$\begin{aligned} \dot{\underline{x}} &= \underline{F}\underline{x} + \underline{g}u + \underline{z} \\ y &= \underline{c}'\underline{x} \end{aligned} \qquad (2.5.5)$$

where $\underline{z}$ is a constant unknown disturbance vector and y is the controlled variable. Assume a proportional-integral-controller with state-vector feedback

$$\begin{aligned} \dot{x}_I &= -\underline{c}'\underline{x}+w \\ u &= k_I x_I + k_P w - \underline{k}'\underline{x} \end{aligned} \qquad (2.5.6)$$

Figure 2.8 illustrates eqs. (2.5.5) and (2.5.6).

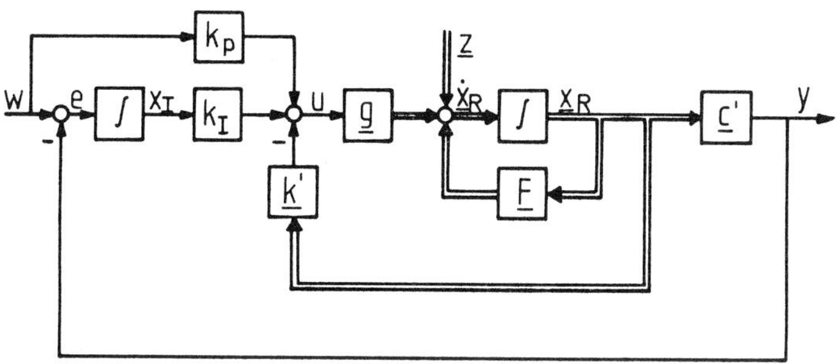

Figure 2.8 PI-controller with state-vector feedback

The controller of figure 2.8 corresponds to a state-vector feedback

$$u = [-\underline{k}' \quad k_I]\begin{bmatrix} \underline{x} \\ x_I \end{bmatrix} + k_P w \qquad (2.5.7)$$

for the system augmented by an integrator

$$\begin{bmatrix} \dot{\underline{x}} \\ \dot{x}_I \end{bmatrix} = \begin{bmatrix} \underline{F} & \underline{0} \\ -\underline{c}' & 0 \end{bmatrix}\begin{bmatrix} \underline{x} \\ x_I \end{bmatrix} + \begin{bmatrix} \underline{g} \\ 0 \end{bmatrix} u + \begin{bmatrix} \underline{I} \\ 0 \end{bmatrix} \underline{z} + \begin{bmatrix} 0 \\ 1 \end{bmatrix} w$$

The feedback gains $\underline{k}'$ and k_I are determined such that the augmented system is stabilized. Then in steady-state

$$\lim_{t\to\infty} \dot{x}_I(t) = \lim_{t\to\infty}[w(t)-y(t)] = 0$$

for $\underline{z}$ and w constant. The property of zero steady-state error is robust with respect to variation of the plant $(\underline{c}', \underline{F}, \underline{g})$ as long as the loop remains stable.

2.5.2 Transients and Location of the Eigenvalues

Besides the stationary response, the transient from $y(0) = 0$ to $y(\infty) = \underline{c}'\underline{x}(\infty)$ for a reference step is of special interest. It may be specified for example by the following requirements:

a) overshoot, less than 10 %, i.e. $\max_t y_{step}(t) \le 1.1$.

b) given time t_1, after which the absolute value of the return difference $|e(t)| = |1-y_{step}(t)|$ amounts to less than 2 %.

c) the avoidance of weakly damped high frequency solution terms,

d) the required actuator input $\max_t |u(t)|$ should be small.

In the structure of figure 2.8 first the eigenvalues are fixed by the feedback gains k_I and $\underline{k}'$. This will be more closely studied in section 2.6. $y_{step}(t)$ is then composed of the input step and solution terms, which correspond to the closed-loop eigenvalues. The essential design step is therefore the assignment of appropriate eigenvalues. Also for other than step inputs and disturbance inputs affecting the plant of eq. (2.5.5) as for instance $\underline{z}(t)$ the appropriate choice of the eigenvalues leads to fast and well-damped responses.

The reference step response can then still be formed by the choice of k_P. Also additional dynamic elements can be provided for in place of k_P or as a prefilter for w.

The solution terms which occur for the transient from $\underline{x}(0) = \underline{0}$ to $\underline{x}(\infty)$ in eq. (2.5.2) result from $e^{\underline{H}t}$, i.e. from the eigenvalues $\underline{s}_i$ of $\underline{H}$. For simple eigenvalues one obtains terms $e^{s_i t}$, $i = 1,2\dots n$ from eq. (2.4.9), for double eigenvalues $te^{s_i t}$ from eq. (2.4.10) etc. Since the characteristic polynomial always has real coefficients, complex eigenvalues only arise in conjugate pairs. They must join together in y to form a real signal, that is to form the term

$$y_i(t)=(\alpha+j\beta)e^{(\sigma_i-j\omega_i)t}+(\alpha-j\beta)e^{(\sigma_i+j\omega_i)t}=2e^{\sigma_i t}(\alpha\cos\omega_i t+\beta\sin\omega_i t) \tag{2.5.8}$$

The corresponding factor of the characteristic polynomial is

$$P_i(s) = (s-\sigma_i+j\omega_i)(s-\sigma_i-j\omega_i) = s^2-2\sigma_i s+\sigma_i^2+\omega_i^2 \tag{2.5.9}$$

If all eigenvalues are placed to the left of a line at $\sigma = -a$ parallel to the imaginary axis of the s-plane, then all solution terms $y_i(t)$ decay at least as e^{-at}.

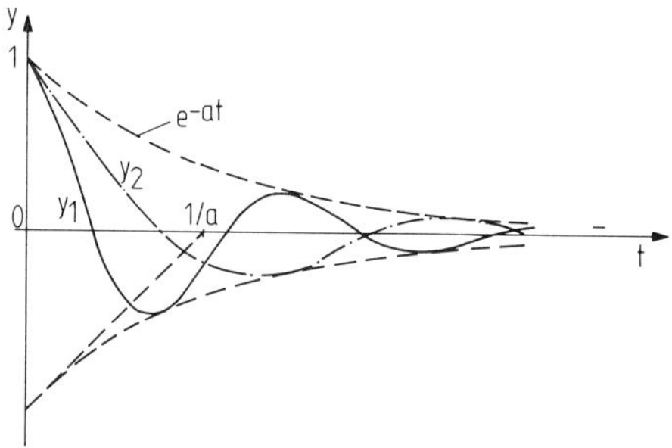

Figure 2.9 Two solution terms with the same negative real part $\sigma = -a$.

Figure 2.9 shows two examples y_1 and y_2 whose eigenvalues have the same negative real part $\sigma = -a$. However, a term of type y_1 is less desirable because it has more overshoot and oscillations within the envelope e^{-at} than a term of type y_2. For higher frequencies the eigenvalues should lie still further left in the s-plane. This is achieved by the assignment of a minimum value for the damping ζ. The term $P_i(s)$ may be written in the following form:

$$P_i(s) = s^2-2\sigma_i s+\sigma_i^2+\omega_i^2 = s^2+2\zeta_i\omega_{ni}s+\omega_{ni}^2 \tag{2.5.10}$$

$\omega_{ni} = \sqrt{\sigma_i^2+\omega_i^2}$ is the "natural frequency", which represents the distance of the eigenvalues from the origin of the s-plane, and $\zeta_i = -\sigma_i/\omega_{ni}$ is the "damping". Solving for σ_i and ω_i yields $\sigma_i = -\zeta_i\omega_{ni}$, $\omega_i = \omega_{ni}\sqrt{1-\zeta_i^2}$. Figure 2.10 illustrates these relations in the s-plane

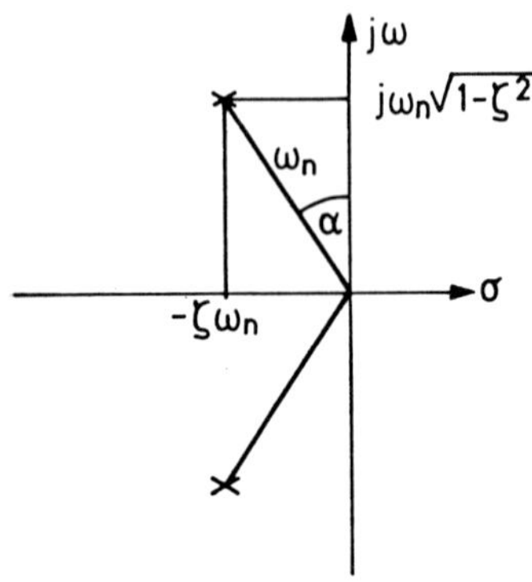

Figure 2.10

Natural frequency ω_n and damping ζ of a conjugate complex pair of poles at $s = \sigma \pm j\omega$

A value ζ_i of the damping corresponds to an angle α_i with respect to the imaginary axis with

$$\tan \alpha_i = \frac{\zeta_i}{\sqrt{1-\zeta_i}} \tag{2.5.11}$$

and eq. (2.5.8) reads

$$y_i(t) = e^{-\zeta_i \omega_{ni} t}\left[\alpha \times \cos\left(\omega_{ni} t \sqrt{1-\zeta_i^2}\right) + \beta \times \sin\left(\omega_{ni} t \sqrt{1-\zeta_i^2}\right)\right] \tag{2.5.12}$$

ω_{ni} only occurs in a product with t and can therefore be interpreted as a scaling of the time. Figure 2.11 shows the step response of the system

$$g_s(s) = \frac{\omega_n^2}{s^2 + 2\zeta\omega_n s + \omega_n^2} \tag{2.5.13}$$

plotted over normalized time $\omega_n t$ for different damping values.

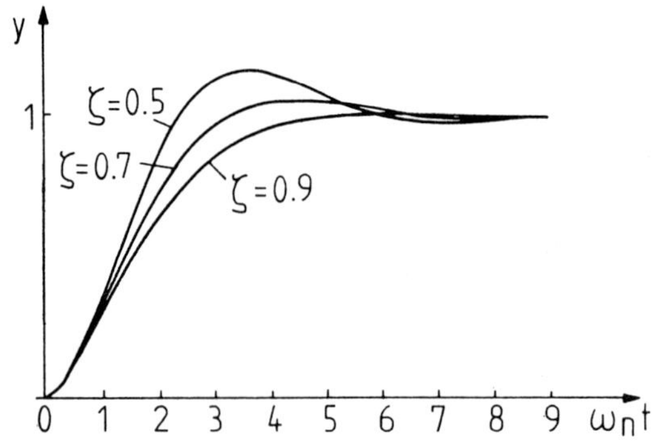

Figure 2.11

Step response of the system of eq. (2.5.13)

The step response for $\zeta = 1/\sqrt{2} \approx 0.7$ is considered as specially favorable. This value is distinguished by the fact that the absolute value of the frequency response $|g_s(j\omega)|$ has a maximum only if the damping is $\zeta < 1/\sqrt{2}$. This maximum occurs at the resonance frequency $\omega_r = \sqrt{1-2\zeta^2}$. In other words, for $\zeta \geq 1/\sqrt{2}$ no resonance occurs. In figure 2.10, for this value of $\zeta = 1/\sqrt{2}$, $\alpha = 45^o$.

The form of the step response essentially does not change if the transfer function (2.5.13) is augmented by far left poles, zeros far from the origin or stable pole-zero pairs lying close together such that they almost cancel. Then the conjugate complex pole pair in eq. (2.5.13) is designated as "dominant". It may also be defined via a partial fractions decomposition of the transfer function into several parallel first and second order terms. A dominating term (2.5.13) has the largest residue. Additive faster solution terms with sufficient damping have a significant effect only in the initial part of the transient and vanish before the maximal overshoot.

The requirement $\sigma \leq -a$ ensures that for real eigenvalues or small natural frequencies ω_n the transients are not too sluggish. Combining the two conditions for damping and real part gives the desirable domain $\sigma \leq \sigma_o$, $\zeta \geq \zeta_o$ represented in figure 2.12 for the position of the eigenvalues of the closed loop.

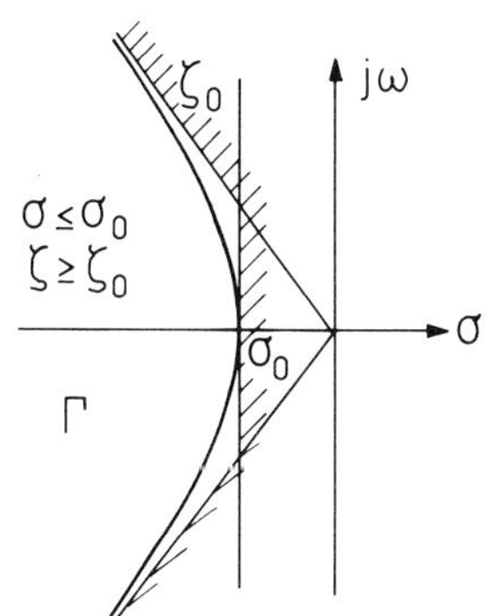

Figure 2.12

Desired position of the closed loop eigenvalues

Since the piecewise definition of the complex boundary curve is awkward in some design methods, the two demands can also be replaced by the requirement that all eigenvalues lie left of the hyperbola in figure 2.12 [80.6], [82.1], [82.6], [83.1]. Its equation is

$$\omega^2 = \frac{1-\zeta_o^2}{\zeta_o^2}(\sigma^2-\sigma_o^2)\ , \quad \sigma \le \sigma_o \le 0 \qquad (2.5.14)$$

The specification of a desired pole region like Γ in figure 2.12 admits the definition of "Γ-stability" or "nice stability" as the property of a system, that all its eigenvalues are located in Γ. An essential design step is Γ-stabilization. For this step it is recommended that the natural frequency of open-loop complex eigenvalues is not changed too much by closing the loop. Unstable or unsufficiently damped eigenvalues are moved to the left into Γ. Simultaneously far left eigenvalues may be moved to the right, such that also the center of gravity of the eigenvalues is not moved too much. The center of gravity of the eigenvalues of an $n\times n$ matrix $\underline{A}$ follows from eq. (A.7.34) as (trace $\underline{A}$)/n. The trace remains unchanged if actuator state variables are not fed back. The system with plant states $\underline{x}$ and actuator states $\underline{x}_a$ is

$$\begin{bmatrix}\dot{\underline{x}}\\ \dot{\underline{x}}_a\end{bmatrix} = \begin{bmatrix}\underline{F} & \underline{g}\underline{c}_a'\\ 0 & \underline{F}_a\end{bmatrix}\begin{bmatrix}\underline{x}\\ \underline{x}_a\end{bmatrix} + \begin{bmatrix}0\\ \underline{g}_a\end{bmatrix}u \qquad (2.5.15)$$

and for feedback of $\underline{x}$ only

$$u = -[\underline{k}' \quad 0]\begin{bmatrix}\underline{x}\\ \underline{x}_a\end{bmatrix}$$

$$\begin{bmatrix}\dot{\underline{x}}\\ \dot{\underline{x}}_a\end{bmatrix} = \begin{bmatrix}\underline{F} & \underline{g}\underline{c}_a'\\ -\underline{g}_a\underline{k}' & \underline{F}_a\end{bmatrix}\begin{bmatrix}\underline{x}\\ \underline{x}_a\end{bmatrix} \qquad (2.5.16)$$

The trace of the closed-loop dynamic matrix does not depend on $\underline{k}'$. The geometric center of all eigenvalues of plant and actuator is

$$S_{cg} = (\text{trace } \underline{F} + \text{trace } \underline{F}_a)/(n+n_a) \qquad (2.5.17)$$

where n_a is the number of actuator states.

S_{cg} can be changed only if the actuator states $\underline{x}_a$ are also fed back. On the other hand, if the actuator dynamics are neglected in the design and the geometric center is shifted far to the left, then the analysis of the system with actuator will show, that its eigenvalues have moved far to the right, because eq. (2.5.17) must be satisfied.

2.6 Pole Shifting

2.6.1 Closed-loop Eigenvalues

For a system

$$\dot{\underline{x}} = \underline{F}\underline{x} + \underline{g}u \tag{2.6.1}$$

the output vector

$$\underline{y} = \underline{C}\underline{x} \tag{2.6.2}$$

is measured and fed back to the input u by proportional feedback

$$u = -\underline{k}_y'\underline{y} + w \tag{2.6.3}$$

Then the closed-loop state equation is

$$\dot{\underline{x}} = (\underline{F}-\underline{g}\underline{k}_y'\underline{C})\underline{x} + \underline{g}w \tag{2.6.4}$$

For the characteristic polynomial the notation

$$\begin{aligned} P(s) &= p_o+p_1s+\ldots p_{n-1}s^{n-1}+s^n = [\underline{p}' \quad 1]\underline{s}_n \\ \underline{p}' &:= [p_o \quad p_1\ldots p_{n-1}], \quad \underline{s}_n' := [1 \quad s\ldots s^n] \end{aligned} \tag{2.6.5}$$

is introduced. A polynomial P(s) is thus characterized by its coefficient vector $\underline{p}'$.

For the open-loop system the characteristic polynomial is

$$Q(s) = \det(s\underline{I}-\underline{F}) = [\underline{q}' \quad 1]\underline{s}_n \tag{2.6.6}$$

and for the closed loop

$$P(s) = \det(s\underline{I}-\underline{F}+\underline{g}\underline{k}_y'\underline{C}) = [\underline{p}' \quad 1]\underline{s}_n \tag{2.6.7}$$

This relation $\underline{p} = \underline{p}(\underline{k}_y)$ will be further examined in this section.

Laplace transform of eq. (2.6.1) with $\underline{x}(0) = \underline{0}$ yields

$$s\underline{x}_s(s) = \underline{F}\underline{x}_s(s) + \underline{g}u_s(s)$$

$$\underline{x}_s(s) = (s\underline{I}-\underline{F})^{-1}\underline{g}u_s(s)$$

and with eqs. (2.6.2) and (2.6.3)

$$u_s(s) = -\underline{k}_y'\underline{C}(s\underline{I}-\underline{F})^{-1}\underline{g}u_s(s) + w_s(s)$$

$$[1+\underline{k}_y'\underline{C}(s\underline{I}-\underline{F})^{-1}\underline{g}]u_s(s) = w_s(s) \tag{2.6.8}$$

The resolvent $(s\underline{I}-\underline{F})^{-1}$ may be evaluated by Leverrier's algorithm (see eqs. (2.3.7) and (A.7.35))

$$(s\underline{I}-\underline{F})^{-1} = \frac{\underline{D}_0+\underline{D}_1 s+\ldots+\underline{D}_{n-1}s^{n-1}}{q_0+q_1 s+\ldots+q_{n-1}s^{n-1}+s^n} = \frac{\underline{D}(s)}{Q(s)} \tag{2.6.9}$$

and eq. (2.6.8) becomes

$$\left(1 + \frac{\underline{k}_y'\underline{C}\underline{D}(s)\underline{g}}{Q(s)}\right)u_s(s) = w_s(s)$$

$$u_s(s) = \frac{Q(s)}{Q(s)+\underline{k}_y'\underline{C}\underline{D}(s)\underline{g}}\, w_s(s) \tag{2.6.10}$$

The poles of the uncancelled transfer function are the eigenvalues of $\underline{F}-\underline{g}\underline{k}_y'\underline{C}$; the characteristic polynomial of the closed loop is therefore

$$P(s) = Q(s) + \underline{k}_y'\underline{CD}(s)\underline{g} \tag{2.6.11}$$

The matrix $\underline{D}(s)\underline{g}$ can be written with its coefficient matrix $\underline{W}$, i.e.

$$\underline{D}(s)\underline{g} = \underline{Ws}_{n-1}$$
$$\underline{W} := [\underline{D}_o\underline{g},\ \underline{D}_1\underline{g}\ldots\underline{D}_{n-1}\underline{g}]\ ,\ \underline{s}_{n-1} = [1 \quad s\ldots s^{n-1}]' \tag{2.6.12}$$

Written in coefficient vectors eq. (2.6.11) obtains the form

$$[\underline{p}' \quad 1]\underline{s}_n = [\underline{q}' \quad 1]\underline{s}_n + \underline{k}_y'\underline{CWs}_{n-1}$$

and eliminating $s^n = s^n$

$$\underline{p}'\underline{s}_{n-1} = \underline{q}'\underline{s}_{n-1} + \underline{k}_y'\underline{CWs}_{n-1}$$

Matching the coefficients of the powers of s on both sides yields

$$\underline{p}' = \underline{q}' + \underline{k}_y'\underline{CW} \tag{2.6.13}$$

For a given characteristic polynomial of the open loop, $\underline{q}'$, and a given $\underline{CW}$, the characteristic polynomial of the closed loop, $\underline{p}'$, can easily be calculated for every given feedback vector $\underline{k}_y'$. Eq. (2.6.13) shows that the relation between $\underline{k}_y'$ and $\underline{p}$ is affine, i.e. it consists of a linear transformation by $\underline{CW}$ and a shift of the origin by $\underline{q}'$. The case of state-vector feedback $u = -\underline{k}'\underline{x}$ is contained in eq. (2.6.13) with $\underline{C} = \underline{I}$. Then

$$\underline{p}' = \underline{q}' + \underline{k}'\underline{W} \tag{2.6.14}$$

Example: Loading bridge

$\underline{Ws}_{n-1} = \underline{D}(s)\underline{g}$ was evaluated already in eq. (2.3.10)

$$\underline{D}(s)\underline{g} = \begin{bmatrix} (s^2+g/\ell)/m_c \\ s(s^2+g/\ell)/m_c \\ -s^2/m_c\ell \\ -s^3/m_c\ell \end{bmatrix} = \frac{1}{\ell m_c}\begin{bmatrix} g & 0 & \ell & 0 \\ 0 & g & 0 & \ell \\ 0 & 0 & -1 & 0 \\ 0 & 0 & 0 & -1 \end{bmatrix}\begin{bmatrix} 1 \\ s \\ s^2 \\ s^3 \end{bmatrix}$$

Leverrier's algorithm also yields

$$\underline{q}' = [0 \quad 0 \quad -f_{43} \quad 0]$$

$$= [0 \quad 0 \quad \frac{(m_L+m_c)g}{m_c \ell} \quad 0]$$

The coefficient vector of the closed-loop characteristic polynomial results from eq. (2.6.14)

$$[p_0 \quad p_1 \quad p_2 \quad p_3] = [0 \quad 0 \quad \frac{(m_L+m_c)g}{m_c \ell} \quad 0]$$

$$+ [k_1 \quad k_2 \quad k_3 \quad k_4] \begin{bmatrix} g & 0 & \ell & 0 \\ 0 & g & 0 & \ell \\ 0 & 0 & -1 & 0 \\ 0 & 0 & 0 & -1 \end{bmatrix} \frac{1}{\ell m_c} \tag{2.6.15}$$

If the closed-loop is stable, then $P(s) = [\underline{p}' \quad 1]\underline{s}_n$ is a Hurwitz polynomial. A necessary stability condition is that all coefficients p_i are positive. For the loading bridge the conditions are

Coefficient	Necessary stability condition
$p_0 = k_1 g/\ell m_c$	$k_1 > 0$
$p_1 = k_2 g/\ell m_c$	$k_2 > 0$
$p_2 = [(m_L+m_c)g+k_1\ell-k_3]/\ell m_c$	$k_3 < (m_L+m_c)g+k_1\ell$
$p_3 = (k_2\ell-k_4)/\ell m_c$	$k_4 < k_2\ell$

According to eqs. (C.1.6) and (C.1.8) a necessary and sufficient condition for stability is that $\Delta_3 > 0$ in addition to the above conditions (which contain also $\Delta_1 > 0$)

$$\Delta_3 = \begin{bmatrix} p_3 & p_1 & 0 \\ 1 & p_2 & p_0 \\ 0 & p_3 & p_1 \end{bmatrix} = p_1 p_2 p_3 - p_0 p_3^2 - p_1^2 > 0$$

Let for example $k_1 > 0$, $k_2 > 0$, $k_3 = k_4 = 0$.

Then $\Delta_3 = \dfrac{k_2^2 g^2 m_L}{\ell^2 m_c^3}$

This expression is positive for all physical parameter values. Thus feedback $u = -[k_1 \quad k_2 \quad 0 \quad 0]\underline{x}$, $k_1 > 0$, $k_2 > 0$ (i.e. negative feedback of trolley position and velocity) stabilizes all loading bridges.

2.6.2 Feedback Gains for Pole Placement

We now turn to the question: when can the relationship eq. (2.6.13) be inverted, or in other words, when can a suitable $\underline{k}_y'$ be found for every desired $\underline{p}'$? Clearly this is the case if and only if $\underline{C}$ and $\underline{W}$ are invertible. $\underline{C}$ being invertible means that n linearly independent sensors exist. Then the state variables may be chosen such that $\underline{C} = \underline{I}$, and a state-vector feedback, $u = -\underline{k}'\underline{x}$ and eq. (2.6.14) for $\underline{p}'$, results.

Eq. (A.7.44) shows that $\underline{W}$ has the same rank as the controllability matrix. Thus $\underline{W}$ is nonsingular if and only if the system $(\underline{F}, \underline{g})$ is controllable. For a controllable system eq. (2.6.14) can be inverted, i.e.

$$\underline{k}' = (\underline{p}' - \underline{q}')\underline{W}^{-1} \qquad (2.6.16)$$

Small gains $\underline{k}$ can be expected for small changes $\underline{p}' - \underline{q}'$ of the characteristic polynomial. For weakly controllable systems the controllability matrix and therefore $\underline{W}$ is almost singular. Then large gains $\underline{k}$ are required [67.1], [68.5].

An interpretation of eq. (2.6.16) can be given by the feedback canonical form, eq. (A.3.8),

$$\begin{aligned} \dot{\underline{x}}_f &= \underline{F}_f\underline{x}_f + \underline{g}_f u \\ y &= [r_0 \quad r_1 \ldots r_{n-1}]\underline{x}_f \end{aligned} \qquad (2.6.17)$$

$$\underline{F}_f := \begin{bmatrix} 0 & 1 & & 0 \\ \vdots & & \ddots & \\ 0 & & & 1 \\ -q_o & -q_1 & \dots & -q_{n-1} \end{bmatrix}, \quad \underline{g}_f := \begin{bmatrix} 0 \\ \vdots \\ 0 \\ 1 \end{bmatrix}$$

with the transfer function (see eq. (A.3.10)).

$$\frac{y_s(s)}{u_s(s)} = \frac{r_o+r_1 s+\dots+r_{n-1}s^{n-1}}{q_o+q_1 s+\dots+q_{n-1}s^{n-1}+s^n_{n-1}} \tag{2.6.18}$$

If the original controllable system (2.6.1) is transformed to this form by $\underline{x}_f = \underline{Tx}$, $\underline{F}_f = \underline{TFT}^{-1}$, $\underline{g}_f = \underline{Tg}$, then the controllability matrices are related by eq. (2.3.19), i.e. $\underline{\mathcal{C}}_f = \underline{T}\,\underline{\mathcal{C}}$, or

$$\underline{T}^{-1} = \underline{\mathcal{C}}\;\underline{\mathcal{C}}_f^{-1} \tag{2.6.19}$$

and by eq. (A.3.9)

$$\underline{T}^{-1} = [\underline{g}, \underline{Fg}\dots\underline{F}^{n-1}\underline{g}] \times \begin{bmatrix} q_1 & q_2 & \dots & q_{n-1} & 1 \\ q_2 & & ⋰ & ⋰ & 0 \\ \vdots & ⋰ & ⋰ & & \\ q_{n-1} & 1 & & & \\ 1 & 0 & & & 0 \end{bmatrix} \tag{2.6.20}$$

Comparison with eq. (A.7.44) shows that $\underline{T}^{-1} = \underline{W}$, thus $\underline{W}^{-1}$ is the matrix which transforms the given system to its feedback canonical form. Feedback $u = -\underline{k}_f\underline{x}_f+w$ changes this form to

$$\dot{\underline{x}}_f = (\underline{F}_f-\underline{g}_f\underline{k}_f')\underline{x}_f = \begin{bmatrix} 0 & 1 & & 0 \\ & & \ddots & \\ & & & 1 \\ -q_o-k_{f1} & \dots & & -q_{n-1}-k_{fn} \end{bmatrix}\underline{x}_f + \begin{bmatrix} 0 \\ \vdots \\ 0 \\ 1 \end{bmatrix} w$$

with the characteristic polynomial

$$P(s) = (q_0+k_{f1})+(q_1+k_{f2})s+\ldots+(q_{n-1}+k_{fn})s^{n-1}+s^n = Q(s)+\underline{k}_f'\underline{s}_{n-1} \quad (2.6.21)$$

and the closed-loop transfer function

$$\frac{y_s(s)}{w_s(s)} = \frac{r_0+r_1s+\ldots+r_{n-1}s}{p_0+p_1s+\ldots+p_{n-1}s^{n-1}+s^n} \quad (2.6.22)$$

We observe:

a) the zeros remain unchanged by state feedback,

b) state feedback of $\underline{x}_f$ changes the characteristic equation coefficient vector to

$$\underline{p}' = \underline{q}'+\underline{k}_f' \quad (2.6.23)$$

Transforming back to the original state variables $u = -\underline{k}_f'\underline{x}_f = -\underline{k}_f'\underline{W}^{-1}\underline{x} = -\underline{k}'\underline{x}$, thus $\underline{k}_f' = \underline{k}'\underline{W}$ and eq. (2.6.23) reads

$$\underline{p}' = \underline{q}'+\underline{k}'\underline{W}$$

which is identical to eq. (2.6.14), with the exception that eq. (2.6.14) has been proved also for systems with uncontrollable eigenvalues, where $\underline{W}$ is singular.

A different form of the pole assignment relation $\underline{k} = \underline{k}(\underline{p})$ was shown by the author [72.1]:

$$\begin{aligned} \underline{k}' &= \underline{e}'P(\underline{F}) \\ \underline{e}' &:= [0 \ \ldots \ 0 \quad 1][\underline{g}, \ \underline{F}\underline{g}\ldots\underline{F}^{n-1}\underline{g}]^{-1} \end{aligned} \quad (2.6.24)$$

This has become known as "Ackermann's formula". (See also eq. (A.7.46).) It is proved first for the feedback canonical form. Substituting $\underline{F}_f$ for s in eq. (2.6.21)

$$P(\underline{F}_f) = Q(\underline{F}_f)+k_{f1}\underline{F}_f^{\,0}+k_{f2}\underline{F}_f^{\,1}+\ldots+k_{fn}\underline{F}_f^{\,n-1}$$

By the Cayley-Hamilton theorem, eq. (A.7.30)

$\underline{Q}(\underline{F}_f) = \underline{0}$. Thus $\underline{e}_f'\underline{P}(\underline{F}_f) = k_{f_1}\underline{e}_f' + k_{f_2}\underline{e}_f'\underline{F}_f + \ldots + k_{fn}\underline{e}_f'\underline{F}_f^{n-1}$

Now by eq. (A.3.9)

$$\begin{aligned} \underline{e}_f' &= [1 \quad 0 \; \ldots \; 0] \\ \underline{e}_f'\underline{F}_f &= [0 \quad 1 \qquad 0] \\ &\vdots \\ \underline{e}_f'\underline{F}_f^{n-1} &= [0 \; \ldots \; 0 \quad 1] \end{aligned}$$

and therefore

$$\underline{e}_f'\underline{P}(\underline{F}_f) = \underline{k}_f' \tag{2.6.25}$$

The quantities $\underline{k}_f'$, $\underline{e}_f'$ and $\underline{F}_f$ are related to the corresponding quantities in the original state variables by

$$\begin{aligned} \underline{k}_f' &= \underline{k}'\underline{W} \\ \underline{e}_f' &= [0 \ldots 0 \quad 1]\,\underline{\mathcal{C}}_f^{-1} = [0 \ldots 0 \quad 1]\,\underline{\mathcal{C}}^{-1}\underline{W} = \underline{e}'\underline{W} \\ \underline{F}_f &= \underline{W}^{-1}\underline{FW} \end{aligned}$$

By substitution into eq. (2.6.25) the original formulation $\underline{k}' = \underline{e}'\underline{P}(\underline{F})$, eq. (2.6.24), follows. The vector $\underline{e}'$ is the last row of the inverted controllability matrix $\underline{\mathcal{C}}$. $\underline{e}'$ may be computed by solving

$$\underline{e}'[\underline{g}, \underline{Fg} \ldots \underline{F}^{n-1}\underline{g}] = [0 \ldots 0 \quad 1] \tag{2.6.26}$$

by Gauss elimination with pivoting.

It should be noted that $\underline{e}'$ is invariant under state feedback. For the pair $(\underline{H}, \underline{g}) = (\underline{F}-\underline{gk}', \underline{g})$ the $\underline{e}'$ vector satisfies

$$\underline{e}_H'[\underline{g}, \underline{Hg}, \ldots \underline{H}^{n-1}\underline{g}] = \underline{e}_H'[\underline{g}, (\underline{F}-\underline{gk}')\underline{g}, \ldots (\underline{F}-\underline{gk}')^{n-1}\underline{g}] = [0 \ldots 0 \quad 1]$$

Substituting the first column into the second one and so on

$$\underline{e}_H'[\underline{g}, \underline{F}\underline{g} \ldots \underline{F}^{n-1}\underline{g}] = [0 \ldots 0 \quad 1]$$

results and by comparison with eq. (2.6.26)

$$\underline{e}_H = \underline{e} \text{ for all } \underline{k}' \qquad (2.6.27)$$

In case of numerical difficulties, $\underline{e}$ can be computed by this feedback solution with arbitrary $\underline{k}'$. Also a transformation to Hessenberg form may be used as in eq. (A.5.13).

An advantage of eq. (2.6.24) is, that the desired characteristic polynomial can be given in factorized form

$$P(s) = (a_0+b_0s+s^2)(a_1+b_1s+s^2)\ldots(c+s) \qquad (2.6.28)$$

where the last term (c+s) occurs only for n odd. Then

$$\underline{k}' = \underline{e}'(a_0\underline{I}+b_0\underline{F}+\underline{F}^2)(a_1\underline{I}+b_1\underline{F}+\underline{F}^2)\ldots(c\underline{I}+\underline{F}) \qquad (2.6.29)$$

In this parameterization the class of all stabilizing feedbacks is characterized by eq. (2.6.29) with $a_i > 0$, $b_i > 0$ for all i, $c > 0$. In the factorized form large differences in the size of the coefficients p_i, as they typically occur for large n, are avoided in the calculations.

Remark 2.1:

The form (2.6.29) is particularly suited for determining sensitivities of $\underline{k}$ with respect to the assigned eigenvalues or the a, b and c coefficients. For $\underline{k}' = \underline{e}'P(\underline{F}) = -\underline{e}'(\underline{F}-s_1\underline{I})(\underline{F}-s_2\underline{I})\ldots(\underline{F}-s_n\underline{I})$ the sensitivity with respect to s_1 is

$$\frac{\partial \underline{k}'}{\partial s_1} = -\underline{e}'(\underline{F}-s_2\underline{I})\ldots(\underline{F}-s_n\underline{I})$$

and for $\underline{k}' = \underline{e}'(a\underline{I}+b\underline{F}+\underline{F}^2)R(\underline{F})$

$$\frac{\partial \underline{k}'}{\partial a} = \underline{e}'R(\underline{F}) \;, \quad \frac{\partial \underline{k}'}{\partial b} = \underline{e}'\underline{F}\,R(\underline{F}).$$

Eq. (2.6.24) may also be written

$$\underline{k}' = \underline{e}'P(\underline{F})$$
$$= p_o\underline{e}'+p_1\underline{e}'\underline{F}+\ldots+p_{n-1}\underline{e}'\underline{F}^{n-1}+\underline{e}'\underline{F}^n$$
$$= [\underline{p}' \quad 1]\underline{E} \qquad (2.6.30)$$

$$\underline{E}:=\begin{bmatrix}\underline{e}' \\ \underline{e}'\underline{F} \\ \vdots \\ \underline{e}'\underline{F}^n\end{bmatrix}$$

The (n+1)×n matrix $\underline{E}$ is called "Pole assignment matrix of the pair ($\underline{F}$, g)". By eq. (2.6.16), $\underline{E}$ is related to $\underline{W}$ by

$$\underline{E} = \begin{bmatrix}\underline{W}^{-1} \\ -\underline{q}\underline{W}^{-1}\end{bmatrix} \qquad (2.6.31)$$

Example: Loading bridge of eq. (2.1.15)

By eqs. (2.6.26) and (2.3.18)

$$[e_1 \quad e_2 \quad e_3 \quad e_4]\begin{bmatrix}0 & g_2 & 0 & g_4f_{23} \\ g_2 & 0 & g_4f_{23} & 0 \\ 0 & g_4 & 0 & g_4f_{43} \\ g_4 & 0 & g_4f_{43} & 0\end{bmatrix} = [0 \quad 0 \quad 0 \quad 1]$$

Solving and substituting the physical variables yields

$$\underline{e}' = \frac{\ell m_c}{g}[1 \quad 0 \quad \ell \quad 0]$$

and the pole placement equation is

$$[k_1 \quad k_2 \quad k_3 \quad k_4] = [p_0 \quad p_1 \quad p_2 \quad p_3 \quad 1] \begin{bmatrix} 1 & 0 & \ell & 0 \\ 0 & 1 & 0 & \ell \\ 0 & 0 & -g & 0 \\ 0 & 0 & 0 & -g \\ 0 & 0 & \frac{(m_c+m_L)g^2}{\ell m_c} & 0 \end{bmatrix} \frac{\ell m_c}{g} \tag{2.6.32}$$

This result permits some interesting conclusions:

1. Stability requires $p_0 > 0$, i.e. $k_1 > 0$.
2. Stability requires $p_1 > 0$, i.e. $k_2 > 0$.
3. The rope angle must not be fed back for stabilization, however $k_3 = 0$ imposes the following condition on $P(s)$

   $$k_3 = p_0\ell^2 m_c/g - p_2\ell m_c + (m_c+m_L)g = 0 \tag{2.6.33}$$

4. Similarly $k_4 = 0$ implies

   $$p_3 = p_1\ell/g \tag{2.6.34}$$

5. The load mass m_L enters only into k_3

   $$k_3 = k_{30} + m_L g, \quad k_{30} = p_0\ell^2 m_c/g - p_2\ell m_c + m_c g \tag{2.6.35}$$

   If the load mass m_L is known, then the gain k_3 can be scheduled such that $\underline{p}'$ and the eigenvalues are unchanged for different loads.

The pole assignment matrix has some interesting properties:

1. For every controllable pair $(\underline{F}, \underline{g})$ there exists one unique pole assignment matrix $\underline{E}$; conversely $\underline{W}$ and $\underline{q}$ can be computed from $\underline{E}$ via eq. (2.6.31). This leads to a unique realization $\underline{F} = \underline{W}\underline{F}_f\underline{W}^{-1}$, $\underline{g} = \underline{W}\underline{g}_f$ with $\underline{F}_f$ and $\underline{g}_f$ as in eq. (2.6.17).

2. A numerical test for the accuracy of $\underline{E}$ is

$$\underline{E}\underline{g} = \begin{bmatrix} \underline{e}'\underline{g} \\ \underline{e}'\underline{F}\underline{g} \\ \vdots \\ \underline{e}'\underline{F}^n\underline{g} \end{bmatrix} = \begin{bmatrix} 0 \\ 0 \\ \vdots \\ 1 \\ -q_{n-1} \end{bmatrix} \tag{2.6.36}$$

The first n rows follow from eq. (2.6.26) and are used in the test. The last row follows from Cayley-Hamilton. It is not necessary to calculate the open-loop characteristic polynomial for the application of (2.6.24). If therefore q_{n-1} is not available, a different test for the last row of $\underline{E}$ is useful: Let $\underline{k}' = \underline{e}'\underline{F}^n$, then $\underline{p} = \underline{0}$ by eq. (2.6.30) and all eigenvalues are shifted to s = 0. Calculate

$$\det(s\underline{I}-\underline{F}+\underline{g}\underline{e}'\underline{F}^n) = [\tilde{\underline{p}}' \quad 1]\underline{s}_n \tag{2.6.37}$$

For exact calculation $\tilde{\underline{p}}'$ should be zero. Its size is a measure for the accuracy of $\underline{e}'\underline{F}^n$, the last row of $\underline{E}$.

3. If the test for numerical accuracy is unsatisfactory, then $\underline{E}$ may be computed by a feedback solution for $(\underline{H}, \underline{g}) = (\underline{F}-\underline{g}\underline{k}', \underline{g})$ instead of $(\underline{F}, \underline{g})$ with arbitrary $\underline{k}'$. The relation between the open and closed-loop pole assignment matrices is

$$\underline{E}_H = \begin{bmatrix} \underline{e}_H' \\ \underline{e}_H'\underline{H} \\ \vdots \\ \underline{e}_H'\underline{H}^{n-1} \\ \underline{e}_H'\underline{H}^n \end{bmatrix} = \begin{bmatrix} \underline{e}' \\ \underline{e}'\underline{F} \\ \vdots \\ \underline{e}'\underline{F}^{n-1} \\ \underline{e}'\underline{F}^n-\underline{k}' \end{bmatrix} \tag{2.6.38}$$

This follows directly from eqs. (2.6.26) and (2.6.27). If $\underline{k}'$ is determined iteratively as $\underline{k}' = \underline{k}^{(1)}+\underline{k}^{(2)}+\underline{k}^{(3)}+\ldots$, then the pole assignment matrix must be changed only in its last row between iterations.

The mappings between the set of eigenvalues $\Lambda = \{s_1, s_2 \ldots s_n\}$, the $\underline{p}'$ vector and the $\underline{k}'$ vector are summarized in the following diagram:

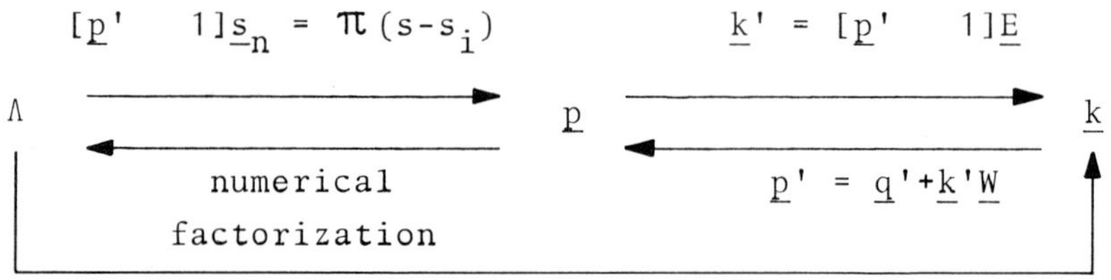

$$\begin{array}{ccccc} & [\underline{p}' \quad 1]\underline{s}_n = \Pi(s-s_i) & & \underline{k}' = [\underline{p}' \quad 1]\underline{E} & \\ \Lambda & \rightleftarrows & \underline{p} & \rightleftarrows & \underline{k} \\ & \text{numerical factorization} & & \underline{p}' = \underline{q}'+\underline{k}'\underline{W} & \end{array}$$

(arrow from Λ to $\underline{k}$:)

$$\underline{k}' = \underline{e}'\,\Pi(\underline{F}-s_i\underline{I}) \tag{2.6.39}$$

2.6.3 Fixed Eigenvalues

So far we have assumed that the desired characteristic polynomial P(s) has been specified already. On the other hand we have seen in section 2.5 that only a desired region Γ in the eigenvalue plane generally can be recommended. The exact location of the eigenvalues in this region should be dictated by other design requirements. An important one is that the required actuator input magnitudes $|u| = |\underline{k}'\underline{x}|$ for realistic states $\underline{x}$ should not be too large. In view of

$$|u| = |\underline{k}'\underline{x}| \leq ||\underline{k}'|| \times ||\underline{x}|| \tag{2.6.40}$$

$||k||$ should be kept small. In other words: In the K space, in which $\underline{k}'$ lives, we do not want to move too far away from the origin $\underline{k} = \underline{0}$, which corresponds to the open loop. For a better understanding it is useful to look simultaneously at movements of $\underline{k}'$ and the corresponding eigenvalue migrations. In particular we are interested in directions in $\underline{k}'$ space such that some of the open-loop eigenvalues remain unchanged by feedback. For $\underline{k}' = \underline{0}$ let

$$\det(s\underline{I}-\underline{F}) = Q(s) = R(s)W(s) \tag{2.6.41}$$

$$W(s) := w_o+w_1 s+\ldots+w_{m-1}s^{m-1}+s^m \tag{2.6.42}$$

$$R(s) := r_o+r_1 s+\ldots+r_{n-m-1}s^{n-m-1}+s^{n-m} = (s-s_{m+1})(s-s_{m+2})\ldots(s-s_n) \tag{2.6.43}$$

Let us shift only the m eigenvalues of W(s) to

$$V(s) = v_o+v_1 s+\ldots+v_{m-1}s^{m-1}+s^m = (s-s_1)(s-s_2)\ldots(s-s_m) \tag{2.6.44}$$

such that

$$\det(s\underline{I}-\underline{F}+\underline{gk}') = P(s) = R(s)V(s) = (s-s_1)(s-s_2)\ldots(s-s_n) \tag{2.6.45}$$

By eq. (2.6.24)

$$\underline{k}' = \underline{e}'P(\underline{F}) = \underline{e}'\underline{R}(\underline{F})V(\underline{F})$$

$$= \underline{e}_R'V(\underline{F}) \quad \text{with} \quad \underline{e}_R' := \underline{e}'R(\underline{F}) \tag{2.6.46}$$

For V(s) = W(s) the open loop with $\underline{k} = \underline{0}$ is obtained, i.e.

$$\underline{0}' = \underline{e}_R'W(\underline{F}) \tag{2.6.47}$$

and eq. (2.6.46) may be written

$$\underline{k}' = \underline{e}_R'[V(\underline{F})-W(\underline{F})] \tag{2.6.48}$$

$$\underline{k}' = (\underline{v}'-\underline{w}') \begin{bmatrix} \underline{e}_R' \\ \underline{e}_R'\underline{F} \\ \vdots \\ \underline{e}_R'\underline{F}^{m-1} \end{bmatrix} \tag{2.6.49}$$

The vectors $\underline{e}_R'$, $\underline{e}_R'\underline{F}$ ... $\underline{e}_R'\underline{F}^{m-1}$ span the m-dimensional subspace of K in which the eigenvalues in R(s) are unobservable from $\underline{k}'\underline{x}$. Therefore they remain unchanged by closing the loop.

Let for example m = 1, $V(s) = v_o+s$, i.e. only one real eigenvalue is shifted. Then

$$\underline{k}' = (v_o-w_o)\underline{e}_R' \tag{2.6.50}$$

The second row of the observability matrix for the output $\underline{k}'\underline{x}$ is

$$\underline{k}'\underline{F} = (v_o-w_o)\underline{e}_R'\underline{F} \tag{2.6.51}$$

and by eq. (2.6.47), $\underline{0} = \underline{e}_R'(w_o\underline{I}+\underline{F})$

$$\underline{k}'\underline{F} = -w_o(v_o-w_o)\underline{e}_R' = -w_o\underline{k}' \tag{2.6.52}$$

i.e. by eqs. (2.3.36) and (2.3.40) only the subsystem with eigenvalue $-w_o$ is observable.

Now let m = 2, $V(s) = v_0+v_1s+s^2$. A pair of eigenvalues is shifted and the n-2 eigenvalues in R(s) remain at their open loop locations. Eq. (2.6.46) becomes

$$\underline{k}' = \underline{e}_R'(v_0\underline{I}+v_1\underline{F}+\underline{F}^2) = [v_0 \quad v_1 \quad 1]\begin{bmatrix}\underline{e}_R' \\ \underline{e}_R'\underline{F} \\ \underline{e}_R'\underline{F}^2\end{bmatrix} \tag{2.6.53}$$

and eq. (2.6.49) becomes

$$\underline{k}' = (v_0-w_0)\underline{e}_R'+(v_1-w_1)\underline{e}_R'\underline{F} = [(v_0-w_0) \quad (v_1-w_1)]\begin{bmatrix}\underline{e}_R' \\ \underline{e}_R'\underline{F}\end{bmatrix} \tag{2.6.54}$$

The eigenvalues in R(s) are unobservable and remain unchanged by a feedback vector in the plane spanned by $\underline{e}_R'$ and $\underline{e}_R'\underline{F}$. Assume that W(s) has a complex pair of roots, then w_0 is the square of the natural frequency and w_1 is proportional to the damping. Moving in the direction $\underline{e}_R'$ in K space means $v_0 > w_0$ and the natural frequency is increased. Moving in the direction $\underline{e}_R'\underline{F}$ increases the damping of the complex pair of eigenvalues.

Example 1:

$$\dot{\underline{x}} = \begin{bmatrix}1 & 3 \\ -4 & -7\end{bmatrix}\underline{x} + \begin{bmatrix}0 \\ 1\end{bmatrix}u \tag{2.6.55}$$

$Q(s) = (5+s)(1+s)$

The closed-loop eigenvalues shall be real and satisfy $s_1 < -2$, $s_2 < -2$.

$\underline{e} \quad = [1/3 \quad 0]$

$\underline{e}'\underline{F} = [1/3 \quad 1]$

First the eigenvalue at s = -1 is shifted to the left, i.e. R(s) = 5+s is fixed

$\underline{e}_R' = \underline{e}'R(\underline{F}) = 5\underline{e}'+\underline{e}'\underline{F} = [2 \quad 1]$

With W(s) = (1+s) eq. (2.6.49) is

$\underline{k}' = (v_o - 1)[2 \quad 1]$

Figure 2.13 shows this line marked $s_2 = -5$.

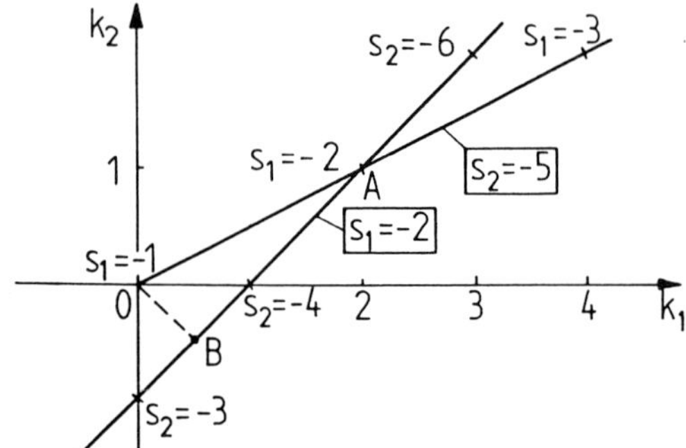

Figure 2.13 Shift s_1 from -1 to -2 at point A, then shift s_2 from -5 to -3.5 at point B.

Choose $s_1 = -2$ i.e. $\underline{k}'^{(1)} = [2 \quad 1]$ at point A, keep this eigenvalue fixed in a second design step and try to reduce $||\underline{k}||^2 = k_1{}^2 + k_2{}^2$. $\underline{k}$ is composed as $\underline{k} = \underline{k}^{(1)} + \underline{k}^{(2)}$. $\underline{k}^{(2)}$ is a feedback around the modified system $\underline{F} - \underline{g}\underline{k}'^{(1)}$, $\underline{g}$. The new origin in figure 2.13 is the point A. By eq. (2.6.38) the vectors $\underline{e}'$ and $\underline{e}'\underline{F}$ are unchanged by $\underline{k}'^{(1)}$. Now $R(s) = 2+s$ is fixed, i.e.

$$\underline{e}'_R = \underline{e}'R(\underline{F}) = 2\underline{e}' + \underline{e}'\underline{F} = [1 \quad 1]$$

with $W(s) = 5+s$ eq. (2.6.49) is

$$\underline{k}'^{(2)} = (v-5)[1 \quad 1].$$

This straight line originating at A is plotted in figure 2.13 and marked $s_1 = -2$. The smallest feedback gain $||k||$ on this line is at point B with $\underline{k}'_B = [0.5 \quad -0.5]$, $P(s) = (s+2)(s+3.5)$. This second shift of s_2 from -5 to -3.5 reduces $||k||$ by a factor 3.2.

A different scaling of the state variables has the same effect as using a different norm $||k||^2 = \alpha k_1^2 + \beta k_2^2$, $\alpha > 0$, $\beta > 0$. Figure 2.13 shows that $-4 < s_2 < -3$ is the set of Pareto-optimal solutions for which $|k_1|$ can be reduced only at the expense of $|k_2|$ and vice versa.

Example 2: Loading bridge

The open-loop eigenvalues are shown in figure 2.3. The double eigenvalue at s = 0 is to be kept and the imaginary eigenvalues are to be stabilized. Here $R(s) = s^2$ and eq. (2.6.53) with $\underline{e}_R = \underline{e}'\underline{F}^2$, $w_o = \omega_L^2$, $w_1 = 0$

$$\underline{k}' = (v_o-\omega_L^2)\underline{e}'\underline{F}^2 + v_1\underline{e}'\underline{F}^3$$

$$= (v_o-\omega_L^2)[0 \quad 0 \quad -\ell m_c \quad 0] + v_1[0 \quad 0 \quad 0 \quad -\ell m_c]$$

Now introduce

$$v_o+v_1 s+s^2 = \omega_o^2+2\zeta\omega_o s+s^2$$

$$\underline{k}' = [0 \quad 0 \quad -(\omega_o^2-\omega_L^2)\ell m_c \quad -2\zeta\omega_o\ell m_c] \qquad (2.6.56)$$

It is seen that a negative k_4 increases the damping. k_3 may be chosen as zero if the natural frequency ω_L is left unchanged as ω_o. Negative k_3 increases the natural frequency of the pendulum, positive k_3 reduces it. For $k_3 = \omega_L^2\ell m_c = (m_L+m_c)g$ the stability boundary is reached.

2.6.4 Fixed Gains

Instead of fixing some eigenvalues in a design step we may fix some gains in $\underline{k}'$.

By eq. (2.6.14)

$$\underline{p}' = \underline{q}' + \underline{k}'\underline{W}$$

Let

$$\underline{W} = \begin{bmatrix} \underline{w}_1' \\ \underline{w}_2' \\ \vdots \\ \underline{w}_n' \end{bmatrix}$$

If k_i is fixed, then the term $k_i\underline{w}_i'$ is constant and can be added to the constant $\underline{q}$. The feedback k_i changes the characteristic polynomial to $\underline{q}^* = \underline{q}'+k_i\underline{w}_i'$ and $\underline{p}$ can be moved only in the (n-1)-dimensional affine subspace

$$\underline{p}' = \underline{q}^* + [k_1 \ldots k_{i-1}, \; k_{i+1} \ldots k_n] \begin{bmatrix} \underline{w}_1' \\ \vdots \\ \underline{w}_{i-1}' \\ \underline{w}_{i+1}' \\ \vdots \\ \underline{w}_n' \end{bmatrix} \tag{2.6.57}$$

Also output feedback

$$u = -\underline{k}_y'\underline{Cx} \;, \; \text{rank } \underline{C} = m < n$$

has a similar effect. According to eq. (2.6.13) $\underline{p}'$ is then restricted to the affine subspace

$$\underline{p}' = \underline{q}'+\underline{k}_y'\underline{CW} \tag{2.6.58}$$

A special case of eq. (2.6.58) is, that only one output variable $y = \underline{c}'\underline{x}$ is fed back. Then

$$\underline{p}' = \underline{q}'+k_y\underline{c}'\underline{W}$$

or, going back to polynomials, see eq. (2.6.11),

$$P(s) = Q(s) + k_y\underline{c}'\underline{D}(s)\underline{g}$$

The roots of $P(s) = 0$ as a function of k_y may be found by a root locus [48.1], [50.1] from

$$k_y \frac{\underline{c}'\underline{D}(s)\underline{g}}{Q(s)} = -1 \tag{2.6.59}$$

$\underline{c}'\underline{D}(s)\underline{g} = R(s) = r_0+r_1s+\ldots+r_ms^m$ is the numerator polynomial of the transfer function and $Q(s)$ is its denominator polynomial.

Both must be factorized, then eq. (2.6.59) becomes

$$K \times \frac{(s-s_{o1})(s-s_{o2})\dots(s-s_{om})}{(s-s_{p1})(s-s_{p2})\dots(s-s_{pn})} = -1 \tag{2.6.60}$$

with $K = k_y r_m$. The root locus satisfies the phase condition of eq. (2.6.60). For each point on the root locus K is determined by the absolute value condition of eq. (2.6.60).

2.6.5 Pole Shifting in Multi-input Systems

Design parameters for multi-input systems are treated in chapter 9. A simple solution is to ignore all these degrees of freedom in the design and to assign only eigenvalues as in the single-input case.

If $\underline{F}$ is cyclic and $(\underline{F}, \underline{G})$ is controllable, then almost every pair $(\underline{F}, \underline{g}) = (\underline{F}, \underline{G\gamma})$ with $\underline{\gamma}$ chosen at random is controllable. (See the discussion after eq. (2.3.24).) Thus the input vector $\underline{u}$ may be generated from a scalar input v by

$$\underline{u} = \underline{\gamma} v \tag{2.6.61}$$

Figure 2.14 shows this case of linearly dependent inputs.

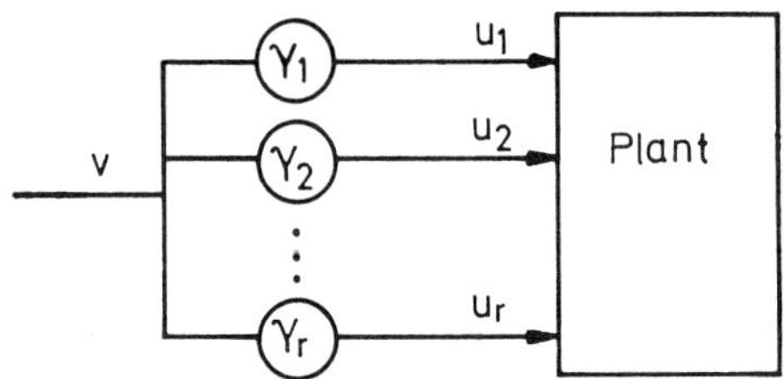

Figure 2.14

Linearly dependent inputs

Now $v = -\underline{k}'\underline{x}$ is determined by single-input pole assignment and the feedback gain matrix in $\underline{u} = -\underline{Kx}$ is the dyadic product (see eq. (A.7.6))

$$\underline{K} = \underline{\gamma k}' \tag{2.6.62}$$

In the case that $\underline{F}$ is noncyclic it can be made cyclic first by almost any feedback before the above procedure is applied. The design of eq. (2.6.62) with rank $\underline{K} = 1$ was proposed by Popov [64.11]. A disadvantage of this solution is, that all actuators require the peak power at the same time.

Also several other techniques for pole assignment for multi-input systems by state feedback have been proposed. Methods which do not give the designer some freedom to choose systematically one of the many feedback matrices which assign the desired poles, are not recommended.

2.7 Exercises

2.1 Transform the state representation (2.1.15) of the loading bridge and the measurement eq. (2.1.18) into block diagonal form by introducing position x_1^* and velocity x_2^* of the common center of gravity of trolley and load as new state variables instead of x_1 and x_2.

2.2 Transform the state representation of the loading bridge, eq. (2.1.15), with output load position, eq. (2.1.17), into observability canonical form, see appendix A.

2.3 The load m_L at the loading bridge varies in the range $m_{Lo} \leq m_L \leq m_{Lmax}$. Assume that a stabilizing state feedback for $m_L = m_{Lo}$ has been determined. Discuss the stability of this loop, if m_L is increased.

2.4 For the loading bridge calculate the transfer function from u to the trolley velocity x_2. What is the characteristic polynomial of the observable subsystem?

2.5 Calculate the transition matrix of the loading bridge via the block-diagonal form of exercise 2.1.

2.6 For the loading bridge example, assume a constant wind force which acts parallel to u on the load. Determine a controller which allows an exact stationary positioning of the load and provides a well damped transient with $\zeta \geq \zeta_o = 1/\sqrt{2}$ and $\sigma < \sigma_o = -1$.

2.7 Determine the stability boundaries in figure 2.13.

2.8 In the example of eq. (2.6.56) only the eigenvalues of the subsystem "pendulum" were shifted. Shift them such that damping $1/\sqrt{2}$ is obtained and the natural frequency remains unchanged. Do a second design step, in which these eigenvalues are fixed now and the double eigenvalue at $s = 0$ can be moved. (Suggested parameter values $m_L = 3000$ kg, $m_C = 1000$ kg, $\ell = 10$ m, $g = 10\ \mathrm{ms}^{-2}$.)

2.9 An incremental feedback $\underline{k}' = [\Delta k_1 \quad \Delta k_2 \quad 0 \quad 0]$ with small $\Delta k_1 > 0$ and $\Delta k_2 > 0$ stabilizes the loading bridge. It shifts the eigenvalues from $Q(s) = s^2(s^2+\omega_L^2)$ to $P(s) = (s^2+as+b)(s^2+cs+\omega_L^2+d)$ with small a,b,c,d. How are a,b,c,d related to Δk_1 and Δk_2?

2.10 Plot the root locus for the loading bridge with $\sqrt{g/\ell} = 1$, $m_L = 3m_C$ and feedback $u = -k_y[1 \quad \alpha \quad 0 \quad 0]\underline{x}$ for

a) $\alpha = 3$, b) $\alpha = 2\sqrt{3}$, c) $\alpha = 4$.

Repeat for x_2 replaced by $\hat{x}_2$ produced by approximate differentiation $\hat{x}_{2s} = \dfrac{5\omega_L s}{s+5\omega_L} \times x_{1s}$.

3 Modelling and Analysis of Sampled-Data Systems

3.1 Discretization of the Plant

Let the plant be a continuous system which can be described by the vector differential equation

$$\underline{\dot{x}}(t) = \underline{Fx}(t) + \underline{g}u(t) \tag{3.1.1}$$

and the measurement equation

$$y(t) = \underline{c}'\underline{x}(t) \tag{3.1.2}$$

The input u(t) is produced by a hold element, i.e.

$$u(t) = u(kT),\ kT < t < kT + T \tag{3.1.3}$$

The general solution of the differential equation is

$$\underline{x}(t) = e^{\underline{F}(t-t_o)}\underline{x}(t_o) + \int_{t_o}^{t} e^{\underline{F}(t-\tau)}\underline{g}u(\tau)d\tau \tag{3.1.4}$$

The particular solution for the interval of constant $u(\tau)$ is

$$\underline{x}(kT+T) = e^{\underline{F}T}\underline{x}(kT) + \int_{kT}^{kT+T} e^{\underline{F}(kT+T-\tau)}d\tau\underline{g}u(kT) \tag{3.1.5}$$

and with the substitution $v = kT+T-\tau$

$$\underline{x}(kT+T) = e^{\underline{F}T}\underline{x}(kT) + \int_{0}^{T} e^{\underline{F}v}dv\underline{g}u(kT) \tag{3.1.6}$$

This is a vector difference equation of the form

$$\underline{x}(kT+T) = \underline{A}(T)\underline{x}(kT) + \underline{b}(T)u(kT) \tag{3.1.7}$$

The suitable choice of the sampling interval T will be discussed in section 4.3. Here T is regarded as given and the notation of eqs. (3.1.2) and (3.1.7) is simplified to

$$\begin{aligned} \underline{x}[k+1] &= \underline{A}\underline{x}[k] + \underline{b}u[k] \\ y[k] &= \underline{c}'\underline{x}[k] \end{aligned} \tag{3.1.8}$$

This state equation of the discrete system was introduced into the control literature by Kalman and Bertram [58.4]. It is illustrated by figure 3.1.

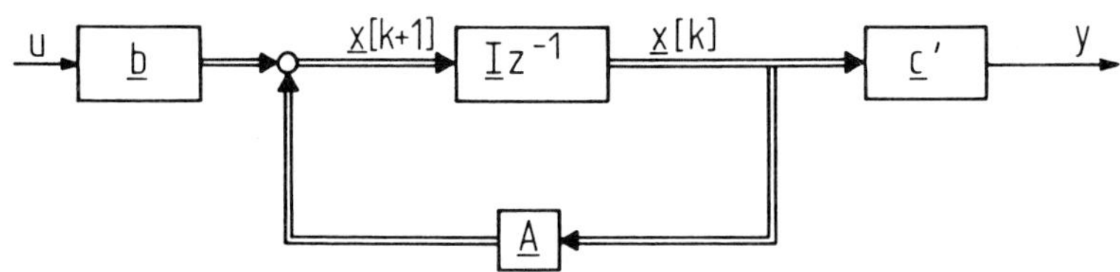

Figure 3.1 State representation of the discrete system

$\underline{I}z^{-1}$ represents n parallel storage elements which delay their input sequences, that is the components of the vector $\underline{x}$, by one sampling interval.

The discrete state representation (3.1.8) is determined by the calculation of $\underline{A}$ and $\underline{b}$. This can be performed by computation of

$$\underline{A}(T) = e^{\underline{F}T} \tag{3.1.9}$$

and the subsequent integration

$$\underline{b} = \underline{R}\underline{g} \ , \ \underline{R} := \int_0^T \underline{A}(v)\,dv \tag{3.1.10}$$

If $\underline{A}$ is calculated by the polynomial form eq. (2.4.8), that is

$$\underline{A} = e^{\underline{F}T} = \sum_{i=0}^{n-1} c_i(T)\underline{F}^i \tag{3.1.11}$$

then the coefficients c_i can be directly integrated

$$\underline{R} = \int_0^T e^{\underline{F}v} dv = \sum_{i=0}^{n-1} \int_0^T c_i(v)\, dv \underline{F}^i \tag{3.1.12}$$

Example:

$$\dot{\underline{x}} = \begin{bmatrix} 0 & 1 \\ 0 & -1 \end{bmatrix} \underline{x} + \begin{bmatrix} 0 \\ 1 \end{bmatrix} u \tag{3.1.13}$$

$$\det(s\underline{I}-\underline{F}) = s(s+1)$$

From eq. (A.7.55) follows that there exist c_0 and c_1 such that

$$e^{\underline{F}t} = c_0\underline{I} + c_1\underline{F}$$

This relation is not only satisfied by the matrix $\underline{F}$, but also by its eigenvalues s_i, i.e.

$$e^{s_i t} = c_0 + c_1 s_i$$

$$\left.\begin{array}{l} s_1 = 0 \rightarrow 1 = c_0 \\ s_2 = -1 \rightarrow e^{-t} = c_0 - c_1 \end{array}\right\} c_1 = 1-e^{-t}$$

$$\underline{A} = e^{\underline{F}T} = \underline{I} + (1-e^{-T})\underline{F} = \begin{bmatrix} 1 & 1-e^{-T} \\ 0 & e^{-T} \end{bmatrix}$$

$$\underline{R} = \int_0^T e^{\underline{F}v} dv = \int_0^T dv \underline{I} + \int_0^T (1-e^{-v}) dv \underline{F}$$

$$= T\underline{I} + (T+e^{-T}-1)\underline{F} = \begin{bmatrix} T & T+e^{-T}-1 \\ 0 & 1-e^{-T} \end{bmatrix}$$

$$\underline{b} = \underline{R}\underline{g} = \begin{bmatrix} T+e^{-T}-1 \\ 1-e^{-T} \end{bmatrix}$$

for example for T = 1

$$\underline{x}[k+1] = \begin{bmatrix} 1 & 0.6321 \\ 0 & 0.3679 \end{bmatrix} \underline{x}[k] + \begin{bmatrix} 0.3679 \\ 0.6321 \end{bmatrix} u[k]$$

For the numerical calculation for systems of higher order, it is advantageous to determine first

$$\underline{R} = \int_0^T e^{\underline{F}v} dv$$

$$= \int_0^T \sum_{m=0}^{\infty} \frac{1}{m!} \underline{F}^m v^m dv$$

$$= \sum_{m=0}^{\infty} \frac{1}{m!} \underline{F}^m \int_0^T v^m dv$$

$$= T \sum_{m=0}^{\infty} \frac{1}{(m+1)!} \underline{F}^m T^m \tag{3.1.14}$$

From

$$\underline{FR} = \int_0^T \underline{F} e^{\underline{F}v} dv = e^{\underline{F}v} \Big|_0^T = e^{\underline{F}T} - \underline{I} = \underline{A} - \underline{I}$$

it follows that

$$\underline{A} = \underline{I} + \underline{FR} \tag{3.1.15}$$

and from eq. (3.1.10) $\underline{b} = \underline{R}\underline{g}$. Note that the matrices $\underline{F}$ and $\underline{R}$ commute, see eq. (3.1.14). Thus we may also write $\underline{A} = \underline{I} + \underline{RF}$.

The numerical calculation of $\underline{R}$ is performed by truncation of the series

$$\underline{R} \approx \underline{R}_N = T \times \sum_{m=0}^{N} \frac{1}{(m+1)!} \underline{F}^m T^m \tag{3.1.16}$$

Källström [73.3] has examined numerically suitable methods for this calculation. For improvement of the convergence, first a short sampling interval $\tau = T/2^p$, p integer, is taken. The result is then formed as

$$\underline{A}(T) = e^{\underline{F}T} = \left[e^{\underline{F}T/2^p}\right]^{2^p} = \Big[\dots \underbrace{\left[\left[e^{\underline{F}\tau}\right]^2\right]^2 \dots\Big]^2}_{p\ \text{times}} \tag{3.1.17}$$

In the work of Källström it turns out to be favorable to fix the number of the series terms at N = 8 and to choose p such that

$$\begin{aligned} &\|(\underline{F}/2^p)^N/N!\| \le \varepsilon \\ &\|(\underline{F}/2^{p-1})^N/N!\| > \varepsilon \end{aligned} \tag{3.1.18}$$

where ε is the machine accuracy. In the examples examined in [73.3] with realistically chosen sampling period T (see section 4.3), p = 3 or 4 has resulted.

Further methods for the numerical calculation of the exponential function of a matrix can be found in the literature, for example [78.3], [80.4].

The relation between the eigenvalues of $\underline{F}$ and $\underline{A}$ can be seen from the Jordan form of $\underline{F}$, eq. (2.4.11). From eqs. (2.4.12) and (2.4.13) it follows that

$$\underline{A} = e^{\underline{F}T} = \begin{bmatrix} e^{\underline{J}_1 T} & \underline{0} & & \\ \underline{0} & e^{\underline{J}_2 T} & & \\ & & \ddots & \\ & & & e^{\underline{J}_m T} \end{bmatrix} \tag{3.1.19}$$

$$e^{\underline{J}_i T} = e^{s_i T} \begin{bmatrix} 1 & T & T^2/2! & \dots & T^{r-1}/(r-1)! \\ 0 & 1 & T & \ddots & \vdots \\ 0 & 0 & 1 & \ddots & T^2/2! \\ & & & \ddots & T \\ 0 & & & & 1 \end{bmatrix} \tag{3.1.20}$$

The character of the solution of the difference equation (3.1.8) is determined essentially by the eigenvalues of $\underline{A}$, that is by the zeros of the characteristic polynomial

$$P_A(z) := \det(z\underline{I}-\underline{A}) \tag{3.1.21}$$

From eq. (3.1.20) it can be seen that for every eigenvalue s_i of the matrix $\underline{F}$ there corresponds an eigenvalue of $\underline{A}$ at

$$z_i = e^{s_i T} \tag{3.1.22}$$

with the same multiplicity.

Remark 3.1

From eqs. (3.1.19) and (3.1.20) follows for the discretized plant

$$\det \underline{A} = \det e^{\underline{F}T} = \prod e^{s_i T} = e^{\sum s_i T} = e^{\mathrm{trace}\underline{F}T} > 0 \tag{3.1.23}$$

Note that the trace of $\underline{F}$, eq. (A.7.34), is real. If eq. (3.1.8) describes a control algorithm or a closed loop, then $\det \underline{A}$ can also be negative. Special cases arise only when $\det \underline{A} = 0$ (see section 4.1).

Remark 3.2:

If the sampling interval T approaches zero, then the staircase function (3.1.3) approximates a continuous input and the difference equation (3.1.7) reduces to a differential equation:

$$\underline{x}(kT+T) - \underline{x}(kT) = (\underline{A}(T)-\underline{I})\underline{x}(kT) + \underline{b}(T)u(kT)$$

and with $kT = t$ and $T \to 0$

$$\lim_{T\to 0} \frac{1}{T}[\underline{x}(t+T)-\underline{x}(t)] = \lim_{T\to 0} \frac{1}{T}[\underline{A}(T)-\underline{I}]\underline{x}(t) + \lim_{T\to 0} \frac{1}{T}\underline{b}(T)u(t)$$

$$\dot{\underline{x}}(t) = \underline{F}\underline{x}(t) + \underline{g}u(t)$$

Therefore

$$\underline{F} = \lim_{T \to 0} \frac{1}{T} [\underline{A}(T) - \underline{I}] \tag{3.1.24}$$

$$\underline{g} = \lim_{T \to 0} \frac{1}{T} \underline{b}(T) \tag{3.1.25}$$

3.2 Homogeneous Solutions: Eigenvalues, Solution Sequences

The difference equation (3.1.8) is directly suitable for the recursive calculation of $\underline{x}[k]$ and y[k] from the initial state $\underline{x}[0]$ and a numerically given input sequence u[j], j = 0,1,2...k-1.

The general solution is obtained by recursive substitution as follows

$$\begin{aligned}
\underline{x}[1] &= \underline{Ax}[0] + \underline{bu}[0] \\
\underline{x}[2] &= \underline{Ax}[1] + \underline{bu}[1] \\
&= \underline{A}^2\underline{x}[0] + \underline{Abu}[0] + \underline{bu}[1] \\
&\vdots \\
\underline{x}[k] &= \underline{A}^k\underline{x}[0] + \underline{A}^{k-1}\underline{bu}[0] + \underline{A}^{k-2}\underline{bu}[1] + \ldots + \underline{bu}[k-1]
\end{aligned} \tag{3.2.1}$$

Eq. (3.2.1) can be written in either matrix notation

$$\underline{x}[k] = \underline{A}^k\underline{x}[0] + [\underline{A}^{k-1}\underline{b}, \underline{A}^{k-2}\underline{b} \ldots \underline{b}] \begin{bmatrix} u[0] \\ u[1] \\ \vdots \\ u[k-1] \end{bmatrix} \tag{3.2.2}$$

or as a convolution sum

$$\underline{x}[k] = \underline{A}^k\underline{x}[0] + \sum_{i=0}^{k-1} \underline{A}^{k-i-1}\underline{bu}[i] \tag{3.2.3}$$

For an output equation $y[k] = \underline{c}'\underline{x}[k] + du[k]$

$$\underline{y}[k] = \underline{c}'\underline{A}^k\underline{x}[0] + \sum_{i=0}^{k-1} \underline{c}'\underline{A}^{k-i-1}\underline{bu}[i] + du[k] \tag{3.2.4}$$

If the difference equation describes an open-loop sampled-data system with $\underline{A} = e^{\underline{F}T}$, then

$$\underline{A}^k(T) = \underline{A}[kT] \tag{3.2.5}$$

is determined simply by substituting kT for T in $\underline{A}$. For arbitrary $\underline{A}$, $\underline{A}^k$ is numerically calculated, by expressing k as a binary number,

$$k = a_o 2^o + a_1 2^1 + a_2 2^2 + \ldots + a_N 2^N, \quad a_i \varepsilon \{0,1\}.$$

For the calculation of

$$\underline{A}^k = \underline{A}^{a_o} \times \underline{A}^{2a_1} \times \underline{A}^{2^2 a_2} \ldots \underline{A}^{2^N a_N} \tag{3.2.6}$$

only the powers A, A^2, A^4, A^8, A^{16}... need be formed by squaring.

First some solutions of the homogeneous (or zero input) system

$$\underline{x}[k+1] = \underline{Ax}(k) \tag{3.2.7}$$

will be studied. The examples are of first and second order. Examples of higher order can be decomposed into simpler subsystems by transformation into Jordan form.

Example 1: n = 1

$$\begin{aligned} x[k+1] &= z_1 x[k] \\ x[k] &= z_1^k x[0] \end{aligned} \tag{3.2.8}$$

For example, for $z_1 > 0$ this can describe a zero-input sampled system with the continuous part

$$\dot{x} = s_1 x \quad , \quad z_1 = e^{s_1 T}$$

Figure 3.2 shows the eigenvalues and the corresponding form of x[k] for x[0] = 0 and

a) $z_1 = 1.2$, b) $z_1 = 1$, c) $z_1 = 0.8$. For $z_1 < 0$ the corresponding sequences with alternating signs arise;

d) $z_1 = -0.8$, e) $z_1 = -1$.

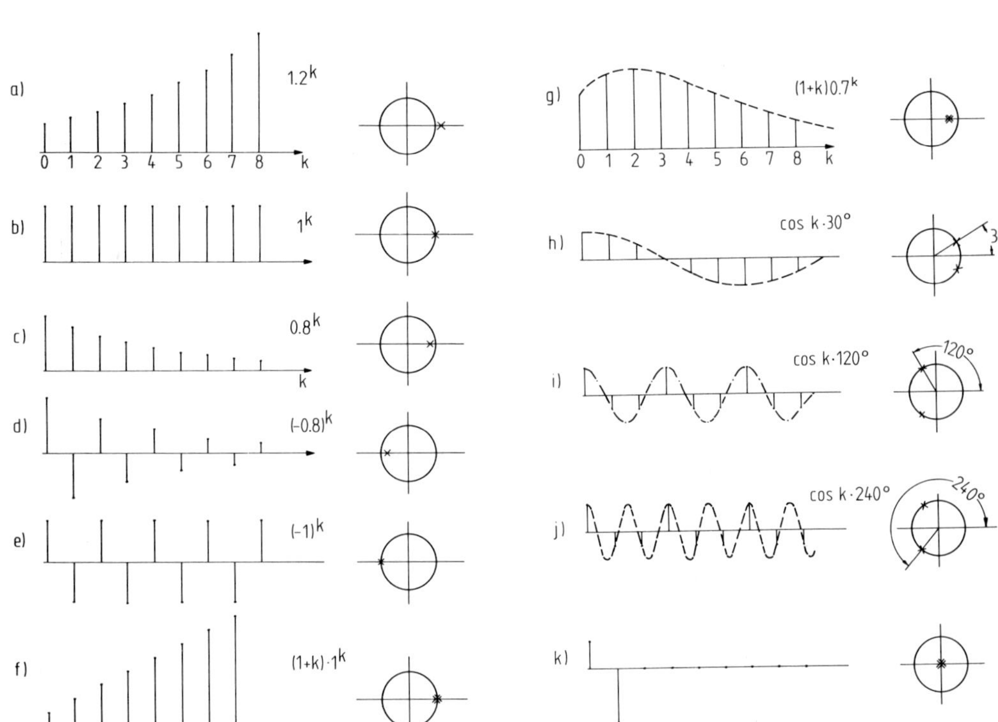

Figure 3.2 Homogeneous solutions and corresponding eigenvalues.

Example 2: n = 2, double eigenvalues

Continuous part

$$\underline{\dot{x}} = \begin{bmatrix} s_1 & 1 \\ 0 & s_1 \end{bmatrix} \underline{x}$$

$$\underline{x}[k+1] = z_1 \begin{bmatrix} 1 & T \\ 0 & 1 \end{bmatrix} \underline{x}[k] \quad , \quad z_1 = e^{s_1 T} \tag{3.2.9}$$

$$\underline{x}[k] = z_1^k \begin{bmatrix} 1 & kT \\ 0 & 1 \end{bmatrix} \underline{x}[0]$$

$$x_1[k] = z_1^k x_1[0] + kTz_1^k x_2[0]$$

Besides the solution term z_1^k, for double eigenvalues a term kTz_1^k arises, for triple eigenvalues also $(kT)^2 z_1^k$, etc.

For $\underline{x}[0] = [1 \quad 1]'$, $x_1[k] = (1+kT)z^k$. Figure 3.2 shows $x_1[k]$ for T = 1 and f) $z_1 = 1$, g) $z_1 = 0.7$. For $z_1 < 0$ the corresponding cases with alternating signs arise.

Example 3: n = 2, complex eigenvalues

Continuous part

$$\dot{\underline{x}} = \begin{bmatrix} \sigma_1 + j\omega_1 & 0 \\ 0 & \sigma_1 - j\omega_1 \end{bmatrix} \underline{x}$$

$$y = \begin{bmatrix} 1 & 1 \end{bmatrix} \underline{x}$$

$$\underline{x}[k+1] = e^{\sigma_1 T} \begin{bmatrix} e^{j\omega_1 T} & 0 \\ 0 & e^{-j\omega_1 T} \end{bmatrix} \underline{x}[k]$$

$$\underline{x}[k] = e^{\sigma_1 kT} \begin{bmatrix} e^{j\omega_1 kT} & 0 \\ 0 & e^{-j\omega_1 kT} \end{bmatrix} \underline{x}[0] \qquad (3.2.10)$$

The eigenvalues $z_{1,2} = e^{\sigma_1 T} \times e^{\pm j\omega_1 T}$ have the absolute value $e^{\sigma_1 T}$ and the angle $\omega_1 T$ to the positive real axis of the z-plane. For $\underline{x}[0] = [0.5 \quad 0.5]'$, $y(kT) = e^{\sigma_1 kT} \cos \omega_1 kT$. In figure 3.2 some cases with $\sigma_1 = 0$ are illustrated. For $\sigma_1 > 0$ $(s < 0)$ the corresponding increasing (decreasing) oscillations arise. Shown are:

h) $\omega_1 T = \pi/6 = 30^{\circ}$, i) $\omega_1 T = 2\pi/3 = 120^{\circ}$,
j) $\omega_1 T = 4\pi/3 = 240^{\circ}$.

Cases i) and j) have the same eigenvalues and the same discrete output $y(kT)$. However, the continuous outputs $y(t)$, shown here by dotted lines, have different frequencies. In practice, for given ω_1 one always chooses the sampling period T so that $\omega_1 T < \pi$ (see section 4.3).

In the special case $\omega_1 = 0$, two Jordan blocks with equal eigenvalues arise in the dynamic matrix

$$\underline{F} = \begin{bmatrix} s_1 & 0 \\ 0 & s_1 \end{bmatrix}$$

of the continuous system, i.e. $\underline{F}$ is not cyclic. This means that there is no vector $\underline{c}'$ such that $(\underline{F},\underline{c}')$ is observable. Here:

$$\text{rank} \begin{bmatrix} \underline{c}' \\ \underline{c}'\underline{F} \end{bmatrix} = \text{rank} \begin{bmatrix} c_1 & c_2 \\ c_1 s_1 & c_2 s_1 \end{bmatrix} = 1 \qquad (3.2.11)$$

There are two identical subsystems of first order whose outputs are added. The two states can no longer be distinguished from the sum signal. From an input-output point of view the system can be reduced to a system of first order. Solution forms corresponding to those in figure 3.2 a, b and c then arise.

A further special case is $\omega_1 T = \pi = 180^{\circ}$. Here the eigenvalues $\sigma_1 + j\omega_1$ and $\sigma_1 - j\omega_1$ of the continuous system are distinct, that is $\underline{F}$ is cyclic, but the eigenvalues of the discrete system both lie at $z_1 = -e^{\sigma_1 T}$, $\underline{A}$ is therefore not cyclic. Solutions such as in figures 3.2d and e arise.

Example 4: Eigenvalues at z = 0

$$P_A(z) = \det(z\underline{I}-\underline{A}) = z^n \tag{3.2.12}$$

From Cayley-Hamilton $\underline{A}^n = \underline{0}$ and therefore

$$\underline{x}[n] = \underline{A}^n\underline{x}[0] = \underline{0} \tag{3.2.13}$$

for all initial states $\underline{x}[0]$. Such a solution cannot arise in an open control loop with $\underline{A} = e^{\underline{F}T}$. However, this case occurs in the closed loop with a deadbeat controller (see section 4.4). As an example, a solution for

$$\underline{x}[k+1] = \begin{bmatrix} 2 & -4 \\ 1 & -2 \end{bmatrix} \underline{x}[k] \tag{3.2.14}$$

with $\underline{x}[0] = [1 \quad 1]'$, $y = [1 \quad 0]\underline{x}$ is represented in figure 3.2 k. Here $y[0] = 1$, $y[1] = -2$, $y[k] = 0$ for $k = 2,3,4...$

It is obvious from the form of the different solution terms that the solutions asymptotically approach zero if the absolute values of all eigenvalues are less than one. In other words: A necessary and sufficient condition for the asymptotic stability of the discrete system $\underline{x}[k+1] = \underline{Ax}[k]$ is that all the zeros of the characteristic polynomial

$$P_A(z) = \det(z\underline{I}-\underline{A}) = a_o + a_1 z + \ldots + a_{n-1}z^{n-1} + z^n \tag{3.2.15}$$

lie in the interior of the unit circle of the z-plane. Algebraic tests on the polynomial coefficients for the satisfaction of this condition are given in appendix C. See exercise 3.2.

3.3 Inhomogeneous Solutions: z-Transfer Function, Impulse and Step Responses

If the input sequence u in the difference equation

$$\underline{x}[k+1] = \underline{Ax}[k] + \underline{b}u[k] \tag{3.3.1}$$

is given analytically, for example as a sine sequence, then the application of the z-transform is recommended. It is defined by

$$f_z(z) = \mathcal{Z}\{f[k]\} := \sum_{k=0}^{\infty} f[k]z^{-k} \tag{3.3.2}$$

Its rules of calculation are derived in appendix B. Applying the left shifting theorem (B.4.1) to eq. (3.3.1)

$$\begin{aligned} z\{\underline{x}_z(z) - \underline{x}[0]\} &= \underline{A}\underline{x}_z(z) + \underline{b}u_z(z) \\ (z\underline{I}-\underline{A})\underline{x}_z(z) &= z\underline{x}[0] + \underline{b}u_z(z) \\ \underline{x}_z &= z(z\underline{I}-\underline{A})^{-1}\underline{x}[0] + (z\underline{I}-\underline{A})^{-1}\underline{b}u_z(z) \end{aligned} \tag{3.3.3}$$

Remark 3.3:

For the homogeneous or zero-input solution ($u \equiv 0$) comparison with eq. (3.2.3) gives the correspondence

$$\begin{aligned} \mathcal{Z}\{\underline{A}^k\} &= z(z\underline{I}-\underline{A})^{-1} \\ \underline{A}^k &= \mathcal{Z}^{-1}\{z(z\underline{I}-\underline{A})^{-1}\} \end{aligned} \tag{3.3.4}$$

Therefore $\underline{A}^k$ can be calculated for general k by the Leverrier-algorithm eq. (A.7.35) (z is inserted for λ), and inverse z-transform.

The output relation is now

$$\begin{aligned} y_z(z) &= \underline{c}'\underline{x}_z(z) + du_z(z) \\ &= z\underline{c}'(z\underline{I}-\underline{A})^{-1}\underline{x}[0] + [\underline{c}'(z\underline{I}-\underline{A})^{-1}\underline{b} + d]u_z(z) \end{aligned} \tag{3.3.5}$$

The expression

$$h_z(z) := \underline{c}'(z\underline{I}-\underline{A})^{-1}\underline{b} + d \tag{3.3.6}$$

is defined as the "z-transfer function". For the initial state

$\underline{x}[0] = \underline{0}$ this gives

$$y_z(z) = h_z(z) \times u_z(z) \tag{3.3.7}$$

For the determination of the output sequence y[k] the inverse z-transform

$$y[k] = \mathcal{Z}^{-1}\{y_z(z)\} \tag{3.3.8}$$

must be applied as described in section 3.6.2 and appendix C. Eq. (3.3.6) shows that the z-transfer function is a rational expression, the system described by $h_z(z)$ is therefore called a "rational system". The denominator of $h_z(z)$ is $\det(z\underline{I}-\underline{A})$, in other words the eigenvalues of the matrix $\underline{A}$ are the poles of the uncancelled z-transfer function. This is the reason why no distinction is made between z in eq. (3.1.21) (the eigenvalue equation of $\underline{A}$), and z as the complex variable of the z-transform eq. (3.3.2).

Example:

The discretized system of example (3.1.13) is augmented by an output equation as follows

$$\underline{x}(kT+T) = \begin{bmatrix} 1 & 1-e^{-T} \\ 0 & e^{-T} \end{bmatrix} \underline{x}(kT) + \begin{bmatrix} T+e^{-T}-1 \\ 1-e^{-T} \end{bmatrix} u$$

$$y(kT) = [1 \quad 0]\ \underline{x}(kT) \tag{3.3.9}$$

$$(z\underline{I}-\underline{A})^{-1} = \begin{bmatrix} z-1 & e^{-T}-1 \\ 0 & z-e^{-T} \end{bmatrix}^{-1} = \frac{1}{(z-1)(z-e^{-T})} \begin{bmatrix} z-e^{-T} & 1-e^{-T} \\ 0 & z-1 \end{bmatrix}$$

$$h_z(z) = \underline{c}'(z\underline{I}-\underline{A})^{-1}\underline{b}$$

$$= \frac{z(T-1+e^{-T}) + (1-e^{-T}-Te^{-T})}{(z-1)(z-e^{-T})} \tag{3.3.10}$$

This example for the sampling period T = 1 will be often referred to in the following sections. Its state representation is

$$\underline{x}[k+1] = \begin{bmatrix} 1 & 0.6321 \\ 0 & 0.3679 \end{bmatrix} \underline{x}[k] + \begin{bmatrix} 0.3679 \\ 0.6321 \end{bmatrix} u[k]$$

$$y[k] = [1 \quad 0]\, \underline{x}[k]$$

and its z-transfer function

$$h_z = \frac{0.3679z + 0.2642}{(z-1)(z-0.3679)}$$

Just as with the s-transfer function, eq. (2.3.6), the z-transfer function is invariant with respect to a basis transformation $\underline{x}^* = \underline{Tx}$ in the state space. The z-transfer function

$$h_z(z) = \frac{b_o+b_1 z+\ldots+b_{n-1}z^{n-1}+b_n z^n}{a_o+a_1 z+\ldots+a_{n-1}z^{n-1}+z^n} \tag{3.3.11}$$

$$= b_n + \frac{(b_o-a_ob_n)+(b_1-a_1b_n)z+\ldots+(b_{n-1}-a_{n-1}b_n)z^{n-1}}{a_o+a_1 z+\ldots+a_{n-1}z^{n-1}+z^n}$$

has 2n+1 numerator and denominator coefficients. If numerator and denominator are coprime, then they define a controllable and observable realization e.g. in the control-canonical form (A.3.8)

$$\underline{x}[k+1] = \begin{bmatrix} 0 & 1 & \ddots & 0 \\ & & \ddots & 1 \\ -a_o & -a_1 & \ldots & -a_{n-1} \end{bmatrix} \underline{x}[k] + \begin{bmatrix} 0 \\ \vdots \\ 0 \\ 1 \end{bmatrix} u[k] \tag{3.3.12}$$

$$y[k] = [b_o-a_ob_n \quad b_1-a_1b_n \ldots b_{n-1}-a_{n-1}b_n]\underline{x}[k] + b_n u[k]$$

All other realizations of order n emerge from (3.3.12) by a basis transformation with $\underline{T}$. Similar results as discussed in section 2.3.7 for continuous systems apply to possible cancellations in h_z. We will come back to this in connection with the controllability and observability of sampling systems.

An elementary solution of the difference eq. (3.1.8) is the response to a unit impulse

$$u[k] = \Delta[k] = \begin{cases} 0 & k \neq 0 \\ 1 & k = 0 \end{cases} \tag{3.3.13}$$

for the initial state $\underline{x}(0) = \underline{0}$. In other words: The input of the continuous systems (3.1.1) is the output of the hold element shown in figure 3.3.

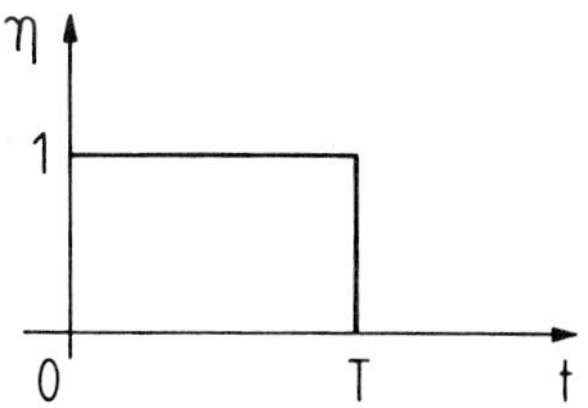

Figure 3.3
Response of the hold element to the impulse $\Delta[k]$

From eq. (3.2.4) the impulse response is

$$y[k] = h[k] = \begin{cases} 0 & k < 0 \quad \text{(causality)} \\ d & k = 0 \\ \underline{c}'\underline{A}^{k-1}\underline{b} & k > 0 \end{cases} \tag{3.3.14}$$

With the help of this impulse response and with $\underline{x}(0) = \underline{0}$, eq. (3.2.4) can also be written as

$$y[k] = \sum_{i=0}^{k} h[k-i]u[i] \tag{3.3.15}$$

Eq. (3.3.15) is the discrete equivalent of the convolution integral, eq. (2.4.15). Correspondingly eq. (3.3.15) is designated as "convolution sum" and $h[k]$ as "weighting sequence".

Eq. (3.3.15) also allows the representation of the response of general discrete linear systems which are not necessarily describable by a vector difference equation; for example the weighting sequence $h[k]$ can be given numerically.

For rational systems the impulse response is describable in the form (3.3.14) and has a simple correspondence with the characteristic polynomial

$$P_A(z) = \det(z\underline{I}-\underline{A}) = a_0 + a_1 z + \ldots + a_{n-1}z^{n-1} + z^n \qquad (3.3.16)$$

From the Cayley-Hamilton theorem

$$P_A(\underline{A}) = a_0\underline{I} + a_1\underline{A} + \ldots + a_{n-1}\underline{A}^{n-1} + \underline{A}^n = \underline{0} \qquad (3.3.17)$$

Multiplying this equation from the left by $\underline{c}'\underline{A}^{k-n-1}$, $k > n$, and from the right by $\underline{b}$, and noting eq. (3.3.14)

$$a_0\underline{c}'\underline{A}^{k-n-1}\underline{b} + a_1\underline{c}'\underline{A}^{k-n}\underline{b} + \ldots + a_{n-1}\underline{c}'\underline{A}^{k-2}\underline{b} + \underline{c}'\underline{A}^{k-1}\underline{b} =$$

$$= a_0 h[k-n] + a_1 h[k-n+1] + \ldots + a_{n-1}h[k-1] + h[k] = 0, \quad k > n \qquad (3.3.18)$$

In other words: If one knows the values h[1], h[2]...h[n] of the weighting sequence as well as the characteristic equation, then the entire weighting sequence can be calculated. Conversely the characteristic polynomial and a state representation can be derived from a measured impulse response h[k] by eq. (3.3.18). This problem which is known as minimal realization is treated for example in [66.3], [71.3], [71.4], [79.5].

For the determination of a mathematical model from a test measurement of the system the unit step input may also be chosen. The step input is not changed by the sample and hold operation, the choice of the sampling period need not be made before the test measurement. For $\underline{x}(0) = \underline{0}$ and

$$u[k] = \begin{cases} 0 & k < 0 \\ 1 & k \geq 0 \end{cases} \qquad (3.3.19)$$

one obtains from eq. (3.3.15)

$$y_{step}[k] = \sum_{i=0}^{k} h[k-i] \qquad (3.3.20)$$

The step response is the sum of the impulse responses. This is plausible from figure 3.3 because the unit step may be represented as the sum of shifted impulse responses of the hold element. Conversely

$$h[k] = h(kT) = y_{step}(kT) - y_{step}(kT-T) \qquad (3.3.21)$$

The impulse response h(kT) can therefore be determined from the step response $y_{step}(t)$ of the continuous part of the system for every chosen sampling interval T. For small T the measurement of h(kT) is more accurate, however, than the measurement of $y_{step}(kT)$ and calculation of h(kT) via eq. (3.3.21).

The z-transfer function can be determined from the impulse response. Applying the z-transform to the convolution sum (3.3.15) gives

$$y_z(z) = \sum_{k=0}^{\infty} \sum_{i=0}^{k} h[k-i]u[i]z^{-k} \qquad (3.3.22)$$

For causal systems h[k-i] = 0 for i > k, thus

$$y_z(z) = \sum_{i=0}^{\infty} \sum_{k=0}^{\infty} h[k-i]u[i]z^{-k}$$

Introducing the variable m = k-i gives

$$y_z(z) = \sum_{i=0}^{\infty} \left(\sum_{m=-i}^{\infty} h[m]u[i]z^{-m} \right) z^{-i}$$

The second sum can begin at m = 0 instead of m = -i because h[m] = 0 for m < 0. Since the individual elements of the double sum depend either on m or on i only, $y_z(z)$ may be written as a product of two simple sums

$$y_z(z) = \sum_{m=0}^{\infty} h[m]z^{-m} \times \sum_{i=0}^{\infty} u[i]z^{-i} \qquad (3.3.23)$$

$$y_z(z) = h_z(z) \times u_z(z) \qquad (3.3.24)$$

The z-transfer function $h_z(z)$ is the z-transform of the impulse response h[k] from eq. (3.3.14):

$$h_z(z) = d + \sum_{k=1}^{\infty} \underline{c}'\underline{A}^{k-1}\underline{b}z^{-k} \tag{3.3.25}$$

This relation can also be derived by inserting eq. (3.3.4) into eq. (3.3.6). It shows in particular that the excess of poles over zeros in $h_z(z)$ is equal to the value of m corresponding to the first non-zero term in the impulse response. If $h[0] = d \neq 0$, $h_z(z)$ has equal numerator and denominator degree and the pole excess is zero. If $h[0] = 0$, $h[1] = \underline{c}'\underline{b} \neq 0$, the pole excess is one, etc.

If a frequency domain description by the s-transfer function is given for the continuous part, it is not necessary to take the detour of using a state description. It is easier to determine the impulse response of the hold element and plant by the Laplace-transform. h[k] is the response of the continuous part $g_s(s)$ to the impulse response of the hold element shown in figure 3.3. It has the Laplace-transform

$$\eta_s(s) = \frac{1-e^{-Ts}}{s} \tag{3.3.26}$$

Therefore

$$h_s(s) = \eta_s(s)g_s(s) = \frac{1-e^{-Ts}}{s} \times g_s(s)$$

$$h(t) = \mathcal{L}^{-1}\left\{\frac{1-e^{-Ts}}{s} g_s(s)\right\}$$

With the step response of the plant

$$v(t) := \mathcal{L}^{-1}\{g_s(s)/s\} \tag{3.3.27}$$

$$h(t) = v(t) - v(t-T).$$

The z-transfer function is

$$h_z(z) = \mathcal{Z}\{h(kT)\} = \mathcal{Z}\{v(kT)\} - \mathcal{Z}\{v(kT-T)\}$$

and from the right shifting theorem (B.3.1)

$$h_z(z) = (1-z^{-1})\,\mathcal{Z}\{v(kT)\}$$

$$h_z(z) = \frac{z-1}{z} \times \mathcal{Z}\left\{\mathcal{L}^{-1}\{g_s(s)/s\}_{t = kT}\right\} \qquad (3.3.28)$$

The notation

$$f_z(z) = \mathcal{Z}\left\{f_s(s)\right\} := \mathcal{Z}\left\{\mathcal{L}^{-1}\{f_s(s)\}_{t = kT}\right\} \qquad (3.3.29)$$

is introduced for the operation arising in eq. (3.3.28). Therefore

$$h_z(z) = \frac{z-1}{z}\mathcal{Z}\left\{g_s/s\right\} \qquad (3.3.30)$$

Figure 3.4 summarizes the relations between the different system descriptions

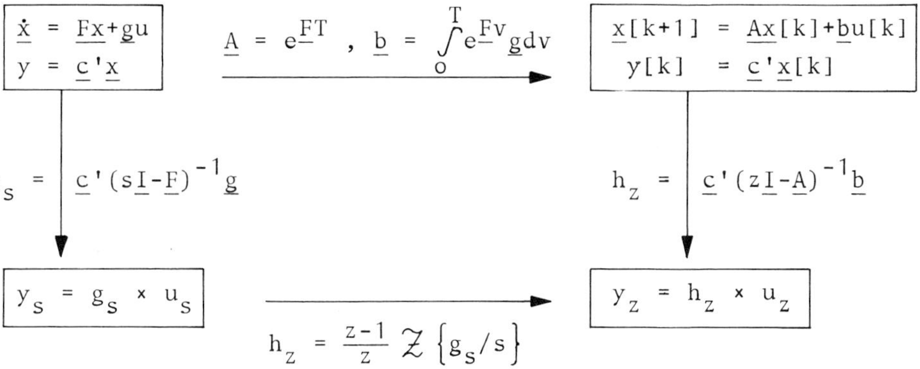

$$\begin{array}{|l|}\hline \dot{\underline{x}} = \underline{F}\underline{x}+\underline{g}u \\ y = \underline{c}'\underline{x} \\ \hline\end{array} \xrightarrow{\underline{A} = e^{\underline{F}T},\ \underline{b} = \int_0^T e^{\underline{F}v}\underline{g}dv} \begin{array}{|l|}\hline \underline{x}[k+1] = \underline{A}\underline{x}[k]+\underline{b}u[k] \\ y[k] = \underline{c}'\underline{x}[k] \\ \hline\end{array}$$

$$g_s = \underline{c}'(s\underline{I}-\underline{F})^{-1}\underline{g} \downarrow \qquad\qquad \downarrow h_z = \underline{c}'(z\underline{I}-\underline{A})^{-1}\underline{b}$$

$$\boxed{y_s = g_s \times u_s} \xrightarrow[h_z = \frac{z-1}{z}\mathcal{Z}\{g_s/s\}]{} \boxed{y_z = h_z \times u_z}$$

Figure 3.4 Relations between the continuous and discrete system representations

In simple cases one can immediately determine the z-transfer function from the table in section B.13.

Example:

For the system of example (3.1.13) and (3.3.9)

$$\underline{\dot{x}} = \begin{bmatrix} 0 & 1 \\ 0 & -1 \end{bmatrix} \underline{x} + \begin{bmatrix} 0 \\ 1 \end{bmatrix} u$$

$$y = [1 \quad 0]\underline{x}$$

The transfer function is

$$g_s(s) = \underline{c}'(s\underline{I}-\underline{F})^{-1}\underline{g} = \frac{1}{s(s+1)}$$

We now suppose that the system is given in this form, see figure 3.5.

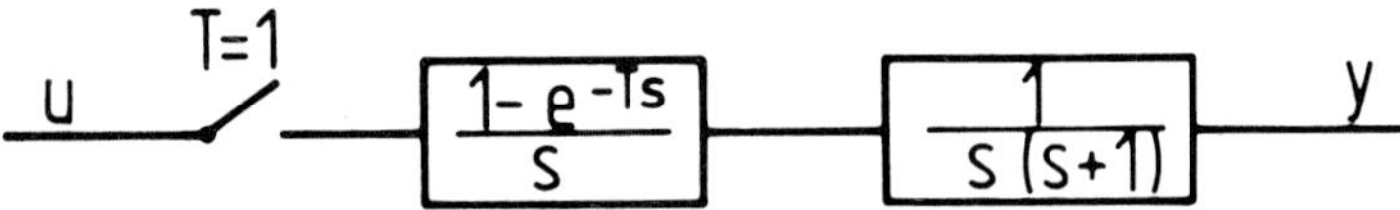

Figure 3.5 Determination of the z-transfer function

$$\frac{g_s(s)}{s} = \frac{1}{s^2(s+1)}$$

In the table in appendix B one finds the row

$f(t)$	$f_s(s)$	$f_z(z)$
$at-1+e^{-at}$	$\dfrac{a^2}{s^2(s+a)}$	$\dfrac{(aT-1+e^{-aT})z^2+(1-aTe^{-aT}-e^{-aT})z}{(z-1)^2(z-e^{-aT})}$

for a = 1 the z-transfer function is therefore

$$h_z(z) = \frac{z-1}{z} \times f_z(z) = \frac{(T-1+e^{-T})z + (1-Te^{-T}-e^{-T})}{(z-1)(z-e^{-T})}$$

This expression agrees with the solution (3.3.10), which was derived from the discrete state representation. For T = 1

$$h_z(z) = \frac{0.3679z+0.2642}{(z-1)(z-0.3679)} \tag{3.3.31}$$

For higher order s-transfer functions the desired expression cannot be found in the table. Rather $g_s(s)$ should be decomposed by partial fractions. If only simple poles arise in $g_s(s)$

$$g_s(s) = \sum_{i=1}^{n} \frac{R_i}{s+a_i}$$

and from eq. (3.3.30)

$$\begin{aligned} h_z(z) &= \frac{z-1}{z} \sum_{i=1}^{n} \mathcal{Z}\left\{\frac{R_i}{s(s+a_i)}\right\} \\ &= \frac{z-1}{z} \sum_{i=1}^{n} \frac{R_i(1-e^{-a_iT})z}{a_i(z-1)(z-e^{-a_i})} \end{aligned} \tag{3.3.32}$$

$$h_z(z) = \sum_{i=1}^{n} \frac{R_i(1-e^{-a_iT})}{a_i(z-e^{-a_iT})}$$

For plants with an integrator one of the coefficients a_i equals zero and the corresponding term is

$$\lim_{a_i \to 0} \frac{R_i(1-e^{-a_iT})}{a_i(z-e^{-a_iT})} = \frac{R_iT}{z-1} \tag{3.3.33}$$

For a double pole in $g_s(s)$ a term $1/(s+1)^2$ arises from the partial fractions decomposition and $h_z(z)$ contains the term

$$\frac{z-1}{z}\mathcal{Z}\left\{\frac{1}{s(s+a)^2}\right\} = \frac{z(1-e^{-aT}-aTe^{-aT})+e^{-aT}(aT-1+e^{-aT})}{a^2(z-e^{-aT})^2} \tag{3.3.34}$$

and for $a \to 0$

$$\frac{z-1}{z}\mathcal{Z}\left\{\frac{1}{s^3}\right\} = \frac{T^2(z+1)}{2(z-1)^2} \tag{3.3.35}$$

In general: Every pole of $g_s(s)$ at s_i corresponds to a pole of the uncancelled z-transfer function $h_z(z)$ at $z = e^{s_i T}$ with the same multiplicity. This statement corresponds fully to the statement formulated in eq. (3.1.22) for the state representation.

A partial check of the calculated z-transfer function is obtained by comparison of the steady-state values of step responses of the continuous and discrete systems. If $\lim_{t\to\infty} v(t)$ exists, then

$$\lim_{t\to\infty} v(t) = \lim_{k\to\infty} v(kT) \tag{3.3.36}$$

From the final value theorem of the Laplace transform applied to eq. (3.3.27), the step response of the continuous system has the steady-state value

$$\lim_{t\to\infty} v(t) = \lim_{s\to 0} s v_s(s) = \lim_{s\to 0} g_s(s) = g_s(0)$$

On the other hand, from the final value theorem of the z-transform, eq. (B.8.1)

$$\lim_{k\to\infty} v(kT) = \lim_{z\to 1} (z-1) v_z(z) = \lim_{z\to 1} h_z(z) = h_z(1)$$

The two expressions must be equal, i.e.

$$g_s(0) = h_z(1) \tag{3.3.37}$$

Another question of interest is whether the zeros of $g_s(s)$ and the zeros of $h_z(z)$ have a simple correspondence. The answer is no. Example (3.3.31) has already shown that $h_z(z)$ can have a zero when $g_s(s)$ has none.

Not even the minimum phase property, i.e. all zeros of $g_s(s)$ in the left half s-plane, leads to the analogous property of the z-transfer function, as the following two examples show. See also [81.9], [81.10], [84.1].

Example 1:

The s-transfer function

$$g_s(s) = 1/s^3 \tag{3.3.38}$$

has no zeros in the right half s-plane. The corresponding z-transfer function

$$h_z(z) = \frac{z-1}{z}\mathcal{Z}\left\{\frac{1}{s^4}\right\} = \frac{T^3(z^2+4z+1)}{6(z-1)^3}$$

has a zero outside of the unit circle at z = -3.732.

Example 2:

In this example the continuous part is non-minimum phase.

$$g_s(s) = \frac{0.5-s}{(1+s)^2} \tag{3.3.39}$$

$$v_s(s) = \frac{g_s(s)}{s} = \frac{0.5 - s}{s(1+s)^2} = \frac{0.5}{s} - \frac{0.5}{1+s} - \frac{1.5}{(1+s)^2}$$

$$v(t) = 0.5 - 0.5\, e^{-t} - 1.5\, te^{-t}$$

v(t) is the step response of the system (3.3.39) as shown in figure 3.6.

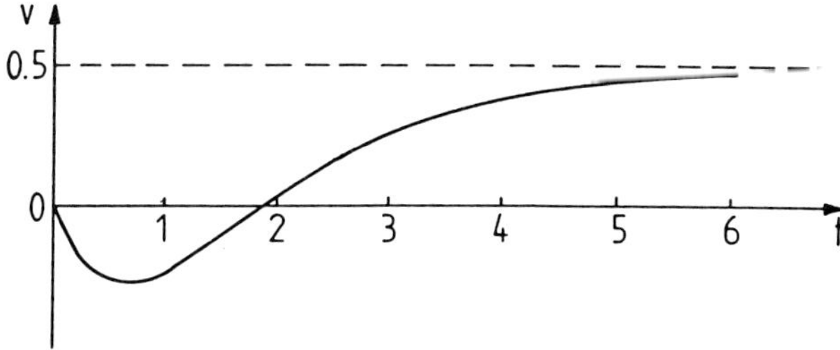

Figure 3.6 Step response of the non-minimum phase system $(0.5-s)/(1+s)^2$

From eq. (3.3.30) the z-transfer function is

$$h_z(z) = \frac{z-1}{z}\,[v(0)+v(T)z^{-1}+v(2T)z^{-2}+\dots]$$

with $v(0) = 0$ and $v(T) \neq 0$ for $T \neq 1.904$, or in other words $h_z(z)$ has a pole excess of one. In the special case of $T = 1.904$, where the pole excess is two, a zero goes to infinity.

The z-transfer function is

$$h_z(z) = \frac{z-1}{z}\mathcal{Z}\left\{\frac{0.5}{s} - \frac{0.5}{1+s} - \frac{1.5}{(1+s)^2}\right\}$$

$$= \frac{0.5z}{z-1} - \frac{0.5z}{z-e^{-T}} - \frac{1.5Tze^{-T}}{(z-e^{-T})^2}$$

$$= 0.5\,\frac{z(1-e^{-T}-3Te^{-T})+e^{-T}(e^{-T}-1+3T)}{(z-e^{-T})^2}$$

The zero at

$$z = \frac{e^{-T}(1-3T-e^{-T})}{1-(1+3T)e^{-T}}$$

lies in the unit circle for $T > 2.8385$, although the zeros of $g_s(s)$ lie in the right s-half plane.

Figure 3.7 shows how the position of the zero depends on T.

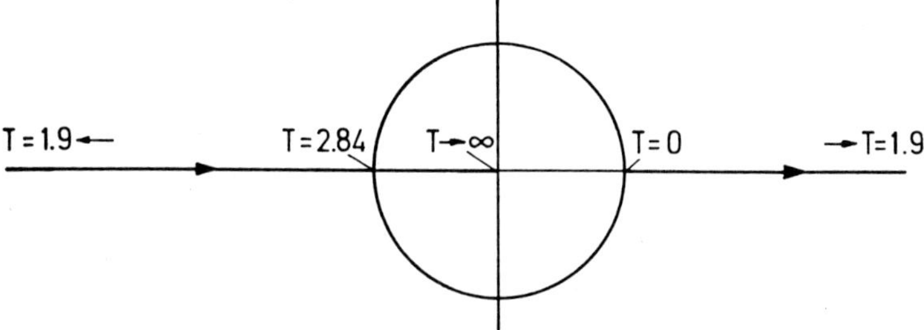

Figure 3.7 Position of the zero as a function of the sampling period T.

3.4 Discrete Controller and Control Loop

3.4.1 Representation by the z-Transfer Function

In the control loop of figure 1.7 a linear control algorithm of order m can be implemented according to eq. (1.1.2)

$$u[k] = d_m e[k] + d_{m-1} e[k-1] + \ldots + d_o e[k-m] - c_{m-1} u[k-1] - \ldots - c_o u[k-m] \tag{3.4.1}$$

The corresponding z-transfer function of the controller according to eq. (1.1.5) is

$$d_z = \frac{u_z}{e_z} = \frac{d_m z^m + d_{m-1} z^{m-1} + \ldots + d_o}{z^m + c_{m-1} z^{m-1} + \ldots + c_o} \tag{3.4.2}$$

The causality condition of eq. (3.4.1) stipulates that no future values can be used for calculation of u[k]. This is reflected in the z-transfer function (3.4.2) by the fact that the numerator degree cannot be greater than the denominator degree.

Eq. (3.4.1) also shows that it would not be reasonable to set the proportional feedback term d_m equal to zero. In this case the latest information concerning the return difference (which for example may have been increased by a step of the reference input or a disturbance input in the last sampling interval) would be processed, only after a delay of one sampling interval. Note that at time t = kT only the term $d_m e[k]$ must be formed and added, all other terms in eq. (3.4.1) can be calculated before this time. The time required for one multiplication and one addition is usually small compared to the sampling interval T. See also figures A.3 and A.5 for the implementation of eq. (3.4.1).

It will always be assumed that the controller has equal numerator and denominator degrees. For example

$$m = 0 \qquad d_z(z) = d_o \tag{3.4.3}$$

$$m = 1 \qquad d_z(z) = \frac{d_1 z + d_o}{z + c_o} \tag{3.4.4}$$

$$m = 2 \qquad d_z(z) = \frac{d_2 z^2 + d_1 z + d_o}{z^2 + c_1 z + c_o} \tag{3.4.5}$$

The number of free controller parameters is 2m+1.

A simple control loop is represented in figure 3.8.

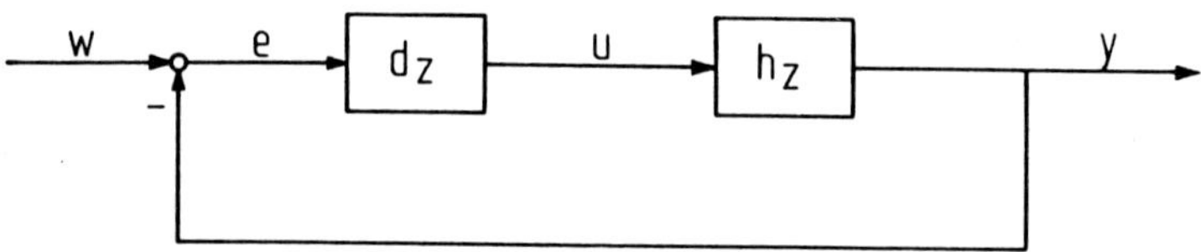

Figure 3.8 Simple control loop

The closed-loop transfer function follows from

$$e_z = w_z - y_z = w_z - h_z d_z e_z$$

$$(1+h_z d_z)e_z = w_z \; , \; e_z = \frac{1}{1+h_z d_z} w_z \tag{3.4.6}$$

$$y_z = \frac{h_z d_z}{1+h_z d_z} w_z \tag{3.4.7}$$

Thus the transfer function is

$$f_z = \frac{h_z d_z}{1+h_z d_z} \tag{3.4.8}$$

The compensator d_z is assumed to have equal numerator and denominator degrees. In feedback notation this corresponds to a "feed-through" term d in the output equation $y = \underline{c}'\underline{x} + du$ as in eqs. (3.2.4) and (3.3.5). It is not realistic to make the same assumption for the plant h_z, because in the closed loop the actuator input u[k] would influence itself immediately via the plant and compensator responses in the closed loop. This would not agree with the operation of the sampler and the digital controller as described above, see also remarks 1.2 and 3.9. Therefore it is assumed that h_z is strictly proper, i.e. there is no feedthrough term d, the plant step response $v(t) = \mathcal{L}^{-1}\{g_s(s)/s\}$ satisfies $v(0) = 0$. It is often true that the pole-zero excess of h_z is equal to one, which is the case when the step response is first nonzero at time instant T, i.e. $v(T) \neq 0$. Then by eq. (3.4.8) also the closed-loop transfer function f_z has pole-zero excess one.

A pole-zero excess greater than one for the open and closed loop arises for systems with transport lag, see section 3.8, or for an especially awkward choice of the sampling period, namely whenever v(t) goes through zero for t = T as in figure 3.6 for T = 1.904.

The type of solution terms which arises in the closed loop is determined by the uncancelled poles of the transfer function, i.e. the zeros of the characteristic equation

$$1 + h_z(z)d_z(z) = 0 \tag{3.4.9}$$

Expressing $h_z(z) = B(z)/A(z)$ and $d_z(z) = D(z)/C(z)$ by their numerator and denominator polynomials, the reference transfer function is

$$f_z = \frac{B(z)D(z)}{A(z)C(z)+B(z)D(z)} \tag{3.4.10}$$

and the characteristic polynomial of the closed loop is

$$P(z) = A(z)C(z) + B(z)D(z) \tag{3.4.11}$$

If A(z) and D(z) (or B(z) and C(z)) have a common zero because of a cancellation compensation, then this zero also arises in P(z); it is therefore an eigenvalue of the closed loop, even if cancelled out in the transfer function f_z.

3.4.2 Pole Assignment by Equating Coefficients

For the control loop of figure 3.8 with the z-transfer function of the plant

$$
h_z = \frac{B(z)}{A(z)} = \frac{b_{n-1}z^{n-1}+b_{n-2}z^{n-2}+\ldots+b_o}{z^n+a_{n-1}z^{n-1}+\ldots+a_o} \tag{3.4.12}
$$

where $B(z)$ and $A(z)$ are coprime, desired poles z_i, $i = 1,2\ldots 2n-1$, can always be assigned by a controller of order $m = n-1$. Let the desired characteristic polynomial be

$$
P(z) = \prod_{i=1}^{2n-1} (z-z_i) = z^{2n-1}+p_{2n-2}z^{2n-2}+\ldots+p_o \tag{3.4.13}
$$

The controller is assumed as

$$
d_z = \frac{D(z)}{C(z)} = \frac{d_{n-1}z^{n-1}+d_{n-2}z^{n-2}+\ldots+d_o}{z^{n-1}+c_{n-2}z^{n-2}+\ldots+c_o} \tag{3.4.14}
$$

By multiplying eq. (3.4.11) out and equating the coefficients of each power of z with the corresponding coefficient in eq. (3.4.13), one obtains the linear system of equations:

$$
2n-1\left\{
\left[
\underbrace{\begin{array}{cccc}
a_o & 0 & & \\
a_1 & a_o & & \\
\vdots & \ddots & \ddots & \\
 & & \ddots & a_o \\
\hline
a_{n-1} & \cdots & & a_1 \\
1 & \ddots & & \vdots \\
0 & \ddots & \ddots & \\
 & \ddots & \ddots & a_{n-1} \\
0 & & & 1
\end{array}}_{n-1}
\middle|
\underbrace{\begin{array}{ccccc}
b_o & 0 & & & \\
b_1 & b_o & & & \\
\vdots & & \ddots & & 0 \\
 & & & \ddots & \\
\hline
b_{n-1} & & & & b_o \\
0 & \ddots & & & \\
 & & \ddots & & \\
 & & & \ddots & \\
0 & & & & b_{n-1}
\end{array}}_{n}
\right]
\right.
\left[\begin{array}{c}
c_o \\ \vdots \\ c_{n-2} \\ \hline d_o \\ \vdots \\ \vdots \\ d_{n-1}
\end{array}\right]
=
\left[\begin{array}{c}
p_o \\ \vdots \\ p_{n-2} \\ \hline p_{n-1}-a_o \\ \vdots \\ \vdots \\ p_{2n-2}-a_{n-1}
\end{array}\right]
\tag{3.4.15}
$$

The $(2n-1)\times(2n-1)$ matrix is nonsingular if and only if the polynomials $A(z)$ and $B(z)$ are coprime [59.4].

Example:

For the plant of figure 3.5 and eq. (3.3.31), h_z is

$$h_z = \frac{B(z)}{A(z)} = \frac{0.3679z+0.2642}{z^2-1.3679z+0.3679} = 0.3679\,\frac{z+0.7181}{(z-1)(z-0.3679)}$$

The controller of order n-1 has the form

$$d_z = \frac{d_1 z+d_o}{z+c_o}$$

Here eq. (3.4.15) is

$$\begin{bmatrix} 0.3679 & 0.2642 & 0 \\ -1.3679 & 0.3679 & 0.2642 \\ 1 & 0 & 0.3679 \end{bmatrix} \begin{bmatrix} c_o \\ d_o \\ d_1 \end{bmatrix} = \begin{bmatrix} p_o \\ p_1-0.3679 \\ p_2+1.3679 \end{bmatrix}$$

$$\begin{bmatrix} c_o \\ d_o \\ d_1 \end{bmatrix} = \begin{bmatrix} 0.5358 & -0.3849 & 0.2764 \\ 3.0385 & 0.5358 & -0.3849 \\ -1.4565 & 1.0462 & 1.9668 \end{bmatrix} \begin{bmatrix} p_o \\ p_1-0.3679 \\ p_2+1.3679 \end{bmatrix}$$

a) For example if all three poles are placed at $z = 0$,

$P(z) = z^3$, $p_o = p_1 = p_2 = 0$, then

$$d_z = \frac{2.3055z-0.7236}{z+0.5197} = 2.3055 \times \frac{z-0.3139}{z+0.5197} \qquad (3.4.16)$$

and the z-transfer function is

$$f_z = \frac{0.8484z^2+0.3430z-0.1912}{z^3} \qquad (3.4.17)$$

b) One can also cancel the pole at $z = 0.3679$ which thus remains as an eigenvalue of the closed loop, i.e.

$P(z) = z^2(z-0.3679)$, $p_o = p_1 = 0$, $p_2 = -0.3679$.

Then

$$d_z = \frac{1.5820z-0.5820}{z+0.4180} = 1.582\,\frac{z-0.3679}{z+0.4180} \qquad (3.4.18)$$

$$f_z = \frac{0.5820z+0.4180}{z^2} \qquad (3.4.19)$$

c) Formally one can also cancel the zero at $z = -0.7181$

$P(z) = z^2(z+0.7181)$, $p_o = p_1 = 0$, $p_2 = 0.7181$. One obtains

$$d_z = \frac{3.7178z-1}{z+0.7181} = 3.7178 \times \frac{z-0.2690}{z+0.7181} \qquad (3.4.20)$$

$$f_z = \frac{1.3678z-0.3679}{z^2} \qquad (3.4.21)$$

It will be shown later that this cancellation is not to be recommended.

d) Finally both cancellations can be performed simultaneously:

$P(z) = z(z+0.7181)(z-0.3679)$, $p_o = 0$, $p_1 = -0.2642$, $p_2 = 0.3502$

and

$$d_z = \frac{2.7179z-1}{z+0.7181} = 2.7179\,\frac{z-0.3679}{z+0.7181} \qquad (3.4.22)$$

$$f_z = \frac{1}{z} \qquad (3.4.23)$$

This solution will also be shown to have severe disadvantages. See figure 3.18.

3.4.3 Integral Controller

The steady-state response of the control loop can be expressed by the steady-state value of the error e for reference inputs such as step or ramp. From the final value theorem of the z-transform (B.8.1), the steady-state value for a stable control loop with e_z as given in eq. (3.4.6) is

$$\lim_{k\to\infty} e(kT) = \lim_{z\to 1}(z-1)e_z(z)$$

$$= \lim_{z\to 1} \frac{z-1}{1+h_z(z)d_z(z)} w_z(z) \qquad (3.4.24)$$

For a step reference input $w(t) = 1(t)$, $w_z(z) = z/(z-1)$

$$\lim_{k\to\infty} e(kT) = \frac{1}{1+h_z(1)d_z(1)} \qquad (3.4.25)$$

This "position error" becomes zero if the open loop z-transfer function $h_z(z)d_z(z)$ contains a pole at $z = 1$, i.e. an integration. (Strictly speaking this is only an approximate integration by an "accumulator".)

For ramp input $w(t) = t$, $w_z(z) = Tz/(z-1)^2$

$$\lim_{k\to\infty} e(kT) = \frac{T}{\lim\limits_{z\to 1}(z-1)h_z(z)d_z(z)} \qquad (3.4.26)$$

This "velocity error" becomes zero if $h_z(z)d_z(z)$ contains a double pole at $z = 1$, i.e. a double integration.

If the steady-state value $h_z(1) = g_s(0)$ (see eq. (3.3.37)), is known exactly, the position error can be corrected by a prefactor V

$$V = \frac{1+h_z(1)d_z(1)}{h_z(1)d_z(1)} = \frac{1+g_s(0)d_z(1)}{g_s(0)d_z(1)} \qquad (3.4.27)$$

for the reference input w.

However, if the steady-state value $g_s(0)$ of the plant is not known exactly, it is recommended that an integral term Σ with a pole at $z = 1$ is used in the controller as in figure 3.9. This makes the zero steady-state error property robust with respect to changes in the plant $h_z(z)$ as long as the control loop remains stable.

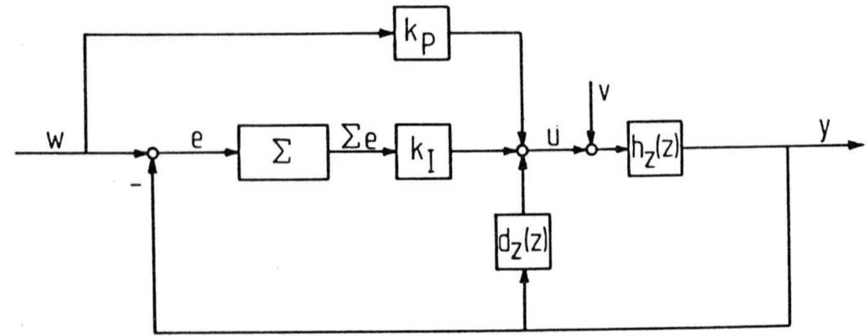

Figure 3.9 Controller with integral part

For Σ an approximate integrator can be chosen. E.g. for the "rectangular rule"

$$u(kT) = u(kT-T) + T \times e(kT-T)$$

the right-shifting theorem (B.3.1) yields

$$u_z(z) = z^{-1} u_z(z) + z^{-1} T e_z(z)$$

and the transfer function

$$\Sigma_1(z) = \frac{u_z(z)}{e_z(z)} = \frac{T}{z-1} \qquad (3.4.28)$$

For the "advanced rectangular rule"

$$u(kT) = u(kT-T) + Te(kT)$$

$$\Sigma_2(z) = \frac{Tz}{z-1} = T + \Sigma_1(z) \qquad (3.4.29)$$

For the trapezoidal rule

$$u(kT) = u(kT-T) + T \frac{e(kT)+e(kT-T)}{2}$$

$$u_z(z) = z^{-1} u_z(z) + \frac{T}{2} (1+z^{-1}) e_z(z)$$

$$\Sigma_3(z) = \frac{T(z+1)}{2(z-1)} = \frac{1}{2}\left[\Sigma_1(z)+\Sigma_2(z)\right] = \frac{T}{2} + \Sigma_1(z) \qquad (3.4.30)$$

The constant terms T or T/2 may be included into k_P and $d_z(z)$ in figure 3.9. Therefore all choices of Σ finally lead to the same controller.

If the order of $d_z(z)$ is n-1 as before, the degree of the characteristic polynomial of the control loop increases by 1 to 2n and an additional control parameter k_I is available. k_P only enters into the numerator of the transfer function from w to y, but not into the characteristic equation. If a stable characteristic polynomial is produced by $d_z(z)$ and k_I, then the summed error $\Sigma e(k)$ tends to a constant and e[k] tends to to zero asymptotically for a step input. Indeed this is also the case whenever a constant disturbance input v affects the plant. The feedback path $d_z(z)$ may also be replaced by a structure with observer and state feedback. In this case Σ is considered as part of the plant (see chapter 6).

A standard control structure for continuous systems is the proportional-integral-differential controller (PID controller). Its transfer function is

$$PID(s) = k_P + \frac{k_I}{s} + k_D \frac{s}{1+T_1 s} \tag{3.4.31}$$

One controller pole is fixed by the integration. For small T_1 the other pole is placed as far left in the s-plane as high frequency disturbances allow. The three numerator coefficients remain as free parameters.

Sometimes practical experience with the tuning of PID controllers is available for plants for which we do not have a good mathematical model. This know-how should be further used if the analog controller is replaced by a digital one. Similar analog and digital controllers may also be desirable if one of them is a back-up for the other and a bumpless switching is required. In these cases the continuous controller transfer function, like the one of eq. (3.4.31), is approximated by a discrete controller. The approximate integration by the trapezoidal rule is especially suited for this problem.

The integrator 1/s is replaced by Σ_3 of eq. (3.4.30). The differentiator can be correspondingly appoximated by the reciprocal of eq. (3.4.30)

$$q = \Sigma_3^{-1} = \frac{2}{T}\frac{z-1}{z+1} \tag{3.4.32}$$

The transfer function q is an approximation for s in the neighborhood of s = 0. This can be seen from eq. (3.4.32) by replacing e^{Ts} approximately by the first two terms of the Taylor series, i.e. $z \approx 1+Ts$

$$q \approx \frac{2}{T}\frac{Ts}{2+Ts} \approx s \quad \text{for small } Ts \tag{3.4.33}$$

Eq. (3.4.32) is also known as "Tustin transformation".

Using eqs. (3.4.30) and (3.4.32) the PID controller (eq. (3.4.31)) can now be approximated by

$$PID(z) = k_P + \frac{k_I T}{2}\frac{z+1}{z-1} + \frac{2k_D}{T}\frac{z-1}{z+1} \tag{3.4.34}$$

This second order controller has poles at z = 1 and z = -1 and the three numerator coefficients remain as free parameters for the design. If it is required that the controller pole at z = -1 is shifted into the unit circle, then the differentiator term of the PID can be formed corresponding to eq. (3.4.31) by the substitution of s = q

$$PID(z) = k_P + \frac{k_I T}{2}\frac{z+1}{z-1} + \frac{2k_D(z-1)}{2T_1(z-1)+T(z+1)} \tag{3.4.35}$$

Similarly other continuous-time compensators can be discretized if T is chosen small. This design ignores the delay of one half sampling interval introduced by the hold element, see eg. (1.2.17). Digital systems designed in this way are always inferior to their analog counterparts. For the design of new control systems, a choice of T as described in section 4.3 and an exact discrete design without approximations is recommended. Special advantages of discrete-time systems (as the deadbeat controller of section 4.4) can be utilized only in this way.

3.4.4 State Representation of Controller and Control Loop

The control loop of figure 3.8 will now be represented in state space, see figure 3.10.

The controller of order m from eq. (3.4.2) can be realized in an arbitrary state representation.

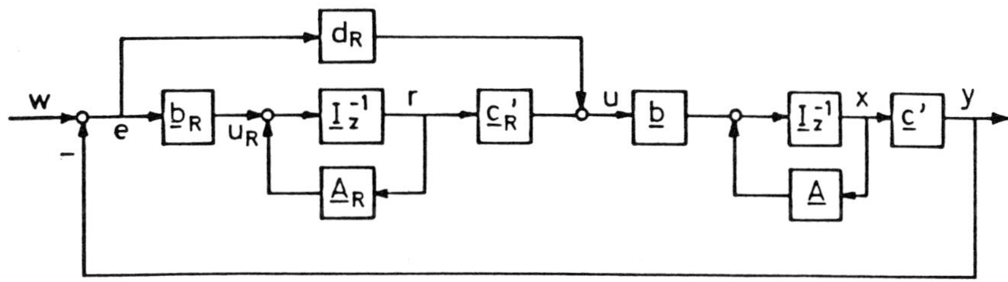

Figure 3.10 State representation of the control loop of figure 3.8

The states $\underline{r}$ of the controller are the contents of the computer memory elements which delay their input sequences by one sampling interval. The memory elements therefore have the transfer function z^{-1}. The figures A.2 to A.5 show different possibilities for controller realization of the form

$$\begin{aligned} \underline{r}[k+1] &= \underline{A}_R \underline{r}[k] + \underline{b}_R e[k] \\ u[k] &= \underline{c}_R' \underline{r}[k] + d_R e[k] \end{aligned} \tag{3.4.36}$$

m^2 of the $(m+1)^2$ coefficients are set to zero or one by the choice of a canonical form so that 2m+1 independent free design parameters remain.

The Jordan form of the realization can be obtained by a partial fractions decomposition of $d_z(z)$. For controllers of higher order with coefficients of greatly varying magnitudes, a factorization of $d_z(z)$ into terms of second order and if necessary one of first order is recommended.

$$d_z(z) = \frac{d_{12}z^2+d_{11}z+d_{10}}{z^2+c_{11}z+c_{10}} \times \frac{d_{22}z^2+d_{21}z+d_{20}}{z^2+c_{21}z+c_{20}} \cdots \frac{d_{i1}z+d_{i0}}{z+c_{i0}} \tag{3.4.37}$$

Here several realizations of second and first order operate in series in order to avoid the numerical difficulties involved with controller coefficients of extremely different orders of magnitude.

The state representation of the closed loop is now obtained by combining the states of plant ($\underline{x}$) and controller ($\underline{r}$) into one state vector

$$\begin{bmatrix} \underline{x}[k+1] \\ \underline{r}[k+1] \end{bmatrix} = \begin{bmatrix} \underline{A} & \underline{0} \\ \underline{0} & \underline{A}_R \end{bmatrix} \begin{bmatrix} \underline{x}[k] \\ \underline{r}[k] \end{bmatrix} + \begin{bmatrix} \underline{b} \\ 0 \end{bmatrix} u[k] + \begin{bmatrix} 0 \\ \underline{b}_R \end{bmatrix} e[k]$$

Since the controller and plant are coupled by $u[k] = \underline{c}_R'\underline{r}[k]+d_R e[k]$ we can write

$$\begin{bmatrix} \underline{x}[k+1] \\ \underline{r}[k+1] \end{bmatrix} = \begin{bmatrix} \underline{A} & \underline{b}\underline{c}_R' \\ 0 & \underline{A}_R \end{bmatrix} \begin{bmatrix} \underline{x}[k] \\ \underline{r}[k] \end{bmatrix} + \begin{bmatrix} \underline{b}d_R \\ \underline{b}_R \end{bmatrix} e[k]$$

Substituting the feedback law $e[k] = w[k] - \underline{c}'\underline{x}[k]$ the difference equation of the closed loop is obtained

$$\begin{bmatrix} \underline{x}[k+1] \\ \underline{r}[k+1] \end{bmatrix} = \begin{bmatrix} \underline{A} - d_R\underline{b}\underline{c}' & \underline{b}\underline{c}_R' \\ -\underline{b}_R\underline{c}' & \underline{A}_R \end{bmatrix} \begin{bmatrix} \underline{x}[k] \\ \underline{r}[k] \end{bmatrix} + \begin{bmatrix} \underline{b}d_R \\ \underline{b}_R \end{bmatrix} w[k] \qquad (3.4.38)$$

Thus the response

$$\begin{bmatrix} y[k] \\ u[k] \end{bmatrix} = \begin{bmatrix} \underline{c}' & 0 \\ -d_R\underline{c}' & \underline{c}_R' \end{bmatrix} \begin{bmatrix} \underline{x}[k] \\ \underline{r}[k] \end{bmatrix} + \begin{bmatrix} 0 \\ d_R \end{bmatrix} w[k] \qquad (3.4.39)$$

to arbitrary reference inputs w[k] can be calculated. w[k] need only be given numerically.

Remark 3.4

> The determination of the controller parameters can also be formulated as proportional output-vector feedback of the system augmented by the controller states $\underline{r}$

$$\begin{bmatrix} \underline{x}[k+1] \\ \underline{r}[k+1] \end{bmatrix} = \begin{bmatrix} \underline{A} & \underline{0} \\ \underline{0} & \underline{0} \end{bmatrix} \begin{bmatrix} \underline{x}[k] \\ \underline{r}[k] \end{bmatrix} + \begin{bmatrix} \underline{b} & \underline{0} \\ \underline{0} & \underline{I} \end{bmatrix} \begin{bmatrix} u \\ \underline{u}_R \end{bmatrix} \tag{3.4.40}$$

$$\begin{bmatrix} y[k] \\ \underline{r}[k] \end{bmatrix} = \begin{bmatrix} \underline{c}' & \underline{0} \\ \underline{0} & \underline{I} \end{bmatrix} \begin{bmatrix} \underline{x}[k] \\ \underline{r}[k] \end{bmatrix}$$

The required controller dynamics are not yet contained in this formulation, only the order. All possible couplings between the subsystems in eq. (3.4.40) and all possible controller dynamics can be introduced by an output-vector feedback of the form

$$\begin{bmatrix} u \\ \underline{u}_R \end{bmatrix} = - \begin{bmatrix} d_R & -\underline{c}_R' \\ \underline{b}_R & -\underline{A}_R \end{bmatrix} \begin{bmatrix} y[k] \\ \underline{r}[k] \end{bmatrix} + \begin{bmatrix} d_R \\ \underline{b}_R \end{bmatrix} w[k] \tag{3.4.41}$$

This is a multi-variable problem of the type

$$\begin{aligned} \underline{x}^*[k+1] &= \underline{A}^*\underline{x}^*[k] + \underline{B}^*\underline{u}^* \\ \underline{y}^* &= \underline{C}^*\underline{x}^*[k] \\ \underline{u}^* &= -\underline{K}^*\underline{y}^* + \underline{b}^*w \end{aligned} \tag{3.4.42}$$

As in eq. (3.4.36) $\underline{K}^*$ has 2m+1 independent free parameters since m^2 coefficients in eq. (3.4.41) can be fixed at zero or one by the choice of a canonical basis for the controller.

See exercise 3.5.

3.5 Root Locus Plots and Pole Specifications in the z-Plane

For the simplest controller $d_z = k_y$ the closed-loop transfer function, eq. (3.4.8) is

$$f_z(z) = \frac{k_y h_z(z)}{1+k_y h_z(z)} \tag{3.5.1}$$

The determination of the closed-loop pole locations as a function of k_y is the same problem as for continuous-time systems, see eq. (2.6.60). As in the continuous case, this can be handled in the z-plane by a root locus plot whose form is determined by the poles and zeros of h_z. All rules for the construction of the root locus curves can be applied without alteration [48.1], [50.1]. The only difference now is that stability is determined by the position of the poles relative to the unit circle. The type of the solution terms follows directly from the position of the roots (see figure 3.2).

The desired pole region for continuous systems (see figure 2.12) with $\sigma \le \sigma_o$, $\zeta \ge \zeta_o$, can now be mapped into the z-plane by $z=e^{Ts}$. If all the eigenvalues $s_i = \sigma_i \pm j\omega_i$ satisfy $\sigma_i \le \sigma_o$, there exists an A such that the solution envelope (see figure 2.9) is $Ae^{\sigma_o t}$. The corresponding s-plane boundary $s = \sigma_o \pm j\omega$ is mapped onto the circle $z = e^{\sigma_o T \pm j\omega}$ with radius $e^{\sigma_o T}$. Points to the left of the s-plane boundary correspond to points inside the circle. In figure 3.11, $e^{\sigma_o T} = 0.6$, or $\sigma_o = -0.5108/T$. In this case it is assured that the solution sequence decays at least as quickly as 0.6^k. Note that the decay rate of the oscillations in continuous time $t = kT$ depends on the sampling interval T. Thus the required time constant σ_o must first be multiplied by T in order to determine the circle radius $e^{\sigma_o T}$. In order that the solution is not too oscillatory within the envelope $\pm e^{\sigma_o kT}$ and that the step response does not have too much overshoot, the solution terms should have a minimum damping $\zeta \ge \zeta_o$. Take the curve $s = -\omega_n\zeta_o \pm j\omega_n\sqrt{1-\zeta_o^2}$, $\omega_n = \sqrt{\sigma^2+\omega^2}$ (see eq. (2.5.10)) in the s-plane for fixed damping ζ_o and any real ω_n, and map it into the z-plane to give the curve

$$z_i = e^{sT} = e^{-\omega_n T\zeta_o} \times e^{\pm j\omega_n T\sqrt{1-\zeta_o^2}} \qquad (3.5.2)$$

The resulting curve in the z-plane is a logarithmic spiral in the z-plane which begins for $\omega_n T = 0$ at $z = 1$. For example for the favorable damping value

$\zeta = 1/\sqrt{2}$

$$z_i = e^{-\alpha} \times e^{\pm j\alpha}, \quad \alpha = \omega_n T/\sqrt{2} \qquad (3.5.3)$$

This spiral for $0 < \alpha < \pi$ is the heart-shaped curve in figure 3.11. The desired pole region Γ lies inside the cross-hatched boundary. Here a favorable k_y-value on the root locus plot is that for which a "dominant" pole pair lies on the boundary curve $\partial\Gamma$ and all others inside.

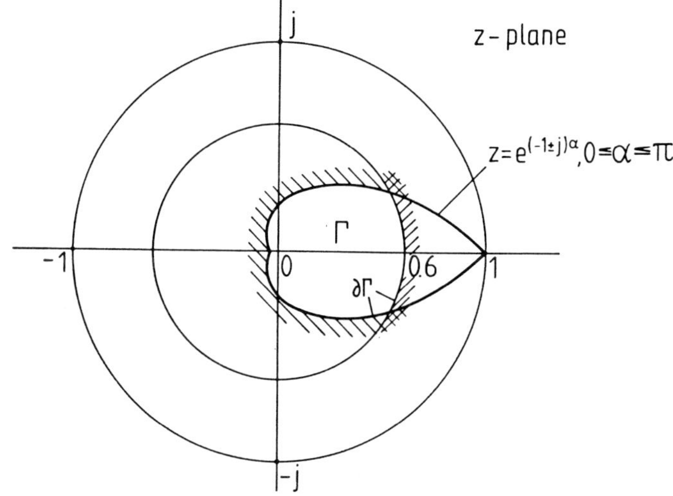

Figure 3.11 The poles of the closed loop should lie in the domain Γ with a pole pair on the boundary curve $\partial\Gamma$. For small sampling periods T the radius of the inner circle, here r = 0.6, may be increased in order to allow a smaller actuator input magnitude.

Example:

For the control loop from figure 3.12 a favorable value k_y should be determined.

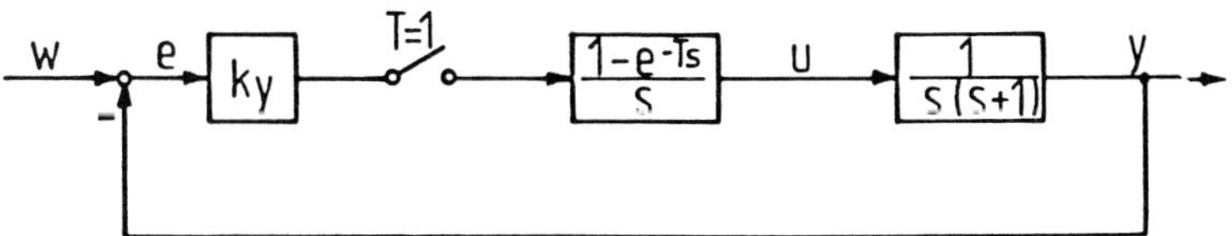

Figure 3.12 Closing of the control loop around the plant of figure 3.5.

From eq. (3.3.31)

$$h_z(z) = \frac{0.3679z+0.2642}{(z-1)(z-0.3679)}$$

The root locus represents the phase relation for

$$k_y h_z(z) = 0.368k_y \times \frac{z+0.718}{(z-1)(z-0.368)} \qquad (3.5.4)$$

It is composed of the parts $0.368 < z < 1$ and $z < 0.718$ of the real axis and a circle around the zero at $z = -0.718$. The circle is fixed by the branching points. There

$$\frac{dh_z(z)}{dz} = 0.368k_y \times \frac{-z^2-1.436+1.350}{(z-1)^2(z-0.368)^2} = 0 \qquad (3.5.5)$$

The branching points lie at $z_{1,2} = -0.718 \pm 1.368$, in other words, the radius of the circle is 1.368. Figure 3.13 shows the root locus curve for $k_y > 0$. The curve $z = e^{(-1\pm j)\alpha}$, $0 \leq \alpha \leq \pi$, from figure 3.11 is also plotted. At the intersection the gain is $k_y = 0.326$.

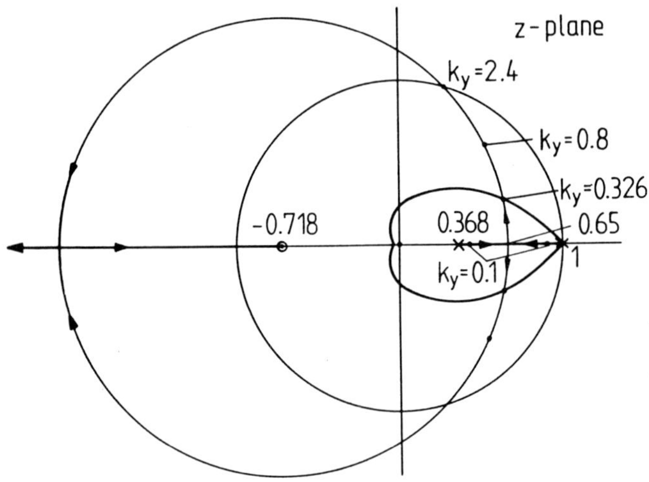

Figure 3.13 Root locus for the control loop of figure 3.12. Favorable value of the gain $k_y = 0.326$. Gain margin factor $2.4/0.326 = 7.36$.

For the same plant the controller $d_z = k_y \times \frac{z-0.3679}{z+0.4180}$ with $k_y = 1.582$ was determined in eq. (3.4.18) with a cancelled pole at $z = 0.3679$ and the two other poles lying at $z = 0$. The root locus for free k_y is represented in figure 3.14.

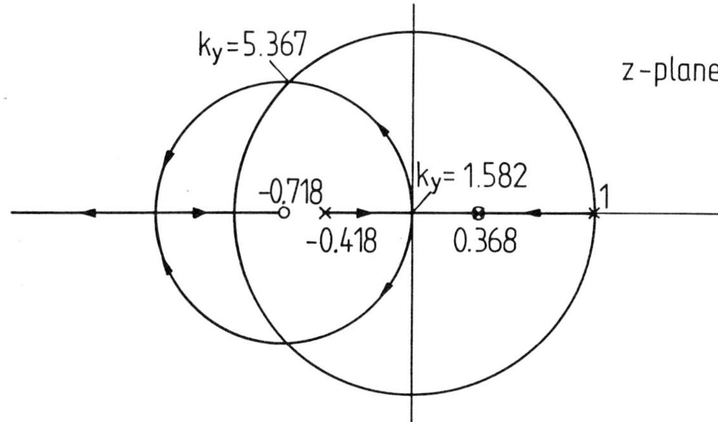

Figure 3.14 Root locus of the compensated system. Deadbeat solution for k_y=1.582. Gain margin factor 5.367/1.582=3.39.

See exercise 3.6.

3.6 Time Domain Solutions and Specifications

3.6.1 Recursive Solution of the Vector Difference Equation

The quality of a control system's performance is usually judged by its time-domain behavior, for example the response to a reference step input. This should neither overshoot too much nor approach the final value too slowly. The maximum actuator input magnitude, which frequently arises in the first sampling interval for the greatest error signal, should adhere to given bounds. Only the specification on the steady-state response can be taken into consideration exactly in the frequency domain, for example by an integral part of the controller or a high loop gain. Whether or not other requirements are met can be judged exactly only by the calculation of solutions in the time domain. In the older design methods a great deal of effort was spent transferring such requirements for systems of second order from the frequency domain to the time domain and applying them approximately for systems of higher order with a "dominant pole pair". With the computers and graphic terminals available today it is no longer a problem to judge the time domain solutions by inspection. Also if linearization, approximation of distributed parameter systems by lumped ones or order reduction has been employed for the design, it is possible to do the simulation using a more realistic and more complex model. This also allows the designer to refine the linear control design empirically in the simulation. These problems will not be covered in this book. However, the generation of solutions for the linear design model plus linear controller will be treated.

The state representation of the control loop, for example in eqs. (3.4.38) and (3.4.39), is directly suited for digital computer simulation. The response in all variables of the system can be produced for arbitrary input sequences w[k] and initial states $\underline{x}[0]$ by recursive application of the difference equation. This is an added advantage of the state representation for sampled systems.

Example:

A state representation for the controller of eq. (3.4.18) is

$$\begin{aligned} r[k+1] &= -0.418r[k] - 1.243e[k] \\ u[k] &= r[k] + 1.582e[k] \end{aligned} \tag{3.6.1}$$

Together with the plant from eq. (3.3.9), T = 1, eq. (3.4.38) is

$$\begin{bmatrix} x_1[k+1] \\ x_2[k+1] \\ r[k+1] \end{bmatrix} = \begin{bmatrix} 0.418 & 0.632 & 0.368 \\ -1 & 0.368 & 0.632 \\ 1.243 & 0 & -0.418 \end{bmatrix} \begin{bmatrix} x_1[k] \\ x_2[k] \\ r[k] \end{bmatrix} + \begin{bmatrix} 0.582 \\ 1 \\ -1.243 \end{bmatrix} w[k] \tag{3.6.2}$$

$$\begin{bmatrix} y[k] \\ u[k] \end{bmatrix} = \begin{bmatrix} 1 & 0 & 0 \\ -1.582 & 0 & 1 \end{bmatrix} \begin{bmatrix} \underline{x}[k] \\ r[k] \end{bmatrix} + \begin{bmatrix} 0 \\ 1.582 \end{bmatrix} w[k]$$

For the initial state $x_1[0] = r[0] = 0$, $x_2[0] = a$ and a step of the reference input $w[k] = 1[k]$ by recursive substitution one obtains

k	$y = x_1[k]$	$x_2[k]$	$r[k]$	$u[k]$
0	0	a	0	1.582
1	0.582 + 0.632a	1 + 0.368a	-1.243	-0.582-a
2	1 + 0.497a	-0.497a	0.786a	0
3	1 + 0.497a×0.368	-0.497a×0.368	0.786a×0.368	0
4	1 + 0.497a×0.368^2	-0.497a×0.368^2	0.786a×0.368^2	0
⋮	⋮	⋮	⋮	⋮

Let $\underline{\tilde{x}}[k] := [x_1[k], x_2[k], r[k]]'$. For a = 0, the steady state $\underline{\tilde{x}}[k] = [1 \quad 0 \quad 0]$, u[k] = 0, y[k] = 1 is reached for k ≥ 2 in two sampling intervals. Such a finite transient is called "deadbeat response". Because of the cancellation of the plant pole at z = 3.679 in the controller of eq. (3.4.18), this pole does not appear in the transfer function from w to y, that is, it is not controllable by w. However, if this eigenvalue is stimulated by a disturbance, for example by a disturbance acting before k = 0 which has led to the initial state $x_2[0] = a \neq 0$, a corresponding solution term 0.3679^k arises.

A finite transient for all initial states results only if all closed loop eigenvalues lie at z = 0, as for the controller of eq. (3.4.16). A state equation for this controller is

$$\begin{aligned} r[k+1] &= -0.520r[k] - 1.922e[k] \\ u[k] &= r[k] + 2.306\, e[k] \end{aligned} \qquad (3.6.3)$$

The state representation of the control loop becomes

$$\begin{bmatrix} x_1[k+1] \\ x_2[k+1] \\ r[k+1] \end{bmatrix} = \begin{bmatrix} 0.152 & 0.632 & 0.368 \\ -1.457 & 0.368 & 0.632 \\ 1.922 & 0 & -0.520 \end{bmatrix} \begin{bmatrix} x_1[k] \\ x_2[k] \\ r[k] \end{bmatrix} + \begin{bmatrix} 0.848 \\ 1.457 \\ -1.922 \end{bmatrix} w[k]$$

$$\begin{bmatrix} y[k] \\ u[k] \end{bmatrix} = \begin{bmatrix} 1 & 0 & 0 \\ -2.306 & 0 & 1 \end{bmatrix} \begin{bmatrix} \underline{x}[k] \\ r[k] \end{bmatrix} + \begin{bmatrix} 0 \\ 2.306 \end{bmatrix} w[k] \qquad (3.6.4)$$

The step response is again calculated for the initial state

$\underline{\tilde{x}}[0] = [x_1[0] \quad x_2[0] \quad r[0]]' = [0 \quad a \quad 0]$

k	$y = x_1[k]$	$x_2[k]$	r[k]	u[k]
0	0	a	0	2.306
1	0.848+0.632a	1.457+0.368a	-1.922	-1.572-1.457a
2	1.191+0.329a	-0.457-0.786a	0.707+1.215a	0.266+0.457a
≥3	1	0	0	0

In this case the dynamic matrix in eq. (3.6.4) is so determined that it is nilpotent, or in other words that a power of this matrix $\underline{\tilde{A}}$ is equal to the zero matrix. Here $\det(zI-\underline{\tilde{A}}) = z^3$. From the Cayley-Hamilton theorem it follows then that $\underline{\tilde{A}}^3 = 0$ and with $w[k] \equiv 0$, $\underline{\tilde{x}}[k+3] = \underline{\tilde{A}}^3\underline{\tilde{x}}[0] = \underline{0}$ for every initial state $\underline{\tilde{x}}[0]$. However, the step response for this solution is not good. In comparison with the step response of the system (3.6.2) three sampling intervals are needed rather than two to achieve zero plant input, and moreover the maximum input amplitude is increased from 1.582 to 2.306 (for sufficiently small a). However, this disadvantage is caused solely by the form of the control loop in figure 3.8. In most cases the variables y and w are available separately and a control structure can be used which allows them to be processed separately. Later we will treat structures in which the controller dynamics are not controllable by w and therefore do not enter into the response to a reference input. The effect is, that the system (3.6.4) is modified such that the reference input w enters at a different point in order to obtain the same step response as in system (3.6.2), but now, as in the last example, a triple eigenvalue lies at $z = 0$.

3.6.2 Numerical Inverse z-Transform

If the controller design was accomplished in the frequency domain, one need not bother with the choice of a basis in state space in order to do the simulation. Rather one can determine the time response of the pertinent variables in the control loop directly by the inverse z-transform. This presupposes that the input variables are z-transformable. Initial states can be taken into consideration in the frequency domain by eq. (3.3.3). However, for the calculation of system responses these states are normally set equal to zero.

For example, if the step response of the closed loop and the corresponding actuator input in the frequency domain, $y_z(z)$ and $u_z(z)$, are calculated in the first design step, one faces the question of the calculation of the corresponding sequences y[k] and u[k]. This is accomplished by the inverse z-transform.

The general inversion is treated in appendix B. Here only the numerical inversion by expansion into a series in z^{-1} is considered. The definition of the z-transform may be written as

$$f_z(z) = \sum_{k=0}^{\infty} f[k]z^{-k} = f[0] + f[1]z^{-1} + f[2]z^{-2} + f[3]z^{-3} + \dots \qquad (3.6.5)$$

If one puts a rational expression $f_z(z)$ into this form by an expansion via the standard division scheme, the coefficients of the sequence f[k] are formed.

Example:

For the control loop of figure 3.12 the response y[k] to a reference step, that is $w_z = z/(z-1)$, will be calculated for two cases

a) if k_y is replaced by the controller d_z of eq. (3.4.18)

b) for the value $k_y = 0.326$ chosen in the root locus of figure 3.13.

For a) with eq. (3.4.19)

$$y_z = f_z w_z = \frac{0.582z+0.418}{z^2} \times \frac{z}{z-1}$$

$$
\begin{array}{l}
\qquad\qquad 0.582z^{-1}+z^{-2}+z^{-3}+z^{-4}\ldots \\
(z^2-z)\,\big|\; 0.582z+0.418 \\
\qquad\quad \underline{-0.582z+0.582} \\
\qquad\qquad\qquad 1 \\
\qquad\qquad\quad \underline{-1 \quad +z^{-1}} \\
\qquad\qquad\qquad\quad z^{-1} \\
\qquad\qquad\qquad \underline{-z^{-1}+z^{-2}} \\
\qquad\qquad\qquad\qquad z^{-2} \\
\qquad\qquad\qquad\quad -z^{-2}+z^{-3} \\
\qquad\qquad\qquad\qquad\quad \vdots
\end{array}
$$

$$
y[0] = 0,\ y[1] = 0.582,\ y[k] = 1 \text{ for } k \geq 2 \tag{3.6.6}
$$

This solution agrees with that of eq. (3.6.2) for the initial state zero.

For stable control loops in which the controller or the plant contain an integration and the steady-state error is therefore zero, it is advantageous to calculate the step response as

$$
y[k] = 1 - e[k] = 1 - \mathfrak{Z}^{-1}\{e_z(z)\} \tag{3.6.7}
$$

because e[k] contains no constant terms and the denominator degree of the expression to be inverted is thereby decreased by one. Here for eq. (3.4.18)

$$
e_z = \frac{1}{1+0.326\times h_z} \times \frac{z}{z-1} = \frac{z^2-0.368z}{z^2-1.248z+0.454}
$$

$$
\begin{array}{l}
\qquad\qquad\qquad\qquad\qquad\quad 1+0.880z^{-1}+0.644z^{-2}+0.404z^{-3}+\ldots \\
e_z = (z^2-1.248z+0.454)\,\big|\; z^2-0.368z \\
\qquad\qquad\qquad\qquad\qquad \underline{-z^2+1.248z-0.454} \\
\qquad\qquad\qquad\qquad\qquad\quad 0.880z-0.454 \\
\qquad\qquad\qquad\qquad\qquad \underline{-0.880z+1.098-0.400z^{-1}} \\
\qquad\qquad\qquad\qquad\qquad\qquad\quad 0.644-0.400z^{-1} \\
\qquad\qquad\qquad\qquad\qquad\qquad \underline{-0.644+0.804z^{-1}-0.292z^{-2}} \\
\qquad\qquad\qquad\qquad\qquad\qquad\qquad 0.404z^{-1}-0.292z^{-2} \\
\qquad\qquad\qquad\qquad\qquad\qquad\qquad\quad \vdots
\end{array}
\tag{3.6.8}
$$

The result is

k	e[k]	y[k] = 1 - e[k]	u[k] = 0.326 e[k]
0	1	0	0.326
1	0.880	0.120	0.287
2	0.644	0.356	0.210
3	0.404	0.596	0.132
	⋮	⋮	⋮

Figure 3.15 illustrates this response for $k_y = 0.326$.

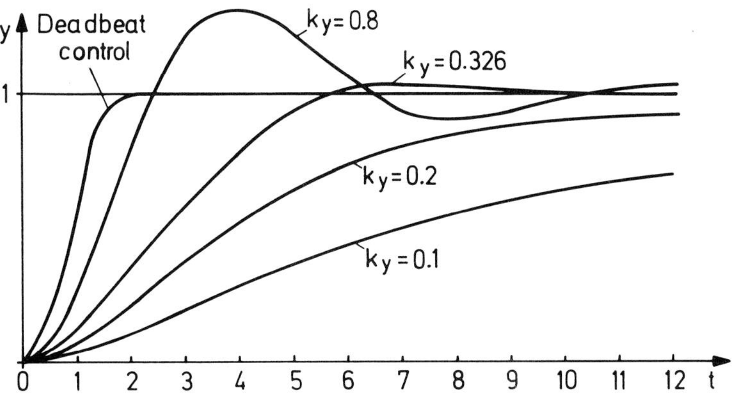

Figure 3.15 Step responses of the control loop from figure 3.12 with the sampling period T = 1.

In the division scheme (3.6.8) rounding errors in the first calculation step can strongly effect later values. In contrast to the notation in eq. (3.6.8) the full accuracy should be used for the expansion. We first put the problem into a suitable form. Consider the following rational expression

$$f_z = \frac{b_n z^n + b_{n-1} z^{n-1} + \ldots + b_o}{z^n + a_{n-1} z^{n-1} + \ldots + a_o} \qquad (3.6.9)$$

which is to be inversly z-transformed. The division scheme with the discrete time k written as index: $f_k = f[k]$, yields

$$
\begin{array}{l}
\qquad\qquad\qquad\qquad\qquad\qquad\qquad f_o \quad + \quad f_1 z^{-1} + \ldots \\
f_z = (z^n + a_{n-1} z^{n-1} + \ldots + a_o) \overline{\Big)\, b_n z^n \quad + \quad b_{n-1} z^{n-1} \quad + \quad b_{n-2} z^{n-2} + \ldots + b_o} \\
\qquad\qquad\qquad\qquad\qquad\qquad\qquad -f_o z^n \quad - \quad a_{n-1} f_o z^{n-1} \quad - \quad a_{n-2} f_o z^{n-2} - \ldots - a_o f_o \\
\qquad\qquad\qquad\qquad\qquad\qquad\qquad \overline{0 \times z^n + (b_{n-1} - a_{n-1} f_o) z^{n-1} \quad + \quad (b_{n-2} - a_{n-2} f_o) z^{n-2} + \ldots + (b_o - a_o f_o)} \\
\qquad\qquad\qquad\qquad\qquad\qquad\qquad\qquad\qquad -f_1 z^{n-1} \quad - \quad a_{n-1} f_1 z^{n-2} - \ldots - a_1 f_1 - a_o f_1 z^{-} \\
\qquad\qquad\qquad\qquad\qquad\qquad\qquad\qquad\qquad \overline{0 \times z^{n-1} + (b_{n-2} - a_{n-2} f_o - a_{n-1} f_1) z^{n-2} + \ldots} \\
\qquad\qquad\qquad\qquad\qquad\qquad\qquad\qquad\qquad\qquad\qquad\qquad f_2 z^{n-2} + \ldots \\
\qquad\qquad\qquad\qquad\qquad\qquad\qquad\qquad\qquad\qquad\qquad\qquad \overline{0 \times z^{n-2} + \ldots}
\end{array}
\tag{3.6.10}
$$

$$
\begin{aligned}
f_o &= b_n \\
f_1 &= b_{n-1} - a_{n-1} f_o \\
f_2 &= b_{n-2} - a_{n-2} f_o - a_{n-1} f_1 \\
&\;\vdots \\
f_n &= b_o - a_o f_o - \ldots - a_{n-1} f_{n-1} \\
f_{n+1} &= \quad -a_o f_1 - \ldots - a_{n-1} f_n
\end{aligned}
$$

This gives the recursion formula

$$
\begin{aligned}
f_k &= b_{n-k} - a_o f_{k-n} - a_1 f_{k-n+1} - \ldots - a_{n-1} f_{k-1} \\
b_i &= 0, \; f_i = 0 \text{ for } i < 0
\end{aligned}
\tag{3.6.11}
$$

This relation can also be derived via a determinant formulation, see [75.4]. It is especially suited for memory-efficient programming with a pocket calculator.

Example:

The division performed in eq. (3.6.8) is for general coefficients

$$
e_z = \frac{z^2 + b_1 z}{z^2 + a_1 z + a_o}
\tag{3.6.12}
$$

From eq. (3.6.11)

$$\begin{aligned} e[0] &= 1 \\ e[1] &= b_1 - a_1 e[0] \\ e[2] &= \quad - a_1 e[1] - a_0 e[0] \\ e[3] &= \quad - a_1 e[2] - a_0 e[1] \\ &\vdots \\ e[k] &= -a_1 e[k-1] - a_0 e[k-2] \end{aligned}$$

For example, this relation can be programmed on a HP-calculator as follows. After inserting the starting values

a_0 STO 1
a_1 STO 2
e[0] STO 3
e[1] STO 4

the recursion relation (3.6.11) is calculated by the following program

```
LBL A
RCL 1
RCL 3
 x
RCL 2
RCL 4
STO 3
 x
 +
CHS
STO 4
RTN
```

(3.6.13)

It is essential that the commands STO 3 and STO 4 put the values e[k-2] into memory 3 and e[k-1] into memory 4. The calculator is thereby prepared for the next recursion step which is started by the command A. This procedure requires only four memory locations. A program for the inversion of z-transforms up to degree n = 9 was written for an HP 67 by this principle.

Example:

For some values of k_y indicated on the root locus of figure 3.13 the step responses are represented in figure 3.15 along with the deadbeat solution of eq. (3.6.6). Here the calculated values for the sampling instants $t = kT$ have been interpolated continuously. The calculation of the form of the solution between the sampling instants will be handled in section 3.7.

For $k_y = 0.1$ one of the two real eigenvalues lies too near to $z = 1$ and leads to a very slow transient. The situation already looks better for $k_y = 0.2$, in other words for two eigenvalues near the branching point $z = 0.65$. A damping of $1/\sqrt{2}$ gives an acceptable value $k_y = 0.326$.

$k_y = 0.8$ is the optimum if $\sum_{k=0}^{\infty} e^2[k]$ is minimized, however this results in considerable overshoot. At $k_y = 2.4$ the stability boundary is reached. The maximal value of u[k] arises in the first sampling interval in all these cases and is equal to k_y. The deadbeat control from eq. (3.6.6) looks most satisfactory in the step response. The cost however is a much larger maximal input amplitude u[0] = 1.582.

If u[k] or other variables in the control loop are of interest also, it is advisable to expand only the common denominator into a sequence in z^{-1} and to multiply this sequence by the respective numerator polynomials.

Remark 3.5:

The numerical inversion by dividing out the expression can also be used for the calculation of solutions of continuous systems. One approximates the continuous system by a corresponding sampled-data system with sufficiently small sampling period T. The smoothing of the sampled signal need not be performed by an extrapolation, that is by a causal element with its phase delay, but rather can be done by noncausal interpolation. A linear interpolation has the impulse response shown in figure 3.16

$$y[k] = \begin{cases} 0 & k \neq 0 \\ 1 & k = 0 \end{cases} \tag{3.6.14}$$

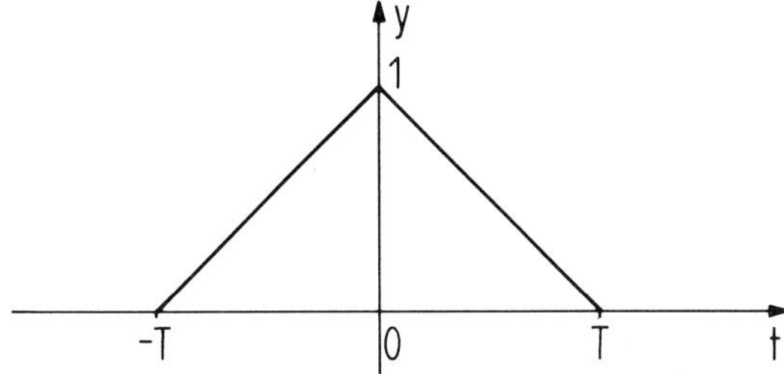

Figure 3.16 Impulse response of the linear interpolator

$$
\begin{aligned}
y(t) &= \left[\frac{1}{T}\ (t+T)\times 1(t+T) \ - \ 2t\times 1(t) \ + \ (t-T)\times 1(t-T)\right] \\
y_s(s) &= \left[\frac{1}{T}\ \frac{1}{s^2}e^{sT} \ - \ \frac{2}{s^2} \ + \ \frac{1}{s^2}e^{-sT}\right] \qquad (3.6.15) \\
y_s(s) &= \frac{(1-e^{-sT})^2\ e^{sT}}{Ts^2}
\end{aligned}
$$

A sampler and interpolator with this transfer function approximates the curve of the input signal by joining successive input values. This approximation should be introduced into the control loop in a position where the signal is as smooth as possible. For typical plants with low-pass characteristics the chord-approximation is placed immediately after the plant and before the position where the reference input enters the loop. In order to calculate the reference step response the configuration of figure 3.17 is used.

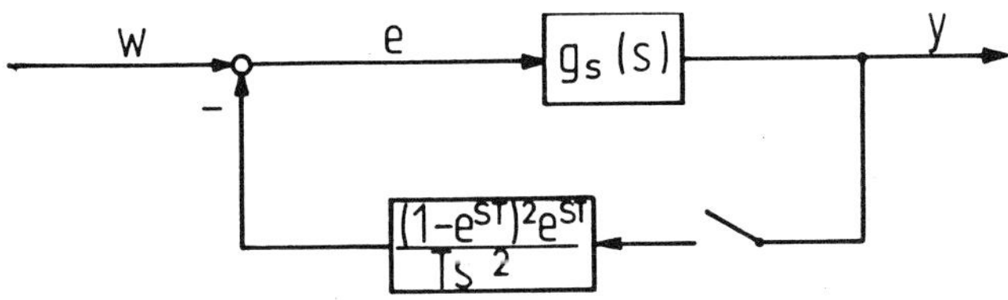

Figure 3.17 Substitute control loop configuration for the numerical calculation of the step response of a continuous control loop.

For the sampling points $e[kT] = w[kT]-y[kT]$, i.e. $e_z = w_z-y_z$

$$
y_z(z) = \mathcal{Z}\{w_s g_s\} \ - \ \frac{(z-1)^2}{Tz} \times \mathcal{Z}\{g_s/s^2\}\ y_z(z) \qquad (3.6.16)
$$

This numerical method is particularly advantageous for time-delay plants, see section 3.8.

3.7 Behavior Between the Sampling Instants

For the representation of solutions in figure 3.15 the points $y(kT)$ have been connected continuously. This can be quite inaccurate, e.g. for the deadbeat solution $y(0) = 0$, $y(T) = 0.582$, $y(kT) = 1$, $k = 2,3,4...$ Therefore the entire solution $y(t)$ will now be calculated. Let

$$t = kT + \gamma T \quad , \quad 0 \leq \gamma < 1 \tag{3.7.1}$$

In eq. (3.1.1) to (3.1.7) the transition from the differential equation

$$\dot{\underline{x}} = \underline{Fx} + \underline{g}u \quad , \quad u = u(kT) \text{ for } kT < t < kT + T \tag{3.7.2}$$

to the difference equation

$$\underline{x}(kT+T) = \underline{Ax}(kT) + \underline{b}u(kT) \tag{3.7.3}$$

has been carried out. The solution

$$\underline{x}(t) = e^{\underline{F}(t-t_o)}\underline{x}(t_o) + \int_{t_o}^{t} e^{\underline{F}(t-\tau)}\underline{g}u(\tau)d\tau \tag{3.7.4}$$

of the differential eq. (3.7.2) is examined once again and $t_o = kT$, $t = kT + \gamma T$ is inserted

$$\underline{x}(kT+\gamma T) = e^{\underline{F}\gamma T}x(kT) + \int_{kT}^{kT+\gamma T} e^{\underline{F}(kT+\gamma T-\tau)}d\tau\underline{g}u(kT) \tag{3.7.5}$$

$$\underline{x}(kT+\gamma T) = \underline{A}_\gamma\underline{x}(kT) + \underline{b}_\gamma u(kT)$$

with

$$\underline{A}_\gamma = e^{\underline{F}\gamma T} \quad , \quad \underline{b}_\gamma = \int_0^{\gamma T} e^{\underline{F}v}dv \times \underline{g} \tag{3.7.6}$$

The form (3.7.3) is the special case with $\gamma = 1$, $\underline{A} = \underline{A}_\gamma\big|_{\gamma=1}$ and $\underline{b} = \underline{b}_\gamma\big|_{\gamma=1}$.

The state of the system contains all information concerning the previous history of the system which has an influence on the future form of the solution. Here the content of the hold element $u(kT+\gamma T) = u(kT)$ obviously also belongs to the state at the time instant $t = kT+\gamma T$, $0 < \gamma < 1$. This equation can be put together with eq. (3.7.5) in order to form a state description valid for $0 < \gamma < 1$:

$$\begin{bmatrix} \underline{x}(kT+\gamma T) \\ u(kT+\gamma T) \end{bmatrix} = \begin{bmatrix} \underline{A}_\gamma & \underline{b}_\gamma \\ \underline{0} & 1 \end{bmatrix} \begin{bmatrix} \underline{x}(kT) \\ u(kT) \end{bmatrix}, \quad 0 < \gamma < 1 \tag{3.7.7}$$

u is discontinuous for $\gamma = 0$ and $\gamma = 1$. $u(kT)$ denotes the limit value when the point of discontinuity is approached from the right side. The output equation $y(t) = \underline{c}'\underline{x}(t)$ becomes

$$y(kT+\gamma T) = [\underline{c}' \quad d] \begin{bmatrix} \underline{x}(kT+\gamma T) \\ u(kT+\gamma T) \end{bmatrix} \tag{3.7.8}$$

For the reading out of $y(kT+\gamma T)$ from the difference eq. (3.7.3), eqs.(3.7.7) and (3.7.8) are combined to form

$$y(kT+\gamma T) = \underline{c}'_\gamma \underline{x}(kT) + d_\gamma u(kT) \tag{3.7.9}$$

with $\underline{c}'_\gamma = \underline{c}'\underline{A}_\gamma$, $\quad d_\gamma = \underline{c}'\underline{b}_\gamma + d$

Example:

We examine the plant given in eq. (3.1.13). For the calculation of $\underline{A}_\gamma$ and $\underline{B}_\gamma$, γT must be substituted for T. Once again, $T = 1$. Eq. (3.7.7) is here

$$\begin{bmatrix} x_1(kT+\gamma T) \\ x_2(kT+\gamma T) \\ u(kT+\gamma T) \end{bmatrix} = \begin{bmatrix} 1 & 1-e^{-\gamma} & \gamma+e^{-\gamma}-1 \\ 0 & e^{-\gamma} & 1-e^{-\gamma} \\ 0 & 0 & 1 \end{bmatrix} \begin{bmatrix} x_1(kT) \\ x_2(kT) \\ u(kT) \end{bmatrix} \tag{3.7.10}$$

If u(kT) is produced by the deadbeat controller (3.6.1), then $\underline{x}(kT)$, u(kT) are given by the solution of eq. (3.6.2). For the initial state zero

t	k	γ	$y(t) = x_1(t)$	$x_2(t)$	u(t)
0	0	0	0	0	1.582
0.25	0	0.25	0.046	0.350	1.582
0.5	0	0.5	0.168	0.623	1.582
0.75	0	0.75	0.352	0.847	1.582
1	1	0	0.582	1	-0.582
1.25	1	0.25	0.786	0.650	-0.582
1.5	1	0.5	0.914	0.377	-0.582
1.75	1	0.75	0.980	0.165	-0.582
2	2	0	1	0	0
2.25	2	0.25	1	0	0

The form of y(t) is represented in figure 3.15.

If $u_z(z)$ is calculated in the frequency domain, then the output $y(t) = y(kT+\gamma T)$ can be calculated as the z-transform of a sequence obtained by shifted sampling. We introduce the following notation:

$$f_{z\gamma}(z) := \mathfrak{Z}\{f(kT+\gamma T)\} = \sum_{k=0}^{\infty} f(kT+\gamma T)z^{-k} \tag{3.7.11}$$

The z-transform of eq. (3.7.9) is then written

$$y_{z\gamma}(z) = \underline{c}_\gamma'\underline{x}_z(z) + d_\gamma u_z(z) \tag{3.7.12}$$

By the substitution of $\underline{x}_z$ from eq. (3.3.3)

$$y_{z\gamma}(z) = \underline{c}_\gamma' z(z\underline{I}-\underline{A})^{-1}\underline{x}[0] + [\underline{c}_\gamma'(z\underline{I}-\underline{A})^{-1}\underline{b} + d_\gamma]u_z(z) \tag{3.7.13}$$

In this equation the expression

$$\begin{aligned} h_{z\gamma}(z) &:= \underline{c}_\gamma'(z\underline{I}-\underline{A})^{-1}\underline{b} + d_\gamma \\ &= \underline{c}'\underline{A}_\gamma(z\underline{I}-\underline{A})^{-1}\underline{b} + \underline{c}'\underline{b}_\gamma + d \end{aligned} \tag{3.7.14}$$

is designated as the "extended z-transfer function". For $\gamma \to 0$ this becomes the usual z-transfer function.

Starting from an s-transfer function $g_s(s)$ of the continuous part, proceed as from eq. (3.3.26) to (3.3.28).

$$h_s(s) = \eta_s(s) \times g_s(s) = \frac{1-e^{-Ts}}{s} g_s(s)$$

With the step response $v(t) = \mathcal{L}^{-1}\{g_s(s)/s\}$ of the plant, the response $h(t)$ of the continuous part to the impulse response $\eta(t) = 1(t) - 1(t-T)$ of the hold element is $h(kT+\gamma T) = v(kT+\gamma T) - v(kT+\gamma T-T)$.

From the right shifting theorem of the z-transform

$$\begin{aligned} h_{z\gamma}(z) &= \mathfrak{Z}\{h(kT+\gamma T)\} = (1-z^{-1})\mathfrak{Z}\{v(kT+\gamma T)\} \\ h_{z\gamma}(z) &= \frac{z-1}{z}\mathfrak{Z}\left\{\mathcal{L}^{-1}\{g_s(s)/s\}_{t\,=\,kT+\gamma T}\right\} \end{aligned} \qquad (3.7.15)$$

For this operation the notation

$$f_{z\gamma}(z) = \mathcal{Z}_\gamma\{f_s(s)\} := \mathfrak{Z}\left\{\mathcal{L}^{-1}\{f_s(s)\}_{t\,=\,kT+\gamma T}\right\} \qquad (3.7.16)$$

is introduced, therefore

$$h_{z\gamma}(z) = \frac{z-1}{z}\mathcal{Z}_\gamma\{g_s(s)/s\} \qquad (3.7.17)$$

$\mathcal{Z}_\gamma\{f_s(s)\}$ can be immediately found in the corresponding column of the table in appendix B. For the initial state $\underline{x}[0] = \underline{0}$, the input-output relation

$$y_{z\gamma}(z) = h_{z\gamma}(z) \times u_z(z) \qquad (3.7.18)$$

is obtained from eq. (3.7.13). Note that the usual z-transform of u is formed because $u(kT+\gamma T)$ is not dependent upon γ.

Example:

Let the cancellation controller of eq. (3.4.22) be substituted for k_y in the control loop of figure 3.12. According to eq. (3.4.23) the closed-loop transfer function is then $f_z = 1/z$. In other words, the discrete output y(kT) tracks the input w(kT) exactly with a delay of one sampling interval. The step response is y(0) = 0, y(kT) = 1 for k = 1,2,3... The complete response y(t) will now be calculated. From eq. (3.7.17)

$$h_{z\gamma}(z) = \frac{z-1}{z} \mathcal{Z}_\gamma \left\{ \frac{1}{s^2(s+1)} \right\}$$

From the table in appendix B the following expression is obtained

$$h_{z\gamma} = \frac{z-1}{z} \times \frac{(\gamma-1+e^{-\gamma})z^3+(2-\gamma-\gamma e^{-1}-2e^{-\gamma}+e^{-1})z^2+(e^{-\gamma}-2e^{-1}+\gamma e^{-1})z}{(z-1)^2(z-e^{-1})}$$

$$\doteq \frac{(\gamma-1+e^{-\gamma})z^2+(2.368-1.368\gamma-2e^{-\gamma})z+(e^{-\gamma}-0.736+0.368\gamma)}{(z-1)(z-0.368)}$$

The controller (3.4.22) produces an actuator input

$$u_z = \frac{d_z}{1+d_z h_z} \times w_z = \frac{(z-0.368)(z-1)}{0.368z(z+0.718)} \; \frac{z}{z-1}$$

$$= \frac{z-0.368}{0.368(z+0.718)}$$

$$= 2.718-2.952z^{-1}+2.120z^{-2}-1.522z^{-3}+1.093z^{-4}-0.785^{-5}\pm...$$

Eq. (3.7.18) is therefore

$$y_{z\gamma} = \frac{(\gamma-1+e^{-\gamma})z^2+(2.368-1.368\gamma-2e^{-\gamma})z+(e^{-\gamma}-0.736+0.368\gamma)}{0.368(z-1)(z+0.718)} \tag{3.7.19}$$

The factor (z+0.718) is cancelled for $\gamma = 0$ and $\gamma = 1$, however, this does not occur between the sampling instants for $0 < \gamma < 1$. The inversion is performed by expanding

$1/(z-1)(z+0.718)$ into a sequence in z^{-1}. Multiplying these terms by the numerator, the expression takes the form

$$y_{z\gamma} = y(\gamma T)+y(T+\gamma T)z^{-1}+y(2T+\gamma T)z^{-2}+\ldots \quad (3.7.20)$$

The transients of u(t) and y(t) are represented in figure 3.18.

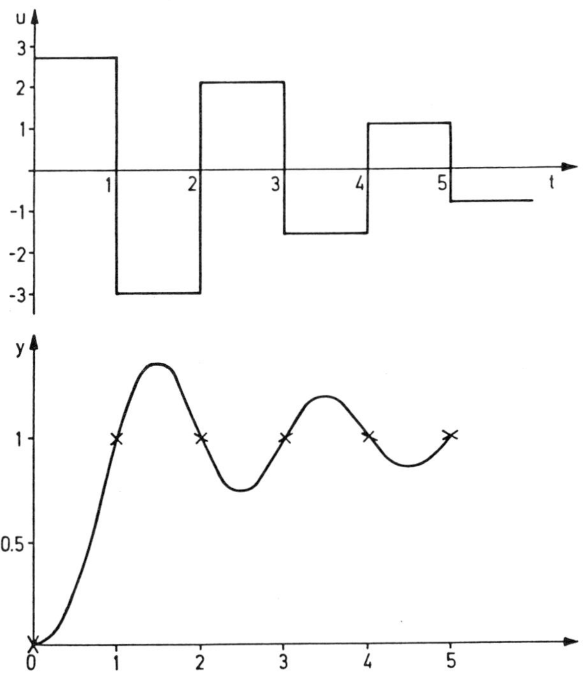

Figure 3.18 Step response at actuator u and output y for the controller of eq. (3.4.22). A reference transfer function 1/z is produced by cancellation.

This complete solution exhibits two disadvantages of this cancellation design:

1) Large actuator amplitudes u arise and last for many sampling intervals.
2) This large amount of actuator power is necessary to force the solution y through the points $y(kT) = 1$, $k = 1,2,3\ldots$ The shape of y(t) between these instants is very unsatisfactory, and there is an overshoot of 37 %. The form of

u and y is essentially determined by the controller pole at z = -0.718. This pole does not appear in $y_z(z)$ due to its cancellation with the corresponding plant zero.

The relationships between the variables formed from f(t) are illustrated by figure 3.19.

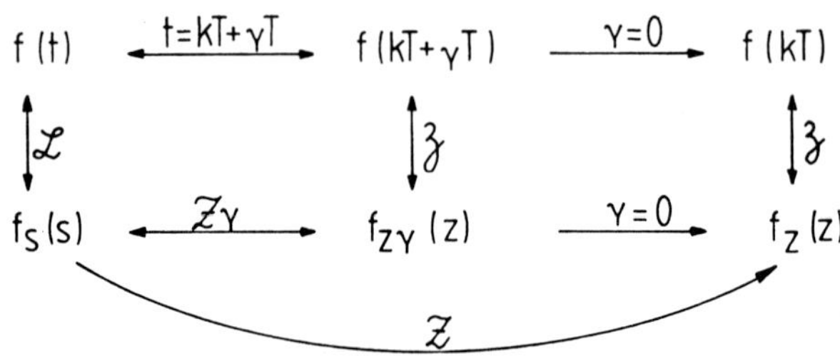

Figure 3.19 Relationship of the variables and transforms formed from f(t).

Information is lost by the insertion of $\gamma = 0$. The transition to the expressions of the last column is therefore not reversible. All other relationships are uniquely reversible.

Remark 3.6:

The Laplace- and z-transforms are usually designated by capital letters in the control literature, i.e., $F(s) = \mathcal{L}\{f(t)\}$, $F(z) = \mathcal{Z}\{f(kT)\}$. It is, however, misleading if the same capital letter is used for both transforms. The confusion gets worse if the z-transform is not being applied to the sequence f(kT) or f(kT+γT), but rather to the signal f(t) and to F(s). One frequently reads the notation

$$F(z) = \mathcal{Z}\{f(t)\} = \mathcal{Z}\{F(s)\} \qquad (3.7.21)$$

This representation has several disadvantages:

1) F(z) and F(s) are different functions for which the same symbol is used.
2) $\mathcal{Z}\{f(t)\}$ and $\mathcal{Z}\{F(s)\}$ are different operations for which the same symbol is used.
3) $\mathcal{Z}\{f(t)\}$ and $\mathcal{Z}\{F(s)\}$ are not uniquely reversible. For a

sequence f(kT) there are arbitrary many functions f(t) which pass through the values f(kT) at the time instants $t = kT$.

4) Because the operation $\mathcal{Z}\{f(t)\}$ already contains the formation of the sequence f(kT), a special transform must be introduced for the sequence $f(kT+\gamma T)$. This is called the "modified z-transform" [52.2] in the sampled-data literature.

$$F(z,\gamma) = \mathcal{Z}_{\gamma}\{f(t)\} := z^{-1} \sum_{k=0}^{\infty} f(kT+\gamma T) z^{-k} \qquad (3.7.22)$$

Because the factor z^{-1} is unnecessary, the "advanced z-transform" for which this factor is omitted, is also used.

It is not necessary, however, to present this "modified z-transform" as another transform and to repeat the proofs for all the rules of the z-transform. Note that the generation of a sequence f_k from a signal f(t) by insertion of $t = kT$ or $t = kT+\gamma T$ has nothing to do with the z-transform of a sequence.

3.8 Time-Delay Systems

In control systems with a transportation process a "time-delay" arises from an input signal $x_i(t)$ to an output signal $x_o(t)$. This can be expressed as

$$x_o(t) = x_i(t-\theta), \quad \theta > 0 \qquad (3.8.1)$$

and by the right shifting theorem of the Laplace transform

$$x_{os}(s) = e^{-\theta s} x_{is}(s) \qquad (3.8.2)$$

The remote control of the movement of the Soviet vehicle Lunochod on the surface of the moon can be taken as an example. More than a second of time-delay is needed for the transmission of commands and the reply back to the earth.

The content of the time-delay element

$$x_o(\tau),\ t - \theta < \tau < t \tag{3.8.3}$$

must be known in order to describe the state at the time instant t. This signal segment influences the future form of the solution. A state of the type of eq. (3.8.3) cannot be described exactly by a finite-dimensional state model. $x_i(\tau)$ may be approximated by a finite number of amplitude values. However, if $x_i(t)$ is produced by a hold element, then it can be represented exactly by one amplitude value per sampling interval, see [77.2], [80.1]. For a time-delay θ at the input of a sampling system

$$\underline{\dot{x}}(t) = \underline{Fx}(t) + \underline{g}u(t-\theta)\ ,\ u(t) = u(kT) \text{ for } kT < t < kT+T \tag{3.8.4}$$

The solution is

$$\underline{x}(t) = e^{\underline{F}(t-t_o)}\underline{x}(t_o) + \int_{t_o}^{t} e^{\underline{F}(t-\tau)}\underline{g}u(\tau-\theta)d\tau$$

and for $t = kT+T,\ t_o = kT$

$$\underline{x}(kT+T) = e^{\underline{F}T}\underline{x}(kT) + \int_{kT}^{kT+T} e^{\underline{F}(kT+T-\tau)}\underline{g}u(\tau-\theta)d\tau \tag{3.8.5}$$

First we examine the case for which the time-delay θ is smaller than the sampling interval T. In the interval of integration

$$u(\tau-\theta) = \begin{cases} u(kT-T) & \text{for } kT < \tau < kT+\theta \\ u(kT) & \text{for } kT+\theta < \tau < kT+T \end{cases} \tag{3.8.6}$$

$$\underline{x}(kT+T) = e^{\underline{F}T}\underline{x}(kT) + \int_{kT}^{kT+\theta} e^{\underline{F}(kT+T-\tau)}\underline{g}d\tau u(kT-T) + \int_{kT+\theta}^{kT+T} e^{\underline{F}(kT+T-\tau)}\underline{g}d\tau u(kT)$$

$$= e^{\underline{F}T}\underline{x}(kT) + \int_{T-\theta}^{T} e^{\underline{F}v}\underline{g}dv u(kT-T) + \int_{0}^{T-\theta} e^{\underline{F}v}\underline{g}dv u(kT)$$

using the notation of eq. (3.1.7)

$$\underline{A}(T) = e^{\underline{F}T} , \quad \underline{b}(T) = \int_0^T e^{\underline{F}v} dv \underline{g} \tag{3.8.7}$$

$$\underline{x}(kT+T) = \underline{A}(T)\underline{x}(kT) + [\underline{b}(T)-\underline{b}(T-\theta)]u(kT-T) + \underline{b}(T-\theta)u(kT)$$

The state variable $u_1(kT) = u(kT-T)$ is introduced for the past value of the input signal which influences the further form of the solution

$$\begin{bmatrix} \underline{x}(kT+T) \\ u_1(kT+T) \end{bmatrix} = \begin{bmatrix} \underline{A} & \underline{b}_1 \\ \underline{0} & 0 \end{bmatrix} \begin{bmatrix} \underline{x}(kT) \\ u_1(kT) \end{bmatrix} + \begin{bmatrix} \underline{b}_2 \\ 1 \end{bmatrix} u(kT) \tag{3.8.8}$$

with $\underline{b}_1 = \underline{b}(T)-\underline{b}(T-\theta)$, $\underline{b}_2 = \underline{b}(T-\theta)$. If the sampling period T is equal to the time-delay θ, then $b(T-\theta) = b(0) = 0$, $\underline{b}_1 = \underline{b}$, $\underline{b}_2 = \underline{0}$. For $T < \theta \leq 2T$ in eq. (3.8.8) $u_1[k] = u(kT-2T)$ and the additional state $u_2(kT) = u(kT-T)$ must be introduced:

$$\begin{bmatrix} \underline{x}(kT+T) \\ u_1(kT+T) \\ u_2(kT+T) \end{bmatrix} = \begin{bmatrix} \underline{A} & \underline{b}_1 & \underline{b}_2 \\ \underline{0} & 0 & 1 \\ \underline{0} & 0 & 0 \end{bmatrix} \begin{bmatrix} \underline{x}(kT) \\ u_1(kT) \\ u_2(kT) \end{bmatrix} + \begin{bmatrix} 0 \\ 0 \\ 1 \end{bmatrix} u(kT) \tag{3.8.9}$$

In general the time-delay θ is expressed by an integer multiple of the sampling period T and a difference term γT:

$$\theta = mT-\gamma T , \quad 0 \leq \gamma < 1 \tag{3.8.10}$$

and eq. (3.8.9) is extended to

$$\begin{bmatrix} \underline{x}(kT+T) \\ u_1(kT+T) \\ \cdot \\ \cdot \\ \cdot \\ u_m(kT+T) \end{bmatrix} = \begin{bmatrix} \underline{A} & \underline{b}_1 & \underline{b}_2 & 0 \ldots 0 \\ \underline{0} & 0 & 1 & \\ \cdot & & & 1 \ddots \\ \cdot & & & \quad \ddots 1 \\ \cdot & & & \\ \underline{0} & 0 & & 0 \end{bmatrix} \begin{bmatrix} \underline{x}(kT) \\ u_1(kT) \\ \cdot \\ \cdot \\ \cdot \\ u_m(kT) \end{bmatrix} + \begin{bmatrix} 0 \\ \cdot \\ \cdot \\ \cdot \\ 0 \\ 1 \end{bmatrix} u(kT)$$

$$\underline{b}_1 = \underline{b}(T)-\underline{b}(\gamma T) , \quad \underline{b}_2 = \underline{b}(\gamma T) \tag{3.8.11}$$

The eigenvalues of the extended dynamic matrix are those of the subsystem $\underline{A}$ and additional m eigenvalues at $z = 0$. Figure 3.20 illustrates eq. (3.8.11).

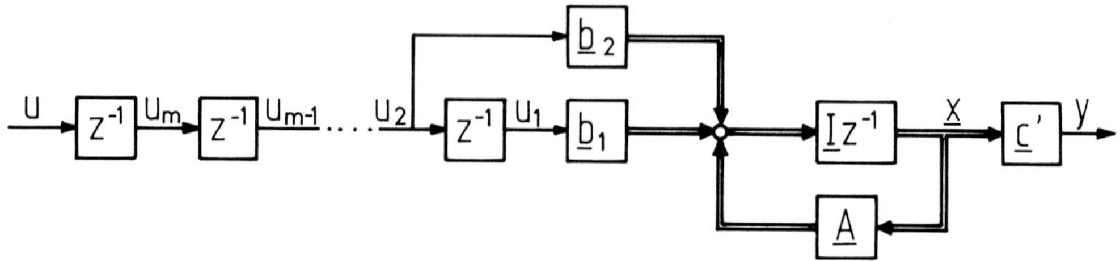

Figure 3.20 Sampling system with time-delay θ at the input, $mT-T < \theta \leq mT$

The z-transfer function is obtained from the z-transform of eq. (3.8.11) with initial state zero

$$
\begin{aligned}
z\underline{x}_z &= \underline{A}\underline{x}_z + \underline{b}_1 u_{1z} + \underline{b}_2 u_{2z} \\
z u_{1z} &= u_{2z} \\
&\vdots \\
z u_{mz} &= u_z \\
(z\underline{I}-\underline{A})\underline{x}_z &= (\underline{b}_1 + z\underline{b}_2) z^{-m} u_z \\
y_z &= \underline{c}'(z\underline{I}-\underline{A})^{-1}(\underline{b}_1 + z\underline{b}_2) z^{-m} u_z
\end{aligned}
\tag{3.8.12}
$$

In addition to the poles of the subsystem $\underline{A}$ the transfer function has m poles at $z = 0$. Therefore, in contrast to the continuous case with time-delay, this is again a rational system with rational transfer function and finite-dimensional state representation.

Example:

For the familiar example of the plant of eq. (3.1.13) the input is delayed by a time-delay of $\theta = 1.25$ seconds

$$
\dot{\underline{x}}(t) = \begin{bmatrix} 0 & 1 \\ 0 & -1 \end{bmatrix} \underline{x}(t) + \begin{bmatrix} 0 \\ 1 \end{bmatrix} u(t-1.25) \tag{3.8.13}
$$

$T = 1$, so $\theta = m-\gamma$, $m = 2$, $\gamma = 0.75$. From the solution of eq. (3.1.13)

$$\underline{A}(T) = \begin{bmatrix} 1 & 1-e^{-T} \\ 0 & e^{-T} \end{bmatrix}$$

$$\underline{b}(T) = \begin{bmatrix} T + e^{-T}-1 \\ 1 - e^{-T} \end{bmatrix}$$

$$\underline{b}_1 = \underline{b}(T) - \underline{b}(\gamma T) = \underline{b}(1) - \underline{b}(0.75) = \begin{bmatrix} e^{-1}-e^{-0.75}+0.25 \\ -e^{-1}+e^{-0.75} \end{bmatrix}$$

$$\underline{b}_2 = \underline{b}(\gamma T) = \underline{b}(0.75) = \begin{bmatrix} e^{-0.75}-0.25 \\ 1-e-0.75 \end{bmatrix}$$

Eq. (3.8.11) is therefore

$$\begin{bmatrix} x_1(kT+T) \\ x_2(kT+T) \\ u_1(kT+T) \\ u_2(kT+T) \end{bmatrix} = \begin{bmatrix} 1 & 0.6321 & 0.1455 & 0.2224 \\ 0 & 0.3679 & 0.1045 & 0.5276 \\ 0 & 0 & 0 & 1 \\ 0 & 0 & 0 & 0 \end{bmatrix} \begin{bmatrix} x_1(kT) \\ x_2(kT) \\ u_1(kT) \\ u_2(kT) \end{bmatrix} + \begin{bmatrix} 0 \\ 0 \\ 0 \\ 1 \end{bmatrix} u(kT) \tag{3.8.14}$$

For the measurement equation $y(kT) = [1 \quad 0]\underline{x}(kT)$ the z-transfer function eq. (3.8.12) becomes

$$h_z = \frac{1}{(z\ 1)(z\ 0.3678)} [1 \quad 0] \begin{bmatrix} z-0.3679 & 0.6321 \\ 0 & z-1 \end{bmatrix} \begin{bmatrix} 0.1455+0.2224z \\ 0.1045+0.5276z \end{bmatrix} z^{-2}$$

$$h_z = \frac{0.2224z^2+0.3972z+0.0125}{z^2(z-1)(z-0.3679)} \tag{3.8.15}$$

If the description of the continuous system is given in the frequency domain, that is by a transfer function

$$g_s(s) = g_{rs}(s) \times e^{-\theta s} \tag{3.8.16}$$

then one proceeds as in eqs. (3.3.26) to (3.3.28)

$$h_s(s) = \frac{1-e^{-Ts}}{s} e^{-\theta s} \times g_{rs}(s) \tag{3.8.17}$$

The step response of the rational part is

$v(t) = \mathcal{L}^{-1}\{g_{rs}(s)/s\}$

With this result

$$h(t) = \mathcal{L}^{-1}\left\{\frac{1-e^{-Ts}}{s} e^{-\theta s} g_{rs}(s)\right\} = v(t-\theta) - v(t-\theta-T)$$

$$h_z(z) = \frac{z-1}{z} \times \mathcal{Z}\left\{v(kT-\theta)\right\} \tag{3.8.18}$$

The time-delay θ is expressed as in eq. (3.8.10) by
$\theta = mT-\gamma T,\ 0 \le \gamma < 1$

$$h_z(z) = \frac{z-1}{z} \times z^{-m} \mathcal{Z}\{v(kT+\gamma T)\}$$

$$= \frac{z-1}{z^{1+m}} \times \mathcal{Z}\left\{\mathcal{L}^{-1}\{g_s(s)/s\}_{t=kT+\gamma T}\right\} \tag{3.8.19}$$

and with the notation of eq. (3.7.16)

$$h_z(z) = \frac{z-1}{z^{1+m}} \mathcal{Z}_\gamma \{g_s(s)/s\} \tag{3.8.20}$$

$\mathcal{Z}_\gamma\{f_s(s)\}$ can be found in the table in appendix B.

Example:

The root locus for the control loop of figure 3.21 will now be sketched.

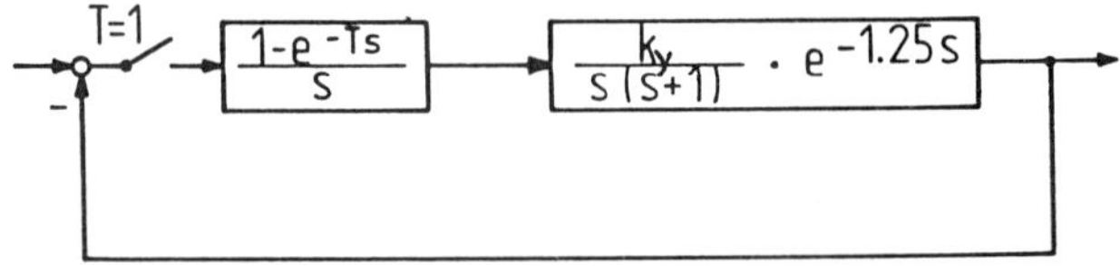

Figure 3.21 Sampled control loop with time-delay

m = 2, $\gamma = 0.75$, and for the determination of

$$h_z(z) = \frac{z-1}{z^3} \mathcal{Z}_\gamma \left\{ \frac{1}{s^2(s+1)} \right\}_{\gamma=0.75}$$

One has - as in the example of eq. (3.7.18) -

$$h_z(z) = \frac{(\gamma-1+e^{-\gamma})z^2+(2.368-1.368\gamma-2e^{-\gamma})z+(e^{-\gamma}-0.736+0.368\gamma)}{z^2(z-1)(z-0.3679)} \qquad \gamma=0.75$$

$$= \frac{0.2224z^2+0.3972z+0.0125}{z^2(z-1)(z-0.3679)}$$

The result agrees with the z-transfer function (3.8.15) determined from the state representation. The numerator has the factorization 0.2224(z+0.032)(z+1.754).

The following considerations and simple calculations suffice to sketch the root locus.

1. In the neighborhood of z=0 the transfer function may be approximated by $(z+0.03)/z^2$ with a phase angle contribution of 2π by the poles at z = 1 and z = 0.368. The root locus curve is therefore a small circle around the point z = -0.03 passing through z = 0. For greater distances from z = 0, z + 0.03 can be approximately cancelled with a pole at z = 0.
2. A further branching point must lie between z = 0.368 and z = 1.
3. $h_z(z)$ has a double zero at $z = \infty$. Two root locus plot branches therefore go off to infinity in the directions $\pm j$. The intersection of the asymptotes with the real axis at z_a is determined from the equilibrium equation

   $$(1-z_a)+(0.368-z_a)-2z_a = (-0.03-z_a)+(-1.754-z_a)$$

   to be $z_a = 1.577$.

   This results in the form of the root locus sketched in figure 3.22, also see eq. (C.2.22).

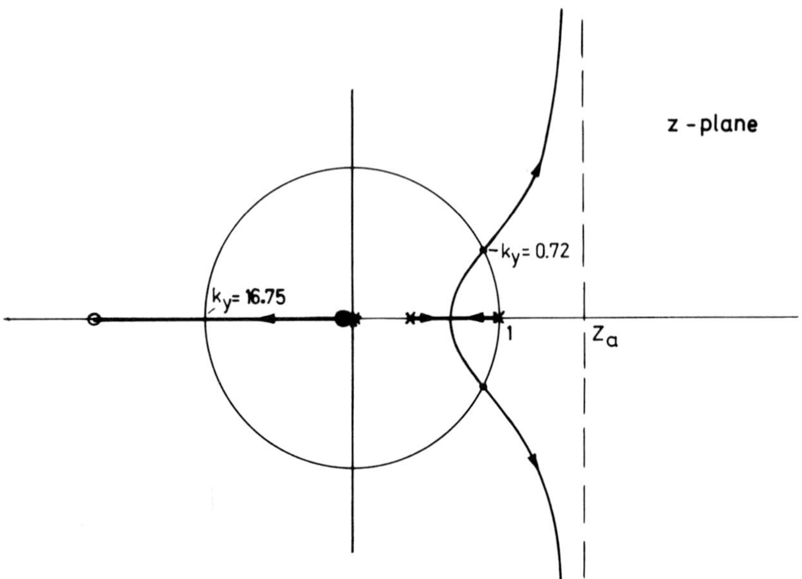

Figure 3.22 Sketch of the root locus of the control loop from figure 3.21.

3.9 Frequency Response Methods

3.9.1 Frequency Response Determination

The frequency response of a linear system is defined by the transfer function $g_s(s)$ on the imaginary axis $s = j\omega$, i.e. $g_s(j\omega)$. For a rational system with the state representation

$$\underline{\dot{x}} = \underline{F}\underline{x} + \underline{g}u$$
$$y = \underline{c}'\underline{x} + du$$

the frequency response according to eq. (2.3.9) is

$$g_s(j\omega) = \underline{c}'(j\omega\underline{I}-\underline{F})^{-1}\underline{g} + d \tag{3.9.1}$$

This frequency response is of special interest because for stable systems, it can be experimentally determined by excitation of the system with a harmonic input variable

$$u(t) = a \times e^{j\omega_1 t} + a^* \times e^{-j\omega_1 t}, \quad t \geq 0 \tag{3.9.2}$$

where the star designates the complex conjugate value. For the initial state $\underline{x}(0) = \underline{0}$ and the input (3.9.2) in the frequency domain

$$y_s(s) = g_s(s) \times u_s(s)$$

$$= g_s(s) \times \left(\frac{a}{s-j\omega_1} + \frac{a^*}{s+j\omega_1}\right) \tag{3.9.3}$$

By a partial fraction expansion one obtains for $\pm j\omega_1 \neq s_i$ (s_i = pole of $g_s(s)$ with the multiplicity m_i)

$$y_s(s) = \frac{a g_s(j\omega_1)}{s-j\omega_1} + \frac{a^* g_s^*(j\omega_1)}{s+j\omega_1} + \sum_i \left(\frac{c_{i1}}{s-s_i} + \ldots + \frac{c_{im_i}}{(s-s_i)^{m_i}}\right) \tag{3.9.4}$$

If the system is stable, i.e. all s_i have a negative real part, then exponentially decreasing time functions correspond to the terms under the sum (these are called "transients") and in the steady state only the following term remains

$$y_\infty(t) = a g_s(j\omega_1) e^{j\omega_1 t} + a^* g_s^*(j\omega_1) e^{-j\omega_1 t} \tag{3.9.5}$$

The frequency response can be expressed in the following form

$$g_s(j\omega_1) = A(\omega_1) e^{j\varphi(\omega_1)} \tag{3.9.6}$$

where $A(\omega_1)$ is the absolute value and $\varphi(\omega_1)$ is the phase angle. Thus eq. (3.9.5) reduces to

$$y_\infty(t) = A(\omega_1)\left(a e^{j[\omega_1 t+\varphi(\omega_1)]} + a^* e^{-j[\omega_1 t+\varphi(\omega_1)]}\right) \tag{3.9.7}$$

For example, for $a = a^* = 1/2$ the input is $u(t) = \cos\omega_1 t$ and the steady state response is

$$y_\infty(t) = A(\omega_1)\cos[\omega_1 t+\varphi(\omega_1)] \tag{3.9.8}$$

After a sufficiently long time a sine-signal of frequency ω_1 remains at the output of a stable system (2.2.1) as a response to a sinusoidal input of the same frequency ω_1. The amplitude of this signal is multiplied by the factor $A(\omega_1)$ and its phase angle has been shifted by $\varphi(\omega_1)$ with respect to the input signal, where

$$A(\omega_1) = |g_S(j\omega_1)| \quad \text{and} \quad \varphi(\omega_1) = \arctan\frac{\text{Im } g_S(j\omega_1)}{\text{Re } g_S(j\omega_1)} \tag{3.9.9}$$

This result is the same even if the stable, linear, time-invariant system does not have a rational transfer function, for example a system with time-delay.

Thus, the modelling of an unknown stable system is possible through a frequency response measurement. The system is stimulated by a sinusoidal input of frequency ω_1. After the fading of all transients the amplitude ratio $A(\omega_1)$ and the phase shift $\varphi(\omega_1)$ are determined. This process is repeated for sufficiently many frequencies ω_1 and the result is plotted as a frequency response curve or -"Nyquist"-plot, i.e. the locus of points $A(\omega_1)e^{j\varphi(\omega_1)}$ in the complex plane.

This Nyquist diagram represents the mapping of the imaginary axis of the s-plane into the g_S-plane. For unstable systems the frequency response cannot be determined experimentally. However, $g_S(j\omega)$ can be calculated by substituting $s = j\omega$ in the transfer function or by eq. (3.9.1) from a state representation.

The importance of the frequency response curve is that it allows the stability of the closed-loop system around $g_S(s)$ to be graphically checked by the Nyquist-criterion, see [55.2], [57.2], [80.8]. This procedure of modelling and stability analysis can also be transferred to sampling systems.

The discrete frequency response is the value of the z-transfer function on the unit circle, i.e. for $z = e^{j\omega T}$. According to eq. (1.2.16), the relation

$$h_z(e^{j\omega T}) = h_s^*(j\omega) = \frac{1}{T} \sum_{m=-\infty}^{\infty} h_s(j\omega + jm\omega_A) \tag{3.9.10}$$

with

$$h_s(j\omega) = \frac{1-e^{-j\omega T}}{j\omega} \times g_s(j\omega) \tag{3.9.11}$$

exists between the frequency response $g_s(j\omega)$ of a continuous system and the discrete frequency response of the same system with sampler and hold connected to the input.

$h_s(j\omega)$ is determined from the measured frequency response $g_s(j\omega)$ and the frequency response of the hold element of eq. (1.2.17) by multiplication of the absolute values and addition of the phase angles, always with respect to the same frequency. Since the plant frequently has low-pass characteristics, in other words $|g_s(j\omega)|$ decreases as $1/\omega$ or even $1/\omega^2$ for high frequencies ω, it is usually sufficient to approximate $h_z(e^{j\omega T})$ by a few terms of the sum

$$h_z(e^{j\omega T}) \approx \frac{1}{T} \sum_{m=-N}^{N} h_s(j\omega + jm\omega_A) \tag{3.9.12}$$

The discrete frequency response $h_z(e^{j\omega T})$ can be graphically constructed by addition of the terms on the right hand side of eq. (3.9.12), see [55.2], [59.1].

There is another possibility of determining the discrete frequency response of the sampler with hold element and continuous subsystem directly. In this case the sampling period T must already be fixed for the determination of the model of the plant.

A sinusoidal input sequence u(kT) with arbitrary phase relative to the sampling instants is expressed as

$$u(kT) = ae^{jk\omega T} + a^* e^{-jk\omega T}, \quad k = 0,1,2... \tag{3.9.13}$$

The z-transform of the output signal is

$$y_z(z) = h_z(z)u_z(z) = h_z(z)\left[\frac{az}{z-e^{j\omega T}} + \frac{a^*z}{z-e^{-j\omega T}}\right] \quad (3.9.14)$$

By a partial fractions decomposition of $y_z(z)/z$ one obtains

$$y_z(z) = \frac{h_z(e^{j\omega T})az}{z-e^{j\omega T}} + \frac{h_z^*(e^{j\omega T})a^*z}{z-e^{-j\omega T}} + \frac{\text{polynomial in } z}{\text{denominator of } h_z(z)} \quad (3.9.15)$$

For a stable system $h_z(z)$, the inverse z-transform of the last term produces a sequence which goes to zero as k goes to infinity. After the transient have vanished the following steady-state solution remains

$$y_\infty(kT) = h_z(e^{j\omega T})ae^{jk\omega T} + h_z^*(e^{j\omega T})a^*e^{-jk\omega T} \quad (3.9.16)$$

With the notation

$$h_z(e^{j\omega T}) = M(\omega T)e^{j\phi(\omega T)} \quad (3.9.17)$$

this can be written as

$$\begin{aligned} y_\infty(kT) &= M(\omega T)\{ae^{j[k\omega T+\phi(\omega T)]} + a^*e^{-j[k\omega T+\phi(\omega T)]}\} \\ &= M(\omega T)u[kT+\phi(\omega T)] \end{aligned} \quad (3.9.18)$$

Therefore as for continuous systems one obtains the absolute value of the discrete frequency response as the ratio of the amplitudes (and the phase as the phase-shift) of the sinusoidal output sequence with respect to the input sequence. Numerical values for the discrete frequency response $h_z(e^{j\omega T})$ of eq. (3.9.17) are obtained by measurements at different frequencies. These values are plotted in the h_z-plane as a discrete frequency response curve which can be used for a stability test.

The discrete frequency response $h_z(e^{j\omega T})$ has the following properties

1. For $\omega T = i\pi$, $i = 0,1,2...$ $h_z(e^{j\omega T}) = h_z(\pm 1)$ is real.

2. From eq. (3.9.10) follows

$$h_z[e^{j(\omega T+2\pi)}] = \frac{1}{T}\sum_{m=-\infty}^{\infty} h_s[\frac{j}{T}(\omega T+2\pi+2m\pi)]$$

$$= \frac{1}{T}\sum_{m=-\infty}^{\infty} h_s[\frac{j}{T}(\omega T+2m\pi)]$$

$$= h_z(e^{j\omega T})$$

$h_z(e^{j\omega T})$ is therefore periodic with the period $\omega T = 2\pi$.

3. The discrete frequency response curve is always symmetric to the real axis. Therefore it suffices to draw the curve for $0 < \omega T < \pi$.

3.9.2 Nyquist Criterion

With the help of the discrete frequency response plot $h_z(e^{j\omega T})$, the application of the Nyquist stability criterion can now be explained.

For the rational expression

$$E(z) = \frac{(z-z_{o1})\dots(z-z_{om})}{(z-z_1)\dots(z-z_n)} \qquad (3.9.19)$$

it is first assumed that neither a pole nor a zero lies on the unit circle. Examine the variation of the phase angle ϕ_E as z moves once counterclockwise around the unit circle beginning at $z = -1$. For every zero in the unit circle ϕ_E is varied by 2π, for every pole by -2π. Poles and zeros outside of the unit circle make no contribution. If E(z) has p_c zeros and p_o poles in the unit circle, then the phase angle ϕ_E varies by

$$\Delta\phi[E(e^{j\omega T})] = (p_c-p_o)2\pi \qquad (3.9.20)$$

for one encirclement ($-\pi < \omega T < \pi$).

Now let

$$E(z) = 1 + h_z(z) = 1 + \frac{B(z)}{A(z)} = \frac{A(z)+B(z)}{A(z)}$$

$$= \frac{\text{closed loop characteristic polynomial}}{\text{open loop characteristic polynomial}}$$

Then p_c is the number of closed poles in the unit circle and p_o the number of open loop poles in the unit circle.

The system is stable if all poles of the closed loop lie in the unit circle, i.e. $p_c = n$. In this case the phase angle must change in one encirclement by

$$\Delta\phi[1+h_z(e^{j\omega T})] = (n-p_o)2\pi \tag{3.9.21}$$

This stability condition is especially simple if the open loop $h_z(z)$ is stable. In this case $p_o = n$ and the system is stable if the phase angle variation is zero. The phase angle variation is determined by plotting $1+h_z(e^{j\omega T})$ for $-\pi < \omega T < \pi$ in the $(1+h_z)$-plane and counting the number of encirclements of the origin by this curve. Equivalently $h_z(e^{j\omega T})$ may be plotted in the h_z-plane. Then the number of encirclements of the Nyquist diagram around the "critical point" $h_z = -1$ must be counted. In the special case of a stable open loop the closed loop is also stable, if and only if the point -1 is not encircled by the Nyquist curve $h_z(e^{j\omega T})$. If the open loop has p_o poles in unit circle, then for a stable loop the critical point is encircled $n-p_o$ times by $h_z(e^{j\omega T})$.

So far no poles of the open and closed loop were assumed to lie on the unit circle. If now a pole of the closed loop lies on the unit circle, the Nyquist curve passes through the critical point $h_z = -1$. If a pole of the open loop lies on the unit circle, then $h_z(e^{j\omega T})$ goes to infinity and returns with a phase angle rotation of $\pm\pi$, i.e. returns from infinity from the opposite direction. Here it must now be established whether this angle variation should be considered as $+\pi$ or $-\pi$. There are two equivalent possibilities which involve bypassing the pole on the unit circle by using a small semicircular indentation on the outside (inside). The mapping of this semicircle into the $h_z(z)$ plane gives a large circle which runs through the angle $-\pi(+\pi)$. The pole is considered to lie inside (outside) the unit circle.

Example:

In the control loop of figure 3.12 for $k_y = 1$ one obtains by a partial fraction decomposition of eq. (3.3.22)

$$h_z(z) = \frac{1}{z-1} - \frac{0.632}{z-0.368}$$

$$= \frac{z^{-1}-1}{(z-1)(z^{-1}-1)} - \frac{0.632(z^{-1}-0.368)}{(z-0.368)(z^{-1}-0.368)}$$

$$= \frac{z^{-1}-1}{2-z-z^{-1}} - \frac{0.632(z^{-1}-0.368)}{1.135-0.368(z+z^{-1})}$$

$$h_z(e^{j\omega T}) = \frac{\cos\omega T-j\sin\omega T-1}{2(1-\cos\omega T)} - \frac{0.632(\cos\omega T-j\sin\omega T-0.368)}{1.135-0.736\cos\omega T}$$

$$= \left(-\frac{1}{2} - \frac{0.632\cos\omega T-0.232}{1.135-0.736\cos\omega T}\right)+j\left(\frac{-1}{2(1-\cos\omega T)} + \frac{0.632}{1.135-0.736\cos\omega T}\right)\sin\omega T$$

The Nyquist curve is shown in figure 3.23 together with the root locus in z-plane, in this case a circle around the zero at $z = -0.718$ together with portions of the real axis. The pole at $z = 1$ is here bypassed on the right and is therefore considered as lying inside the unit circle, i.e. $p_o = n = 2$. Correspondingly, the Nyquist curve is completed by a large circle to the right (indicated by the dotted lines). The curve does not encircle the critical point -1, i.e. $\Delta\phi = 0$ and condition (3.9.21) is satisfied. Both poles of the closed loop lie in the unit circle and the system is stable.

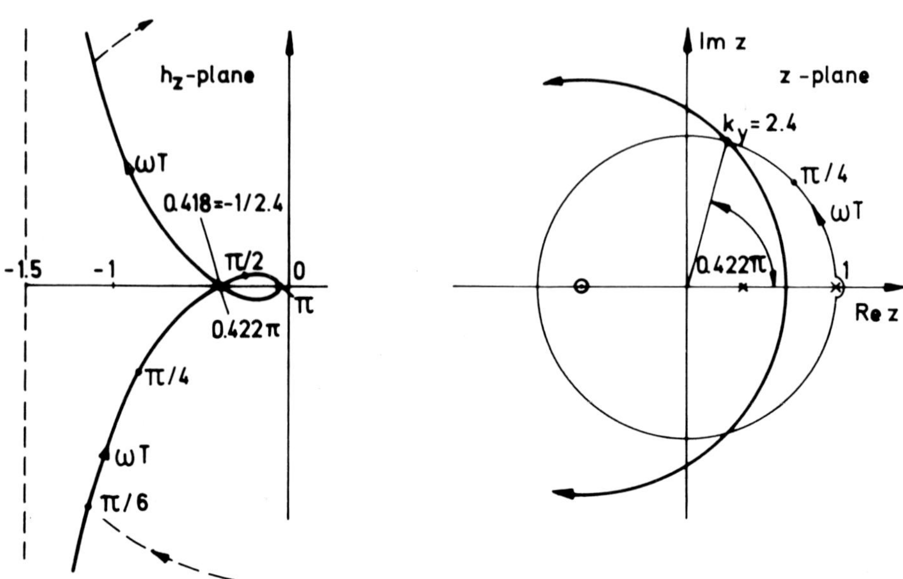

Figure 3.23 Nyquist diagram and root locus for the control loop of figure 3.12

For $k_y \neq 1$ it is not necessary to redraw the Nyquist diagram with a different scale. Writing eq. (3.5.4) as follows

$$h_z(z) = -1/k_y \tag{3.9.22}$$

shows that it is only necessary to view $-1/k_y$ as the critical point instead of -1. In this example the stability boundary is reached if $-1/k_y = 0.418$, i.e. $k_y = 2.4$. The root locus also crosses over the unit circle for this gain at a point with an angle $\omega T = 0.422\pi$ with respect to the positive real axis. This angle, or for T = 1 this frequency ω, arises on the Nyquist curve at the intersection with the negative real axis. $h_z(z)$ maps the unit circle from z-plane to the Nyquist curve and the root locus to the negative real axis. Stability is decided by the relative position of these two curves in either representation.

A correction term $d_z(e^{j\omega T})$ can be used to improve the form of the Nyquist curve $f_{oz}(e^{j\omega T}) = d_z(e^{j\omega T})h_z(e^{j\omega T})$. The discrete frequency response curve of the correction term is plotted and every two points of $d_z(z)$ and $h_z(z)$ having the same ωT are multiplied. This is performed by the multiplication of the absolute values and the addition of the phase angles. The correction term must first ensure stability, i.e. the Nyquist plot of $f_{oz}(e^{j\omega T})$ must satisfy the encirclement condition (3.9.21). This means that the zeros of

$$1 + f_{oz}(z) = 0 \qquad (3.9.23)$$

lie in the unit circle of the z-plane. If such a zero is located on the unit circle with an angle $\omega_1 T$ to the positive real axis, then

$$1 + f_{oz}(e^{j\omega_1 T}) = 0 \qquad (3.9.24)$$

One can judge how far away one is from this critical case by the "stability margin"

$$\rho = \min_{\omega} |1 + f_{oz}(e^{j\omega T})| \qquad (3.9.25)$$

ρ is the radius of the largest circle around the point -1 in the f_{oz}-plane, which is avoided by the plot of $f_{oz}(e^{j\omega T})$, see figure 3.24.

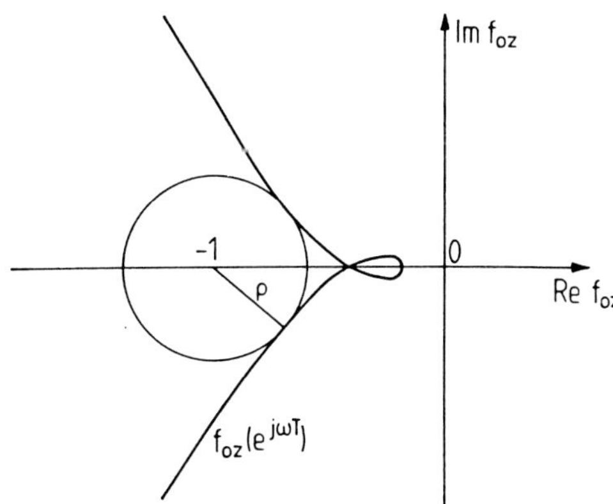

Figure 3.24

The stability margin ρ is the radius of the largest circle around $f_{oz} = -1$ which is avoided by the Nyquist curve $f_{oz}(e^{j\omega T})$

This raises the question of how this stability margin is related to the desired position of the eigenvalues in the z-plane, see figure 3.11. There is no simple correspondence here, but a qualitative idea can be obtained as follows:

Assume that not only the unit circle $z = e^{j\omega T}$ is conformally mapped by $f_{oz}(z)$ but also concentric circles $r \times e^{j\omega T}$, r constant, and radii with varying r and fixed ω. Their image is an orthogonal net of curves in the f_{oz}-plane. The mapping from the z-plane into the $f_{oz}(z)$-plane is unique. Inversely, however, n points in the z-plane correspond to one in the $f_{oz}(z)$-plane if the numerator of $f_{oz}(z)$ is of degree n. The n eigenvalues of the closed loop correspond to the point $f_{oz}(z) = -1$. If an eigenvalue now lies very close to the unit circle, then the point -1 also lies near to the Nyquist plot. The mapping from z to $f_{oz}(z)$ is conformal almost everywhere, i.e. the small angles and distances are unchanged. Therefore the small rectangle abcd at the unit circle, see figure 3.25, is mapped into the similar rectangle ABCD at the Nyquist plot.

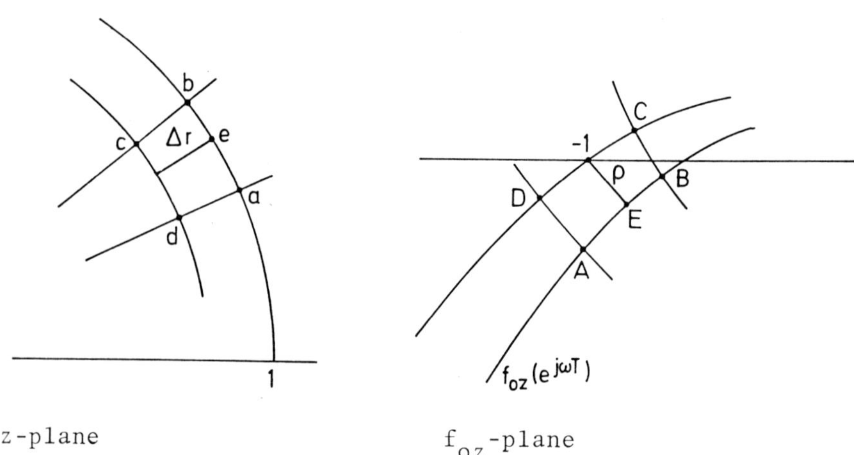

Figure 3.25 Conformal mapping of the "rectangle" abcd into the "rectangle" ABCD.

Now let E in figure 3.25 be the point on the Nyquist plot closest to the critical point -1. Select as A and B the neighboring points on the Nyquist plot, for which the frequencies $\omega_A T$ and $\omega_B T$ are known from the calculation or experimental determination of $f_{oz}(e^{j\omega T})$. Then the ratio of section AB to the stability margin ρ is the same as the ratio of the section ab to Δr in the z-plane. Thus

$$\Delta r \approx \frac{\omega_B T - \omega_A T}{|f_{oz}(e^{j\omega_B T}) - f_{oz}(e^{j\omega_A T})|} \times \rho \tag{3.9.26}$$

In the neighborhood of E we can at least qualitatively say that for given stability margin ρ and wide spread frequency pattern on $f_{oz}(e^{j\omega T})$ the pole distance Δr from the unit circle is smaller than for densely packed frequencies. A large stability margin ρ therefore still does not guarantee a favorable position of the eigenvalues.

A further relation between ρ and the eigenvalues of the closed loop follows from eq. (3.9.23) for

$$1+f_{oz}(z) = \frac{A(z)+B(z)}{A(z)} = \frac{P(z)}{A(z)} \tag{3.9.27}$$

$$= \frac{\text{closed loop characteristic polynomial}}{\text{open loop characteristic polynomial}}$$

The stability margin is therefore

$$\rho = \min_{\omega} |P(e^{j\omega T})/A(e^{j\omega T})| \tag{3.9.28}$$

Even for suitable positions of the zeros of P(z), the case can arise where A(z) becomes relatively large on the unit circle and therefore causes a small stability margin ρ. In this case the system may become unstable for small gain or phase changes, for example by neglected actuator or sensor dynamics. This case can be avoided by closing the loop so that all the poles of A(z) are not shifted too far.

It is therefore important that the positions of the resulting poles are checked for design processes which guarantee a stability margin - such as Nyquist plots, Bode diagrams and (implicitely) Riccati design (see section 9.4). Vice versa it must be checked that a sufficient stability margin is ensured for design procedures which guarantee a suitable pole position - such as root locus and pole assignment.

Remark 3.7:

An indirect description of the stability margin is given by the gain margin A_r and the phase margin ϕ_r. The values A_r and ϕ_r are found by measuring the distance between $f_{oz}(e^{j\omega T})$ and the critical point -1 along the negative real axis and along the unit circle. A given gain margin and phase margin can lead to very different stability margins, as illustrated by figure 3.26.

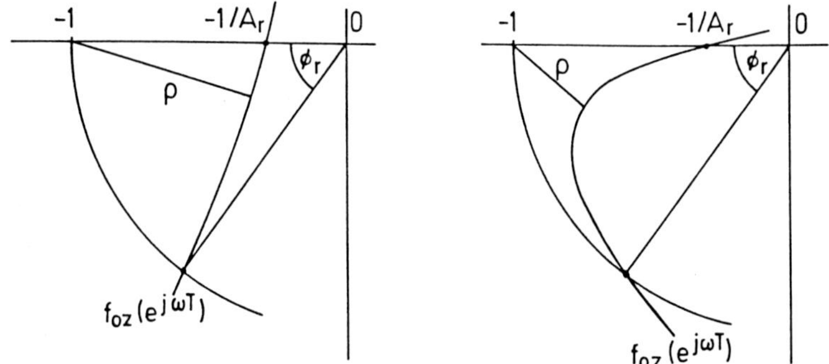

Figure 3.26 Gain margin A_r and phase margin ϕ_r are not good measures of the stability margin ρ.

3.9.3 Absolute Stability, Tsypkin and Circle Criteria

The stability margin ρ is also important for the examination of absolute stability. The statement of this problem is illustrated in figure 3.27.

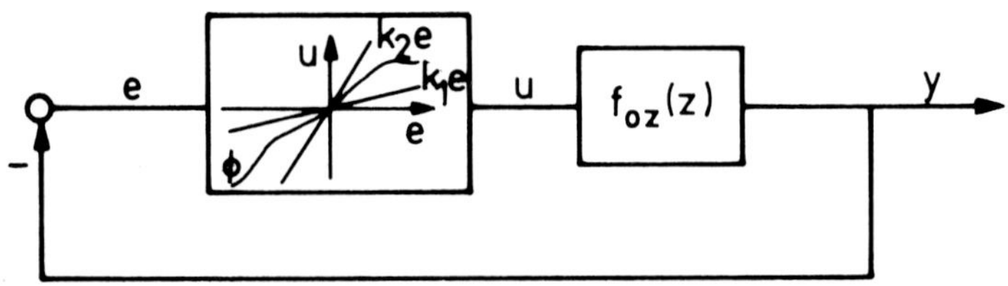

Figure 3.27 Control loop with nonlinearity in a sector

Besides the linear elements, which are described by the open loop z-transfer function $f_{oz}(z)$, the control loop contains a nonlinear term which is represented by a characteristic $u = \phi(e)$. In many applications this is a nonlinearity of the actuator. Suppose it is known only that the nonlinearity lies in a sector

$$k_1 < \phi(e)/e < k_2 \qquad (3.9.29)$$

The control loop is called "absolutely stable with respect to the sector (k_1,k_2)" if it is asymptotically stable in the large for all nonlinearities satisfying (3.9.29). For $k_1 = 0$, $k_2 = k$, $f_{oz}(z)$ stable, Tsypkin [63.5], [63.6] has proven the following sufficient stability condition: The control loop of figure 3.27 is absolutely stable with respect to the sector (0,k), if

$$\mathrm{Re} f_{oz}(e^{j\omega T}) > -1/k \quad \text{for} \quad 0 \leq \omega T \leq \pi \qquad (3.9.30)$$

Graphically this means that the total Nyquist plot $f_{oz}(e^{j\omega T})$ must be located to the right of a line parallel to the imaginary axis through the point $f_{oz}(z) = -1/k$. It is easy to determine graphically the largest value of k for which we can guarantee absolute stability by choosing this line so that it touches the Nyquist plot, see figure 3.28.

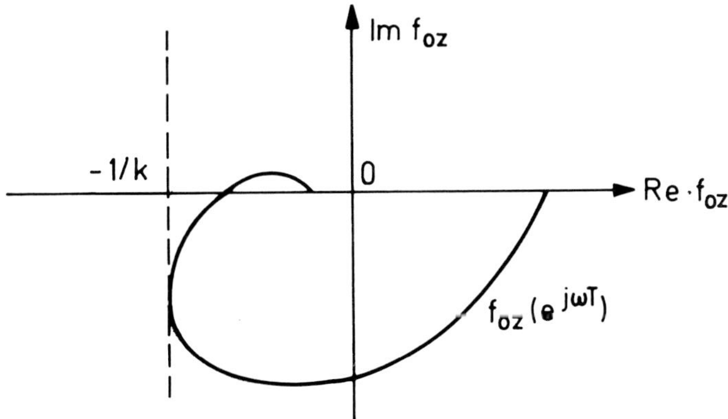

Figure 3.28 Determination of the largest sector of absolute stability

If the plant is unstable or if $k_1 \neq 0$, the problem may be reduced to the previous one by conversion of the nonlinearity and the linear dynamic system as indicated in the block diagram 3.29.

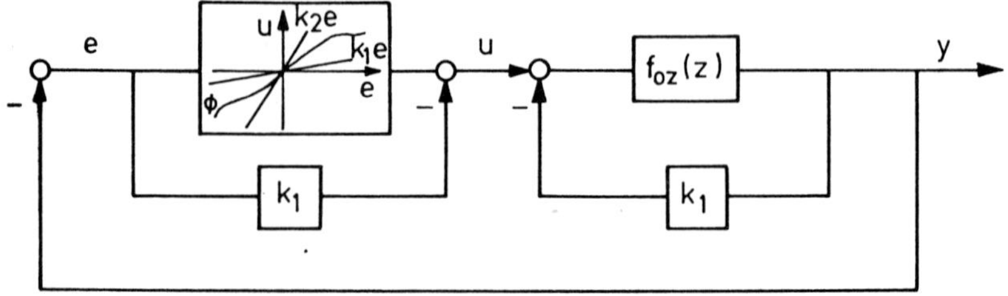

Figure 3.29 Rotation of the sector by $k_1 e$

The effects of two additional connections with the gain k_1 cancel each other. A new nonlinearity in the sector $(0,\ k_2-k_1)$ occurs from e to u. The z-transfer function of the open loop is

$$g_{oz}(z) = \frac{f_{oz}(z)}{1+k_1 f_{oz}(z)} \tag{3.9.31}$$

If $g_{oz}(z)$ is stable, then Tsypkin's result can be applied to the Nyquist plot of $g_{oz}(e^{j\omega T})$. Practically no new plot of the Nyquist plot need be made. Rather the stability condition can be expressed in terms of $f_{oz}(e^{j\omega T})$. The condition (3.9.30) is here

$$\begin{aligned} &\mathrm{Re}\, g_{oz}(e^{j\omega T}) > \frac{-1}{k_2-k_1} \\ &\mathrm{Re}\, \frac{f_{oz}(e^{j\omega T})}{1+k_1 f_{oz}(e^{j\omega T})} > \frac{-1}{k_2-k_1} \end{aligned} \tag{3.9.32}$$

For $f_{oz}(e^{j\omega T}) = \mathrm{Ref} + j\mathrm{Imf}$, then

$$\mathrm{Re}\, \frac{(\mathrm{Ref}+j\mathrm{Imf})(1+k_1\mathrm{Ref}-jk_1\mathrm{Imf})}{(1+k_1\mathrm{Ref})^2+k_1^2\mathrm{Imf}^2} > \frac{-1}{k_2-k_1}$$

$$\frac{\mathrm{Ref}(1+k_1\mathrm{Ref})+k_1\mathrm{Imf}^2}{(1+k_1\mathrm{Ref})^2+k_1^2\mathrm{Imf}^2} > \frac{-1}{k_2-k_1}, \quad k_2-k_1 > 0$$

$$(k_2-k_1)\mathrm{Ref}(1+k_1\mathrm{Ref}) + k_1(k_2-k_1)\mathrm{Imf}^2 > -[1+2k_1\mathrm{Ref}+k_1^2\mathrm{Ref}^2+k_1^2\mathrm{Imf}^2]$$

$$(k_2-k_1)\mathrm{Ref} + k_1k_2(\mathrm{Ref}^2+\mathrm{Imf}^2) > -[1+2k_1\mathrm{Ref}]$$

$$(k_2+k_1)\mathrm{Ref} + k_1k_2(\mathrm{Ref}^2+\mathrm{Imf}^2) > -1$$

For $k_1 > 0$, i.e. $k_1k_2 > 0$ the inequality can be divided by k_1k_2:

$$\frac{k_2+k_1}{k_1k_2}\,\mathrm{Ref} + \mathrm{Ref}^2 + \mathrm{Imf}^2 > -\frac{1}{k_1k_2}$$

Completing the square in Ref leads to

$$\left(\frac{k_2+k_1}{2k_1k_2} + \mathrm{Ref}\right)^2 + \mathrm{Imf}^2 > \frac{-1}{k_1k_2} + \left(\frac{k_2+k_1}{2k_2k_1}\right)^2 \qquad (3.9.33)$$

$$\left(\frac{k_2+k_1}{2k_1k_2}\right) + \mathrm{Ref}^{\;2} + \mathrm{Imf}^2 > \left(\frac{k_2-k_1}{2k_1k_2}\right)^2$$

The boundary of the region determined by the inequality is a circle with center $-(k_2+k_1)/2k_1k_2$ and radius $(k_2-k_1)/2k_1k_2$, i.e. a circle with a real center and intersection points with the real axis at $-1/k_1$ and $-1/k_2$. The inequality is satisfied if the Nyquist plot $f_{oz}(e^{j\omega T})$ avoids this disc, figure 3.30 (circle criterion).

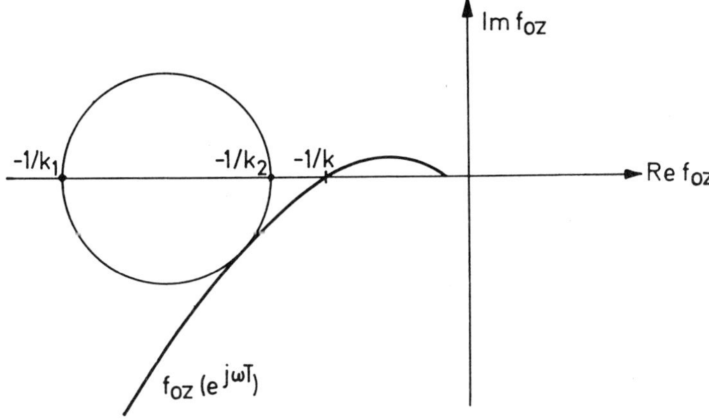

Figure 3.30 Sufficient condition for the absolute stability sector (k_1,k_2)

The assumption that $g_{oz}(z)$ in eq. (3.9.31) is stable can be checked in the same figure by the Nyquist criterion. If $f_{oz}(z)$ does not have a pole outside the unit circle, then the point $-1/k_1$ may not be encircled by $f_{oz}(e^{j\omega T})$. Therefore in this case, as in figure 3.30, the Nyquist plot must not encircle the disc. For unstable open loops the encirclement condition (3.9.21) must be satisfied not only with respect to $f_{oz}(z) = -1$ but also with respect to the disc around it.

Figure 3.30 shows that different sectors (k_1,k_2) can be chosen. The closer k_2 comes to the stability boundary $\bar{k}$ of the linear system, the smaller the circle and thereby the sector (k_1,k_2) becomes.

An extended criterion for nonlinearities with bounded slope $|d\phi(e)/de| < K'$ was derived by Jury and Lee [64.12].

Remark 3.8:

> Describing functions for nonlinearities in discrete-time systems were studied in [66.4]. These extend from $-1/k_1$ to $-1/k_2$, but not as far in the imaginary direction, as the circle in figure 3.30. The application of the describing function is more tedious than the circle criterion. Also it does not lead to sufficient stability conditions but predicts approximately the frequency and amplitude of a possible limit cycle.

3.9.4 Other Graphical Frequency Response Methods

It should be mentioned that alternative graphical representations for the condition

$$1+f_{oz}(z) = 1+d_z(z)h_z(z) = 0 \tag{3.9.34}$$

exist in the literature. These will only be briefly discussed here.

a) Eq. (3.9.34) can be written as

$$h_z(z) = -\frac{1}{d_z(z)} \tag{3.9.35}$$

The negative inverse frequency response plot of an assumed controller $-1/d_z(e^{j\omega T})$ is plotted together with the Nyquist plot $h_z(e^{j\omega T})$ of the plant and the figure is then examined to see, for which value of ωT both curves have the same phase angle. The amplitude margin is calculated from the absolute value ratio for this ωT.

b) The inversion of (3.9.35) is

$$d_z(z) = -\frac{1}{h_z(z)} \tag{3.9.36}$$

Here the negative-inverse Nyquist plot of the plant is combined with the controller frequency response plot.

c) If represented in logarithms, eq. (3.9.34) is

$$\ell n f_{oz}(z) = \ell n d_z(z)+\ell n h_z(z) = j(2k+1)\pi,\ k = 0,\pm 1,\pm 2... \tag{3.9.37}$$

The advantage here is that

$$h_z(e^{j\omega T}) = H(\omega T)e^{j\alpha(\omega T)} \quad \text{and}$$

$$d_z(e^{j\omega T}) = D(\omega T)e^{j\beta(\omega T)} \tag{3.9.38}$$

can easily be multiplied:

$$\ell n f_{oz}(e^{j\omega T}) = \ell n d_z(e^{j\omega T})+\ell n h_z(e^{j\omega T})$$

$$= [\ell n D(\omega T)+\ell n H(\omega T)]+j[\beta(\omega T)+\alpha(\omega T)] \tag{3.9.39}$$

The frequency response can be represented in the Nichols-diagram [47.1]. Here the magnitude and phase of the closed loop discrete frequency response can be concluded from the corresponding values for the open loop. The stability margin can also be seen from this logarithmic curve. If ρ' is the smallest distance of the critical point $j\pi$ (respectively $j3\pi$ etc. for systems with very large phase angles) from the logarithmic frequency response curve and ω_A and ω_B are the neighboring frequency marks on this curve, then corresponding to eq. (3.9.26), one approximately obtains the distance of a dominant pole pair from the unit circle as

$$\Delta r = \frac{\omega_B T - \omega_A T}{\ln|f_{oz}(e^{j\omega_B T})| - \ln|f_{oz}(e^{j\omega_A T})|} \times \rho' \qquad (3.9.40)$$

Since ρ' is not the exact image of ρ from figure 3.24, small differences result for the approximations of eqs. (3.9.26) and (3.9.40). It is important that the absolute value and phase are measured in the same units. An absolute value of $e = 2.72$ (= 1 neper) must be plotted in the same magnitude as an angle of 1 radian = 57.3^o.

If decibels are used as an absolute value measure (1db = 8.68 neper) magnitudes of 90° and 13.6 db are equivalent. Unfortunately this scale ratio is often not observed and the resulting image is not conformal.

d) If the logarithmic representation is plotted separately for absolute value $\ln D(\omega T) + \ln H(\omega T)$ and phase $\beta(\omega T) + \alpha(\omega T)$ over the frequency, then the Bode diagram [1945] is obtained. For rational transfer functions in factorized form, this provides simple asymptotic approximations. In order to use this, the unit circle is mapped by the bilinear transformation

$$q = \frac{2}{T}\frac{z-1}{z+1}, \quad z = \frac{2+Tq}{2-Tq} \qquad (3.9.41)$$

on the imaginary axis of the q-plane. The factor 2/T is often omitted in the literature, but only with this factor $q \approx s$ for small s, see eq. (3.4.33). For $z = e^{j\omega T}$ one obtains the corresponding q on the imaginary axis

$$q = \frac{2}{T}\frac{e^{j\omega T}}{e^{j\omega T}+1} = j\frac{2}{T}\tan\frac{\omega T}{2} \qquad (3.9.42)$$

The frequency ω is associated with the transformed frequency

$$\Omega = \frac{2}{T}\tan\frac{\omega T}{2} \qquad (3.9.43)$$

which for small ω yields

$$\Omega \approx \omega \tag{3.9.44}$$

With these modifications design with Bode diagrams can be performed as for continuous systems [77.1], [78.2], [80.1].

3.10 Special Sampling Problems

So far only systems with one sampler or synchronous samplers have been treated. Kalman and Bertram [59.2] have shown that the state representation is also applicable to other sampling problems. Some typical cases are given below.

It has been shown in eq. (3.7.7) that the variables $\underline{u}(t)$ stored in the hold elements must be introduced as state variables in order to describe the state between the sampling points. This is especially necessary for the description of systems with several non-synchronized samplers. The extended state vector is

$$\underline{z}(t) = \begin{bmatrix} \underline{x}(t) \\ \underline{u}(t) \end{bmatrix} \tag{3.10.1}$$

The time instants at which any sampler closes, are denoted by t_k, $k = 0,1,2...$, with $t_{k+1} > t_k$. The first difference of this sequence gives the intervals $t_{k+1}-t_k = T_k$. One now distinguishes two types of transitions:

a) the sampling transition, during which $\underline{x}$ remains constant but one or more hold element values in $\underline{u}(t)$ are reset

$$\underline{z}(t_k^+) = \begin{bmatrix} \underline{I} & 0 \\ 0 & 0 \end{bmatrix} \underline{z}(t_k^-) + \begin{bmatrix} 0 \\ \underline{I} \end{bmatrix} \underline{u}(k) \tag{3.10.2}$$

and

b) the interval transition, for which all hold element values $\underline{u}(t)$ remain constant:

$$\underline{z}(t_{k+1}^-) = \begin{bmatrix} \underline{A}(T_k) & \underline{b}(T_k) \\ \underline{0} & \underline{I} \end{bmatrix} \underline{z}(t_k^+) \qquad (3.10.3)$$

$z(t_k^+)$ denotes the right-sided and $z(t_k^-)$ the left-sided boundary value of $z(t)$ at the instants of discontinuity $t = t_k$.

The two forms of transitions alternate with each other and the two types of transition matrices (3.10.2) and (3.10.3) are multiplied correspondingly. One chooses a main interval for which the entire process is periodic. In practice, the beginning of the main interval is fixed so that as many hold elements as possible are set to new values at this time instant. The corresponding state variables can finally be eliminated from the state-description for the main interval. It is necessary, however, to carry these state variables along in order to allow a systematic bookkeeping for all transitions in the main interval.

Example: Nonsynchronous sampling

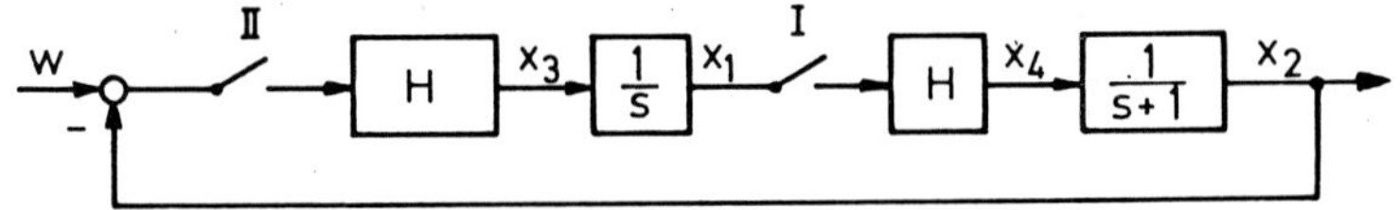

Figure 3.31 Control loop with two nonsynchronized samplers. Sampler I samples at $t = kT$, sampler II at $t = kT+\tau$, $k = 0,1,2...$

$\dot{x}_1 = x_3$

$\dot{x}_2 = -x_2 + x_4$

$\dot{x}_3 = \dot{x}_4 = 0$ in the interval transition

$$\underline{\dot{x}} = \begin{bmatrix} 0 & 0 & 1 & 0 \\ 0 & -1 & 0 & 1 \\ 0 & 0 & 0 & 0 \\ 0 & 0 & 0 & 0 \end{bmatrix} \underline{x} = \underline{Fx}$$

$$\underline{\Phi}(t) = e^{\underline{F}t} = \begin{bmatrix} 1 & 0 & t & 0 \\ 0 & e^{-t} & 0 & 1-e^{-t} \\ 0 & 0 & 1 & 0 \\ 0 & 0 & 0 & 1 \end{bmatrix}$$

At the time instant $t = kT$ the first sampling transition $x_4(kT^+) = x_1(kT^-)$ i.e. $\underline{x}(kT^+) = \Phi_I\underline{x}(kT^-)$ occurs, where

$$\underline{\Phi}_I = \begin{bmatrix} 1 & 0 & 0 & 0 \\ 0 & 1 & 0 & 0 \\ 0 & 0 & 1 & 0 \\ 1 & 0 & 0 & 0 \end{bmatrix}$$

The second sampling transition at the time $t = kT+\tau$ is

$x_3(kT+\tau^+) = -x_2(kT+\tau^-) + w(kT+\tau)$ i.e.
$\underline{x}(kT+\tau^+) = \Phi_{II}\underline{x}(kT+\tau^-) + \underline{b}w(kT+\tau)$ with

$$\underline{\Phi}_{II} = \begin{bmatrix} 1 & 0 & 0 & 0 \\ 0 & 1 & 0 & 0 \\ 0 & -1 & 0 & 0 \\ 0 & 0 & 0 & 1 \end{bmatrix}, \quad \underline{b} = \begin{bmatrix} 0 \\ 0 \\ 1 \\ 0 \end{bmatrix}$$

By multiplication of the four transitions the difference equation for the instants $t = kT^+$ is obtained as follows

$$\begin{aligned} \underline{x}(kT+T^+) &= \underline{\Phi}_I\underline{x}(kT+T^-) \\ &= \underline{\Phi}_I\underline{\Phi}(T-\tau)\underline{x}(kT+\tau^+) \\ &= \underline{\Phi}_I\underline{\Phi}(T-\tau)[\underline{\Phi}_{II}\underline{x}(kT+\tau^-) + \underline{b}w(kT+\tau)] \\ \underline{x}(kT+T^+) &= \underline{\Phi}_I\underline{\Phi}(T-\tau)[\underline{\Phi}_{II}\underline{\Phi}(\tau)\underline{x}(kT^+) + \underline{b}w(kT+\tau)] \end{aligned} \tag{3.10.4}$$

$$\underline{x}(kT+T^+) = \begin{bmatrix} 1 & (\tau-T)e^{-\tau} & \tau & (\tau-T)(1-e^{-\tau}) \\ 0 & e^{-T} & 0 & 1-e^{-T} \\ 0 & -e^{-\tau} & 0 & -1+e^{-\tau} \\ 1 & (\tau-T)e^{-\tau} & \tau & (\tau-T)(1-e^{-\tau}) \end{bmatrix} \underline{x}(kT^+) + \begin{bmatrix} T-\tau \\ 0 \\ 1 \\ T-\tau \end{bmatrix} w(kT+\tau) \tag{3.10.5}$$

At the observed time instants $x_4(kT^+) = x_1(kT^+)$; one can therefore eliminate x_4 from eq. (3.10.5). The reduced state vector is

$$\underline{x}^* = \begin{bmatrix} x_1 \\ x_2 \\ x_3 \end{bmatrix}$$

$$\underline{x}^*(kT+T^+) = \begin{bmatrix} 1+(\tau-T)(1-e^{-\tau}) & (\tau-T)e^{-\tau} & \tau \\ 1-e^{-T} & e^{-T} & 0 \\ -1+e^{-\tau} & -e^{-\tau} & 0 \end{bmatrix} \underline{x}^*(kT^+) + \begin{bmatrix} T-\tau \\ 0 \\ 1 \end{bmatrix} w(kT+\tau)$$

(3.10.6)

Example: Non-instantaneous sampling

A description by a difference equation is also possible for a periodic switching process in a control loop. The control loop of figure 3.32 will be taken as an example

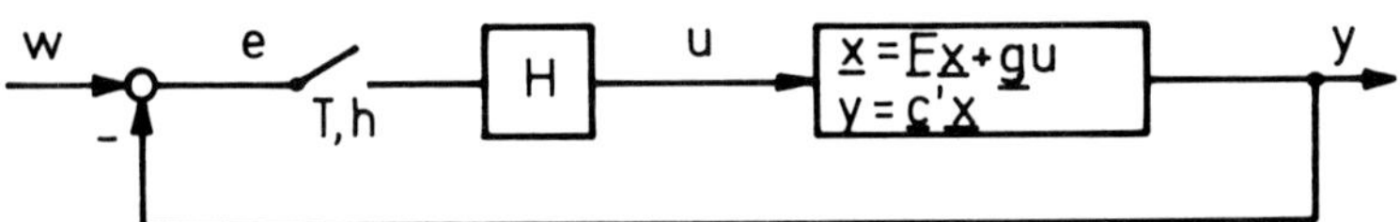

Figure 3.32 Control loop with periodically operating switch

The switch with hold element can for example be implemented as in figure 1.6 with h >> RC.

$$u(t) = \begin{cases} e(t) & kT < t < kT+h \\ e(kT+h) & kT+h < t < kT+T \end{cases} \qquad (3.10.7)$$

The switch is closed in the first interval $kT < t < kT+T$, therefore

$$\dot{\underline{x}} = (\underline{F}-\underline{g}\underline{c}')\underline{x} + \underline{g}w = \underline{F}^*\underline{x} + \underline{g}w \qquad (3.10.8)$$

For $\underline{\Phi}_1(\tau) = e^{\underline{F}^*\tau}$

$$\underline{x}(kT+h) = \underline{\Phi}_1(h)\underline{x}(kT) + \int_{kT}^{kT+h} \underline{\Phi}_1(kT+h-v)\underline{g}w(v)dv \qquad (3.10.9)$$

In the second interval $kT+h < t < kT+T$ the equation for the open loop is

$$\dot{\underline{x}} = \underline{F}\underline{x} + \underline{g}(w(kT+h) - \underline{c}'\underline{x}(kT+h))$$

For $\underline{\phi}_2(v) = e^{\underline{F}v}$ and $\underline{b}_2(v) = \int_0^v \underline{\phi}_2(\tau)\underline{g}d\tau$

$$\underline{x}(kT+T) = \underline{\phi}_2(T-h)\underline{x}(kT+h)+\underline{b}_2(T-h)[w(kT+h)-\underline{c}'\underline{x}(kT+h)]$$

$$= [\underline{\phi}_2(T-h)-\underline{b}_2(T-h)\underline{c}']\underline{x}(kT+h)+\underline{b}_2(T-h)w(kT+h) \qquad (3.10.10)$$

By substituting eq. (3.10.9)

$$\underline{x}(kT+T) = [\underline{\phi}_2(T-h)-\underline{b}_2(T-h)\underline{c}']\underline{\phi}_1(h)\underline{x}(kT)+\underline{b}_2(T-h)w(kT+h)+$$

$$+ [\underline{\phi}_2(T-h)-\underline{b}_2(T-h)\underline{c}'] \int_{kT}^{kT+h} \underline{\phi}_1(kT+h-v)\underline{g}w(v)dv \qquad (3.10.11)$$

Remark 3.9:

This procedure is also applicable for continuous parts with feedthrough from input to output, i.e. $y = \underline{c}'\underline{x}+du$, and a switch with $h \ll T$ as sampler. In this case an idealization of the sampler with hold element would lead to false results. If the physical sampling process is ideal, however, then an idealization of the continuous part as a system with direct input-output connection is no longer allowable. A very small time constant, which can be easily neglected for an unsampled system, here leads to an essential alteration of the solution. Remember, that it was assumed that an ideal sampler samples the right side value at an instant of discontinuity.

Example: Cyclic sampling for multivariable systems

In multivariable systems it occurs that the inputs $u_1(t)$, $u_2(t)\ldots u_r(t)$ are not sampled synchronously. Instead the variables are sampled cyclically one after another. For example, they could be produced with a single analog digital converter. Then

$$
\begin{aligned}
u_1(t) &= u_1(kT) && kT < t < kT+T \\
u_2(t) &= u_2(kT+T/r) && kT+T/r < t < kT+T+T/r \\
&\vdots \\
u_r(t) &= u_r(kT+T(r-1)/r) && kT+T(r-1)/r < t < kT+T+T(r-1)/r
\end{aligned}
\qquad (3.10.12)
$$

In order to record the values in the hold elements, the extended state vector $\underline{z}$ is introduced

$$
\underline{z} = \begin{bmatrix} \underline{x} \\ u_1 \\ \vdots \\ u_r \end{bmatrix} \qquad (3.10.13)
$$

Two types of transitions occur alternately:

1. Reset of the hold element to a new $\underline{u}$-value in the interval $(kT+qT/r)^- < t < (kT+qT/r)^+$ and

2. Dynamic transition in the interval $(kT+qT/r)^+ < t < (kT+(q+1)T/r)^-$.

One obtains a difference equation for the interval from kT^- to $kT+T^-$ by multiplication of the transitions. The variable u_1 can then be eliminated because $u_1(kT)$ is a current input variable and therefore not a part of the state at time kT.

The calculations are performed as follows for a system with two inputs.

$$
\underline{\dot{x}} = \underline{Fx} + \underline{Gu}
$$

$$
\underline{y} = \underline{Cx} = \begin{bmatrix} \underline{c}_1' \\ \underline{c}_2' \end{bmatrix} \underline{x} \qquad (3.10.14)
$$

The system has two input variables

$$
\begin{aligned}
u_1(t) &= e_1(kT) && kT < t < kT+T \\
u_2(t) &= e_2(kT+T/2) && kT+T/2 < t < kT+T+T/2
\end{aligned}
$$

and two output variables y_1 and y_2. Let $\underline{e} = \underline{w}-\underline{y}$, where $\underline{w}$ consists of the two reference variables of the control system. Figure 3.33 displays the system

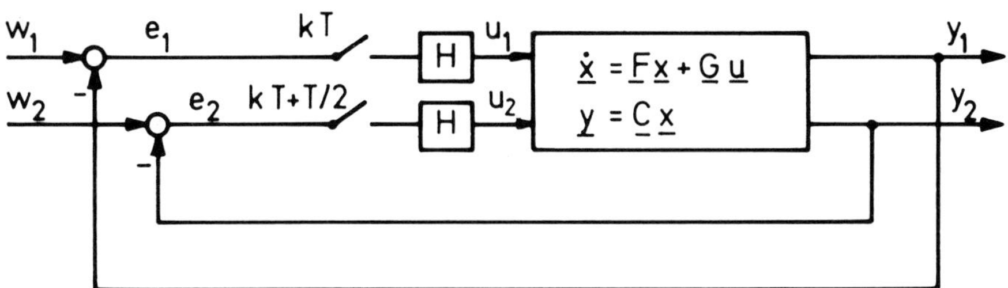

Figure 3.33 Two-loop control system with cyclic sampling of the inputs.

For constant $\underline{u}$ the solution of (3.10.14) is

$$\underline{x}(t) = e^{\underline{F}(t-t_o)}\underline{x}(t_o) + \int_{t_o}^{t-t_o} e^{\underline{F}\tau}d\tau\underline{Gu} \tag{3.10.15}$$

$\underline{u}$ is constant in each case for an interval of length T/2, therefore one calculates

$$\underline{A} = e^{\underline{F}T/2} \text{ and } \underline{B} = \int_0^{T/2} e^{\underline{F}\tau}d\tau\underline{G} = [\underline{b}_1 \quad \underline{b}_2] \tag{3.10.16}$$

The state vector is $z = [\underline{x}',u_1,u_2]'$.

Four different transitions arise:

1. Reset of the first hold element to $e_1(kT)$, i.e., $u_1(kT) = e_1(kT)$. $\underline{I}_n$ is the n×n identity matrix.

$$\underline{z}(kT^+) = \left[\begin{array}{c|cc} \underline{I}_n & \multicolumn{2}{c}{\underline{0}} \\ \hline \underline{0} & 0 & 0 \\ & 0 & 1 \end{array}\right] \underline{z}(kT^-) + \left[\begin{array}{cc} \multicolumn{2}{c}{\underline{0}} \\ \hline 1 & 0 \\ 0 & 0 \end{array}\right] \underline{e}(kT)$$

2. Dynamic transition in the interval $kT < t < kT+T/2$

$$\underline{z}(kT+T/2^-) = \left[\begin{array}{c|c} \underline{A} & \underline{B} \\ \hline \underline{0} & \underline{I}_2 \end{array}\right] \underline{z}(kT^+)$$

3. Reset of the second hold element to $e_2(kT+T/2)$, i.e. $u_2(kT+T/2) = e_2(kT+T/2)$

$$\underline{z}(kT+T/2^+) = \left[\begin{array}{c|cc} \underline{I}_n & \multicolumn{2}{c}{\underline{0}} \\ \hline \underline{0} & 1 & 0 \\ & 0 & 0 \end{array}\right] \underline{z}(kT+T/2^-) + \left[\begin{array}{cc} \multicolumn{2}{c}{\underline{0}} \\ \hline 0 & 0 \\ 0 & 1 \end{array}\right] \underline{e}(kT+T/2)$$

4. Dynamic transition in the interval $kT+T/2 < t < kT+T$

$$\underline{z}(kT+T^-) = \left[\begin{array}{c|c} \underline{A} & \underline{B} \\ \hline \underline{0} & \underline{I}_2 \end{array}\right] \underline{z}(kT+T/2^+)$$

Inserting the first, second, and third equation into the fourth one, yields

$$\underline{z}(kT+T^-) = \left[\begin{array}{c|cc} \underline{A}^2 & \underline{0} & \underline{Ab}_2 \\ \hline \underline{0} & \multicolumn{2}{c}{\underline{0}} \end{array}\right] \underline{z}(kT^-) + \left[\begin{array}{cc} \underline{Ab}_1+\underline{b}_1 & \underline{0} \\ \hline 1 & 0 \\ 0 & 0 \end{array}\right] \underline{e}(kT) + \left[\begin{array}{cc} \underline{0} & \underline{b}_2 \\ \hline 0 & 0 \\ 0 & 1 \end{array}\right] \underline{e}(kT+T/2)$$

The state variable u_1 can be eliminated from this equation because it is equal to $e_1(kT)$ at the examined time point $t = kT$. For the state vector $\underline{x}^* = [\underline{x}', u_2]'$ the difference equation is

$$\underline{x}^*(kT+T^-) = \left[\begin{array}{c|c} \underline{A}^2 & \underline{Ab}_2 \\ \hline \underline{0} & 0 \end{array}\right] \underline{x}^*(kT^-) + \left[\begin{array}{cc} (\underline{A}+\underline{I})\underline{b}_1 & \underline{0} \\ \hline 0 & 0 \end{array}\right] \underline{e}(kT) + \left[\begin{array}{cc} \underline{0} & \underline{b}_2 \\ \hline 0 & 1 \end{array}\right] \underline{e}(kT+T/2) \qquad (3.10.17)$$

The loop is closed by

$$\underline{e}(kT) = \underline{w}(kT) - [\underline{C} \mid 0]\underline{x}^*(kT^-)$$

$$\underline{e}(kT+T/2) = \underline{w}(kT+T/2) - [\underline{C} \mid 0]\underline{x}^*(kT+T/2^-)$$

$$= \underline{w}(kT+T/2) - [C \mid 0]\left\{\left[\begin{array}{c|c} \underline{A} & \underline{b}_2 \\ \hline \underline{0} & 1 \end{array}\right] \underline{x}^*(kT^-) + \left[\begin{array}{c|c} \underline{b}_1 & \underline{0} \\ \hline 0 & 0 \end{array}\right] \underline{e}(kT)\right\}$$

$$= \underline{w}(kT+T/2) - [\underline{CA} \mid \underline{Cb}_2]\underline{x}^*(kT^-) -$$

$$- [\underline{Cb}_1 \mid 0]\{\underline{w}(kT) - [\underline{C} \mid 0]\underline{x}^*(kT^-)\}$$

$$= \underline{w}(kT+T/2) - \underline{Cb}_1 w_1(kT) - [\underline{C}(\underline{A}-\underline{b}_1\underline{c}_1') \mid Cb_2]\underline{x}^*(kT^-)$$

Then from eq. (3.10.17)

$$\underline{x}^*(kT+T^-) = \left[\begin{array}{c|c} \underline{A}^2-(\underline{A}+\underline{I})\underline{b}_1\underline{c}_1'-\underline{b}_2\underline{c}_2'(\underline{A}-\underline{b}_1\underline{c}_1') & (\underline{A}-\underline{b}_2\underline{c}_2')\underline{b}_2 \\ \hline -\underline{c}_2'(\underline{A}-\underline{b}_1\underline{c}_1') & -\underline{c}_2'\underline{b}_2 \end{array}\right]\underline{x}^*(kT^-) +$$

$$+ \left[\begin{array}{c|c} (\underline{A}+\underline{I}-\underline{b}_2\underline{c}_2')\underline{b}_1 & \underline{b}_2 \\ \hline -\underline{c}_2'\underline{b}_1 & 1 \end{array}\right]\left[\begin{array}{c} w_1(kT) \\ w_2(kT+T/2) \end{array}\right] \tag{3.10.18}$$

With the exception of the fact that w_1 and w_2 are sampled at different time instants, the system is again in the basic form of the state representation.

Example: Cascade control with different sampling frequencies

Let the control loop of figure 3.8 and eq. (3.4.38) be part of an outer control loop which operates with a sampling period $T_N = NT$, i.e. an integer multiple of the basic sampling rate. In eq. (3.4.38) let

$$\underline{x}^* = \begin{bmatrix} \underline{x} \\ \underline{r} \end{bmatrix}, \quad \underline{A}^* = \begin{bmatrix} \underline{A} - d_R\underline{bc}' & \underline{bc}_R' \\ -\underline{b}_R\underline{c}' & \underline{A}_R \end{bmatrix}, \quad \underline{b}^* = \begin{bmatrix} \underline{b}d_R \\ \underline{b}_R \end{bmatrix}$$

$$\underline{x}^*(kT+T) = \underline{A}^*\underline{x}^*(kT) + \underline{b}^*w(kT) \tag{3.10.19}$$

Because w(kT) was produced by a sampling interval NT,

$$w(mT) = w(mT+T) = \ldots w(mT+NT-T), \quad m = iN, \quad i = 1,2,3\ldots \tag{3.10.20}$$

$$\begin{aligned} \underline{x}^*(iNT+T) &= \underline{A}^*\underline{x}^*(iNT) + \underline{b}^*w(iNT) \\ \underline{x}^*(iNT+2T) &= \underline{A}^{*2}\underline{x}^*(iNT) + (\underline{A}^*\underline{b}^*+\underline{b}^*)w(iNT) \\ &\vdots \\ \underline{x}^*(iNT+NT) &= \underline{A}^{*N}\underline{x}^*(iNT) + (\underline{A}^{*N-1}\underline{b}^*+\underline{A}^{*N-2}\underline{b}^*+\ldots+\underline{b}^*)w(iNT) \end{aligned} \tag{3.10.21}$$

and for $T_N = NT$

$$\underline{x}^*(iT_N+T_N) = \underline{A}_N\underline{x}^*(iT_N) + \underline{B}_N w(iT_N) \tag{3.10.22}$$

This difference equation can be incorporated into the description of the outer control loop with sampling period T_N.

In this section only the time-domain analysis of multirate sampled-data systems was treated. A frequency-domain approach can be found in [67.5]. The extension of optimal regulator and Kalman filtering techniques to the multirate case and some aerospace applications are discussed in [83.6].

3.11 Excercises

3.1 Determine the discrete state representation of the loading bridge, eq. (2.1.15). Hint: Simplify using the results of exercise 2.1.

3.2 Determine the sampling period T for the loading bridge such that the complex eigenvalues of the discrete system lie at an angle $\omega T = 45^o$ with respect to the positive real axis. How often per period of the pendulum will the system be sampled? Calculate and sketch the form of the solution for $\underline{x}(0) = \underline{0}$ and a unit step $u(kT) = 1(kT)$ as input.

3.3 Calculate the z-transfer function $h_z(z)$ of the loading bridge for input u and output x_1 = position of the trolley

a) from the state representation, see exercise 3.1,

b) from the s-transfer function (2.3.12).

3.4 Calculate the z-transfer function of the following system

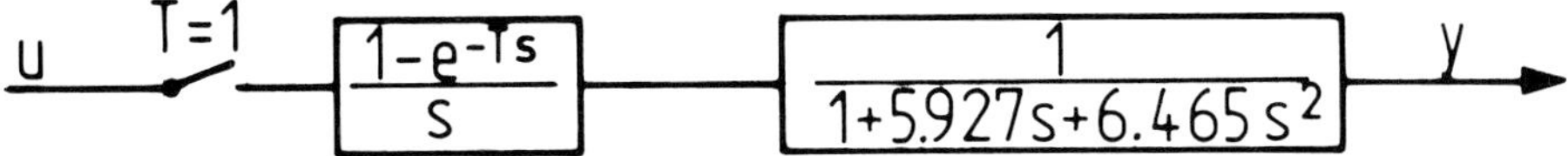

Figure 3.34 Calculation of the z-transfer function from s-transfer function

3.5 For the plant of figure 3.34, determine a controller which places all eigenvalues at $z = 0$ and makes the stationary error zero in the step response.

3.6 Plot the root locus curve for the loading bridge with the output x_1 = trolley position and the parameter values $m_C = 1000$ kg, $m_L = 3000$ kg, $\ell = 10$ m, $g = 10$ m/sec^2, $T = \pi/8$ seconds.

a) for $k_y h_z(z) = -1$. Is there a k_y for which the closed control loop is stable?

b) for a controller $k_y \times \frac{z-0.9}{z+0.5}$. From this choose a suitable value $k_y = k_1$.

3.7 Calculate the step response at the sampling instants of the control loop of exercise 3.6, for $e_z = w_z - x_{1z}$, $u_z = e_z \times k_1 (z-0.9)/(z+0.5)$ with a) $k_1 = 5000$, b) $k_1 = 7000$.

3.8 Calculate the step response for exercise 3.7 also at the time instants $kT+\gamma T$, $\gamma - 1/3$ and $\gamma = 2/3$.

3.9 Let the plant from figure 3.34 contain an additional time-delay of $\theta = 1.2$ seconds. The loop is closed by $u = 0.7(w-y)$. For $w(t) = 1(t)$ determine the step response $y(kT)$ and its steady-state value (for the final value theorem of the z-transform see appendix B).

3.10 Examine the same control system as in exercise 3.9, however without sampler and hold element. Approximately determine the step response by the introduction of a sampler and a linear interpolator in the feedback. Begin with the sampling period T = 0.6 sec and check the accuracy of the solution by repeating it for T = 0.3 sec.

3.11 Plot the Nyquist diagram for the control loop of exercise 3.6b and determine the stability margin ρ.

4 Controllability, Choice of Sampling Period and Pole Assignment

In this chapter the properties of controllability and reachability will be introduced for discrete systems. Also their relation with the corresponding properties of the continuous subsystem will be presented.

Considering actuator constraints there are regions in state space from which an initial state can be transferred to the zero state within a given time interval. The size of this region serves as a measure for the choice of the sampling period. As in the continuous case, all controllable eigenvalues can be arbitrarily shifted by a state vector feedback. An interesting special case for discrete systems is the pole assignment at $z = 0$, which leads to finite transients (deadbeat response).

4.1 Controllability and Reachability

A state $\underline{x}(t_1)$ of a linear system is "reachable" if there exists a finite time instant $t_0 < t_1$ and an input $u(t)$ for $t_0 < t < t_1$ such that the system is transferred from the zero state $\underline{x}(t_0) = \underline{0}$ to the desired state $\underline{x}(t_1)$ [69.1]. A state $x(t_1)$ of a linear system is "controllable" if there exists a finite time instant $t_2 > t_1$ and an input $u(t)$ for $t_1 < t < t_2$ such that $\underline{x}(t_1)$ is transferred to the zero state $\underline{x}(t_2) = \underline{0}$. "Complete reachability (or complete controllability) of the system at time t_1" means that every state $x(t_1)$ is reachable (or controllable). For the time invariant systems examined here, the properties of controllability and reachability are independent of the choice of t_1. If the system is moreover continuous, the two properties are equivalent, i.e. a state is controllable if and only if it is reachable. The two properties are therefore often not distinguished in the control engineering literature, rather one speaks only of controllability. If a time invariant continuous system is controllable every initial state $\underline{x}(t_o)$ can be transferred to every desired final state by a suitable input signal in finite time. For a discrete system

$$\underline{x}[k+1] = \underline{A}\underline{x}[k] + \underline{b}u[k] \tag{4.1.1}$$

the possibility of satisfying the two requirements can be seen in the solution of the difference equation. That is

$$\begin{aligned}
\underline{x}[k+1] &= \underline{A}\underline{x}[k] + \underline{b}u[k] \\
\underline{x}[k+2] &= \underline{A}^2\underline{x}[k] + \underline{A}\underline{b}u[k] + \underline{b}u[k+1] \\
&\vdots \\
\underline{x}[k+N] &= \underline{A}^N\underline{x}[k] + \underline{A}^{N-1}\underline{b}u[k] + \ldots + \underline{A}\underline{b}u[k+N-2] + \underline{b}u[k+N-1]
\end{aligned}$$

$$\underline{x}[k+N] = \underline{A}^N\underline{x}[k] + [\underline{A}^{N-1}\underline{b},\ \underline{A}^{N-2}\underline{b}\ \ldots\ \underline{b}]\underline{u}_{N-1}[k] \tag{4.1.2}$$

Here N consecutive values of the control sequence u are combined into a vector

$$\underline{u}_{N-1}[k] := \begin{bmatrix} u[k] \\ u[k+1] \\ \vdots \\ u[k+N-1] \end{bmatrix} \tag{4.1.3}$$

The system can be transferred from any given initial state $\underline{x}[k]$ to every desired final state $\underline{x}[k+N] = \underline{x}_E$ if

$$\text{rank } [\underline{A}^{N-1}\underline{b},\ \underline{A}^{N-2}\underline{b}\ldots\underline{b}] = n \tag{4.1.4}$$

where n is the order of the system. Eq. (4.1.4) can obviously not be satisfied for $N < n$. It is now asserted that eq. (4.1.4) is true for $N = n$ if it can be satisfied for some $N \geq n$. For the proof, assume that a vector $\underline{A}^i\underline{b}$ is linearly dependent upon the vectors $\underline{b}$, $\underline{A}\underline{b}\ldots\underline{A}^{i-1}\underline{b}$, i.e.

$$\underline{A}^i\underline{b} = \sum_{j=0}^{i-1} c_j\underline{A}^j\underline{b} \tag{4.1.5}$$

By multiplication with $\underline{A}$ on the left, it follows for the next vector that

$$\underline{A}^{i+1}\underline{b} = \sum_{j=0}^{i-1} c_j\underline{A}^{j+1}\underline{b} = \sum_{j=0}^{i-2} c_j\underline{A}^{j+1}\underline{b} + c_{i-1}\sum_{j=0}^{i-1} c_j\underline{A}^j\underline{b} \qquad (4.1.6)$$

$\underline{A}^{i+1}$ is thus linearly dependent upon $\underline{b}$, $\underline{Ab}\ldots\underline{A}^{i-1}\underline{b}$. The same holds for all further vectors $\underline{A}^k\underline{b}$, $k > i$. The sequence of vectors $\underline{b}$, $\underline{AB}$, $\underline{A}^2\underline{b}\ldots$, which is also called the "orbit" of the vector $\underline{b}$ with respect to the matrix $\underline{A}$, is so constructed that once a linearly dependent vector has arisen all further vectors will also be linearly dependent.

Therefore if a transition from the state $\underline{x}[k]$ to the state $\underline{x}_E$ is possible at all, then it is also possible in n sampling intervals and leads to $\underline{x}[k+n] = x_E$. This condition is satisfied for all states if and only if

$$\det[\underline{b}, \underline{Ab}\ldots\underline{A}^{n-1}\underline{b}] \neq 0 \qquad (4.1.7)$$

As for the question of reachability ($\underline{x}[k] = \underline{0}$ and $\underline{x}[k+n] = \underline{x}_E$ for arbitrary $\underline{x}_E$) eq. (4.1.2) can be solved for the desired control sequence

$$\underline{u}_{n-1}[k] = [\underline{A}^{n-1}\underline{b}, \underline{A}^{n-2}\underline{b}\ldots\underline{b}]^{-1}\underline{x}_E \qquad (4.1.8)$$

if and only if condition (4.1.7) is satisfied. Eq. (4.1.7) represents a necessary and sufficient condition for the reachability of the system.

The reachability test for the pair ($\underline{A}$, $\underline{b}$) in eq. (4.1.7) completely corresponds to the controllability test for the pair ($\underline{F}$, $\underline{g}$) in eq. (2.3.17). Thus also the Hautus criterion, eq. (2.3.25), can be applied to the pair ($\underline{A}$, $\underline{b}$) [72.2], i.e. an eigenvalue z_i of $\underline{A}$ is reachable if and only if

$$\text{rank}\,[\underline{A}-z_i\underline{I}, \underline{b}] = n \qquad (4.1.9)$$

For the controllability problem $\underline{x}[k+n] = \underline{0}$ and eq. (4.1.2) is

$$\underline{A}^n\underline{x}[k] = -[\underline{A}^{n-1}\underline{b}, \underline{A}^{n-2}\underline{b}\ldots\underline{b}]\underline{u}_{n-1}[k] \qquad (4.1.10)$$

and the control sequence is

$$\underline{u}_{n-1}[k] = -[\underline{A}^{n-1}\underline{b}, \underline{A}^{n-2}\underline{b}\ldots\underline{b}]^{-1}\underline{A}^n\underline{x}[k] \qquad (4.1.11)$$

The controllability therefore always follows from the reachability condition (4.1.7). However, the converse only holds if all zero eigenvalues of $\underline{A}$ are reachable, i.e. if

$$\text{rank}\,[\underline{A},\ \underline{b}] = n \tag{4.1.12}$$

This is illustrated by the following two examples.

1. $$\underline{A} = \begin{bmatrix} 0 & 1 \\ 0 & 0 \end{bmatrix}, \quad \underline{b} = \begin{bmatrix} 0 \\ 1 \end{bmatrix}, \quad \text{rank}\,[\underline{A},\ \underline{b}] = 2 \tag{4.1.13}$$

$$[\underline{b},\ \underline{Ab}] = \begin{bmatrix} 0 & 1 \\ 1 & 0 \end{bmatrix}, \quad \text{reachable.}$$

$\underline{x}[k+2] = 0$ for $u[k] \equiv 0$, controllable.

2. $$\underline{A} = \begin{bmatrix} 0 & 1 \\ 0 & 0 \end{bmatrix}, \quad \underline{b} = \begin{bmatrix} 1 \\ 0 \end{bmatrix}, \quad \text{rank}\,[\underline{A},\ \underline{b}] = 1 \tag{4.1.14}$$

$$[\underline{b},\ \underline{Ab}] = \begin{bmatrix} 1 & 0 \\ 0 & 0 \end{bmatrix}, \quad \text{not reachable.}$$

$\underline{x}[k+2] = 0$ for $u[k] \equiv 0$, controllable.

In practice, the interesting cases are either $\underline{A} = e^{\underline{F}T}$ which implies $\det\underline{A} \neq 0$, or the zero eigenvalues, resulting from time-delays are reachable. Thus we will not stress the subtle difference between controllability and reachability and - as is usual in the control engineering literature - only speak of controllability. In this sense eqs. (4.1.7) and (4.1.9) are also designated as controllability conditions.

In the frequent case of $\det\underline{A} \neq 0$, eq. (4.1.10) becomes

$$\underline{x}[k] = -[\underline{A}^{-1}\underline{b},\ \underline{A}^{-2}\underline{b}\ldots\underline{A}^{-n}\underline{b}]\underline{u}_{n-1}[k] \tag{4.1.15}$$

It is obvious here that the states in the subspace spanned by the vectors $\underline{A}^{-1}\underline{b},\ \underline{A}^{-2}\underline{b}\ldots\underline{A}^{-n}\underline{b}$ can be transferred to the zero state in finite time. For controllable systems this is the entire state space.

For continuous systems the corresponding controllability conditions hold with $\underline{A}$ and $\underline{b}$ replaced by $\underline{F}$ and $\underline{g}$. This brings up the question whether the property of controllability changes, if a sampler and hold is connected to the input of a continuous system. This problem was examined by Jury [57.3]. Kalman [60.4] has formulated the following connection with controllability:

The discrete system is controllable if and only if

1. the continuous part is controllable and

2. $\exp s_i T \neq \exp s_j T$ for $s_i \neq s_j$ (4.1.16)

i.e. if distinct eigenvalues s_i, s_j of the continuous system lead to distinct eigenvalues $\exp s_i T$, $\exp s_j T$ of the discrete system.

The second condition is violated if the continuous system has complex eigenvalues $s_1 = \sigma + j\omega_1$ and $s_2 = \sigma + j\omega_2$ with the same real part and the sampling period is chosen such that $(\omega_1 - \omega_2)T = \pm q2\pi$, $q = 1,2,3...$ For a complex conjugate pair $s_{1,2} = \sigma \pm j\omega$ this is true for $T = +q\pi/\omega$. In the neighborhood of this sampling period the system is hard to control.

An explanation for this phenomenon is that the eigenvalues $\exp s_i T$ and $\exp s_j T$ cannot belong to the same Jordan block of the $\underline{A}$ matrix for $s_i \neq s_j$. If these eigenvalues of $\underline{A}$ are equal, then $\underline{A}$ is not cyclic i.e. not controllable by one input, see section 2.3.4.

Example: Loading bridge

The loading bridge is controllable by continuous inputs according to eq. (2.3.18). For a staircase input this property can be lost. For which load is this the case? The eigenvalues are $s_{1,2} = 0$ and $s_{3,4} = \pm j\omega_L$, $\omega_L^2 = g(m_L + m_C)/\ell m_C$, see eq. (2.3.2). The controllability condition (4.1.16) is violated for

$$T = q\pi/\omega_L \,, \quad q = 1,2,3 \ldots$$

$$\omega_L^2 = \left(\frac{q\pi}{T}\right)^2 = \frac{g(m_L+m_C)}{\ell m_C} \tag{4.1.17}$$

$$m_L = m_C \left[\frac{\ell}{g}\left(\frac{q\pi}{T}\right)^2 - 1\right]$$

In practice one always chooses $T < \pi/\omega_{Lmax}$, where ω_{Lmax} corresponds to the maximal load m_{Lmax}. Then controllability cannot be lost for any admissible load. The choice of T will be discussed in section 4.3.

4.2 Controllability Regions for Constrained Inputs

In many control systems the input amplitudes are constrained by the power of the actuator, for example in the form

$$|u[k]| \leq 1 \quad \text{for all } k \tag{4.2.1}$$

We now examine the problem of controllability in N steps for a nonsingular $\underline{A}$-matrix. For $k = 0$, $\underline{x}[k+N] = 0$, eq. (4.1.2) produces

$$\underline{x}[0] = -[\underline{A}^{-1}\underline{b},\ \underline{A}^{-2}\underline{b}\ldots\underline{A}^{-N}\underline{b}]\underline{u}_{N-1}[0] \tag{4.2.2}$$

The constraint (4.2.1) represents a hypercube in the space of the control sequences $\underline{u}_{N-1}$, eq. (4.1.3). It is mapped by eq. (4.2.2) into the state space and generates there the "controllability region for N sampling steps", i.e. the set of all initial states $\underline{x}[0]$ which can be transferred to the zero state in N steps with a constrained input.

Example: Second order system with complex eigenvalues

$$\dot{\underline{z}} = \underline{Fz} + \underline{g}u = \begin{bmatrix} \sigma & -\omega \\ \omega & \sigma \end{bmatrix} \underline{z} + \begin{bmatrix} g_1 \\ g_2 \end{bmatrix} u \tag{4.2.3}$$

For simplification this system is first transformed by

$$\underline{x} = \underline{T}^{-1}\underline{z} = \frac{1}{g_1^2+g_2^2}\begin{bmatrix} g_1 & g_2 \\ -g_2 & g_1 \end{bmatrix}\underline{z} \tag{4.2.4}$$

to the form

$$\dot{\underline{x}} = \underline{T}^{-1}\underline{FTx} + \underline{T}^{-1}\underline{g}u = \begin{bmatrix} \sigma & -\omega \\ \omega & \sigma \end{bmatrix}\underline{x} + \begin{bmatrix} 1 \\ 0 \end{bmatrix} u \tag{4.2.5}$$

The transition to discrete time by eq. (3.1.6) yields

$$\underline{x}[k+1] = \underline{A}(T)\underline{x}[k] + \underline{b}(T)u[k]$$

$$\underline{A}(T) = e^{\underline{F}T} = e^{\sigma T}\begin{bmatrix} \cos\omega T & -\sin\omega T \\ \sin\omega T & \cos\omega T \end{bmatrix}$$

$$\underline{b}(T) = \frac{1}{\sigma^2+\omega^2}\begin{bmatrix} e^{\sigma T}(\sigma\cos\omega T + \omega\sin\omega T) - \sigma \\ e^{\sigma T}(\sigma\sin\omega T - \omega\cos\omega T) + \omega \end{bmatrix} \tag{4.2.6}$$

How does the controllability region increase with the number of sampling steps? First examine two steps

$$\underline{x}[1] = \underline{Ax}[0] + \underline{b}u[0]$$

$$\underline{x}[2] = A^2\underline{x}[0] + \underline{Ab}u[0] + \underline{b}u[1]$$

The initial states

$$\underline{x}[0] = -\underline{A}^{-1}\underline{b}u[0] - \underline{A}^{-2}\underline{b}u[1] \tag{4.2.7}$$

with $|u[0]| \leq 1$, $|u[1]| \leq 1$ can be transferred to the zero state in two steps. These initial states lie in the parallelogram with the corners $\pm\underline{A}^{-1}\underline{b} \pm\underline{A}^{-2}\underline{b}$. This is crosshatched in figure 4.1.

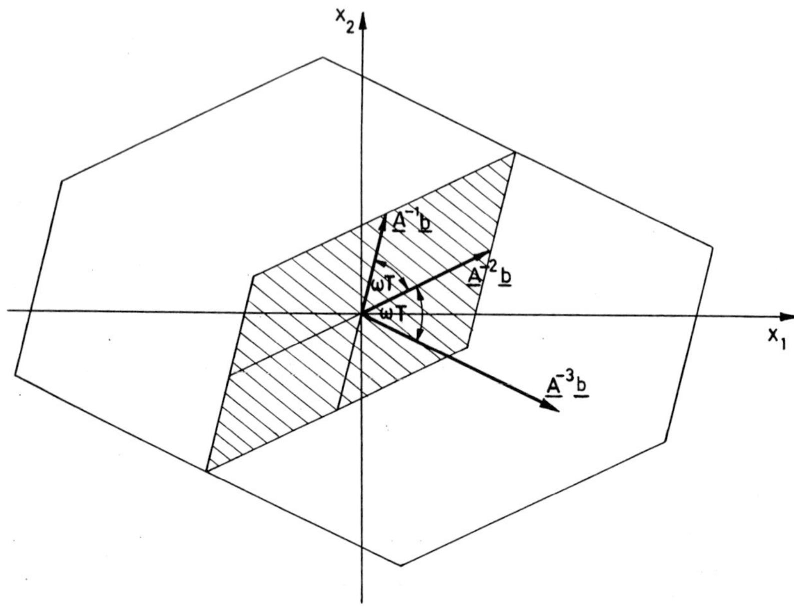

Figure 4.1 Initial states in the crosshatched parallelogram can be transferred to the zero state in two sampling steps, initial states in the hexagon in three steps.

The vector $\underline{A}^{-1}\underline{b}$ is generated from $\underline{b}$ by a rotation through the angle ωT and multiplication of the length by the factor $e^{-\sigma T}$

If three sampling steps are allowed, then

$$\underline{x}[0] = -\underline{A}^{-1}\underline{b}u[0] - \underline{A}^{-2}\underline{b}u[1] - \underline{A}^{-3}\underline{b}u[2] \tag{4.2.8}$$

The initial states which can be transferred to the zero state in three steps lie in the hexagon represented in figure 4.1 with the corners

$$\underline{A}^{-1}\underline{b} + \underline{A}^{-2}\underline{b} \pm \underline{A}^{-3}\underline{b}, \quad -\underline{A}^{-1}\underline{b} - \underline{A}^{-2}\underline{b} \pm \underline{A}^{-3}\underline{b},$$
$$\underline{A}^{-1}\underline{b} - \underline{A}^{-2}\underline{b} - \underline{A}^{-3}\underline{b}, \quad -\underline{A}^{-1}\underline{b} + \underline{A}^{-2}\underline{b} + \underline{A}^{-3}\underline{b}.$$

For four steps an octogon correspondingly arises, etc. A measure for the size of the controllability region for N sampling steps is its area. This was calculated in [72.5] as

$$F_N(T) = 4\,\frac{\|\underline{A}^{-1}\underline{b}\|^2}{1-e^{-2\sigma T}}\,[e^{-\sigma T}(1-e^{-2\sigma T\times(N-1)})\times|\sin\omega T|$$
$$+\ e^{-2\sigma T}(1-e^{-2\sigma T\times(N-2)})\times|\sin 2\omega T|$$
$$\vdots$$
$$+\ e^{-(N-2)\sigma T}(1-e^{-2\sigma T\times 2})\times|\sin(N-2)\omega T|$$
$$+\ e^{-(N-1)\sigma T}(1-e^{-2\sigma T})\quad\times|\sin(N-1)\omega T|\,] \tag{4.2.9}$$

Figure 4.1 has been drawn for a stable system with $\sigma < 0$. Here the length of the vectors $\underline{A}^{-i}\underline{b}$ increases with i; as $N \to \infty$ the entire x_1-x_2-plane is covered, the series (4.2.9) diverges for increasing N. For unstable systems with $\sigma > 0$ the lengths of the vectors $\underline{A}^{-m}\underline{b}$ decrease essentially as $e^{-\sigma T}$. $\underline{A}^{-N}\underline{b}$ goes to zero and F_N in eq. (4.2.9) converges to a finite value. The states which lie outside this controllability region are moving so fast away from the origin that the constrained input is not able to bring them back inside.

For n = 2 the two vectors $\underline{A}^{-1}\underline{b}$, $\underline{A}^{-2}\underline{b}$ are required in order to span the state plane. Further vectors, $\underline{A}^{-3}\underline{b}$, $\underline{A}^{-4}\underline{b}$, etc. only enlarge the controllability region. In general n linearly independent vectors $\underline{A}^{-1}$, $\underline{A}^{-2}\underline{b}\ldots\underline{A}^{-n}\underline{b}$ are required to span the state space. For a constraint $|u| \leq 1$, the controllability region is a polyhedron with 2n corners which lie symmetric to the point $\underline{x} = \underline{0}$. Further vectors $\underline{A}^{-(n+1)}\underline{b}\ldots\underline{A}^{-N}\underline{b}$, etc. enlarge the controllability region to form a polyhedron with 2N corners.

The loss of controllability for $\omega T = q\pi$, q = 1,2,3... as discussed in eq. (4.1.16) is now obvious: All vectors $\underline{A}^{-1}\underline{b}$, $\underline{A}^{-2}\underline{b}$, $\underline{A}^{-3}\underline{b}\ldots$ become parallel and therefore $F_N = 0$. A reasonable choice of T is always constrained by $T < \pi/\omega$ (see figure 3.2). But it would be unsatisfactory to choose T in the neighborhood of π/ω because the system cannot be controlled well there. The choice of the sampling period is discussed in more detail for systems of higher order in the following section.

4.3 Choice of the Sampling Interval

The choice of sampling period requires a compromise between different demands. An argument for a long sampling interval is the cost, which increases with the necessary speed of computing and analog-digital conversion. Moreover, the region in the state space in which a linear controller adheres to the existing constraints of the input, for example the crosshatched domain in figure 4.1, shrinks as T decreases. Therefore the smaller T is chosen the more significant the constraints on $|u|$ become.

On the other hand T is also bounded from above, for example by a required minimum bandwidth of the control loop which is necessary for a good response or by the loss of controllability for complex eigenvalues. In the following sections the choice of the sampling interval will be discussed both from a controllability and a bandwidth point of view.

4.3.1 Controllability Aspects [72.5], [75.1]

The examination of the controllability regions for constrained input amplitude form the basis for the following considerations regarding the choice of sampling period. Again the area, volume or hypervolume are taken as a measure of the controllability region. In order to compare the effect of different sampling periods, a fixed total control time τ is assumed. It is then divided into N sampling intervals $T = \tau/N$. Obviously the condition $N \geq n$ must hold in order that the hypervolume can become greater than zero. The measure F essentially increases with increasing N, at least

$$F_{mN}(\tau/mN) \geq F_N(\tau/N) \quad , \; m = 2,3,4... \tag{4.3.1}$$

This follows because a control sequence of step length τ/N represents a special case of the control sequence of step length τ/mN, in particular this is the case where m consecutive amplitude values are equal. On the other hand, $F_N(\tau/N)$ cannot increase without bound with N, i.e. for $T \to 0$. Rather it converges to the controllability region of the continuous system whose optimal input

$u(t)$ is approximated better and better by the staircase curve $u(kT)$, $kT < t < kT+T$ as $T \to 0$. Since the number of steps N is a direct measure for the calculation time, it can be decided whether the increase of $F_N(\tau/N)$ for increasing N is worth the necessary greater cost of calculation time. In our analysis we choose τ such that the first nontrivial case $T = \tau/n$ is an obviously too long sampling interval.

Example 1:

For illustration consider the example of eq. (4.2.5). Let $\omega = \pi$, $\sigma = -1$, and $\tau = 1$ (i.e. we begin with $T = \tau/n = 1/2$ or $\omega T = \pi/2$; the sinusoidal oscillation is sampled only four times per period). The controllability regions are plotted for some values of N in figure 4.2. Also $F_N(\tau/N)$ is calculated from eq. (4.2.9) and represented in figure 4.2 f. Figures a to e represent the following cases:

a) $N = 2$, $T = 0.5$, $\omega T = \pi/2$, $F_2(0.5) = 2.26$

b) $N = 3$, $T = 0.333$, $\omega T = \pi/3$, $F_3(0.333) = 2.98$

c) $N = 4$, $T = 0.25$, $\omega T = \pi/4$, $F_4(0.25) = 3.28$

d) $N = 5$, $T = 0.2$, $\omega T = \pi/5$, $F_5(0.2) = 3.46$

e) $N = 6$, $T = 0.167$, $\omega T = \pi/6$, $F_6(0.167) = 3.54$

The point A with the coordinates $x_1 = -0.342$, $x_2 = -1.075$ is common to all polygons. At this point the trajectories begin which lead to the origin in time τ for $u[0] = u[1] = \ldots u[N-1] = 1$. For $u = 1$ the heavily shaded vectors $\underline{A}^{-1}\underline{b} + \underline{A}^{-2}\underline{b} + \ldots + \underline{A}^{-N}\underline{b}$ merge as $N \to \infty$ into the trajectory of the continuous system for $u \equiv 1$ i.e. the switching curve of the time optimal system. For $u = -1$ the other part of the switching curve lies symmetric to the origin and runs from point B to zero. As N increases, the polygon makes a better and better approximation of the curve of optimal transfer time τ for the continuous system. Figures 4.2 a to e

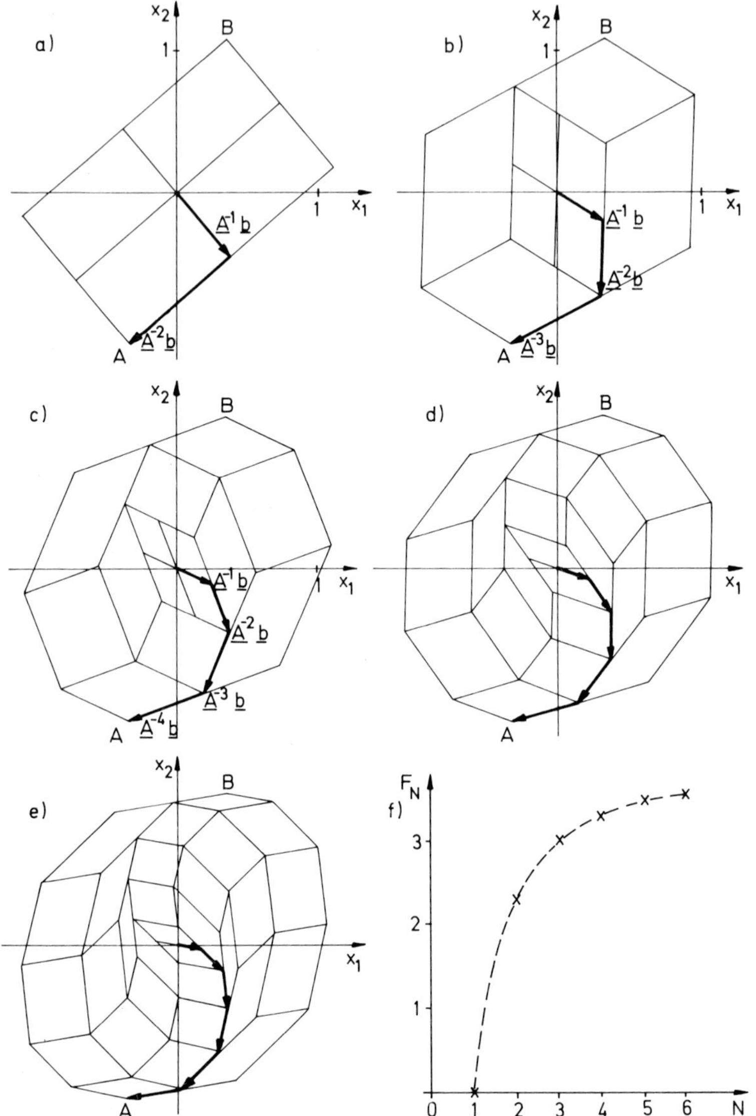

Figure 4.2 Controllability regions for a) $\omega T = \pi/2$, b) $\omega T = \pi/3$, c) $\omega T = \pi/4$, d) $\omega T = \pi/5$, e) $\omega T = \pi/6$, area F_N as a function of step number N.

already indicate the general shape of the envelope; great changes are not to be expected for N > 6. On the other hand the computing cost rises with N because the control algorithm must run N times within the fixed time τ. Figure 4.2 f shows that there is no great increase in controllability for $N > 4$, i.e. if the sampling period is smaller than a quarter of that for which controllability is entirely lost. This would be the case for $\omega T = \pi$ where $\underline{A}^{-1}\underline{b}$ and $\underline{A}^{-2}\underline{b}$ are parallel and the area is zero. This yields the recommended choice of T

$$T \leq \pi/4\omega \qquad (4.3.2)$$

i.e. the eigenvalues z_i lie at an angle $\omega T \leq \pi/4$ to the positive real axis of the z-plane.

The controllability region does not depend only on the angle ωT between the vectors $\underline{A}^{-m}\underline{b}$, but also on their lengths. From example (4.2.6)

$$|b| = \sqrt{\frac{e^{2\sigma T}-2e^{\sigma T}\cos\,\omega T+1}{\sigma^2+\omega^2}} = \frac{|z_i-1|}{|s_i|}, \quad s_i = \sigma+j\omega, \quad z_i = e^{s_iT}$$

$$|A^{-m}b| = e^{-m\sigma T} \times |b|, \qquad m = 1,2,3\ \dots \qquad (4.3.3)$$

If the eigenvalues of F have a large negative real part σ, then $e^{-\sigma T}$ becomes large and the controllability region is narrow but elongated in the direction $\underline{A}^{-N}\underline{b}$. Correspondingly, $e^{-\sigma T}$ is small for unstable systems with large σ and the controllability region essentially extends in the direction $\underline{A}^{-1}\underline{b}$. In order to avoid this effect, T must be chosen so that neither $|\omega T|$ nor $|\sigma T|$ becomes too large.

Since the controllability region according to eq. (4.3.3) depends on the absolute value of the eigenvalue $|s_i| = \sqrt{\sigma^2+\omega^2}$, it is reasonable to modify the rule of thumb eq. (4.3.2) to

$$T \leq \pi/4|s_i| \qquad (4.3.4)$$

In this form the rule is also applicable to real eigenvalues $s_i = \sigma$ as will be illustrated by the next example.

Example 2:

$$\underline{\dot{x}} = \begin{bmatrix} -a & 0 \\ 0 & 0 \end{bmatrix} \underline{x} + \begin{bmatrix} 1 \\ 1 \end{bmatrix} u \quad , \; a > 0 \qquad (4.3.5)$$

$$y = \begin{bmatrix} -1 & 1 \end{bmatrix} x$$

(The corresponding transfer function is

$$\frac{y_s}{u_s} = \frac{a}{s(s+a)} = \frac{1}{s(1+s/a)} \; , \text{ the time constant is } 1/a.)$$

The discretized state equation is

$$\underline{x}(kT+T) = \underline{A}(t)\underline{x}(kT)u + \underline{b}(T)u(kT)$$

$$\underline{A}(T) = \begin{bmatrix} e^{-aT} & 0 \\ 0 & 1 \end{bmatrix} , \quad \underline{b}(T) = \begin{bmatrix} (1-e^{-aT})/a \\ T \end{bmatrix}$$

$$a\underline{A}^{-1}\underline{b} = \begin{bmatrix} e^{aT}-1 \\ aT \end{bmatrix} , \quad a\underline{A}^{-2}\underline{b} = \begin{bmatrix} e^{aT}(e^{aT}-1) \\ aT \end{bmatrix} \quad \text{etc.}$$

The total time for the control is chosen as $\tau = 3/a$ i.e. three time constants of the exponential solution term. After this time the transient has settled without control to e^{-3}, i.e. 5 % of the initial value. Figure 4.3 shows the set of initial states that can be transferred to zero within a time τ and with the bound $|u| \leq 1$. (Upper half plane only, lower half plane is symmetric with respect to $\underline{x} = \underline{0}$). Figures 4.3 a to d represent the following cases:

a) $N = 2$, $aT = 1.5$ b) $N = 3$, $aT = 1$

c) $N = 4$, $aT = 0.75$ d) $N = 5$, $aT = 0.6$

This example illustrates, that the controllability region increases remarkably as we go from $aT = 1.5$ to $a = 1$ and then to $a = 0.75$. However, a further reduction of T to $aT = 0.6$ has only a less significant effect. This is in agreement with the rule of eq. (4.3.4) with

$$aT = |s_i|T \leq \pi/4 = 0.78.$$

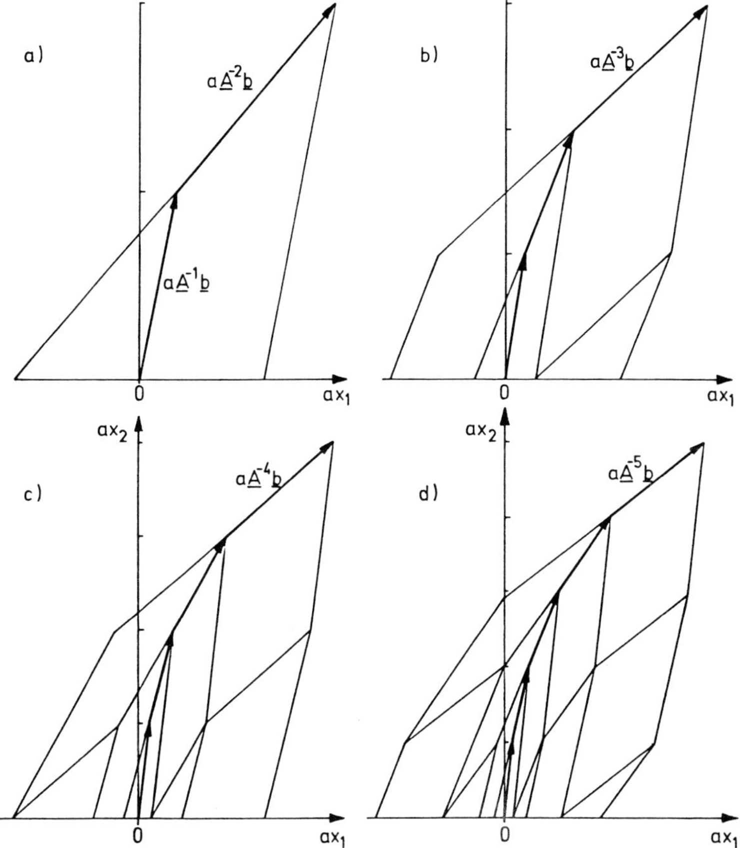

Figure 4.3 Controllability regions for a) a) aT = 1.5, b) aT = 1, c) aT = 0.75, d) aT = 0.6.

For n > 2 one can imagine the system transformed to Jordan form. The preceeding considerations must then hold for each subsystem. Condition (4.3.4) must therefore be satisfied for all the eigenvalues s_i of F. The fastest modes with the maximum value of $|s_i|$ determine the choice of T. The resulting recipe is: In the s-plane draw the smallest circle around s = 0 such that all eigenvalues are enclosed, let its radius be r. Then choose

$$T = \pi/4r \quad \text{i.e.} \quad \omega_A = \frac{2\pi}{T} = 8r \qquad (4.3.6)$$

The sampling frequency ω_A is eight times the radius of the above circle. It is thereby assured that all eigenvalues in the z-plane lie inside the curve

$$z = e^{Ts} = e^{Tr} \times e^{j\alpha} = e^{\frac{\pi}{4}} e^{j\alpha} \quad -\pi \leq \alpha \leq \pi \tag{4.3.7}$$

as represented in figure 4.4.

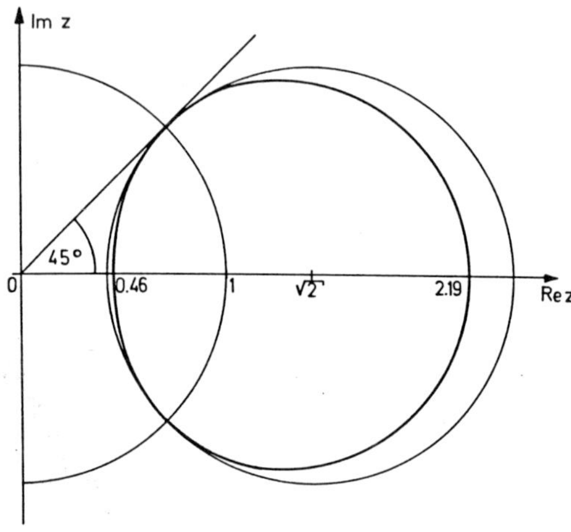

Figure 4.4

The sampling period T must be chosen such that the eigenvalues of $\underline{A}$ lie inside the heavily shaded curve. The lightly shaded unit circle shifted by $\sqrt{2}$ to the right can be used as an approximation.

Remark 4.1:

Because only a rule of thumb has been made plausible here, the curve of eq. (4.3.7) can also be approximated by a circle with radius 1 around the point $z = \sqrt{2}$, see figure 4.4.

As $T \to 0$ all eigenvalues $z_i = e^{s_i T}$ approach $z = 1$. A neighborhood of the point $z = 1$ is fixed by the curve from eq. (4.3.7). All eigenvalues of the open system should lie in this neighborhood for a suitable choice of T. If eigenvalues lie outside, then T has been chosen too large and the controllability is noticably impaired by sampling compared to the continuous system. If all eigenvalues lie close to $z = 1$, the system is sampled too frequently and an unnecessarily fast computer is required.

So far we have discussed the choice of the sampling interval T only from an open-loop point of view, i.e. we have not considered the generation of the time-optimal input sequence in a feedback control system. This could be done by a nonlinear controller. It has to compare the actual state and the state space partition as in figure 4.2 and 4.3 in order to decide on the required actuator signal u(kT) for time-optimal control. For small T the partition is finer, therefore a more complicated decision logic and consequently more time is needed for the decision, but less time T is available. Thus T should not be too small.

In this book we are concerned with the design of linear controllers. The system can be transferred to the zero state in n sampling intervals by a linear feedback controller. This "deadbeat control" is equivalent to an assignment of poles at $z = 0$ and will be discussed in section 4.4. For the second order systems of figures 4.2 and 4.3 with bound $|u| \leq 1$ the deadbeat controller gives the same input sequence as the time-optimal open-loop solution if the state is inside the innermost parallelogram which is fixed by $\underline{A}^{-1}\underline{b}$ and $\underline{A}^{-2}\underline{b}$. For states outside this region the bound $|u| \leq$ causes a different behavior. The loop designed by neglecting the bound may even become unstable by the existing actuator saturation. Thus the linear design is justified only if T is not chosen too small, even if a fast enough computer is available. This point may also be illustrated by a typical step response with $\max|u(kT)|$ at $k = 0$. For small T this may be a short initial impulse of large magnitude. For a longer T approximately the same effect is achieved by an equivalent impulse with smaller magnitude but longer duration T. In this latter case the neglected actuator saturation has a less significant effect.

Another argument in favor of a not too short sampling interval T is the following: If slowly varying signals are sampled fast, then there is little difference in two consecutive samples. Therefore the control algorithm must be processed with high accuracy. For example a 16 or 32 bit microprocessor may be required instead of an 8 bit one.

Remark 4.2:

The "hypercube constraint" $|u[k]| \leq 1$ for $\underline{u}_N$ may be replaced by "hypersphere constraint"

$$T\underline{u}'_{N-1}\underline{u}_{N-1} \leq 1 \qquad (4.3.8)$$

In some applications $\underline{u}'_{N-1}$ represents an input energy. The inequality (4.3.8) is mapped by

$$\underline{x}[0] = \underline{A}\underline{u}_{N-1}, \ \underline{Q} = [\underline{A}^{-1}\underline{b}, \ \underline{A}^{-2}\underline{b}\ldots\underline{A}^{-N}\underline{b}], \ N \geq n \qquad (4.3.9)$$

into the state space, resulting in a hyperellipsoid. Its volume

$$V = C(n)\sqrt{T^{-n}\det(\underline{Q}\underline{Q}')} \qquad (4.3.10)$$

can be used as a measure for the controllability region [75.1].

4.3.2 Bandwidth Aspects, Anti-Aliasing Filters [80.1], [81.17]

Consider the control system of figure 4.5.

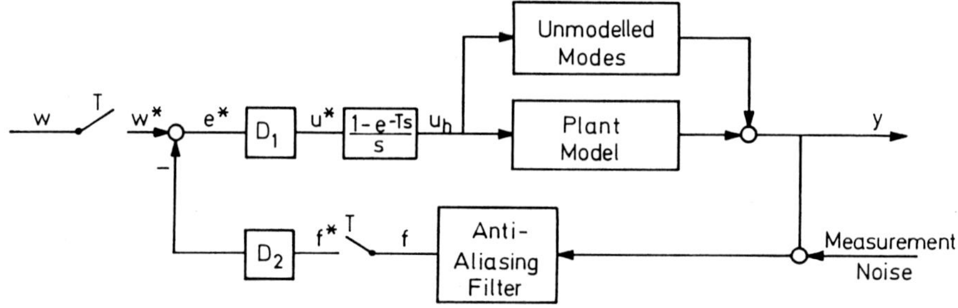

Figure 4.5 A controlled plant with unmodelled modes and measurement noise.

The "star notation" (w^*, e^*, u^*, f^*) refers to the frequency domain description of the sampler as in figure 1.4 and eq. (1.2.8). D_1 and D_2 represent the controller transfer function implemented in a digital computer. Within a design bandwidth the output signal y should follow the reference input w. Within this bandwidth w must be well represented in w^* by appropriate choice of the sampling interval T. The bandwidth is essentially determined by the closed-loop poles. The poles can be arbitrarily shifted if the actuator limitations are ignored. Practically however, only the damping is adjusted and some of the slower modes, in particular integrations, are speeded up, i.e. the eigenvalues closest to the origin are moved away from $s = 0$. There are three limitations to a substantial increase of the bandwidth by closing the loop:

a) actuator saturation and bandwidth,
b) destabilization of unmodelled high-frequency modes (e.g. structural vibrations in mechanical systems),
c) high-frequency measurement noise.

The effects b) and c) are of particular importance in sampled-data systems. Figure 1.5 shows the spectrum of a sampled signal, e.g. f^* in figure 4.5. If the continuous signal f contains a component of frequency ω_1 with $\omega_A/2 < \omega_1 < 3\omega_A/2$, then it shows up in the first lower side band at $\omega_A - \omega_1$ in the frequency range $-\omega_A/2 < \omega_A-\omega_1 < \omega_A/2$, i.e. it is reflected into the original frequency band. Similarly components of frequency ω_2 with $3\omega_A/2 < \omega_2 < 5\omega_A/2$ appear as $2\omega_A-\omega_2$ $(-\omega_A/2 < 2\omega_A-\omega_2 < \omega_A/2)$ in the second lower side band and so on. The frequency ω_1 is called an "alias" of the frequency $\omega_A-\omega_1$. It is not possible from f^* to distinguish whether f was of frequency ω_1 or $\omega_A-\omega_1$ (compare figures 3.2 i) and j) for a time-domain illustration). In order to avoid this "aliasing" an "anti-aliasing filter" is used before the sampler. This is an analog low-pass filter which attenuates the undesired high frequency components. The bandwidth ω_E of the filter is chosen equal to $\omega_A/2$. For example a second-order anti-aliasing filter with damping $1/\sqrt{2}$ has the transfer function

$$f_E(s) = \frac{\omega_E^2}{s^2+\sqrt{2}\omega_E s+\omega_E^2} = \frac{\omega_A^2 4}{s^2+s\omega_A/\sqrt{2}+\omega_A^2/4} \qquad (4.3.11)$$

This filter will have negligible effect in the closed-loop bandwidth if the latter is much smaller than $\omega_A/2$. If the undesired components are not so clearly separated in frequency from the poles of the plant model, then the anti-aliasing filter must be seen as part of the plant and be considered in the design, i.e. the digital controller must compensate for the phase lag introduced by the filter. Anti-aliasing filtering is not possible for signals from a digital sensor.

The recommendation for the choice of the sampling interval is now combined into the following recipe:

1. Plot the open-loop poles s_i of the plant in the s-plane together with the signal poles for the reference input corresponding to the required bandwidth of the closed loop (A in figure 4.6).

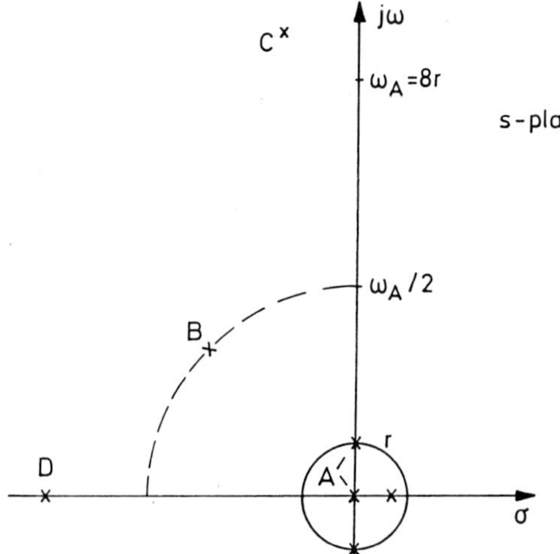

Figure 4.6

On the choice of the sampling frequency $\omega_A = 2\pi/T$

2. Draw the smallest circle around s = 0 such that the above poles are encircled. Its radius is r. Choose $\omega_A = 8r$, i.e. $T = \pi/4r$.

3. If there are unmodelled structural vibrations (e.g. C), neglected actuator dynamics or measurement-noise poles (e.g. D), then include an anti-aliasing filter (e.g. poles at B).

4. If the undesired components (C, D) are not far away from the origin, then a correspondingly small bandwidth of the anti-aliasing filter must be chosen and the filter must be treated as part of the plant in the controller design.

Remark 4.3:

The connection with the sampling theorem of Shannon [49.2] should be noted. It says: If $F(j\omega) = 0$ for $|\omega| \geq \omega_m$ holds for the Fourier transform $F(j\omega)$ of a signal f(t), then f(t) is uniquely determined by the samples f(kT), k = 0, ±1, ±2... if and only if $T \leq \pi/\omega_m$, i.e. $\omega_A \geq 2\omega_m$ is chosen. As shown in figure 1.5, the subspectra in $F(j\omega)$ no longer overlap in this case. The original band identical to the input spectrum can therefore be filtered out by an ideal low-pass filter with the frequency response

$$F_i(j\omega) = \begin{cases} 1 & |\omega| < \omega_m \\ 0 & |\omega| \geq \omega_m \end{cases} \tag{4.3.12}$$

However, such a low-pass filter cannot be realized as a causal transfer function. This can be seen if the inverse Fourier transform of the frequency response $F_i(j\omega)$ is formed. One obtains the following weighting function i.e. the impulse response of the system

$$f_i(t) = \mathcal{F}^{-1}\{F_i(j\omega)\} = \frac{\sin \omega_m t}{\omega_m t} \tag{4.3.13}$$

The response begins at $t = -\infty$, i.e. such a system is not causal. This filter is suitable for the off-line interpolation of a sequence f(kT), but not for real time extrapolation by a causal filter as is required in control systems.

For examples and further discussion of bandwidth aspects in the choice of T see [80.1], [81.17]. Stochastic disturbances and their influence on the choice of T are discussed in [76.3]. The influence of T on a quadratic performance criterion is studied in [70.6].

4.4 Pole Assignment

We next consider the following problem: given a controllable discrete system

$$\underline{x}[k+1] = \underline{A}\underline{x}[k] + \underline{b}u[k] \tag{4.4.1}$$

find a state-vector feedback

$$u[k] = -\underline{k}'\underline{x}[k] \tag{4.4.2}$$

which produces a desired characteristic polynomial

$$P(z) = \det(z\underline{I}-\underline{A}+\underline{b}\underline{k}') = p_o + p_1 z + \ldots + p_{n-1}z^{n-1} + z^n \tag{4.4.3}$$

This corresponds fully to the pole assignment problem for continuous systems, as solved in section 2.6. Corresponding to eq. (2.6.14) the closed loop characteristic polynomial is obtained from

$$\underline{p}' = \underline{a}' + \underline{k}'\underline{W} \tag{4.4.4}$$

$\underline{a}$ and $\underline{W}$ are now formed from $\underline{A}$ and $\underline{b}$ in the same way as $\underline{q}$ and $\underline{W}$ were formed from $\underline{F}$ and $\underline{g}$ in the continuous case, see eqs. (2.6.9) and (2.6.12). Corresponding to (2.6.24), the solution of eq. (4.4.3) for $\underline{k}'$ is

$$\underline{k}' = \underline{e}'P(\underline{A}) \tag{4.4.5}$$

where $\underline{e}'$ is the last row of the inverse of the controllability matrix, i.e.

$$\underline{e}' = [0 \ \ldots \ 0 \quad 1][\underline{b} \quad \underline{Ab} \ \ldots \ \underline{A}^{n-1}\underline{b}]^{-1}$$

Eq. (4.4.5) can also be written in the form

$$\underline{k}' = [\underline{p}' \quad 1]\underline{E} \tag{4.4.6}$$

where $\underline{p}' = [p_o \quad p_1 \ \ldots \ p_{n-1}]$ is the coefficient vector of the desired characteristic polynomial and

$$\underline{E} = \begin{bmatrix} \underline{e}' \\ \underline{e}'\underline{A} \\ \vdots \\ \underline{e}'\underline{A}^n \end{bmatrix} \tag{4.4.7}$$

is the pole assignment matrix of the discrete system (4.4.1).

An interesting and important special case of pole assignment is to place all eigenvalues to $z = 0$, i.e. $P(z) = z^n$, $\underline{p}' = \underline{0}'$ and

$$\underline{k}_D' = \underline{e}'\underline{A}^n \tag{4.4.8}$$

Only finite transients arise for this "deadbeat" solution, namely

$$\underline{x}[k+n] = (\underline{A}-\underline{b}\underline{k}_D')^n\underline{x}[k] = \underline{0} \quad \text{for all } \underline{x}[k] \tag{4.4.9}$$

because from the Cayley-Hamilton theorem the characteristic equation $P(z) = z^n = 0$ is also satisfied by the matrix itself. Therefore $P(\underline{A}-\underline{b}\underline{k}_D') = (\underline{A}-\underline{b}\underline{k}_D')^n = \underline{0}$.

This feedback deadbeat solution is now compared to the feedforward control sequence which transfers a given initial state to the zero state $x[k+n] = 0$ in n steps, see eq. (4.1.11). By uniqueness of this solution this same control sequence must arise in the closed loop with the feedback (4.4.8). This cannot be immediately seen from the equations, but it can be shown in the following way:

Eq. (4.1.11) is formed for $k = 0,1,2...n-1$. For each value of k a column of the following matrix equation is obtained

$$\begin{bmatrix} u[0] & u[1] & \dots & u[n-1] \\ u[1] & u[2] & & 0 \\ \vdots & \vdots & & \vdots \\ u[n-1] & 0 & & 0 \end{bmatrix} = -[\underline{A}^{n-1}\underline{b},\ \underline{A}^{n-2}\underline{b}\dots\underline{b}]^{-1}\underline{A}^n[\underline{x}[0],\ \underline{x}[1]\dots\underline{x}[n-1]] \tag{4.4.10}$$

Here $u[k] = 0$ for $k \geq n$ because then the system is already in the zero state.

The left-hand side of this equation is a Hankel matrix which is its own transpose. Therefore the control sequence can be read from the first row instead of the first column. The first row on the right-hand side has the left factor

$$[1\ \ 0\ldots0][\underline{A}^{n-1}\underline{b},\ \underline{A}^{n-2}\underline{b}\ldots\underline{b}]^{-1} = [0\ldots0\ \ 1][\underline{b}\ \ \underline{A}\underline{b}\ldots\underline{A}^{n-1}\underline{b}]^{-1} = \underline{e}' \qquad (4.4.11)$$

and therefore

$$[u[0],\ u[1]\ldots u[n-1]] = -\underline{e}'\underline{A}^n[\underline{x}[0],\ \underline{x}[1]\ldots\underline{x}[n-1]]$$

This agrees with the state-vector feedback

$$u[k] = -\underline{e}'\underline{A}^n\underline{x}[k] \qquad (4.4.12)$$

The distinction between a feedforward control and a feedback control can be discussed through an examination of eq. (4.4.10). For feedforward control only the initial state $\underline{x}[0]$ and the desired state are presumed to be known. The complete control sequence u[0], u[1]...u[n-1] is precalculated at time t = kT according to the first column of eq. (4.4.10). It then runs as feedforward control. The state $\underline{x}[n] = \underline{0}$ is reached if no disturbance occurs in the control interval and $\underline{A}$ and $\underline{b}$ were exactly known. If $\underline{A}$ is unstable, then the solution diverges from the ideal solution with any disturbance. For state-vector feedback, $\underline{x}[k]$ is presumed to be known at all time points t = kT and only the current u[k] is calculated from the first row of eq. (4.4.10), i.e. eq. (4.4.12) If $\underline{x}[k]$ is altered by a disturbance, then the actual value is fed back. The transient is correspondingly prolonged. If disturbances are acting continuously, then a finite transient is no longer produced. However, all disturbances which have acted more than n sampling intervals before are always completely controlled out. Also for inexactly known unstable $\underline{A}$ the solution goes asymptotically to zero, provided that all eigenvalues of $\underline{A}-\underline{b}\underline{k}_D'$ lie in the unit circle.

The deadbeat solution is especially satisfactory for slow plants with real eigenvalues, also for systems with delay, as found in process technology, and for problems for which the actuator constraints do not play a determining role. On the other hand, the example of the loading bridge, exercise 4.1, shows that this solution can lead to unrealistically large forces and in the example to extremely oscillatory motions of the trolley.

The deadbeat solution can also be unsatisfactory with respect to the required stability margin ρ. From eq. (3.9.28) with $P(z) = z^n$

$$\rho = \min_{\omega} |e^{jn\omega T}/A(e^{j\omega T})| \tag{4.4.13}$$

If the poles of the open loop $A(z) = 0$ lie close to $z = 1$ or to the right of this point, then ρ becomes small at the point $z = -1$, i.e. $\omega T = \pi$.

Example:

For the triple integrator $1/s^3$ of eq. (3.3.38) $A(z) = (z-1)^3$ and

$$\rho = \min_{\omega} |e^{j3\omega T}/(e^{j\omega T}-1)^3| = 1/8 = 0.125 \tag{4.4.14}$$

The system has only a small stability margin although no eigenvalue lies near to the unit circle. If on the contrary one chooses $P(z) = (z-0.5)^3$ as the characteristic polynomial of the closed loop, the stability margin is

$$\rho = \left|\min_{\omega} (e^{j\omega T}-0.5)^3/(e^{j\omega T}-1)^3\right| = 27/64 \approx 0.422 \tag{4.4.15}$$

The stability of this solution is therefore more robust with respect to small changes in gain and phase in the control loop.

If the eigenvalues of the open loop, i.e. the zeros of $A(z)$, lie in the circle represented in figure 4.4, then with respect to the stability margin ρ it is generally advisable to place the eigenvalues of the closed loop to the right of the point $z = 0$. Also, taking the required actuator amplitudes into account, it must be recommended that the eigenvalues be shifted only so far in the direction $z = 0$ as necessary to obtain a quick, well damped transient.

The above observations on the relation between pole position and stability margin remain valid if the pole assignment is not the result of state-vector feedback, eq. (4.4.2), but rather of dynamic output vector feedback with the compensator calculated by eq. (3.4.15).

Example:

Consider the triple integrator of eq. (3.3.38) with the given z-transfer function for T = 1. For a deadbeat-control, $P(z) = z^5$, the z-transfer function of the controller according to eq. (3.4.15) is

$$d_z(z) = \frac{\frac{17}{6} + \frac{23}{3} z + \frac{35}{6} z^2}{\frac{17}{36} + \frac{73}{36} z + z^2} \tag{4.4.16}$$

The controller is unstable with a pole at $z = -1.759$. The second pole at $z = -0.2684$ leads to an "almost cancellation" of the plant zeros at $z = -2+\sqrt{3} = -0.2679$. The stability margin is only $\rho = 0.1402$, this minimal value resulting at $\omega T = 117^o$. This control loop is also sensitive to small phase and amplitude changes in the open loop.

Remark 2.1 on the sensitivity of the state-vector feedback with respect to pole changes and the results of sections 2.6.3 and 2.6.4 on partial pole and gain assignment can be immediately transferred to discrete systems. See exercise 4.5.

4.5 Exercises

4.1 For the loading bridge let $m_C = 1.000$ kg, $m_L = 3.000$ kg, $\ell = 10$ m, $g = 10$ m/s^2, $T = \pi/8$ seconds. Calculate the control sequence which brings a load from the rest position 1 meter from the origin to the new rest position at the origin.

4.2 With the numerical values m_C, m_L, ℓ and g from exercise 4.1, the subsystem pendulum has the state respresentation

$$\begin{bmatrix} x_3[k+1] \\ x_4[k+1] \end{bmatrix} = \begin{bmatrix} \cos 2T & \frac{1}{2}\sin 2T \\ -2\sin 2T & \cos 2T \end{bmatrix} \begin{bmatrix} x_3[k] \\ x_4[k] \end{bmatrix} + 10^4 \begin{bmatrix} \frac{1}{4}(\cos 2T-1) \\ -\frac{1}{2}\sin 2T \end{bmatrix} u[k]$$

Construct the region of the initial states $x_3[0]$, $x_4[0]$ which can be transferred to the zero state within a time of $\tau = \pi/2$ seconds for a constraint $|u(t)| \le 10^4$ Newtons (= kg m/s²). Let a) $T = \tau/2$, b) $T = \tau/4$, c) $T = \tau/6$.

4.3 Determine a state-vector feedback by which every initial state of the loading bridge from exercise 4.1 is transferred to the zero state in four steps and compare the input sequence $u[k]$ with the control sequence of exercise 4.1 for the initial state given there.

4.4 For the control loop of exercise 4.3, the load m_L is changed from 3.000 kg to 1.000 kg for an unaltered feedback. Determine $u[k]$ for the initial state $\underline{x}' = [1 \quad 0 \quad 0 \quad 0]$.

4.5 Perform a partial pole assignment for the loading bridge of exercise 4.1, where the pendulum receives a damping of $1/\sqrt{2}$ for an unchanged natural frequency (mapping this requirement from the s-plane to the z-plane). Assume a double real pole at $z = z_3$ for the remaining two poles. Sketch and discuss $\underline{k}'(z_3)$ for $0 \le z_3 \le 1$. Are there values in this range for which only cable angle, trolley position and trolley velocity must be fed back, but not the cable angular velocity?

4.6 Calculate the transient of the control loop from exercise 4.5 for a) $z_3 = 0.5$, b) $z_3 = 0.75$, c) $z_3 = 0.843$ and the initial state $\underline{x}'[0] = [1 \quad 0 \quad 0 \quad 0]$. Compare the maximum input amplitude and settling time to within 5 % remaining error in the load position for the three cases and for the deadbeat solution of exercise 4.3.

4.7 Consider a plant

$$\underline{x}[k+1] = \begin{bmatrix} 0 & 0 & 1 \\ 0 & 1 & 0 \\ 2 & 0 & 0 \end{bmatrix} \underline{x}[k] + \begin{bmatrix} 1 \\ -1 \\ 1 \end{bmatrix} u[k]$$

Find the state-feedback law $u[k] = -[k_1 \quad k_2 \quad k_3]\underline{x}[k]$ which shifts all eigenvalues to $z = 0$.

5 Observability and Observers

In chapter 4 a state-vector feedback $u = -\underline{k}'\underline{x}$ was assumed for the controller structure. In most control systems not all components of the state vector are measured. Some variables may be difficult to measure; for others the sensors may be expensive or unreliable. If a vector $\underline{y}' = [y_1 \; y_2 \; \dots \; y_s]$ of s linearly independent outputs is measured where

$$\underline{y} = \underline{Cx} + \underline{d}u \;, \quad \text{rank } \underline{C} = s < n \tag{5.1}$$

then an output-vector feedback

$$u = -\underline{k}_y'\underline{y} = -\underline{k}_y'\underline{Cx} - \underline{k}_y'\underline{d}u \tag{5.2}$$

$$u = -\frac{1}{1+\underline{k}_y'\underline{d}} \times \underline{k}_y'\underline{Cx} \tag{5.3}$$

can be implemented. Neglecting the scalar factor $1/(1+\underline{k}_y'\underline{d})$ the connection with state-vector feedback is

$$\underline{k}' = \underline{k}_y'\underline{C} \tag{5.4}$$

or, writing in full

$$[k_1 \dots k_n]' = [k_{y1} \dots k_{ys}]' \begin{bmatrix} c_{11} \dots c_{1n} \\ \vdots \\ c_{s1} \dots c_{sn} \end{bmatrix}$$

Only those state-feedback vectors $\underline{k}'$ can be achieved which may be expressed as a linear combination of the s rows of the $\underline{C}$ matrix, i.e. $\underline{k}$ must lie in the subspace of K space which is spanned by the rows of $\underline{C}$. Thus the positions of the eigenvalues cannot be independently assigned. For example, if $\underline{C}$ only has one row $\underline{c}'$, i.e. only one measured variable is fed back, then only pole combinations on the root locus can be assigned. If this leads to unsatisfactory solutions in classical controller design procedures, a dynamical controller is assumed, for example eqs. (3.4.4), (3.4.5). If this has the order n-1 for one output variable y, then every desired characteristic polynomial can be assigned in the control loop by eq. (3.4.15). One is also led to this result if a special dynamic system called an "observer", with inputs u and y is designed. This system produces a reconstructed value $\hat{\underline{x}}$ which is then used in the state-vector feedback instead of $\underline{x}$, i.e. $u = -\underline{k}'\hat{\underline{x}}$. In the resulting controller structure the additional observer poles do not enter into the closed-loop transfer function. Therefore they can be chosen in view of good agreement $\hat{\underline{x}} \approx \underline{x}$ and noise reduction without compromise with regard to the response. A further advantage is that this structure can be systematically extended to multi-input, multi-output systems.

It is frequently considered a disadvantage of this structure that for high order plants the controller is also of high order. However, these costs are not important for the discrete-time systems examined here if the controller is realized electronically by a digital computer, for example a microprocessor.

The condition for the existence of an observer is that the system is observable from the measurement vector $\underline{y}$. We will first take up the question of observability from discrete values y(kT), respectively from y(kT+γT), $0 < \gamma < 1$.

5.1 Observability and Constructability

A state $\underline{x}(\tau)$ is "observable", if it can be determined from the input $\underline{u}(t)$ and the output $\underline{y}(t)$ over a finite interval $\tau \le t \le t_1$. For a causal operation it is of interest to construct $\underline{x}(\tau)$ from past measurements. A state $\underline{x}(\tau)$ is called "constructable", if it can be determined from $\underline{u}(t)$, $\underline{y}(t)$ over a finite interval $t_2 \le t \le \tau$ [69.1].

In a single-input, single-output discrete system let the following data vectors be given:

$$\underline{u}_{N-1} := \begin{bmatrix} u[0] \\ u[1] \\ \vdots \\ u[N-1] \end{bmatrix} \qquad \underline{y}_{N-1} := \begin{bmatrix} y[0] \\ y[1] \\ \vdots \\ y[N-1] \end{bmatrix} \tag{5.1.1}$$

They are related by eq. (4.1.2) and $y[k] = \underline{c}'\underline{x}[k] + du[k]$

$$\begin{aligned} y[0] &= \underline{c}'\underline{x}[0] + du[0] \\ y[1] &= \underline{c}'\underline{A}\underline{x}[0] + \underline{c}'\underline{b}u[0] + du[1] \\ y[2] &= \underline{c}'\underline{A}^2\underline{x}[0] + \underline{c}'\underline{A}\underline{b}u[0] + \underline{c}'\underline{b}u[1] + du[2] \\ &\vdots \\ y[N-1] &= \underline{c}'\underline{A}^{N-1}\underline{x}[0] + \underline{c}'\underline{A}^{N-2}\underline{b}u[0] + \ldots + du[N-1] \end{aligned} \tag{5.1.2}$$

In matrix notation

$$\underline{y}_{N-1} = \underline{Q}_N\underline{x}[0] + \underline{H}_N\underline{u}_{N-1} \tag{5.1.3}$$

with

$$\underline{Q}_N := \begin{bmatrix} \underline{c}' \\ \underline{c}'\underline{A} \\ \vdots \\ \underline{c}'\underline{A}^{N-1} \end{bmatrix}, \qquad \underline{H}_N := \begin{bmatrix} d & 0 & \cdots & \cdots & 0 \\ \underline{c}'\underline{b} & d & \ddots & & \vdots \\ \vdots & & \ddots & \ddots & \vdots \\ & & & \ddots & 0 \\ \underline{c}'\underline{A}^{N-2}\underline{b} & \cdots & & \underline{c}'\underline{b} & d \end{bmatrix} \tag{5.1.4}$$

The state $\underline{x}[0]$ is observable if eq. (5.1.3) can be solved for $\underline{x}[0]$. This is true for any $\underline{x}[0]$ if

$$\text{rank } \underline{Q}_N = n \tag{5.1.5}$$

In this case all states of the system are observable, in short: the system is observable. The rank is not changed by transposing $\underline{Q}_N$, i.e.

$\underline{Q}'_N = [\underline{c}\ ,\ \underline{A}'\underline{c} \ldots \underline{A}'^{N-1}\underline{c}]$

Mathematically this is the same problem as that of controllability in eq. (4.1.4). There it was shown that the linearly independent columns of $\underline{Q}'_N$ are dense, i.e. the rank does not further increase with N after the first linearly dependent row of $\underline{Q}_N$ has been found.

In other words: If it is at all possible to determine $\underline{x}[0]$ from the data in eq. (5.1.1), then a data window of n samples at input and output is sufficient and the solution of eq. (5.1.3) is

$$\underline{x}[0] = \underline{Q}_n^{-1}(\underline{y}_{n-1} - \underline{H}_n \underline{u}_{n-1}) \tag{5.1.6}$$

This solution exists, i.e. the system is observable, if and only if the observability matrix $\underline{\mathcal{O}} = \underline{Q}_n$ is regular

$$\det \underline{\mathcal{O}} \neq 0 \tag{5.1.7}$$

This condition corresponds to eq. (2.3.38) for the observability of continuous systems. Correspondingly also the Hautus test may be applied: the eigenvalue z_i of $\underline{A}$ is oberservable if and only if

$$\text{rank} \begin{bmatrix} \underline{A} - z_i \underline{I} \\ \underline{c}' \end{bmatrix} = n \tag{5.1.8}$$

For the causal construction of $\underline{x}[n-1]$ at the end of the data window, $\underline{x}[0]$ is substituted into eq. (4.1.2)

$$\underline{x}[n-1] = \underline{A}^{n-1}\underline{x}[0] + [\underline{A}^{n-2}\underline{b},\ \underline{A}^{n-3}\underline{b} \ldots \underline{b}]\underline{u}_{n-1} \tag{5.1.9}$$

Thus every observable system is constructible. The inverse conclusion is true only if all zero eigenvalues of $\underline{A}$ are observable, i.e. if

$$\text{rank} \begin{bmatrix} \underline{A} \\ \underline{c}' \end{bmatrix} = n \tag{5.1.10}$$

Example:

Let $\underline{A} = \underline{0}$.

From eq. (5.1.9) $\underline{x}[n-1] = \underline{b}u[n-2]$, i.e. the system is constructable, but not observable.

In control applications, the system to be observed is the plant. If it is discretized, then $\underline{A} = e^{\underline{F}T}$ does not have zero eigenvalues and zero eigenvalues of time-delay systems, as in figure 3.20, are observable. In these cases observability and constructability are equivalent. Therefore we will only speak of observability in the following.

Corresponding to controllability, the following holds for the continuous and discretized system [60.4]: a discrete system, generated from a continuous system by sampling the output, is observable if the continuous part is observable and distinct eigenvalues s_i, s_j of the continuous system lead to distinct eigenvalues e^{s_iT}, e^{s_jT} of the discrete system. A system with a conjugate complex pole pair $s_{1,2} = \sigma \pm j\omega$ looses its observability if it is sampled with a period $T = \pm q\pi/\omega$, $q = 1,2,3\ldots$.

Clearly the observability is not lost if the continuous output $y(kT+\gamma T)$ is available. According to eq. (3.7.9), here written for multi-variable systems with output $\underline{y} = \underline{Cx} + \underline{Du}$

$$\underline{y}(kT+\gamma T) = \underline{CA}_\gamma \underline{x}(kT) + (\underline{CB}_\gamma + \underline{D})\underline{u}(kT) \tag{5.1.11}$$

with

$$\underline{A}_\gamma = e^{\underline{F}\gamma T}, \quad \underline{B}_\gamma = \int_0^{\gamma T} e^{\underline{F}v}\underline{G}\, dv \tag{5.1.12}$$

The following relation can therefore be deduced, corresponding to eq. (5.1.3)

$$\underline{y}_{n-1}[k+\gamma] = \begin{bmatrix} y[k+\gamma] \\ y[k+1+\gamma] \\ \vdots \\ y[k+n-1+\gamma] \end{bmatrix} = \begin{bmatrix} \underline{CA}_\gamma \\ \underline{CA}_\gamma\underline{A} \\ \vdots \\ \underline{CA}_\gamma\underline{A}^{n-1} \end{bmatrix} \underline{x}[k] + \underline{Mu}_{n-1}[k] \tag{5.1.13}$$

The system is observable from $\underline{y}[k+\gamma]$ if and only if

$$\operatorname{rank}\begin{bmatrix} \underline{CA}_\gamma \\ \underline{CA}_\gamma\underline{A} \\ \vdots \\ \underline{CA}_\gamma\underline{A}^{n-1} \end{bmatrix} = n \quad \text{for some } \gamma\in[0,1] \tag{5.1.14}$$

Correspondingly an eigenvalue z_i is observable from $\underline{y}[k+\gamma]$ if and only if

$$\operatorname{rank}\begin{bmatrix} \underline{A} - z_i\underline{I} \\ \underline{CA}_\gamma \end{bmatrix} = n \quad \text{for some } \gamma\in[0,1] \tag{5.1.15}$$

The eigenvalue z_i is unobservable from the continuous output variable $y(t)$ only if the rank is smaller than n for all γ in the interval $0 \leq \gamma < 1$. Practically it is easier to test this observability from the output $y(t) = y(kT+\gamma T)$ for the continuous system directly. This concept was introduced however, in order to facilitate the understanding of continuous and discrete pole-zero cancellations in section 6.6.

An example is given in figure 3.18 and eq. (3.7.19): the controller pole at $z = -0.718$ cancels the zero of the plant transfer function only for $\gamma = 0$. This eigenvalue is not observable from $y[k]$, but it is observable from $y[k+\gamma]$. Generally it is not advisable in the design of sampled-data control systems to cancel plant zeros by poles of the discrete controller.

5.2 The Observer of Order n

A causal construction of the actual state from n input and n output samples is possible by eqs. (5.1.9) and (5.1.6) providing the system is observable. If the actual state is required at all sampling instants, then many computations are necessary. This computational effort even increases if more than the minimum set of data is used, i.e. $N > n$ in eq. (5.1.3) and a least squares calculation is made in order to smooth noisy data. It is computationally more efficient to use a recursive algorithm which is implemented as a dynamical system called an "observer". (In view of the definitions it should be called "constructor", but "observer" is the standard terminology.)

For a plant

$$\begin{aligned} \underline{x}[k+1] &= \underline{A}\underline{x}[k] + \underline{b}\underline{u}[k] \\ \underline{y}[k] &= \underline{C}\underline{x}[k] \end{aligned} \qquad (5.2.1)$$

an observer should now be designed, i.e. a linear dynamic system with the measurable variables $\underline{u}[k]$ and $\underline{y}[k]$ as input such that this system constructs an estimate $\hat{\underline{x}}$ for the state $\underline{x}$. Let this be the n-th order system

$$\hat{\underline{x}}[k+1] = \underline{F}\hat{\underline{x}}[k] + \underline{G}\underline{u}[k] + \underline{H}\underline{y}[k] \qquad (5.2.2)$$

As Luenberger has shown [64.8], [66.2], $\underline{F}$, $\underline{G}$ and $\underline{H}$ can be chosen such that $\hat{\underline{x}}[k]$ approaches $\underline{x}[k]$ asymptotically. Suppose that a disturbance has affected the plant (5.2.1) but not the system (5.2.2). Therefore initially $\hat{\underline{x}}[0] \neq \underline{x}[0]$. Introduce the construction error

$$\tilde{\underline{x}} := \underline{x} - \hat{\underline{x}} \qquad (5.2.3)$$

and obtain by subtraction of eq. (5.2.2) from (5.2.1) the difference equation for $\tilde{\underline{x}}$

$$\tilde{\underline{x}}[k+1] = \underline{F}\tilde{\underline{x}}[k] + (\underline{A}-\underline{F}-\underline{H}\underline{C})\underline{x}[k] + (\underline{B}-\underline{G})\underline{u}[k] \qquad (5.2.4)$$

The construction error $\tilde{\underline{x}}[k]$ should go asymptotically to zero. This can be achieved by making eq. (5.2.4) a stable, homogeneous difference equation, i.e.

1. $\underline{F} = \underline{A} - \underline{H}\underline{C}$ (5.2.5)

2. $\underline{G} = \underline{B}$ (5.2.6)

3. The system resulting from eq. (5.2.4)

   $$\tilde{\underline{x}}[k+1] = (\underline{A}-\underline{H}\underline{C})\tilde{\underline{x}}[k] \qquad (5.2.7)$$

   must be asymptotically stable.

The choice of $\underline{G} = \underline{B}$ ensures that the construction error $\underline{\tilde{x}}$ is not controllable from $\underline{u}$, i.e. if $\underline{\tilde{x}}[k]$ has become zero, then the influence of the input $\underline{u}$ does not produce a construction error. With eqs. (5.2.5) and (5.2.6) the system (5.2.2) takes on the form

$$\begin{aligned}\underline{\hat{x}}[k+1] &= (\underline{A}-\underline{H}\underline{C})\underline{\hat{x}}[k] + \underline{B}\underline{u}[k] + \underline{H}\underline{y}[k] \\ &= \underline{A}\underline{\hat{x}}[k] + \underline{B}\underline{u}[k] + \underline{H}(\underline{y}[k]-\underline{C}\underline{\hat{x}}[k]) \qquad (5.2.8) \\ &= \underline{A}\underline{\hat{x}}[k] + \underline{B}\underline{u}[k] + \underline{H}\underline{C}\underline{\tilde{x}}[k]\end{aligned}$$

This system is an observer of order n for the plant (5.2.1). The observer is a model of the plant which is driven by the measurable error $\underline{C}\underline{\tilde{x}}$, figure 5.1. Such an observer was first introduced by Kalman [60.4].

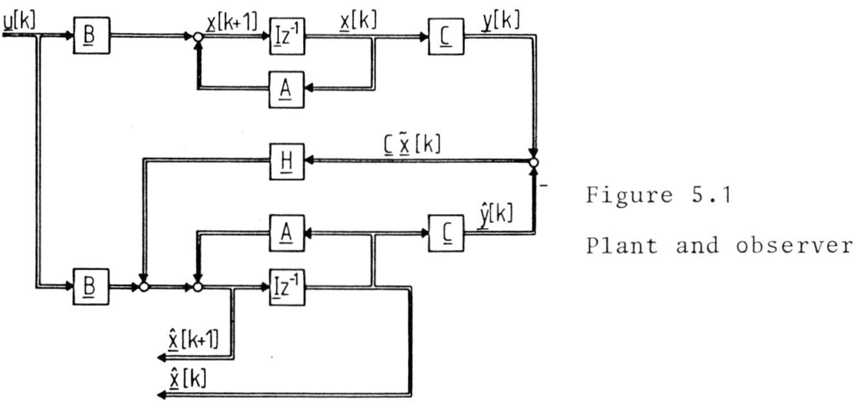

Figure 5.1
Plant and observer

The difference $\underline{C}\underline{\tilde{x}}$ between the precalculated estimate $\underline{\hat{y}} = \underline{C}\underline{\hat{x}}$ and the actually measured value $\underline{y}$ becomes zero if the states of observer and plant are identical. The observer is then merely a model of the plant which is stimulated by $\underline{u}$.

If the system is observable, then desired dynamics can be given to the error difference equation (5.2.7) by the choice of the matrix $\underline{H}$. One assigns n real or complex "observer poles" $z_1 \ldots z_n$ as zeros of the characteristic polynomial of the error equation

$$Q(z) = q_0+q_1z+\ldots+q_{n-1}z^{n-1}+z^n = (z-z_1)(z-z_2)\ldots(z-z_n) \qquad (5.2.9)$$

i.e. for given $\underline{A}$, $\underline{C}$ and Q(z) one determines the gain matrix $\underline{H}$ of the observer such that

$$\det(z\underline{I} - \underline{A} + \underline{HC}) = Q(z) \tag{5.2.10}$$

The question as to how the observer poles z_i should be properly chosen, will be discussed in section 5.4. The solution of eq. (5.2.10) for $\underline{H}$ corresponds basically to pole assignment in the multi-variable case, i.e. for a state vector feedback $\underline{u} = -\underline{Kx}$. Here $\underline{K}$ must be chosen so that

$$\det(z\underline{I} - \underline{A} + \underline{BK}) = P(z) \tag{5.2.11}$$

For a given P(z) the solution is unique for the single-input case $\underline{K} = \underline{k}'$, $\underline{B} = \underline{b}$ and a controllable pair $(\underline{A}, \underline{b})$, see eqs. (4.4.4) and (4.4.6). In the multi-variable case there are more unknowns than equations in (5.2.11) and the solution is no longer unique. This case will be handled in detail in chapter 9. This will also solve the problem in eq. (5.2.10). This is seen by transposing

$$\det(z\underline{I} - \underline{A}' + \underline{C}'\underline{H}') = Q(z) \tag{5.2.12}$$

Comparison with eq. (5.2.11) shows that $\underline{A}$ must be replaced by $\underline{A}'$, $\underline{B}$ by $\underline{C}'$, $\underline{K}$ by $\underline{H}'$, and P(z) by Q(z) in order to obtain eq. (5.2.12).

In the case of only one output variable $y = \underline{c}'\underline{x}$, $\underline{H}$ is a column vector $\underline{h}$ and eqs. (5.2.10) and (5.2.12) are

$$\det(z\underline{I} - \underline{A} + \underline{hc}') = \det(z\underline{I} - \underline{A}' + \underline{ch}') = Q(z) \tag{5.2.13}$$

From eq. (4.4.5) the solution is

$$\underline{h}' = \underline{f}'Q(\underline{A}') = [0 \dots 0 \quad 1][\underline{c}, \underline{A}'\underline{c} \dots (\underline{A}')^{n-1}\underline{c}]^{-1}Q(\underline{A}')$$

$$\underline{h} = Q(\underline{A})\underline{f} = Q(\underline{A})\begin{bmatrix} \underline{c}' \\ \underline{c}'\underline{A} \\ \vdots \\ \underline{c}'\underline{A}^{n-1} \end{bmatrix}^{-1} \begin{bmatrix} 0 \\ \vdots \\ 0 \\ 1 \end{bmatrix} \tag{5.2.14}$$

The parameter vector $\underline{h}$ of the observer is obtained by inserting the dynamic matrix $\underline{A}$ into the desired characteristic polynomial Q and then multiplying by $\underline{f}$, the last column of the inverted observability matrix. Numerically this is calculated like $\underline{e}'$ in eq. (A.5.14).

Corresponding to eq. (4.4.6), $\underline{h}$ can also be expressed by the coefficient vector $\underline{q} = [q_o \quad q_1 \quad \dots \quad q_{n-1}]'$ of the desired characteristic polynomial as

$$\underline{h} = \underline{F}\begin{bmatrix}\underline{q}\\1\end{bmatrix}, \qquad \underline{F} = [\underline{f}, \underline{Af}\dots\underline{A}^n\underline{f}] \tag{5.2.15}$$

Remark 5.1:

The observer-canonical form, eq. (A.4.11), for the state equations of the plant can also be used for assigning the observer poles. This corresponds to the determination of a control law by feedback-canonical form. In this form the equation of the observer is

$$\hat{\underline{x}}_F[k+1] = \begin{bmatrix}0 & & & -a_o-h_{F1}\\ 1 & \ddots & & \vdots\\ & & \ddots & \vdots\\ 0 & & 1 & -a_{n-1}-h_{Fn}\end{bmatrix}\hat{\underline{x}}_F[k]+\underline{b}_F u[k]+\begin{bmatrix}h_{F1}\\ \vdots\\ h_{Fn}\end{bmatrix}y[k] \tag{5.2.16}$$

Its characteristic equation is

$$(a_o+h_{F1})+(a_1+h_{F2})z+\dots+(a_{n-1}+h_{Fn})z^{n-1}+z^n = 0 \tag{5.2.17}$$

By matching the coefficients with those of eq. (5.2.9), one obtains the components of the vector $\underline{h}_F$, namely

$$h_{Fi} = q_{i-1}-a_{i-1}, \quad i = 1,2\dots n \tag{5.2.18}$$

Example: Loading bridge

For the loading bridge with the numerical values from exercise 4.1, the subsystem "pendulum" has the state representation

$$\begin{bmatrix} x_3[k+1] \\ x_4[k+1] \end{bmatrix} = \begin{bmatrix} 1/\sqrt{2} & 1/2\sqrt{2} \\ -\sqrt{2} & 1/\sqrt{2} \end{bmatrix} \begin{bmatrix} x_3[k] \\ x_4[k] \end{bmatrix} + 10^4 \begin{bmatrix} (1/\sqrt{2}-1)/4 \\ -1/2\sqrt{2} \end{bmatrix} u[k]$$

$$. \tag{5.2.19}$$

The angular velocity $x_4[k]$ should be constructed from the measurement of u[k] and the cable angle $x_3[k]$ by an observer with the characteristic polynomial $Q(z) = q_0+q_1z+z^2$. By eq. (5.2.14)

$$\begin{bmatrix} \underline{c}' \\ \underline{c}'\underline{A} \end{bmatrix} \underline{f} = \begin{bmatrix} 0 \\ 1 \end{bmatrix}$$

$$\begin{bmatrix} 1 & 0 \\ 1/\sqrt{2} & 1/2\sqrt{2} \end{bmatrix} \begin{bmatrix} f_1 \\ f_2 \end{bmatrix} = \begin{bmatrix} 0 \\ 1 \end{bmatrix}, \quad \underline{f} = \begin{bmatrix} 0 \\ 2\sqrt{2} \end{bmatrix}$$

Eq. (5.2.15) is here

$$\underline{h} = \begin{bmatrix} h_1 \\ h_2 \end{bmatrix} = \begin{bmatrix} 0 & 1 & \sqrt{2} \\ 2\sqrt{2} & 2 & 0 \end{bmatrix} \begin{bmatrix} q_0 \\ q_1 \\ 1 \end{bmatrix} = \begin{bmatrix} q_1 + \sqrt{2} \\ q_0 2\sqrt{2} + 2q_1 \end{bmatrix}$$

The equation of the observer is therefore

$$\begin{bmatrix} \hat{x}_3[k+1] \\ \hat{x}_4[k+1] \end{bmatrix} = \begin{bmatrix} 1/\sqrt{2} & 1/2\sqrt{2} \\ -\sqrt{2} & 1/\sqrt{2} \end{bmatrix} \begin{bmatrix} \hat{x}_3[k] \\ \hat{x}_4[k] \end{bmatrix} + 10^4 \begin{bmatrix} (1/\sqrt{2}-1)/4 \\ -1/2\sqrt{2} \end{bmatrix} u[k] +$$

$$+ \begin{bmatrix} q_1 + \sqrt{2} \\ q_0 2\sqrt{2} + 2q_1 \end{bmatrix} (x_3[k] - \hat{x}_3[k]) \tag{5.2.20}$$

For systems with several outputs it is advantageous to assign one subsystem to each output. This may be done by transformation of the plant model into a canonical form which exhibits these subsystems - see for example the observer-canonical version of eq. (A.4.1). Then a single-output observer can construct the states of one subsystem from the corresponding output and the coupling terms with other subsystems.

Example:

The correspondence of subsystems and outputs is obvious for the loading bridge. The subobserver determined above for the pendulum can be supplemented by a second subobserver for the trolley with the output x_1 = trolley position. Since exercise 2.1 the coordinates x_1^* = position of the center of gravity and x_2^* = velocity of the center of gravity have been used. x_1^* and x_2^* cannot be measured, rather only x_1 and x_2 for the trolley itself. Therefore the state equations in $\underline{x}^*$ are first transformed back into the trolley coordinates in $\underline{x}$

$$\underline{x}[k+1] = \underline{A}\underline{x}[k] + \underline{b}u[k]$$

$$\underline{A} = \begin{bmatrix} 1 & \pi/8 & 7.5(1-1/\sqrt{2}) & 7.5(\pi/8-1/2\sqrt{2}) \\ 0 & 1 & 7.5\sqrt{2} & 7.5(1-1/\sqrt{2}) \\ 0 & 0 & 1/\sqrt{2} & 1/2\sqrt{2} \\ 0 & 0 & -\sqrt{2} & 1/\sqrt{2} \end{bmatrix}$$

$$\underline{b} = 10^{-3}\begin{bmatrix} \pi^2-96(1\sqrt{2}-1)/512 \\ (\pi+6\sqrt{2})/32 \\ (1/\sqrt{2}-1)/40 \\ -1/20\sqrt{2} \end{bmatrix} \tag{5.2.21}$$

For the subsystem trolley with the state variables x_1 and x_2 and the measurement of x_1, the observability matrix yields

$$\begin{bmatrix} \underline{c}' \\ \underline{c}'\underline{A} \end{bmatrix}\underline{f} = \begin{bmatrix} 1 & 0 \\ 1 & \pi/8 \end{bmatrix}\begin{bmatrix} f_1 \\ f_2 \end{bmatrix} = \begin{bmatrix} 0 \\ 1 \end{bmatrix}, \quad \underline{f} = \begin{bmatrix} 0 \\ 8/\pi \end{bmatrix}$$

The observer produces the characteristic polynomial $\tilde{Q}(z) = \tilde{q}_0 + \tilde{q}_1 z + z^2$ by the feedback vector

$$\underline{h} = \begin{bmatrix} 0 & 1 & 2 \\ 8/\pi & 8/\pi & 8/\pi \end{bmatrix}\begin{bmatrix} \tilde{q}_0 \\ \tilde{q}_1 \\ 1 \end{bmatrix} = \begin{bmatrix} \tilde{q}_1+2 \\ (\tilde{q}_0+\tilde{q}_1+1)8/\pi \end{bmatrix}$$

The coupling terms of the state variables x_3 and x_4 are formed with the correspondingly constructed values $\hat{x}_3$ and $\hat{x}_4$. For the design of the subobserver these values are handled like u:

$$\begin{bmatrix}\hat{x}_1[k+1]\\ \hat{x}_2[k+1]\end{bmatrix} = \begin{bmatrix}1 & \pi/8\\ 0 & 1\end{bmatrix}\begin{bmatrix}\hat{x}_1[k]\\ \hat{x}_2[k]\end{bmatrix} + \begin{bmatrix}7.5(1-1/\sqrt{2}) & 7.5(\pi/8-1/2\sqrt{2})\\ 7.5\sqrt{2} & 7.5(1-1/\sqrt{2}\end{bmatrix}\begin{bmatrix}\hat{x}_3[k]\\ \hat{x}_4[k]\end{bmatrix}$$

$$+ 10^{-3}\begin{bmatrix}[\pi^2-96(1/\sqrt{2}-1)]/512\\ (\pi+6\sqrt{2})/32\end{bmatrix}u[k] + \begin{bmatrix}\tilde{q}_1+2\\ (\tilde{q}_0+\tilde{q}_1+1)8/\pi\end{bmatrix}(x_1[k]-\hat{x}_1[k]) \tag{5.2.22}$$

The two subobservers (5.2.20) and (5.2.22) can now be joined together to form a complete observer:

$$\underline{\hat{x}}[k+1] = \underline{A}\underline{\hat{x}}[k] + \underline{b}u[k] + \begin{bmatrix}\tilde{q}_0+2 & 0\\ (\tilde{q}_0+\tilde{q}_1+1)8/\pi & 0\\ 0 & q_1+\sqrt{2}\\ 0 & q_0 2\sqrt{2}+2q_1\end{bmatrix}\begin{bmatrix}x_1[k]-\hat{x}_1[k]\\ x_3[k]-\hat{x}_3[k]\end{bmatrix} \tag{5.2.23}$$

The characteristic polynomial of the observer above is $Q(z)\times\tilde{Q}(z) = (q_0+q_1z+z^2)(\tilde{q}_0+\tilde{q}_1z+z^2)$.

A practical advantage of the observer of figure 5.1 is that it is also a predictor, providing an estimated value $\underline{\hat{x}}[k+1]$ one sampling interval ahead. Therefore an adjustable parameter α can be introduced into the system according to figure 5.2.

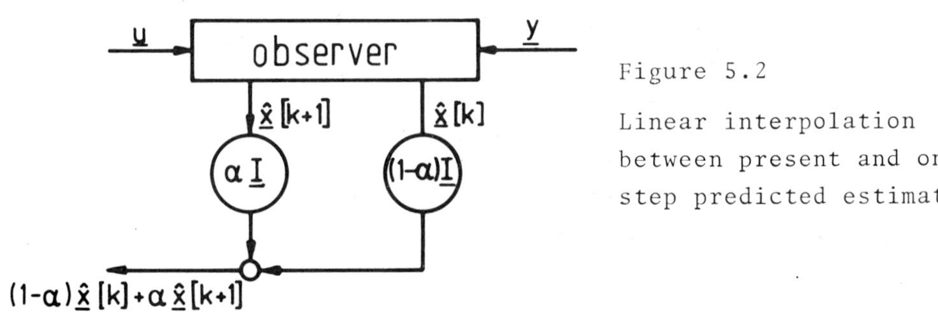

Figure 5.2
Linear interpolation between present and one step predicted estimates

The prediction time is changed by adjusting α between 0 (observer) and 1 (predictor) during the operation of the plant. The prediction produces an additional phase margin of an angle ωT for the frequency ω in the closed loop. This compensates for the phase delay of an anti-aliasing filter eq. (4.3.11), neglected calculation and conversion times or neglected time constants of the plant, the actuator, and the sensors. The amplitude of the frequency response remains unchanged.

5.3 The Reduced Order Observer

If $\underline{\hat{x}}[k]$ from the n-th order observer in figure 5.1 is fed back, then plant disturbances in y[k] do not have an immediate effect on $\underline{\hat{x}}[k]$ and the feedback $\underline{u}[k] = -\underline{K}\underline{\hat{x}}[k]$. Its effect is first noticeable in $\underline{u}[k+1]$. Thus one sampling interval has been lost which could have been used to compensate for the disturbance. Taking the observer (5.2.23) of the loading bridge for instance, $\hat{x}_1$ and $\hat{x}_3$ are constructed in the observer, although the values x_1 and x_3 are measured by

$$\underline{y}[k] = \underline{C}\underline{x}[k] \ , \ \underline{C} = \begin{bmatrix} 1 & 0 & 0 & 0 \\ 0 & 0 & 1 & 0 \end{bmatrix}$$

and can be fed back immediately. In this example an observer of second order would have been sufficient for the construction of $\hat{x}_2$ and $\hat{x}_4$. This would also save calculation time in the observer. As shown in appendix A.6, a system with s output variables $\underline{y} = [y_1 \dots y_s]'$

$$\begin{aligned} \underline{x}[k+1] &= \underline{A}\underline{x}[k] + \underline{B}u[k] \\ \underline{y}[k] &= \underline{C}\underline{x}[k] \ , \ \text{rank } \underline{C} = s \leq n, \ (\underline{A}, \underline{C}) \text{ observable} \end{aligned} \tag{5.3.1}$$

can always be transformed into so-called "sensor coordinates"

$$\begin{bmatrix} \underline{z} \\ \underline{y} \end{bmatrix} = \underline{T}\underline{x} \ , \quad \det \underline{T} \neq 0 \tag{5.3.2}$$

where $\underline{y}$ is part of the state vector and $\underline{z}$ is an (n-s)-vector. Then only an observer of order n-s is needed for the construction of $\hat{\underline{z}}$ (referred to as the reduced-order observer), and $\hat{\underline{x}}$ is obtained as

$$\hat{\underline{x}} = \underline{T}^{-1} \begin{bmatrix} \hat{\underline{z}} \\ \underline{y} \end{bmatrix} \tag{5.3.3}$$

In sensor coordinates, eq. (A.6.3), the state equations of the plant are

$$\underline{z}[k+1] = \underline{Pz}[k] + \underline{Qy}[k] + \underline{Du}[k] \tag{5.3.4}$$

$$\underline{y}[k+1] = \underline{Rz}[k] + \underline{Sy}[k] + \underline{Eu}[k] \tag{5.3.5}$$

where

$$\underline{TAT}^{-1} = \begin{bmatrix} \underline{P} & \underline{Q} \\ \underline{R} & \underline{S} \end{bmatrix}, \quad \underline{TB} = \begin{bmatrix} \underline{D} \\ \underline{E} \end{bmatrix}$$

($\underline{P}$, $\underline{R}$) is observable, and a reduced order observer for $\underline{z}$ can be developed as a linear system of order n-s [64.8], [66.2]:

$$\hat{\underline{v}}[k+1] = \underline{L}\hat{\underline{v}}[k] + \underline{My}[k] + \underline{Nu}[k] \tag{5.3.6}$$

$$\hat{\underline{z}}[k] = \hat{\underline{v}}[k] + \underline{Hy}[k] \tag{5.3.7}$$

Again the matrices $\underline{L}$, $\underline{M}$, $\underline{N}$, and $\underline{H}$ must be determined such that the construction error $\tilde{\underline{z}} = \underline{z}-\hat{\underline{z}}$ goes to zero asymptotically [76.1]. Form $\hat{\underline{z}}[k+1]$ according to eq. (5.3.7) and substitute eqs. (5.3.5) and (5.3.6) to give

$$\hat{\underline{z}}[k+1] = \underline{L}\hat{\underline{v}}[k] + \underline{My}[k] + \underline{Nu}[k] + \underline{HRz}[k] + \underline{HSy}[k] + \underline{HEu}[k] \tag{5.3.8}$$

substituting for $\hat{\underline{v}}$ from eq. (5.3.7) leads to

$$\hat{\underline{z}}[k+1] = \underline{L}\hat{\underline{z}}[k] + \underline{HRz}[k] + (\underline{M}+\underline{HS}-\underline{LH})\underline{y}[k] + (\underline{N}+\underline{HE})\underline{u}[k] \tag{5.3.9}$$

The difference equation of the construction error is obtained by augmenting eq. (5.3.4) by $+\underline{L}\underline{z}$ $-\underline{L}\underline{z}$ and subtracting eq. (5.3.9)

$$\underline{z}[k+1] = \underline{L}\underline{z}[k] + (\underline{P}-\underline{L})\underline{z}[k] + \underline{Q}\underline{y}[k] + \underline{D}\underline{u}[k]$$

$$\hat{\underline{z}}[k+1] = \underline{L}\hat{\underline{z}}[k] + \underline{H}\underline{R}z[k] + (\underline{M}+\underline{H}\underline{S}-\underline{L}\underline{H})\underline{y}[k] + (\underline{N}+\underline{H}\underline{E})\underline{u}[k]$$

$$\tilde{\underline{z}}[k+1] = \underline{L}\tilde{\underline{z}}[k] + (\underline{P}-\underline{H}\underline{R}-\underline{L})\underline{z}[k] + (\underline{Q}-\underline{M}-\underline{H}\underline{S}+\underline{L}\underline{H})\underline{y}[k] + (\underline{D}-\underline{N}-\underline{H}\underline{E})\underline{u}[k] \qquad (5.3.10)$$

$\tilde{\underline{z}}$ goes asymptotically to zero if all input variables $\underline{z}$, $\underline{y}$ and $\underline{u}$ are multiplied by zero and the eigenvalues of $\underline{L}$ lie in the unit circle. This yields

1. $\underline{L} = \underline{P}-\underline{H}\underline{R}$ (5.3.11)

2. $\underline{M} = \underline{Q}-\underline{H}\underline{S}+\underline{L}\underline{H}$ (5.3.12)

3. $\underline{N} = \underline{D}-\underline{H}\underline{E}$ (5.3.13)

4. $\underline{H}$ is chosen so that the error equation

   $$\tilde{z}[k+1] = (\underline{P}-\underline{H}\underline{R})\tilde{z}[k] \qquad (5.3.14)$$

   is stable.

Here again the observer design reduces to the solution of the equation

$$\det(z\underline{I}-\underline{P}+\underline{H}\underline{R}) = Q(z) \qquad (5.3.15)$$

for $\underline{H}$, i.e. the problem of eq. (5.2.10). The only difference is the reduced dimension of the matrices $\underline{P}$, $\underline{H}$ and $\underline{R}$ and the reduced degree of the polynomial Q(z).

The plant of eqs. (5.3.4), (5.3.5) and the reduced order observer of eqs. (5.3.6), (5.3.7), (5.3.11) to (5.3.13) are illustrated by figure 5.3.

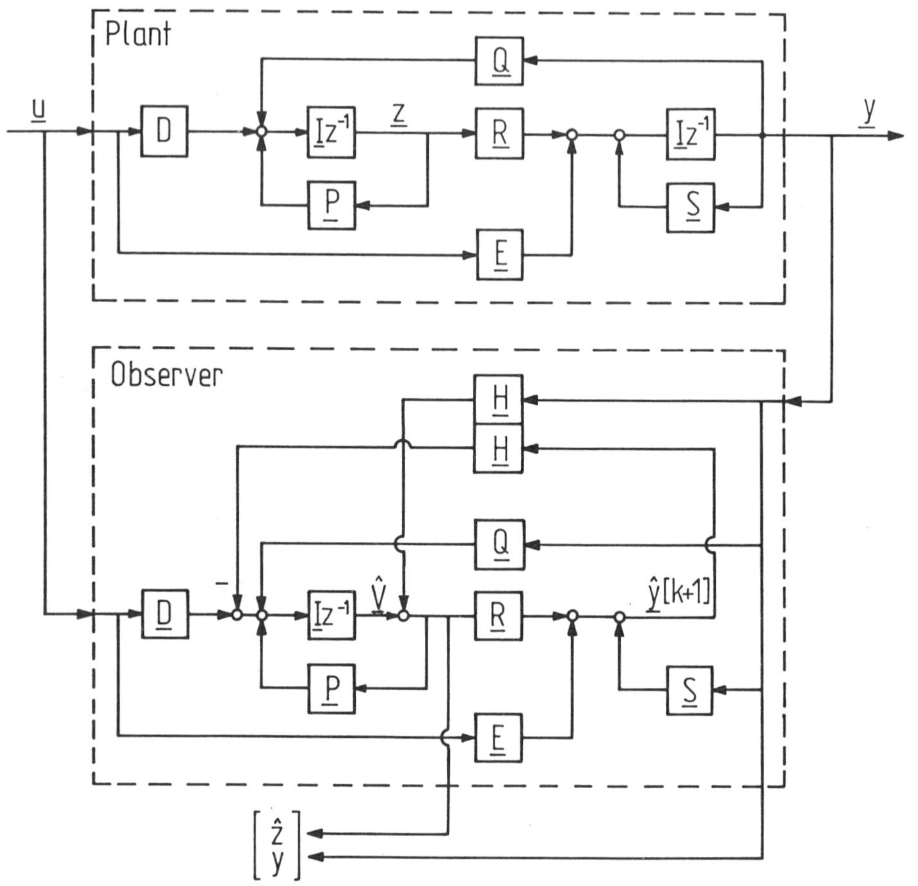

Figure 5.3 Plant and reduced-order observer

An example of the practical application of this reduced observer of eqs. (5.3.6) and (5.3.7) and its determination by eqs. (5.3.11) to (5.3.14) is reported in [80.5]. For a track-guided bus the steering-angle sensor is replaced by an observer for reliability reasons.

5.4 Choice of the Observer Poles

The transients of the construction error of eqs. (5.2.7) and (5.3.14) respectively are determined by the observer poles which are fixed by the choice of the matrix $\underline{H}$.

An especially quick decay of the construction error in a finite transient is obtained with a deadbeat observer, i.e. an observer with all poles at z = 0. The operation of such an observer can be illustrated best in the observer-canonical form (A.3.22). Let the state representation of a plant be transformed into this form

$$\underline{x}_r[k+1] = \begin{bmatrix} 0 & & & & -a_o \\ 1 & \cdot & & & \cdot \\ & & \cdot & & \cdot \\ & & & \cdot & \cdot \\ 0 & & & 1 & -a_{n-1} \end{bmatrix} \underline{x}_r[k] + \begin{bmatrix} b_o \\ \cdot \\ \cdot \\ \cdot \\ b_{n-1} \end{bmatrix} u[k] \qquad (5.4.1)$$

$$y[k] = [0 \; \ldots \; 0 \quad 1] \; \underline{x}_r[k]$$

Then in eqs. (5.3.4) and (5.3.5)

$$\underline{z} = \begin{bmatrix} x_{r1} \\ \vdots \\ x_{rn-1} \end{bmatrix}, \quad \underline{P} = \begin{bmatrix} 0 & \cdot & & 0 \\ 1 & \cdot & \cdot & \\ & \cdot & \cdot & \cdot \\ 0 & & 1 & 0 \end{bmatrix}, \quad \underline{Q} = -\begin{bmatrix} a_o \\ \vdots \\ a_{n-2} \end{bmatrix}, \quad \underline{D} = \begin{bmatrix} b_o \\ \vdots \\ b_{n-2} \end{bmatrix}$$

$$y = x_{rn}, \quad \underline{R} = \underline{r}' = [0 \; \ldots \; 0 \quad 1], \quad S = -a_{n-1}, \quad E = b_{n-1}$$

and

$$\underline{P} - \underline{hr}' = \begin{bmatrix} 0 & & & & -h_1 \\ 1 & \cdot & & & \cdot \\ & & \cdot & & \cdot \\ & & & \cdot & \cdot \\ 0 & & & 1 & -h_{n-1} \end{bmatrix} \qquad (5.4.2)$$

In this form the elements of the vector $\underline{h}$ are directly the coefficients of the observer characteristic polynomial. Here one obtains the deadbeat observer from eqs. (5.3.6), (5.3.7) for $\underline{h} = \underline{0}$ and with $\underline{L}$, $\underline{M}$, $\underline{N}$ according to eqs. (5.3.11) to (5.3.13).

$$\begin{bmatrix} \hat{x}_{r1}[k+1] \\ \vdots \\ \hat{x}_{rn-1}[k+1] \end{bmatrix} = \begin{bmatrix} 0 & & & 0 \\ 1 & \ddots & & \\ & & \ddots & \\ 0 & & 1 & 0 \end{bmatrix} \begin{bmatrix} \hat{x}_{r1}[k] \\ \vdots \\ \hat{x}_{rn-1}[k] \end{bmatrix} + \begin{bmatrix} b_o \\ \vdots \\ b_{n-2} \end{bmatrix} u[k] - \begin{bmatrix} a_o \\ \vdots \\ a_{n-2} \end{bmatrix} y[k]$$

$$\underline{\hat{x}}_r = \begin{bmatrix} \hat{x}_{r1} \\ \vdots \\ \hat{x}_{rn-1} \\ y \end{bmatrix} \qquad (5.4.3)$$

Figure 5.4 illustrates the operation of this observer.

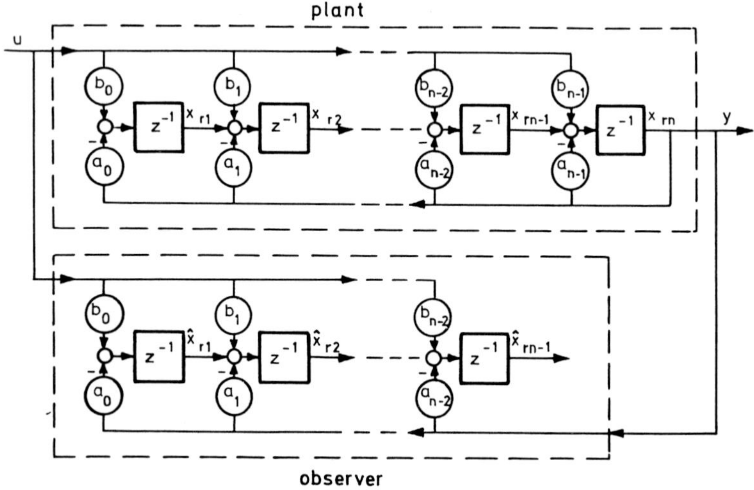

Figure 5.4 Reduced-order deadbeat observer

The observer-canonical form of the plant is represented in the upper portion of the figure. This is an especially convenient model representation which describes the relationship between the input sequence u[k] and the output sequence y[k]. The observer in the lower portion of the figure is actually implemented as shown in the diagram such that the constructed values are available for feedback.

The same variable $b_o u - a_o y$ arises at the input of the first storage element in the observer as in the plant. If initially $\hat{x}_{r1}[0] \neq x_{r1}[0]$, then after one sampling interval

$$\hat{x}_{r1}[1] = x_{r1}[1] \tag{5.4.4}$$

After two sampling intervals

$$\hat{x}_{r1}[2] = x_{r1}[2], \ \hat{x}_{r2}[2] = x_{r2}[2] \tag{5.4.5}$$

and so on until after n-1 steps

$$\hat{x}_{ri}[n-1] = x_{ri}[n-1], \ i = 1,2...n-1, \tag{5.4.6}$$

i.e. the states of observer and plant become identical. Here it was assumed that no disturbance has influenced the plant in the given interval. The reduction of observer order is apparent in figure 5.4: One could also form $\hat{x}_{rn}$ in the observer by an n-th storage element, thus making it a complete model. But then $\hat{x}_{rn}[n] = x_{rn}[n]$ only after n steps whereas in eq. (5.4.3) $\hat{x}_{rn}[k] = y[k] = x_{rn}[k]$ for all k.

It is obviously impossible for $\hat{x}_{r1}[0] \neq x_{r1}[0]$ to obtain a complete estimate value of the state in less than n - 1 sampling intervals. A faster construction of the state is possible, however, if more linearly independent outputs $y_1 ... y_s$ are available. Then each output is used to construct the state of one subsystem. The necessary number of sampling intervals is equal to the order of the largest subsystem minus one. It is apparent here that the orders of the subsystems should be as nearly equal as possible. The observer structure is thereby fixed which allows the fastest construction. In generalization of eq. (5.4.1) the observation dual of the Luenberger canonical form, eq. (A.4.1) can be used.

In the presence of plant disturbances $\underline{z}_s(t)$ and measurement noise $\underline{z}_m(t)$ the continuous time plant equations are

$$\dot{\underline{x}}(t) = \underline{Fx}(t) + \underline{Gu}(t) + \underline{z}_s(t)$$
$$\underline{y}(t) = \underline{Cx}(t) + \underline{z}_m(t)$$

and after discretization with $\underline{s}[k] = \int_0^T e^{Fv}\underline{z}_s(kT+T-v)dv$

$$\begin{aligned} \underline{x}[k+1] &= \underline{Ax}[k] + \underline{Bu}[k] + \underline{s}[k] \\ \underline{y}[k] &= \underline{Cx}[k] + \underline{z}_m[k] \end{aligned} \qquad (5.4.7)$$

A Kalman-Bucy filter [60.2], [61.1] can be used for estimating $\hat{\underline{x}}$ if the disturbances are described as stochastic process with known covariance matrices. This filter has the same structure as an observer of order n, figure 5.1, but the optimal time-variable $\underline{H}[k]$ is determined as the solution of a Riccati difference equation. For steady-state conditions a constant $\underline{H}$-matrix is obtained which results in a favorable position of the observer poles. In practice, however, the stochastic processes, generating $\underline{s}[k]$ and $\underline{z}_m[k]$ are frequently not so well known. Instead of experimenting with assumptions on these processes, one can also use the observer poles as design parameters. Here eq. (5.4.7) provides a qualitative foothold.

It is assumed, that high frequency effects of plant and measurement noise outside the bandwidth of the control system have been reduced by an anti-aliasing filter before the sampler for $\underline{y}(t)$. The following discussion refers to the effects of $\underline{s}[k]$ and $\underline{z}_m[k]$ inside the bandwidth of the control system, where a compromise between fast state construction for the determination of the effect of $\underline{s}[k]$ and good filtering of $\underline{z}_m[k]$ is necessary.

Suppose $\underline{s}[k] = \underline{0}$ and let the measuring noise $\underline{z}_m$ be so large that the measurement $\underline{y}$ is almost useless. In this case a construction of $\hat{\underline{x}}$ by an n-th order observer is recommended. Then only the filtered $\hat{\underline{x}}$, not $\underline{y}$, is fed back. The observer, see figure 5.1, works with very small feedback in $\underline{H}$, almost as a parallel model. $\underline{H}$ is necessary only in order to stabilize the construction error dynamics. If $\underline{A}$ is already stable enough, then the eigenvalues of $\underline{A} - \underline{HC}$ need not be changed i.e. $\underline{H} = \underline{0}$ can be set and feedback of $\hat{\underline{x}}[k]$ results in a pure feedforward control of the plant.

Another extreme case is given for $\underline{z}_m = \underline{0}$, but large disturbances $\underline{s}$. Here it is important that the observer quickly tracks the influence of the disturbances on the plant states. A deadbeat-observer can be used. Because no measurement noise exists, $\underline{y}$ can be fed back unfiltered. For fast tracking the reduced observer is applied. This produces a system as in figure 5.4. An exact agreement of the states of plant and observer can never be reached because of the continually acting plant disturbances $\underline{s}$. But in a deadbeat observer at least the effect of plant disturbances, which have acted some sampling intervals ago, is completely eliminated. The number of these sampling intervals is equal to the observer order.

Practical cases will lie between these two extremes. This means that the observer eigenvalues are shifted closer to the origin than those of the plant $\underline{A}$, but the origin may not be exactly reached. However, all eigenvalues should at least be shifted into the region shown in figure 3.11. The parameter α introduced in figure 5.2 admits a continuous transition from the n-th order observer to the reduced-order observer.

So far the sampling interval T has not been discussed, rather only the number of sampling intervals k. In a noise-free system the state could be constructed arbitrarily fast by letting $T \to 0$ and using a deadbeat observer. An extreme case should indicate the limitations of the construction speed: if $\underline{y}$ changes slowly according to the dynamics of the plant but is sampled very frequently, then consecutive values of $\underline{y}[k]$ differ only very little. The result of the calculations can then be greatly falsified by a small superimposed measurement noise. In the limiting case $T \to 0$ the deadbeat observer approximates a multiple differentiation which greatly accentuates high-frequency noise.

In systems with long sampling period and no measurement noise one can use an observer which operates with a multiple of the sampling frequency for the construction of $\underline{\hat{x}}(kT)$, for example an observer with a sampling period T/N where every N-th value of $\underline{\hat{x}}(kT/N)$ would be used for the feedback. Through the reduction of the sampling period, $\underline{A} = e^{\underline{F}T}$ is replaced by $\underline{A}^{1/N} = e^{\underline{F}T/N}$ and the construction error eq. (5.2.7) becomes

$$\underline{\tilde{x}}(kT+T) = (\underline{A}^{1/N} - \underline{H}^{**}\underline{C})^N \underline{\tilde{x}}(kT) \tag{5.4.8}$$

The N-th power of the eigenvalues assigned to $\underline{A}^{1/N} - \underline{H}^*\underline{C}$ is therefore important for the dynamics of the construction error. In order not to have the N-th power of the eigenvalues in the left half z-plane, it is preferable to assign only positive real eigenvalues for $\underline{A}^{1/N} - \underline{H}^*\underline{C}$. This case can be of interest, if, for example, one does not want to reduce the control sampling period T because of the actuator input constraints, but on the other hand low measurement noise admits a fast observation process.

5.5 Disturbance Observer

So far we have assumed that nothing is known about the kind of disturbances. This corresponds to the case of white noise, for which no conclusions for s[k] can be drawn from s[i], i=k-1, k-2... In practice, however, cases arise for which better assumptions on the disturbance spectrum can be made. These can be deterministic disturbances, for example, step load changes in a rolling mill, or periodic disturbances with known frequency, e.g. vibrations exerted on a helicopter frame by the rotor, or vertical acceleration on a magnetically levitated train travelling over a sagging track. In the latter example the frequency is determined by the velocity of the train and the distance between the supporting columns of the track. Also stochastic disturbances may be considered as being generated by a known dynamic system driven by white noise. Example: wind gusts acting on an airplane.

Because these disturbances are not sampled, they act as continuous disturbance inputs $\underline{z}_s(t)$ on the plant. Their equation in continuous time is

$$\begin{aligned} \dot{\underline{x}} &= \underline{Fx} + \underline{Gu} + \underline{z}_s \\ \underline{y} &= \underline{Cx} \end{aligned} \qquad (5.5.1)$$

It is now assumed that the disturbance inputs $\underline{z}_s$ can be represented as a solution of a known differential equation

$$\begin{aligned} \dot{\underline{q}} &= \underline{Dq} + \underline{\delta}_q \\ \underline{z}_s &= \underline{Mq} \end{aligned} \qquad (5.5.2)$$

where its input $\underline{\delta}_q$ is white noise, i.e. a sequence of independent impulses. Because nothing further is known about $\underline{\delta}_q$, it is set equal to its expected value zero. One is now only interested in the determination of the present state $\underline{q}$ which has developed under the influence of the previous form of $\underline{\delta}_q$. For a periodic disturbance, eq. (5.5.2) is an oscillator of known frequency. This means that $\underline{D}$ has pure imaginary eigenvalues. The amplitude and phase of the components of $\underline{z}_s$ are then dependent on the still unknown state $\underline{q}$. The observer can now be extended so that it also constructs the state vector $\underline{q}$. The continuous plant of eq. (5.5.1) is therefore connected with the disturbance generator, eq. (5.5.2), to form

$$\begin{bmatrix} \dot{\underline{x}} \\ \dot{\underline{q}} \end{bmatrix} = \begin{bmatrix} \underline{F} & \underline{M} \\ \underline{0} & \underline{D} \end{bmatrix} \begin{bmatrix} \underline{x} \\ \underline{q} \end{bmatrix} + \begin{bmatrix} \underline{G} \\ \underline{0} \end{bmatrix} \underline{u} + \begin{bmatrix} \underline{0} \\ \underline{I} \end{bmatrix} \underline{\delta}_q \qquad (5.5.3)$$

$$\underline{y} = \begin{bmatrix} \underline{C} & \underline{0} \end{bmatrix} \begin{bmatrix} \underline{x} \\ \underline{q} \end{bmatrix}$$

It is assumed that this system is observable from $\underline{y}$. The system is discretized and the corresponding observer is determined as before with $\underline{\delta}_q = \underline{0}$.

The construction of $\underline{x}$ is improved by this disturbance observer and moreover a value $\hat{\underline{q}}$ is constructed which can be fed back to $\underline{u}$ for compensation. This has no effect on the dynamics of the disturbance generator, because it is not controllable from $\underline{u}$. However, the effect of $\underline{z}_s$ on the plant can be reduced.

Example:

Step disturbances act at unknown time instants and with unknown amplitude on the input of the plant $1/s(s+1)$. The block diagram is

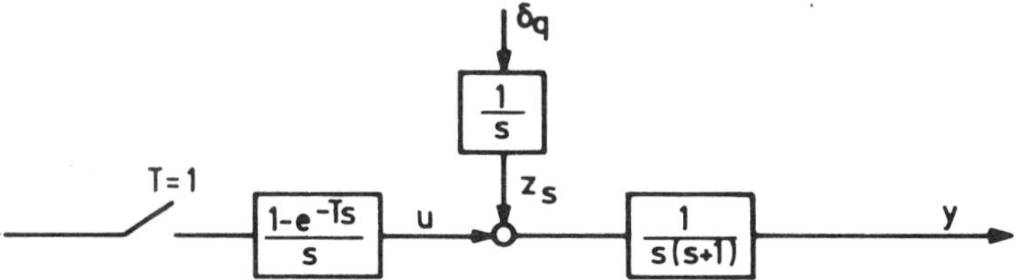

Figure 5.5 Consideration of a step disturbance input z_s.

With

$$D = 0 \ , \ \underline{M} = \begin{bmatrix} 0 \\ 1 \end{bmatrix}$$

and eq. (3.1.13) the augmented system (5.5.3) is

$$\begin{bmatrix} \dot{\underline{x}} \\ \dot{z}_s \end{bmatrix} = \begin{bmatrix} 0 & 1 & 0 \\ 0 & -1 & 1 \\ 0 & 0 & 0 \end{bmatrix} \begin{bmatrix} \underline{x} \\ z_s \end{bmatrix} + \begin{bmatrix} 0 \\ 1 \\ 0 \end{bmatrix} u + \begin{bmatrix} 0 \\ 0 \\ 1 \end{bmatrix} \delta_q$$

$$y = [1 \quad 0 \quad 0] \begin{bmatrix} \underline{x} \\ z_s \end{bmatrix} \tag{5.5.4}$$

Discretizing with T = 1 and substituting the expected value $E\{\delta_q\} = 0$ for δ_q

$$\begin{bmatrix} \underline{x}[k+1] \\ z_s[k+1] \end{bmatrix} = e^{\underline{F}T} \begin{bmatrix} \underline{x}[k] \\ z_s[k] \end{bmatrix} + \int_0^T e^{F\tau} g d\tau u[k]$$

$$= \begin{bmatrix} 1 & 0.632 & 0.368 \\ 0 & 0.368 & 0.632 \\ 0 & 0 & 1 \end{bmatrix} \begin{bmatrix} \underline{x}[k] \\ z_s[k] \end{bmatrix} + \begin{bmatrix} 0.368 \\ 0.632 \\ 0 \end{bmatrix} u[k] \tag{5.5.5}$$

The reduced-order observer for the unmeasured state x_2 and for the disturbance generator state z_s is now determined according to section 5.3. In eqs. (5.3.4) and (5.3.5) $y = x_1$, $\underline{z} = [x_2 \quad z_s]$ and

$$\underline{P} = \begin{bmatrix} 0.368 & 0.632 \\ 0 & 1 \end{bmatrix} \qquad \underline{Q} = \begin{bmatrix} 0 \\ 0 \end{bmatrix} \qquad \underline{D} = \begin{bmatrix} 0.632 \\ 0 \end{bmatrix}$$

$$\underline{R} = [0.632 \quad 0.368] \qquad S = 1 \qquad E = 0.368$$

A deadbeat observer is obtained from eq. (5.3.15) for

$$\det(z\underline{I}-\underline{P}+\underline{hR}) = Q(z) = z^2$$

From eq. (5.2.14)

$$\underline{h} = Q(\underline{P})\begin{bmatrix} \underline{R} \\ \underline{RP} \end{bmatrix}^{-1}\begin{bmatrix} 0 \\ 1 \end{bmatrix} = \underline{P}^2\begin{bmatrix} 0.632 & 0.368 \\ 0.233 & 0.768 \end{bmatrix}^{-1}\begin{bmatrix} 0 \\ 1 \end{bmatrix} = \begin{bmatrix} 1.244 \\ 1.582 \end{bmatrix}$$

Then, following from eqs. (5.3.6), (5.3.7), (5.3.11), (5.3.12) and (5.3.13) the reduced-order observer is

$$\underline{\hat{v}}[k+1] = \begin{bmatrix} -0.418 & 0.175 \\ -1 & 0.418 \end{bmatrix}\hat{v}[k] - \begin{bmatrix} 1.488 \\ 2.165 \end{bmatrix}y[k] + \begin{bmatrix} 0.175 \\ 0.582 \end{bmatrix}u[k]$$

$$\underline{\hat{z}}[k] = \begin{bmatrix} \hat{x}_2[k] \\ \hat{z}_s[k] \end{bmatrix} = \hat{v}[k] + \begin{bmatrix} 1.244 \\ 1.582 \end{bmatrix}y[k] \qquad (5.5.6)$$

5.6 Exercises

5.1 The following values of the trolley position x_1 and the force u were measured for the loading bridge with parameter values according to exercise 4.1

k	$x_1[k]$	u[k]
0	1	-980
1	0.927289831	- 59.9081373
2	0.808951182	-206.3674830
3	0.748895449	-411.7307805

Calculate $\underline{x}[0]$ and $\underline{x}[3]$.

5.2 Calculate a reduced observer in the form of two subobservers for the subsystems pendulum and trolley for the loading bridge of eq. (5.2.21) with

$$\underline{y} = \begin{bmatrix} 1 & 0 & 0 & 0 \\ 0 & 0 & 1 & 0 \end{bmatrix} \underline{x}$$

5.3 Assign zero eigenvalues to the observer of exercise 5.2 and calculate $\underline{\tilde{x}}[k]$ for the initial state $\hat{x}_2[0] = 0$, $\hat{x}_4[0] = 0$, $\underline{x}[0] = 0.1 \times [1 \quad 1 \quad 1 \quad 1]'$, $u[k] \equiv 0$.

5.4 Let a constant wind force affect the loading bridge in the plane of movement of figure 2.1. This acts as a force f on the load and a force 0.2 × f on the trolley. Calculate a disturbance observer which constructs the variables $\underline{\hat{x}}$ and $\hat{f}$ from x_1 and x_3.

6 Control Loop Synthesis

6.1 Design Methodology

In the chapters 1 to 5 several aspects of control system design have been discussed separately; we will now put these elements together.

First the variables which arise in typical multivariable control systems and their desired interaction will be described. Assume that the plant is linear and is described by the following differential equation:

$$\begin{aligned} \dot{\underline{x}}(t) &= \underline{Fx}(t) + \underline{Gu}(t) + \underline{z}_s(t) \\ \underline{y}(t) &= \underline{Cx}(t) + \underline{z}_m(t) \end{aligned} \qquad (6.1.1)$$

The design goal is to have the controlled variable

$$\bar{\underline{y}}(t) = \underline{Lx}(t) \qquad (6.1.2)$$

follow a reference variable $\underline{w}(t)$ as closely as possible despite the influence of the disturbance input $\underline{z}_s$ on the plant and the measurement noise $\underline{z}_m$. According to the definitions, $\bar{\underline{y}}$ and $\underline{w}$ have the same number of elements. This is obviously also the minimal number of actuating inputs in $\underline{u}$, otherwise one would already in the steady-state case have more requirements on $\bar{\underline{y}}$ than steady-state values in $\underline{u}$. A frequent requirement is that the goal of the control should be reached with small actuator input amplitudes $|u_i|$. In the discretization of the plant (6.1.1) it must be noted that the disturbance inputs $\underline{z}_s$ are not sampled or sampled and held. The discretized plant model is written with a disturbance input s[k] as

$$\underline{x}[k+1] = \underline{A}\underline{x}[k] + \underline{B}\underline{u}[k] + \underline{s}[k]$$

$$\underline{y}[k] = \underline{C}\underline{x}[k] + \underline{z}_m[k] \qquad (6.1.3)$$

$$\bar{\underline{y}}[k] = \underline{L}\underline{x}[k]$$

s[k] is formed as in eq. (3.1.5)

$$\underline{s}[k] = \int_0^T e^{\underline{F}v} \underline{z}_s(kT+T-v)\,dv \qquad (6.1.4)$$

Let ($\underline{A}$, $\underline{B}$) be controllable and ($\underline{A}$, $\underline{C}$) be observable.

Remark 6.1:

The plant was assumed to have no feedthrough y = Du in eqs. (6.1.1) and (6.1.3), i.e. the influence of $\underline{u}[k]$ cannot be seen in the output $\underline{y}[k]$, rather at $\underline{y}[k+1]$ at the earliest. If the continuous plant was approximately modelled with feedthrough - for example by neglecting a small time constant - then the actuator input $\underline{u}[k]$ would influence itself via the plant response and feedback. Such a model could only be reasonable as a limiting case for nonideal sampling (see remark 3.9).

There is no general design procedure for all control systems, but the designer may use the following checklist to taylor a procedure for his particular problem.

1. Model or identify the plant. Find a good simulation model. Simplify it, if necessary, for the design. Describe the model uncertainty as accurately as possible. This is particularly important for the low frequency behavior inside the design bandwidth. A very good model contains the uncertain physical parameters explicitely within a known structure (Examples: loading bridge, eq. (2.1.15), track guided bus, eq. (D.2.8)). A good model may also consist of a family of linearized plant models for typical and extremal operating conditions (Example: aircraft, eq. (D.1.2)). A satisfactory model includes at least a

crude description of a parallel, series, or feedback perturbation (D in figure 8.1). A parallel perturbation can describe unmodelled high-frequency modes (see figure 4.5). A series perturbation may describe neglected actuator or sensor dynamics. A feedback perturbation may be assumed to model the uncertainty of unstable poles. A poor model only describes a nominal plant and only formally defines perturbations as $\underline{A}+\Delta\underline{A}$ in a state space model or $g_S(s)+\Delta g_S(s)$ in a plant transfer function.

2. Calculate the plant poles and zeros. What is necessary to improve the dynamics? E.g. stabilization of unstable or unsufficiently damped eigenvalues, or speeding up a sluggish system with time delay and negative real roots close to s = 0. Zeros in the right half plane cannot be removed. Also cancellations by unsufficiently damped poles are not recommended. This is a constraint on the closed-loop responses.

3. Analyze the observability. Select appropriate sensors. Examples are

   a) The loading bridge is observable from the trolley position. However, eq. (2.3.41) shows that it is weakly observable for small m_L. The pendulum with the empty hook has little effect on the trolley motion. This little effect may be completely lost by friction. For a fast positioning of the empty hook it is necessary to measure both the trolley position and the cable angle. The unmeasured state variables (x_2 = trolley velocity and x_4 = cable angular velocity) can be reconstructed by an observer for known m_L or by a differentiating filter for unknown m_L.

   b) For the track-guided bus it is possible to measure the front and rear displacement from the guideline and the steering angle (see eq. (D.2.11)). The unmeasured derivatives of the displacement can be reconstructed by the differentiating filter of figure 8.11. If the automatic system is not used for driving backwards, then it suffices to use only a front antenna and to reconstruct the other four state variables by an observer [80.5].

c) In the aircraft of eq. (D.1.2) it is not easy to measure the rudder deflection because this term in the linearized equations is a deviation from an unknown trim value. Accelerometer and gyro are usually available. (Here a later redesign for robustness against sensor failures in section 8.5.3 results in a triplex gyro and no accelerometer. However, this is not forseeable in the present design step.)

These examples show that a new study must be made for each application. Essential factors in this decision are: observability of the plant, accuracy, bandwidth, noise, dynamic range, reliability, cost and availability of the sensors, possibility of replacement of sensors by feedback dynamics. Digital sensors do not need an A/D conversion in a digital control system. Of course, the accuracy of the control system cannot be better than the accuracy of the sensors.

4. Analyze the controllability. Select appropriate actuators. Essential factors in this decision are: controllability of the plant, bandwidth, saturation, linearity, dynamic range, reliability, cost, energy consumption and availability of the actuators. Digital actuators do not need a D/A conversion in a digital control system. Simulate some trajectories of plant state variables under the open-loop influence of extremal inputs ($|u| = |u|_{max}$). If they are not fast enough, then even the best control law cannot make them faster and a more powerful actuator is needed. If high-frequency structural modes of the plant are too much excited, then a lower bandwidth actuator is the better choice, or the closed-loop bandwidth must be constrained in the design.

5. What is the desired closed-loop bandwidth. What is the frequency range of measurement noise and unmodelled structural vibrations. Consider also the plant eigenvalue locations and actuator bandwidth. Choose the sampling interval T and the anti-aliasing filter (section 4.3). Discretize the plant mo-

del (section 3.1 for state space models, section 3.3 for transfer function models).

6. Which variables are measurable? If only the error e = w-y is measured (e.g. the relative distance from a moving target), then the simple controller structure of figure 1.7 is the only choice. In practice, w and y are frequently available separately and can be processed separately. Then an observer (section 5.2) or reduced order observer (section 5.3) can be used to reconstruct the plant state. If this reconstructed state is used for feedback instead of the real state, then the eigenvalues assigned in the two design steps remain unchanged. This separation principle is derived in section 6.3. A third channel is possible, if also a noise input to the plant can be measured. Then the actuator input is formed as $\underline{u} = \underline{u}_F + \underline{u}_D$ where the feedback part $\underline{u}_F$ is responsible for the closed-loop dynamics. The additional term $\underline{u}_D$ is formed from the measured input via a disturbance compensation filter. This minimizes the total effect of the measured plant noise on the controlled output $\underline{y}$ (see section 6.8). This three-channel structure is further discussed in section 6.2.

7. What is known about the external input variables? The dividing line between a "system" and its "environment" is generally not well defined. Nevertheless we have to draw the line in mathematical-physical modelling. For the design it is convenient to consider all known dynamic elements as part of the plant. This augmentation of the plant model is possible if external reference and noise inputs to the plant can be modelled by a noise generator, that is a dynamic system with completely unknown excitation (e.g. by white noise or by a δ-function input or by unknown initial conditions). This input signal generator may for example be a low pass filter that generates a typical wind gust spectrum from white noise, or it may be an integrator that produces a step function as plant disturbance $\underline{z}_S$ or reference input $\underline{w}$, or it may be an oscillator of known frequency as in the examples of section 5.5. There are two design approaches for this problem. In the first one the states of the noise generator are reconstructed by a disturbance ob-

server (section 5.5) and fed back into a disturbance compensation (section 6.8). In the second approach an internal model of the noise generator is included into the control loop. It produces infinite gain and therefore zero error at this particular frequency. The most frequent special case is the integral controller for step inputs (section 3.4.3). This is not required as a separate integrator if the plant already has a pole at s = 0 and a disturbance step input enters the plant after this integration. The continuous-time or discrete-time implementation of the internal model is discussed in section 6.2.

8. Suppose the controller structure is fixed in consideration of the preceeding aspects. The controller parameters which arise in this structure can be freely chosen and are determined so that requirements on control system behavior are satisfied. The choice of the design method depends to a large extent on the formulation of specifications for the control system performance. The most intuitive form of specification refers to typical trajectories (e.g. response to a reference step or ramp input or response to a disturbance impulse or step input). The steady-state accuracy is primarily a matter of the controller structure (see section 3.4.3). For the transients from the initial state zero to the steady-state the following parameters may be used for specifications of admissible reference step responses y(t):

   i) The magnitude after one sampling interval, i.e. y(T). For fixed T, the maximum achievable value of y(T) depends on the available actuator signal magnitude $|u|$. If the limitation on $|u|$ is not stringent, then a deadbeat response may be specified (see figure 3.15).

   ii) The overshoot $\max_t y(t)$ depends primarily on the damping of dominant eigenvalues. The second order system of eq. (2.5.13) has a small overshoot for damping $\zeta = 1/\sqrt{2} \approx 0.7$ (see figure 2.11). The same holds for discrete systems, see the curve for $k_y = 0.326$ in figure 3.15. A higher damping is required only if no overshoot is allowed (e.g. the temperature in a jet engine must not exceed limits given by materials used in the engine). For the short period longitudinal mode of an aircraft a damping $\zeta \geq 0.35$ is required ([69.7], see figure 8.34).

iii) The settling time t_s (i.e. the time required for the response to settle to an error tolerance of one percent from the final value) primarily depends on the minimum negative real part of all eigenvalues as illustrated by figure 2.9.

iv) The time scale of the step response of a second order system is inversely proportional to ω_n, that is the distance of the poles from the origin. (See figure 2.10 and the time scale $\omega_n t$ in figure 2.11. For the aircraft example see figure 8.34.) Sometimes the time scale is specified in form of the rise time t_r (e.g. the time for the response to rise from 10 % to 90 % of the final value) or as the time to reach the first overshoot.

For the second order system of eq. (2.5.13) the following crude approximate correspondence is given in [80.1]:

$$\begin{aligned} \zeta &\approx 0.6[2-\max_t y(t)] \quad (\text{for } \max_t y(t) > 1.1) \\ \omega_n &\approx 2.5/t_r \\ \zeta\omega_n &\approx 4.6/t_s \end{aligned} \qquad (6.1.5)$$

Figure 2.10 shows that $\zeta\omega_n$ is the negative real part of the poles. For higher order systems ζ and ω_n are considered as minimum values for all eigenvalues. Eq. (6.1.5) only provides an initial guess for the admissible eigenvalue location. After a first design step, the true values of $\max_t y(t)$, t_r and t_s are easily determined by simulation with the design model. The tendencies indicated in ii), iii) and iv) above are then used to improve the closed-loop eigenvalue location in further design steps. Another useful first design step may be the deadbeat solution and modifications which reduce $\max|u|$. Other possible formulations of design requirements use stability margins in the frequency domain. The Bode plot and its discrete-time version of eqs. (3.9.42) and (3.9.43) allows an easy determination of gain and phase margins. However, figure 3.26 shows that the stability margin ρ is a better measure for the distance of the Nyquist plot from the critical point -1. In multivariable systems the stability margin is generalized to the minimum over ω of the singular values of the return difference (see eq. (A.7.32) and [81.18]). Stability margins in the frequency domain also result from a linear quadratic (LQ) design by minimizing the quadratic criterion [60.6], [71.8], [72.3]

$$J = \int_0^\infty (\underline{x}'\underline{Q}\underline{x} + \underline{u}'\underline{R}\underline{u})dt \tag{6.1.6}$$

The resulting optimal control can be obtained directly from the solution of a Riccati equation. However, the stability margins do not result for discrete-time systems (see [78.5] and section 9.4). Design parameters in the form of elements of the matrices $\underline{Q}$ and $\underline{R}$ are introduced in eq. (6.1.6). The effect of the main diagonal elements of $\underline{R}$ is obvious. If one of these is made bigger, then the corresponding input variable in $\underline{u}(t)$ becomes smaller. The main diagonal elements of $\underline{Q}$ have a similar effect on the states $\underline{x}(t)$. The off-diagonal elements of $\underline{Q}$ allow further modifications, however, their effect is not obvious, and they are frequently used in a trial and error procedure. The Riccati design is recommended for design problems which naturally lead to the formulation of the quadratic criterion of eq. (6.1.6). However, if the design objectives are only indirectly related to such a criterion, then the pole assignment method (section 4.4) and its multivariable generalization (sections 9.2 and 9.3) are better suited because each trial requires less computational effort. The pole region assignment method (section 7.5) is suited for tradeoffs between eigenvalue locations, actuator signal magnitudes, and requirements of robustness against large parameter variations, sensor failures, implementation inaccuracies, gain reduction (see section 8.4). In particular the designer learns by this design method during the design process where the conflicts between specifications are and what he can get for free.

9. Each design method emphasizes some requirements and ignores others. Therefore the design must be analyzed for the ignored properties or requirements. A control system resulting from a frequency domain or Riccati design must be analyzed for example for its eigenvalue locations, step responses and robustness. Riccati design may lead to a large bandwidth which is undesirable in view of model uncertainties at high frequencies. A control system resulting from pole region assignment must be examined with respect to its stability margins in the frequency domain and actuator signal magnitude. The effect of actuator nonlinearities on the stability can be analyzed by Tsypkin and Jury-Lee methods (see section 3.9). More indirect design requirements (e.g. ride comfort of a vehicle) must be evaluated for typical disturbances.

10. As a result of the previous design steps the control system will usually be satisfactory with respect to some criteria, but unsatisfactory in view of other requirements. The refinement of the design by modifying the previous design steps leads to a tedious trial-and-error procedure. It is recommended to improve the preliminary solution in a systematic way. This can be achieved by Pareto optimization of vector-valued performance criteria (section 8.6). The goal of this procedure is to improve certain criteria as far as possible while preserving the desired properties that have already been achieved in the preceeding design steps.

11. So far the design and analysis of the resulting control system was performed with a simplified (e.g. linearized) design model. For the evaluation of the resulting control system a more realistic simulation model may be used as described in paragraph 1. Also hardware (sensors, actuators, the controller implementation by microelectronics) may be included into the simulation in order to make it more realistic and to test and qualify the hardware components and software. A further fine

adjustment of the controller parameters by performance-vector optimization is still possible.

12. Finally the controller can be implemented so that some parameters can still be tuned by the operator during the operation of the control loop. Feasible structures are for example the PID controller (section 3.4), tunable observer poles (section 5.4 and figure 6.8), tunable prediction time of the full order observer (figure 5.2), tunable anti-aliasing filters (section 4.3), and tunable prefilters (section 6.7).

6.2 Controller Structures

6.2.1 Feedback, Prefilter and Disturbance Compensation

In some cases disturbance inputs $\underline{z}_S$ in eq. (6.1.1) are partially measurable, for example the outdoor temperature for the control of a heated room. Let $\underline{z}_S = \underline{z}_{S1} + \underline{z}_{S2}$ where $\underline{z}_{S1}$ is measurable and therefore $\underline{s}_1$ can be formed by eq. (6.1.4) for known $\underline{F}$. In contrast $\underline{z}_{S2}$ and the corresponding $\underline{s}_2$ are not measurable. The most general causal control law then uses all known signals $\underline{w}(t-\tau)$, $\underline{y}(t-\tau)$, $\underline{z}_{S1}(t-\tau)$, $\underline{u}(t-\tau)$, $\tau \geq 0$ for the formation of the input variables $\underline{u}(t)$.

It is useful to assume that the controller structure is also linear for the linear plant examined here. In this way the theory of linear systems can be applied to the closed loop. Three transmission paths in the controller of figure 6.1 can now be examined.

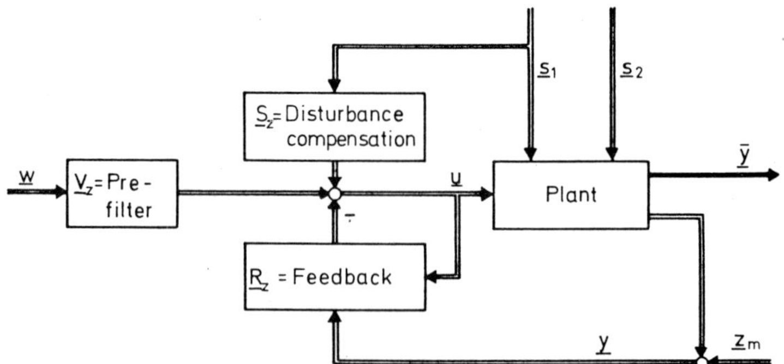

Figure 6.1 Linear controller structure

1. The feedback from $\underline{y}$ (and possibly $\underline{u}$) to $\underline{u}$ has three purposes:
   a) to provide for the desired positions of the eigenvalues of the control loop ("nice stabilization"),

   b) to reduce the influence of the nonmeasurable disturbance inputs $\underline{s}_2$ and $\underline{z}_m$ on $\bar{\underline{y}}$ as much as possible ("disturbance reduction", "filtering"),

   c) to keep the influence of parameter variations, neglected dynamics and nonlinearity of the plant small ("robust control").

2. A prefilter from $\underline{w}$ to $\underline{u}$ may be used. After the controller has been nicely stabilized, i.e. $\bar{\underline{y}}$ is represented as a sum of quickly decaying and well damped solution terms, a desired response to a typical test signal can be obtained from such terms by the choice of the prefilter. The prefilter may reduce the maximum required $|u|$ in the response to a reference step at $\underline{w}$. This is particularly recommended if a deadbeat pole assignment has been made (section 6.7).

   A typical requirement in the multivariable case is decoupling i.e. a stimulation at the i-th component of $\underline{w}$ should have an effect only on the i-th component of $\bar{\underline{y}}$, if possible.

3. The disturbance compensation from $\underline{s}_1$ to $\underline{u}$ is designed to eliminate as much as possible the influence of measurable disturbances $\underline{s}_1$ on $\bar{\underline{y}}$. Since most disturbance inputs are unmeasurable, this can only seldom be applied directly. However, if $\underline{s}_2$ can be modelled by a disturbance generator and its state is observed by a disturbance observer, then the reconstructed value $\hat{\underline{s}}_2$ may be used for a disturbance compensation.

In some control systems $\underline{w}$ is not available separately, but only the return difference $\underline{e} = \underline{w}-\bar{\underline{y}}$. In this case the control loop structure corresponding to figure 3.8 must be chosen. Also reference $\underline{w}$ and disturbance $\underline{s}$ may enter the system together. For an active vehicle suspension for example, the low frequency content of the road surface is reference signal and the high-frequency content is disturbance. Both parts may be separated by filtering with velocity-dependent bandwidth.

The digital controller of order m with state vector $\underline{r}$ may be written as

$$\begin{aligned}\underline{r}[k+1] &= \underline{A}_R\underline{r}[k] - \underline{B}_y\underline{y}[k] + \underline{B}_w\underline{w}[k] + \underline{B}_s\underline{s}_1[k] \\ \underline{u}[k] &= \underline{C}_R\underline{r}[k] - \underline{D}_y\underline{y}[k] + \underline{D}_w\underline{w}[k] + \underline{D}_s\underline{s}_1[k]\end{aligned} \qquad (6.2.1)$$

m^2 coefficients of the matrices can be fixed to zero or one by the choice of a canonical form for $\underline{A}_R$ and one of the other matrices. The usual sign change in the control loop is expressed by feedback of $-\underline{y}$.

The state representation of the closed loop of order n+m is developed by combining eqs. (6.1.3) and (6.2.1). All controller matrices stated in eq. (6.2.1) can be combined into a proportional feedback matrix for the system extended by the controller states:

$$\begin{bmatrix}\underline{x}[k+1]\\ \underline{r}[k+1]\end{bmatrix} = \begin{bmatrix}\underline{A} & \underline{O}\\ 0 & 0\end{bmatrix}\begin{bmatrix}\underline{x}[k]\\ \underline{r}[k]\end{bmatrix} + \begin{bmatrix}\underline{B} & \underline{O}\\ 0 & I\end{bmatrix}\begin{bmatrix}\underline{u}[k]\\ \underline{u}_R[k]\end{bmatrix} + \begin{bmatrix}\underline{I} & \underline{I}\\ \underline{O} & \underline{O}\end{bmatrix}\begin{bmatrix}\underline{s}_1[k]\\ \underline{s}_2[k]\end{bmatrix} \qquad (6.2.2)$$

$$\begin{bmatrix}\underline{u}[k]\\ \underline{u}_R[k]\end{bmatrix} = \begin{bmatrix}-\underline{D}_y\underline{C} & \underline{C}_R\\ -\underline{B}_y\underline{C} & \underline{A}_R\end{bmatrix}\begin{bmatrix}\underline{x}[k]\\ \underline{r}[k]\end{bmatrix} + \begin{bmatrix}\underline{D}_w & -\underline{D}_y & \underline{D}_s\\ \underline{B}_w & -\underline{B}_y & \underline{B}_s\end{bmatrix}\begin{bmatrix}\underline{w}[k]\\ \underline{z}_m[k]\\ \underline{s}_1[k]\end{bmatrix} \qquad (6.2.3)$$

The problem of dynamic output-vector feedback of order m is reduced here to a problem of proportional output-vector feedback for an extended system of order n+m. It is therefore sufficient to examine proportional feedback in the following chapters.

Substitution of eq. (6.2.3) into (6.2.2) yields

$$\begin{bmatrix}\underline{x}[k+1]\\ \underline{r}[k+1]\end{bmatrix} = \begin{bmatrix}\underline{A}-\underline{B}\underline{D}_y\underline{C} & \underline{B}\underline{C}_R\\ -\underline{B}_y\underline{C} & \underline{A}_R\end{bmatrix}\begin{bmatrix}\underline{x}[k]\\ \underline{r}[k]\end{bmatrix} + \begin{bmatrix}\underline{B}\underline{D}_w & -\underline{B}\underline{D}_y & \underline{B}\underline{D}_s+\underline{I} & \underline{I}\\ \underline{B}_w & -\underline{B}_y & \underline{B}_s & \underline{O}\end{bmatrix}\begin{bmatrix}\underline{w}[k]\\ \underline{z}_m[k]\\ \underline{s}_1[k]\\ \underline{s}_2[k]\end{bmatrix} \qquad (6.2.4)$$

One is especially interested in the response of the controlled variables $\bar{\underline{y}}$ and of the actuator inputs $\underline{u}$ in view of their constraints. Therefore the output equation is chosen as

$$\begin{bmatrix} \bar{\underline{y}}[k] \\ \\ \underline{u}[k] \end{bmatrix} = \begin{bmatrix} \underline{L} & \underline{0} \\ \\ -\underline{D}_y\underline{C} & \underline{C}_R \end{bmatrix} \begin{bmatrix} \underline{x}[k] \\ \\ \underline{r}[k] \end{bmatrix} + \begin{bmatrix} \underline{0} & \underline{0} & \underline{0} \\ \\ \underline{D}_w & -\underline{D}_y & \underline{D}_s \end{bmatrix} \begin{bmatrix} \underline{w}[k] \\ \underline{z}_m[k] \\ \underline{s}_1[k] \end{bmatrix} \quad (6.2.5)$$

6.2.2 Discrete and Continuous Compensation

So far the controller structure of figure 6.1 corresponds exactly to the case of continuous systems. We now turn to some special features of sampled systems which refer to the continuous output $\bar{y}(t) = \bar{y}(kT+\gamma T)$, $0 < \gamma < 1$. The input $f^*(t)$ of the hold element is composed of δ-impulses, see figure 1.4 and eq. (1.2.12). Therefore $\bar{y}(t)$ consists of time-shifted impulse responses of the plant including the integrator of the hold element, in addition to the responses to the continuously acting disturbance inputs $z_s(t)$. If the discrete output $\bar{\underline{y}}(kT)$ is forced to follow other functions by use of the control - these functions resulting from the poles of the closed loop and the reference inputs $\underline{w}$ - then this is only achieved at the sampling points, not in between. This is obvious in the stationary case, if $\bar{\underline{y}}(t)$ should follow reference variables $\underline{w}(t)$ with poles of $\underline{w}_s(s)$ on the imaginary axis asymptotically. This only succeeds if the continuous part (including the hold element integrator) also has these poles. Some examples will illustrate this.

Example 1:

Let the transfer function of the plant be $g_s(s) = 1/(s+1)$. Its input is sampled with a sampling period $T = 2$ and held. In the control loop configuration of figure 3.8 let $d_z(z)$ have a double pole at $z = 1$. Let the reference input be $w(t) = t/2$, i.e. it has a double pole at $s = 0$, i.e. discretized at $z = 1$. Provided that the control loop is stable, $\lim_{k\to\infty} e(kT) = 0$ from the final value theorem of the z-transform.

Figure 6.2 shows the steady-state form of u(t) and y(t). y(t) is equal to w(t) only at the sampling instants.

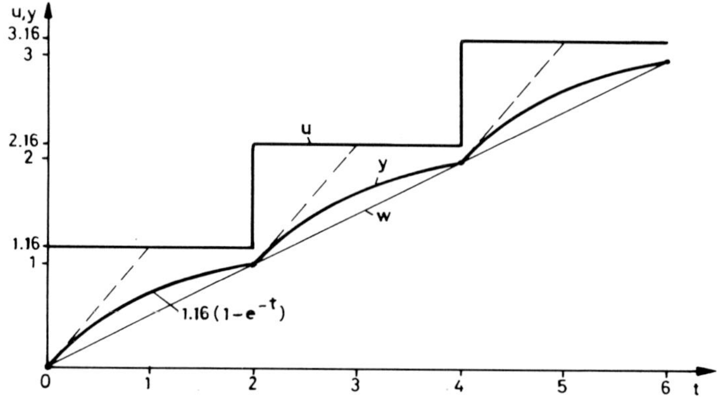

Figure 6.2 The steady-state error $e(t) = w(t)-y(t)$

Due to the integration in the hold element, however, the same system can follow a reference variable $w(t) = 1(t)$ without steady-state errors for $t \neq kT$. u(t) is constant in this case.

For $w(t) = t/2$ the steady-state error can only be made zero if at least one of the two integrations is performed continuously, for example with $\hat{g}_S(s) = (b_0+b_1 s)/s(s+1)$ and a simple pole of $d_z(z)$ at $z = 1$.

Example 2:

In the control loop of figure 6.3 a step disturbance signal z_S is added to the plant input. The system should follow a sinusoidal reference variable $w = \sin\omega t$ of known frequency $\omega = 1$.

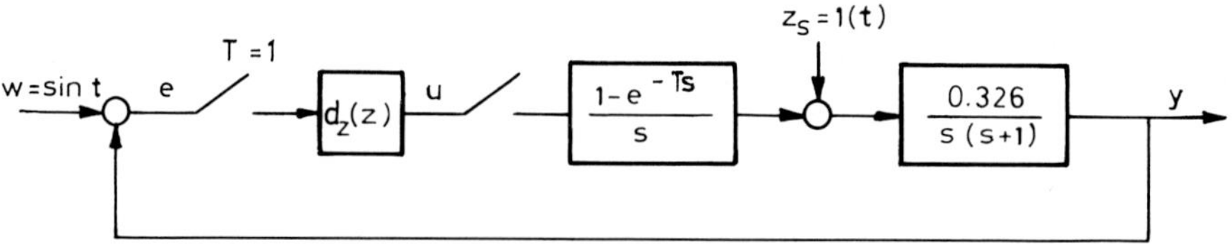

Figure 6.3 Sampled-data control loop with known inputs

Shifting the disturbance z_s (for $z_s(s) = 1/s$) to the output y gives the following z-transform of the disturbance input at y

$$v_z(z) = \mathcal{Z}\left\{\frac{0.326}{s^2(s+1)}\right\} = \frac{0.120z(z+0.718)}{(z-1)^2(z-0.368)}$$

Furthermore

$$w_z(z) = \mathcal{Z}\left\{\sin k\right\} = \frac{0.84z}{z^2-1.083z+1}$$

and

$$h_z(z) = \mathcal{Z}\left\{\frac{1-e^{-Ts}}{s} \times \frac{0.326}{s(s+1)}\right\} = \frac{0.120(z+0.718)}{(z-1)(z-0.368)}$$

The open loop $g_{oz}(z) = d_z(z)h_z(z)$ must contain an internal model of the known inputs. It already has simple poles at $z = 1$ and $z = 0.368$ from $h_z(z)$. The controller $d_z(z)$ is therefore assumed with the three additional poles arising in $v_z(z)$ and $w_z(z)$, i.e. the denominator of $d_z(z)$ must contain the factor $(z-1)(z^2-1.083z+1)$. If d_z is of order m, then three of its 2m+1 free parameters are thereby fixed. The order of the entire system is 2+m. If one wants to fix all eigenvalues, then 2+m free parameters are required, see section 3.4.2. From $2m+1-3 = 2+m$ it follows that $m = 4$, i.e. $d_z(z)$ has the form

$$d_z(z) = \frac{b_0+b_1z+b_2z^2+b_3z^3+b_4z^4}{(z^2-1.083z+1)(z-1)(z-a_0)} \tag{6.2.6}$$

The characteristic equation of the control loop is

$$\begin{aligned}&z^6+(0.12b_4-a_0-3.451)z^5+(0.12b_3+0.86b_4+3.451a_0+5.301)z^4+\\&+(0.12b_2+0.086b_3-5.301a_0-4.616)z^3+(0.12b_1+0.086b_2+4.616a_0+2.135)z^2+\\&+(0.12b_0+0.086b\ -2.135a_0-0.368)z+(0.086b_0+0.368a_0) = 0\end{aligned} \tag{6.2.7}$$

The assignment of the six coefficients of the characteristic equation leads to a nonsingular linear system of equations for the free control parameters. For example, if it is required that the characteristic equation $z^6 = 0$, then $a_0 = -0.712$, $b_0 = 3.047$, $b_1 = -16.018$, $b_2 = 35.74$, $b_3 = -40.082$, $b_4 = 22.82$. If the sinusoidal reference input is switched on at time $t = 0$, then a finite return difference of six sampling intervals occurs. Its z-transform is

$$e_z(z) = \frac{w_z(z)}{1+d_z(z)h_z(z)} = 0.84z^{-1}-1.391z^{-2}+0.042z^{-3}+0.729z^{-4}-0.222z^{-5} \tag{6.2.8}$$

For a step disturbance $z = 1(t)$ the return difference of six sampling intervals is

$$e_z(z) = \frac{-v_z(z)}{1+d_z(z)h_z(z)} = -0.12z^{-1}-0.042z^{-2}+0.004z^{-3}-0.105z^{-4}-0.061z^{-5} \tag{6.2.9}$$

This design is tailored for the special inputs. After the termination of the transient process, the system can follow all input variables $w_z(z)$ exactly at the sampling instants, with poles contained in the z-transfer function of the open loop $g_{oz}(z) = d_z(z)h_z(z)$. In the example these are the input sequences $1(kT)$, kT, e^{-kT}, $\sin kT$, and $(-0.712)^k$. These variables can also be exactly compensated at the sampling instants if they arise as disturbance inputs at the output of the plant.

Since the characteristic equation was fixed at $z^{n+m} = z^6 = 0$, the transients are terminated after six sampling intervals. The return difference $e(kT)$ then remains zero for the continuing sinusoidal reference input $w(t) = \sin t$, i.e. $d_z(z)$ has zero input and works as a discrete sine generator for this frequency. This result also holds for $w(t) = A \sin(t+\varphi)$. If $w(t)$ is switched to a different value of A or φ, then the output $y(t)$ is synchronized in a transient of six sampling steps.

If the steady-state oscillation has been reached, $e(kT) = 0$ at the sampling instants, an error, however, arises in between these points. In the steady-state oscillation a staircase signal occurs at the output of the hold element. The spectrum of the actuator signal, and therefore of the output variable y as well, contains the sideband frequencies of the pulse-amplitude-modulation, i.e. y contains higher frequency components in addition to the desired sine. This error between the sampling instants cannot be removed by a discrete controller alone.

Example 3:

For example 2 a partly continuous controller will be designed which allows exact steady-state tracking.

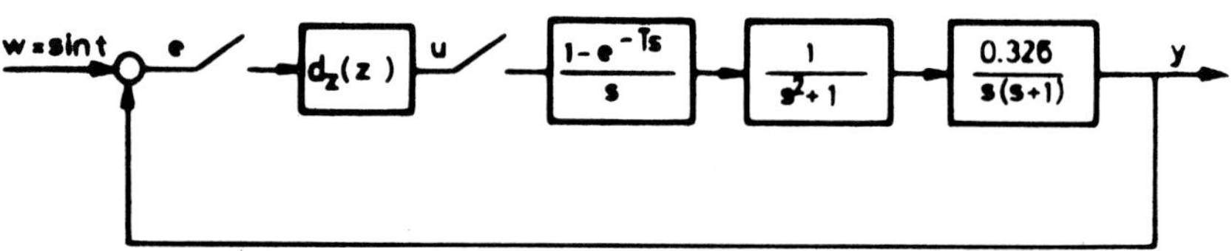

Figure 6.4 Discrete and continuous controller part

For simplification the disturbance input z_s is omitted. A continuous internal model in the form of an oscillator with transfer function $1/(s^2+1)$ is inserted between hold and plant. The z-transfer function $h_z(z)$ of the continuous part is

$$\frac{z-1}{z}\,\mathcal{Z}\left\{\frac{0.326(1-e^{-Ts})}{(s^2+1)s^2(s+1)}\right\} = \frac{0.326[0.028z^3+0.284z^2+0.243z+0.018]}{(z^2-1.083z+1)(z-1)(z-0.368)}$$

The order of the entire system is 4+m for a controller order of m, i.e. 4+m coefficients of the characteristic equation must be fixed. The controller has 2m+1 free parameters. From $2m+1 = 4+m$, it follows that $m = 3$. A finite oscillation process is reached for the characteristic equation $z^7 = 0$. This condition once again produces a linear system of equations. The resulting controller transfer function is

$$d_z(z) = \frac{-6.29+26.9z-33.13z^2+17.81z^3}{0.099+1.402z+2.286z^2+z^3} \qquad (6.2.10)$$

If the characteristic equation for a continuous part $g_s(s)$ of order n and a discrete controller of order m is fixed at $z^{n+m} = 0$ this gives

$$u_z(z) = \frac{\text{numerator } d_z(z) \times \text{denominator } h_z(z)}{z^{n+m}} w_z(z) \qquad (6.2.11)$$

for the output signal of the hold element. If the poles of $w_s(s)$ are reproduced in $g_s(s)$ they also cancel each other in $h_z(z)$ and $w_z(z)$ in eq. (6.2.11). In other words, $u_z(z)$ is a finite polynomial. After termination of the transient $u(t) \equiv 0$. The input of the plant is now generated by the continuous part of the controller. Therefore there are no side-band frequencies, i.e. in the steady-state oscillation the output y(t) follows not only at the sampling instants, but continuously and exactly. The system of figure 6.4 tracks continuously the reference variables $w(t) = 1(t)$, t, e^{-t}, and sin t after a finite transient. The continuous part of the controller, i.e. $1/(s^2+1)$, is called a "notch filter" which represents an "internal model" of the reference signal w.

The three examples show, how $\lim_{k\to\infty}[w(kT) - \bar{y}(kT)] = 0$ can be achieved with a discrete internal model and how $\lim_{k\to\infty}[w(kT-\gamma T) - \bar{y}(kT+\gamma T)] = 0$ can be obtained for all γ with a continuous internal model.

It should be noted that also for transients the formal application of design methods for continuous systems to the discrete case can lead to unfavorable behavior between the sampling instants. This especially holds for design methods in which some of the n+m eigenvalues of eq. (6.2.4) are made unobservable from $\bar{y}(kT)$, for example via decoupling methods. These eigenvalues remain observable from $\bar{\underline{y}}(t) = \underline{y}(kT+\gamma T)$, $0 < \gamma < 1$ and lead to undesired "hidden oscillations" [57.3].

In contrast, controller structures, in which only the states $\underline{x}$ of the plant are controllable from the reference input $\underline{w}$, can be used in sampled-data systems like in continuous-time systems. This arises in the combination of state-vector feedback and state construction by an observer, as is shown in the following section.

6.3 Separation

In chapter 5 the question of how the state $\hat{\underline{x}}[k]$ can be constructed by an observer of order n, eq. (5.2.8) was examined. By the addition of the disturbance variables $\underline{s}[k]$ and $\underline{z}_m[k]$ the error difference eq. (5.2.7) becomes

$$\tilde{\underline{x}}[k+1] = (\underline{A}-\underline{HC})\tilde{\underline{x}}[k] + \underline{s}[k] - \underline{Hz}_m[k] \tag{6.3.1}$$

In chapter 4, results concerning the pole assignment by state-vector feedback $\underline{u} = -\underline{Kx}$ were obtained. Because $\underline{x}$ is not measurable, it is replaced by the constructed value $\hat{\underline{x}} = \underline{x}-\tilde{\underline{x}}$ from the observer. Also the reference variable $\underline{w}$ is introduced by $\underline{u} = -\underline{K}\hat{\underline{x}}+\underline{w}$. Then

$$\underline{u}[k] = -\underline{Kx}[k]+\underline{K}\tilde{\underline{x}}[k]+\underline{w}[k] \tag{6.3.2}$$

Now the combined state equations of the control loop are

$$\begin{bmatrix}\underline{x}[k+1]\\ \tilde{\underline{x}}[k+1]\end{bmatrix} = \begin{bmatrix}\underline{A}-\underline{BK} & \underline{BK}\\ \underline{0} & \underline{A}-\underline{HC}\end{bmatrix}\begin{bmatrix}\underline{x}[k]\\ \tilde{\underline{x}}[k]\end{bmatrix} + \begin{bmatrix}\underline{B} & \underline{0} & \underline{I}\\ \underline{0} & -\underline{H} & \underline{I}\end{bmatrix}\begin{bmatrix}\underline{w}[k]\\ \underline{z}_m[k]\\ \underline{s}[k]\end{bmatrix} \tag{6.3.3}$$

$$\bar{\underline{y}} = [\ \underline{L} \quad \underline{0}\ ]\begin{bmatrix}\underline{x}[k]\\ \tilde{\underline{x}}[k]\end{bmatrix}$$

This equation allows two important conclusions:

1. The dynamics matrix is block triangular, its characteristic equation is therefore

$$\det(z\underline{I}-\underline{A}+\underline{BK}) \times \det(z\underline{I}-\underline{A}+\underline{HC}) = P(z) \times Q(z) \qquad (6.3.4)$$

In other words, the eigenvalues fixed by the state feedback $\underline{K}$ and the observer design $\underline{H}$ are eigenvalues of the entire system. This "separation" property justifies the substitution of $\underline{x}$ by $\hat{\underline{x}}$ in the control law. The separation theorem for a Wiener or Kalman filter in place of the observer was established in [56.2], [61.2].

2. The construction error $\tilde{\underline{x}}$ is not controllable from the reference input $\underline{w}$. Therefore the observer poles in Q(z) do not enter into the reference transfer function. For $\underline{s}[k] \equiv \underline{0}$, $\underline{z}_m[k] \equiv \underline{0}$,

$$\underline{y}_z = \underline{L}(z\underline{I}-\underline{A}+\underline{BK})^{-1}\underline{Bw}_z \qquad (6.3.5)$$

The poles of the reference transfer matrix therefore depend only on $\underline{K}$, not on $\underline{H}$. Consequently, the observer matrix $\underline{H}$ can be tuned to the actual disturbances during the operation of the control system without changing the response to a reference input.

A corresponding state equation of the control loop is also obtained by the application of the reduced order observer. For simplicity let $\underline{z}_m = \underline{s} = \underline{0}$. According to eqs. (5.3.4) and (5.3.5)

$$\begin{bmatrix} \underline{z}[k+1] \\ \underline{y}[k+1] \end{bmatrix} = \begin{bmatrix} \underline{P} & \underline{Q} \\ \underline{R} & \underline{S} \end{bmatrix} \begin{bmatrix} \underline{z}[k] \\ \underline{y}[k] \end{bmatrix} + \begin{bmatrix} \underline{D} \\ \underline{E} \end{bmatrix} \underline{u}[k] \qquad (6.3.6)$$

In the state-vector feedback

$$\underline{u}[k] = -[\underline{K}_z \quad \underline{K}_y] \begin{bmatrix} \hat{\underline{z}}[k] \\ \underline{y}[k] \end{bmatrix} + \underline{w}[k] \qquad (6.3.7)$$

$\hat{\underline{z}} = \underline{z}-\tilde{\underline{z}}$ is fed back from the reduced observer, eqs. (5.3.6) and (5.3.7), instead of the unknown state $\underline{z}$, i.e. $\underline{u}[k]$ is formed as

$$\underline{u}[k] = -\underline{K}_z\underline{z}[k] + \underline{K}_z\tilde{\underline{z}}[k] - \underline{K}_y\underline{y}[k] + \underline{w}[k] \qquad (6.3.8)$$

The construction error $\tilde{\underline{z}}$ satisfies eq. (5.3.14)

$$\tilde{\underline{z}}[k+1] = (\underline{P}-\underline{HR})\tilde{\underline{z}}[k] \tag{6.3.9}$$

Combining eqs. (6.3.6), (6.3.8) and (6.3.9), the closed loop equation is obtained

$$\begin{bmatrix} \underline{z}[k+1] \\ \underline{y}[k+1] \\ \tilde{\underline{z}}[k+1] \end{bmatrix} = \begin{bmatrix} \underline{P}-\underline{DK}_z & \underline{Q}-\underline{DK}_y & \underline{DK}_z \\ \underline{R}-\underline{EK}_y & \underline{S}-\underline{EK}_y & \underline{EK}_z \\ \underline{0} & \underline{0} & \underline{P}-\underline{HR} \end{bmatrix} \begin{bmatrix} \underline{z}[k] \\ \underline{y}[k] \\ \tilde{\underline{z}}[k] \end{bmatrix} + \begin{bmatrix} \underline{D} \\ \underline{E} \\ \underline{0} \end{bmatrix} \underline{w}[k] \tag{6.3.10}$$

The eigenvalues of the entire system are therefore once again the poles fixed by state vector feedback ($\underline{K}_z$, $\underline{K}_y$) and the observer poles as fixed by $\underline{H}$. $\underline{P}-\underline{HR}$ does not enter into the reference transfer function because $\tilde{\underline{z}}$ is not controllable from $\underline{w}$.

The structure of a control system with an observer of full or reduced order is shown in figure 6.5. Here an additional prefilter $\underline{V}_z(z)$ is assumed at the reference input. Its poles are not controllable from $\underline{y}$. On the other hand the observer structure of the feedback $\underline{R}_z$ of figure 6.1 is such that the eigenvalues of the construction error are not controllable from $\underline{s}_1$ and $\underline{w}$.

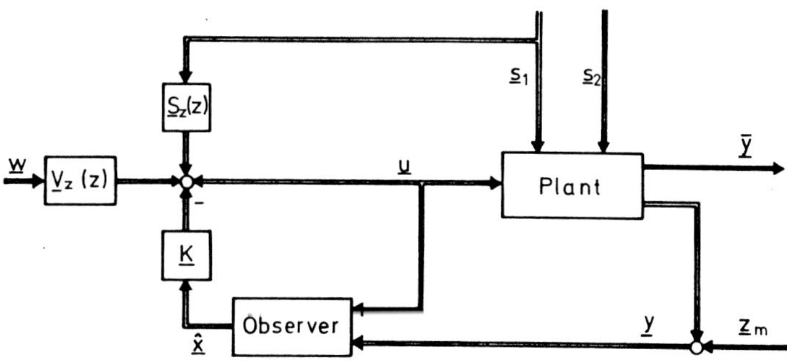

Figure 6.5 Control system with observer

It was pointed out in section 4.3 that large initial peaks in $\underline{u}$ for the step response can be avoided by choosing the sampling period T not too small. If the calculation time in the computer is not critical, however, one can operate the observer with a sampling period

$$T_{observer} = T/N \tag{6.3.11}$$

and then choose only every Nth term from the constructed values $\underline{\hat{x}}(iT/N)$, i.e. $i = K \times N$, in order to obtain $\underline{\hat{x}}(kT)$. If no measurement noise $\underline{z}_m$ is present, N can be chosen so that a deadbeat observer can completely construct the state within one sampling interval T.

6.4 Construction of a Linear Function of the States

In section 6.3 we have connected the observer with the state-vector feedback to a dynamic controller. In this control loop it is not necessary to construct all states in the controller. Only the linear combinations $\underline{K}\underline{\hat{x}}$ or $\underline{K}_z\underline{\hat{z}}$ are necessary. Since we will not deal with the structural properties of multi-variable systems before chapter 9, only the principle for plants with one actuator and s measured independent variables $\underline{y} = [y_1 \ldots y_s]'$ will be explained here.

As Luenberger [71.7] has shown, the plant

$$\begin{aligned} \underline{x}[k+1] &= \underline{A}\underline{x}[k] + \underline{b}u[k] \\ \underline{y}[k] &= \underline{C}\underline{x}[k] \end{aligned} \tag{6.4.1}$$

can be decomposed into s subsystems, each of which is observable from one linear combination of the y_i. This can be done by transformation to the observer-canonical form, i.e. the observability dual of eq. (A.4.1). Here the order of the subsystems is equal to the observability indices, $\nu_1 \ldots \nu_s$. At this point it is sufficient to know that the sum of the observability indices for an observable system is equal to the system order, i.e. $\nu_1 + \nu_2 + \ldots \nu_s = n$ and that the order of the largest subsystem is equal to the largest observability index

$$\nu = \max_i \nu_i \tag{6.4.2}$$

The latter is defined for an observable pair ($\underline{A}$, $\underline{C}$) as the smallest integer ν such that

$$\text{rank}\begin{bmatrix} \underline{C} \\ \underline{CA} \\ \vdots \\ \underline{CA}^{\nu-1} \end{bmatrix} = n \qquad (6.4.3)$$

This decomposition of the plant into subsystems allows s individual observers to work in parallel. Each of the observers requires only one linear combination of measured variables and constructs the state of the corresponding subsystem, see figure 6.6.

Reduced observers have orders ν_1-1, $\nu_2-1 \ldots \nu_s-1$, the largest one therefore has order $\nu-1$.

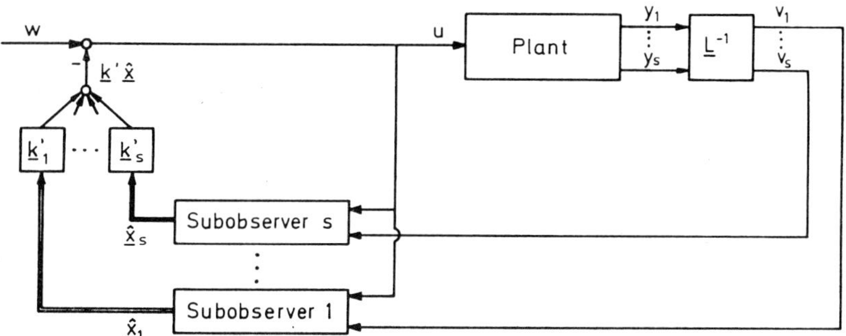

Figure 6.6 Subobservers for the plant decomposed into subsystems

If all subobservers have observer poles which are different from each other, then the complete observer of order n-s cannot be further reduced. One can, however, also assume $\nu-1$ eigenvalues for the largest subobserver in the form of a polynomial Q(s) and assign subsets of these eigenvalues to the other subobservers. From the s+1 transfer functions from u and v_i to $\underline{k}'\hat{\underline{x}}$, i = 1,2...s, the smallest common multiple of all denominators can be extracted at $\underline{k}'\hat{\underline{x}}$. This is exactly Q(s), so the complete observer can be realized as a system of order $\nu-1$. The different numerator polynomials may be realized in the observer-canonical form, figure A.5. See exercise 6.4.

The result is formulated as:

> A linear combination $\underline{k}'\underline{x}$ of state variables can be constructed with an observer of order $\nu-1$, where ν is the observability index of $(\underline{A}, \underline{C})$.
>
> (6.4.4)

For the application of the observer with sampling period T/N, $N = \nu-1$ can be chosen in eq. (6.3.11).

6.5 Synthesis by Polynomial Equations

We have already seen in section 3.4 that pole assignment for single-input systems can also be achieved by assuming a dynamic controller, calculating the characteristic polynomial, and equating the coefficients with the desired characteristic polynomial. This leads to the linear equation (3.4.15) for the controller coefficients. The controller order was assumed to be n-1, corresponding to the reduced observer for plants with one output; but the control loop had the conventional structure of figure 3.8.

The investigations in this chapter have shown that a suitable structure of the control system guarantees that only n eigenvalues of the closed loop are controllable from the reference input. We now want to combine the conceptual simplicity of the equating coefficients method with the advantages of a control loop structure with observer. This approach is especially attractive if the plant description is given in form of a transfer function. A detailed investigation of the calculation of "control-observers" was conducted by Grübel [77.3] using a linear system of equations. In these investigations the question of constraints on the pole assignment is examined if a smaller controller order is assumed. The following presentation is drawn from the work of Chen [69.4] and Wolovich [71.6], [74.1], [81.1]. The analysis given here is for the single variable case. The generalization to the multivariable case will be discussed in remark 6.3.

Let a single-input, single-output plant be described by its z-transfer function

$$y_z(z) = h_z(z)u_z(z)$$

$$h_z(z) = \frac{B(z)}{A(z)} = \frac{b_o+b_1z+\ldots+b_{n-1}z^{n-1}}{a_o+a_1z+\ldots+a_{n-1}z^{n-1}+z^n} \tag{6.5.1}$$

The controller structure is fixed according to figure 6.7, where K, H, Q, L and M are polynomials in z. The disturbance input s enters into the loop at the plant output.

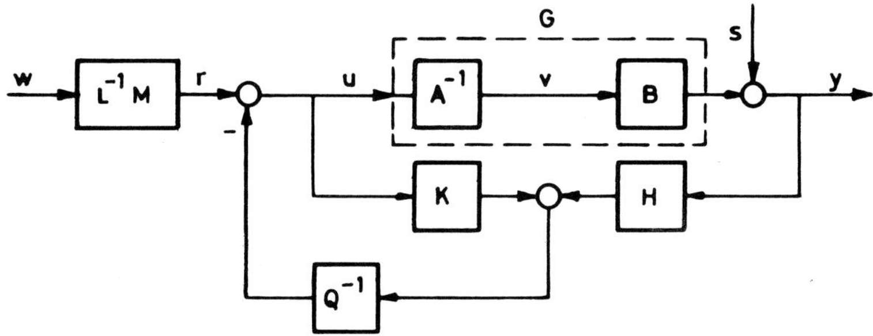

Figure 6.7 Polynomial representation of a structure with observer and state vector feedback.

The variables in figure 6.7 are related in the frequency domain by

$$y = Bv + s \tag{6.5.2}$$

where v is determined from the relation

$$u = Av = r-Q^{-1}(KA+HB)v-Q^{-1}Hs$$

$$(QA+KA+HB)v = Qr-Hs \tag{6.5.3}$$

$$v = (QA+KA+HB)^{-1}(Qr-Hs)$$

and substituted into eq. (6.5.2) to give

$$y = B(QA+KA+HB)^{-1}Qr+[I-B(QA+KA+HB)^{-1}H]s \tag{6.5.4}$$

The transfer function from r to y is

$$\frac{y}{r} = B(QA+KA+HB)^{-1}Q \tag{6.5.5}$$

If this transfer function is made independent of Q, then the additional eigenvalues introduced by Q become observer poles. For this purpose K and H must be chosen such that KA+HB contains the factor Q, i.e.

$$KA + HB = QD \tag{6.5.6}$$

with a polynomial D still to be chosen. Then

$$\frac{y}{r} = B(A+D)^{-1}$$

The denominator of this transfer function corresponds to the polynomial to be determined by pole assignment

$$P = A+D \tag{6.5.7}$$

Because both P and A are of order n, an arbitrary pole placement is possible by choosing

$$D(z) = d_o+d_1z+...+d_{n-1}z^{n-1} \tag{6.5.8}$$

where $d_i = p_i-a_i$ with the a_i's from eq. (6.5.1). D(z) is the difference between the characteristic polynomials of the closed and open loops. By use of the prefilter $L^{-1}M$ one eventually obtains the transfer function

$$f = \frac{y}{w} = B(A+D)^{-1}L^{-1}M \tag{6.5.9}$$

The controller design with pole assignment P(z) and observer characteristic polynomial Q(z) reduces to the solution of the synthesis eq. (6.5.6). First it must be assumed that A and B are relatively prime. Otherwise Q or D must contain the common factor of A and B and would no longer be freely assignable. A common zero of A and B means that a subsystem with these eigenvalues is uncontrollable or unobservable. This eigenvalue cannot be shifted by feedback. It arises as a zero of QD = Q(P-A) and therefore as a zero of the characteristic polynomial QP of the closed loop. The assumption "A, B relatively prime" is not only necessary but also sufficient for the solvability of the synthesis equation. This results from the Euclidean algorithm, see [53.3, theorem 12], [80.2, section 2.4.4]. It shows that for relatively prime polynomials A, B there always exist polynomials K^*, H^*, such that

$$K^*A + H^*B = 1 \tag{6.5.10}$$

Multiplying this equation by QD and putting $K = QDK^*$, $H = QDH^*$, eq. (6.5.6) results.

Remark 6.2:

> The polynomial equation K(z)A(z) + H(z)B(z) = C(z) for given A, B, and C is called a Diophantine equation. This is named after Diophantos of Alexandria, who considered a similar mathematical problem in the third century: find integer solutions for the equation ka + hb = c, where a, b, and c are given integers. Kučera [79.2] uses the theory of Diophantine equations for the synthesis of discrete control systems.

The theory of the reduced observer shows that a solution with an observer characteristic polynomial of degree n-1 exists. Therefore the observer poles can be fixed by the choice of a polynomial

$$Q(z) = q_0+q_1z+\ldots+q_{n-2}z^{n-2}+z^{n-1} \tag{6.5.11}$$

The numerator polynomial H(z) should be of the same degree (see the discussion in section 3.4.1), i.e. it is assumed that

$$H(z) = h_0+h_1 z+\ldots+h_{n-1}z^{n-1} \tag{6.5.12}$$

The expanded polynomials HB and QD are therefore of degree 2n-2. The same holds for KA if

$$K(z) = k_0+k_1 z+\ldots+k_{n-2}z^{n-2} \tag{6.5.13}$$

is assumed. Then the synthesis equation is

$$(k_0+\ldots+k_{n-2}z^{n-2})(a_0+\ldots+z^n) + (h_0+\ldots+h_{n-1}z^{n-1})(b_0+\ldots+b_{n-1}z^{n-1}) =$$

$$= (q_0+\ldots+z^{n-1})(d_0+\ldots d_{n-1}z^{n-1}) \tag{6.5.14}$$

By expansion and equating coefficients the matrix equation

$$\underline{S}_1(A,B) \times \underline{v}(K,H) = \underline{w}(Q,D) \tag{6.5.15}$$

results, from which the 2n-1 controller coefficients in K and H can be calculated.

Written in detail, eq. (6.5.15) reads

$$\left[\begin{array}{cccc|cccc}
1 & & & 0 & 0 & & 0 & b_{n-1} \\
a_{n-1} & \cdot & & & & & \cdot & b_{n-2} \\
\cdot & \cdot & & & & \cdot & & \\
\cdot & \cdot & \cdot & & & \cdot & & \\
 & \cdot & \cdot & & \cdot & & & \\
 & & \cdot & 1 & 0 & \cdot & & b_1 \\
a_1 & & & a_{n-1} & b_{n-1} & & \cdot & b_0 \\
a_0 & \cdot & & & & \cdot & \cdot & 0 \\
0 & \cdot & \cdot & & & \cdot & \cdot & \\
 & \cdot & \cdot & & \cdot & \cdot & & \\
 & & \cdot & \cdot & \cdot & \cdot & & \\
 & & a_0 & a_1 & b_1 & b_0 & & \\
0 & & 0 & a_0 & b_0 & 0 & \cdot\;\cdot\;\cdot & 0
\end{array}\right]
\left[\begin{array}{c}
k_{n-2} \\ k_{n-3} \\ \cdot \\ \cdot \\ \cdot \\ k_0 \\ \hline h_0 \\ h_1 \\ \cdot \\ \cdot \\ \cdot \\ \cdot \\ h_{n-1}
\end{array}\right]
=
\left[\begin{array}{cccccc}
0 & & & & & 1 \\
 & & & & \cdot & q_{n-2} \\
 & & & \cdot & \cdot & \cdot \\
 & & \cdot & \cdot & & \\
 & \cdot & \cdot & & & \\
0 & 1 & \cdot & & & q_1 \\
1 & \cdot & & & & q_0 \\
q_{n-2} & & & & \cdot & 0 \\
 & & & & \cdot & \\
 & & & \cdot & & \\
 & & \cdot & & & \\
q_1 & q_0 & & & & \\
q_0 & 0 & & & & 0
\end{array}\right]
\left[\begin{array}{c}
d_0 \\ d_1 \\ \cdot \\ \cdot \\ \cdot \\ \cdot \\ \cdot \\ d_{n-1}
\end{array}\right]$$

$\underline{S}(A,B)$ is regular if and only if A and B are relatively prime. This is exactly Sylvester's condition for testing two polynomials for a common factor. See exercise 6.6.

Now substitute eq. (6.5.6) into eqs. (6.5.4) and (6.5.3)

$$y = B(A+D)^{-1}r+[1-B(A+D)^{-1}Q^{-1}H]s \quad (6.5.16)$$

$$u = A(A+D)^{-1}r+A(A+D)^{-1}Q^{-1}Hs \quad (6.5.17)$$

Q appears in the denominator of the disturbance-transfer function from s to y. The zeros of Q should therefore correspond to well damped quickly decaying transients, i.e. they should be located in the region represented in figure 3.11. The disturbance and reference input responses can be designed such that they are partially independent of each other, because the observer characteristic polynomial Q only enters into the disturbance response and the prefilter $r = L^{-1}Mw$ only influences the reference input response. The zeros of the plant, i.e. of B, also arise in the transfer function from r to y. They can only be removed by a cancellation by A+D or L. This requires larger input amplitudes, however, because the plant is forced to follow signals in a frequency band where it "does not want to" due to the zeros. New zeros can be introduced by M.

For systems with one input and one output the polynomial synthesis can be summarized as follows:

1. The denominator polynomial P of the transfer function from r to y is assigned and the difference polynomial D = P-A is formed with the denominator polynomial A of the plant.

2. The observer poles are fixed by Q.

3. The polynomials K and H are calculated from eq. (6.5.15).

4. The transfer function from w to y can be further influenced by the prefilter $L^{-1}M$.

The possible cancellations in the polynomial synthesis will be discussed in section 6.6.

Example:

For the plant of eq. (3.3.10) with T = 1

$$h_z(z) = \frac{B(z)}{A(z)} = \frac{0.3679(z+0.7183)}{(z-1)(z-0.3679)} = \frac{0.3679+0.2642}{z^2-1.3679+0.3679} \qquad (6.5.18)$$

A deadbeat-controller is designed with $P(z) = z^2$, i.e. $D(z) = P(z)-A(z) = 1.3679z-0.3679$. Let the observer pole at $z = -q_o$, $-1 < q_o < 1$ be kept free. Eq. (6.5.15) is then

$$\begin{bmatrix} 1 & 0 & 0.3679 \\ -1.3679 & 0.3679 & 0.2642 \\ 0.3679 & 0.2642 & 0 \end{bmatrix} \begin{bmatrix} k_o \\ h_o \\ h_1 \end{bmatrix} = \begin{bmatrix} 0 & 1 \\ 1 & q_o \\ q_o & 0 \end{bmatrix} \begin{bmatrix} -0.3679 \\ 1.3679 \end{bmatrix}$$

$$\begin{bmatrix} k_o \\ h_o \\ h_1 \end{bmatrix} = \begin{bmatrix} 0.2764 & -0.3849 & 0.5359 \\ -0.3849 & 0.5359 & 3.0387 \\ 1.9669 & 1.0461 & -1.4567 \end{bmatrix} \begin{bmatrix} 0 & 1 \\ 1 & q_o \\ q_o & 0 \end{bmatrix} \begin{bmatrix} -0.3679 \\ 1.3679 \end{bmatrix}$$

$$\begin{bmatrix} k_o \\ h_o \\ h_1 \end{bmatrix} = \begin{bmatrix} 0.518 - 0.724q_o \\ -0.724 - 0.385q_o \\ 2.306 + 1.967q_o \end{bmatrix} \qquad (6.5.19)$$

Figure 6.8 shows the system with the controller in observer-canonical form.

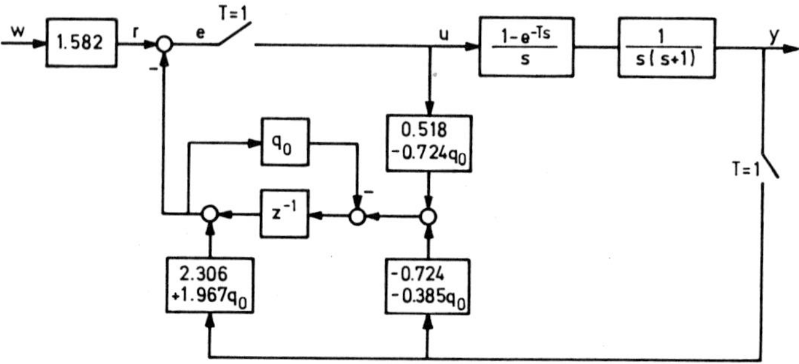

Figure 6.8 Control system with "deadbeat"-reference response and an observer pole tunable in the region $-1 < q_o < 1$, which only influences the disturbance response.

The z-transfer function from r to u follows from eq. (6.5.17) as

$$A(A+D)^{-1} = \frac{(z-1)(z-0.368)}{z^2} \tag{6.5.20}$$

The choice of the prefilter $L^{-1}M$ will be discussed in section 6.7.

Remark 6.3:

The polynomial synthesis can also be extended to multivariable systems with r inputs and s outputs. The derivation in eqs. (6.5.2) to (6.5.8) was written such that it holds unaltered for polynomial matrices instead of polynomials.

One starts with a factorization of the z-transfer matrix

$$\begin{aligned} \underline{y}_z(z) &= \underline{H}_z(z)\underline{u}_z(z) \\ \underline{H}_z(z) &= \underline{B}(z)\underline{A}^{-1}(z) \end{aligned} \tag{6.5.21}$$

The polynomial matrices $\underline{B}(z)$ and $\underline{A}(z)$ have the dimension s×r and r×r, respectively. Here $\underline{A}$ and $\underline{B}$ must be "right coprime", i.e. that a factorization into

$$\underline{A}(z) = \underline{A}^*(z)\underline{G}(z), \quad \underline{B}(z) = B^*(z)\underline{G}(z) \tag{6.5.22}$$

with a common "right divisor" $\underline{G}(z)$ is only possible for a unimodular matrix $\underline{G}(z)$. This is a polynomial matrix whose determinant is a nonzero constant, i.e. independent from z.

Such a factorization is called a minimal matrix-fraction representation. This factorization is not unique, more precisely: If $(\underline{A}(z), \underline{B}(z))$ is a minimal fraction representation of $H_z(z)$, then all other possible minimal fraction representations follow by right-multiplication by an arbitrary r×r unimodular polynomial matrix $\underline{T}(z)$: $(\underline{A}(z)\underline{T}(z), \underline{B}(z)\underline{T}(z))$. The question of how the condition "$\underline{A}(z)$ and $\underline{B}(z)$ right coprime" can be tested and, if necessary, how the greatest common divisor can be divided out, is considered in the following references: [70.5], [74.1], [75.2], [80.2], [81.2].

The factorization in eq. (6.5.21) formally produces a synthesis equation between polynomial matrices corresponding to eq. (6.5.6). Their dimensions are indicated in the following representation of the equation:

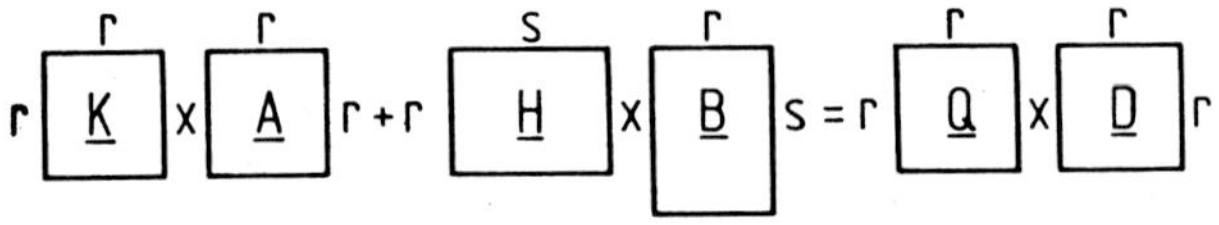

$$\underset{r\times r}{\underline{K}} \times \underset{r\times r}{\underline{A}} + \underset{r\times s}{\underline{H}} \times \underset{s\times r}{\underline{B}} = \underset{r\times r}{\underline{Q}} \times \underset{r\times r}{\underline{D}} \qquad (6.5.23)$$

Thus the synthesis problem can be reduced to the solution of this polynomial equation [71.6]. Internal models for the disturbance and reference inputs can also be incorporated into the design [81.1]. The degree of the individual polynomials fixed in $\underline{K}$, $\underline{H}$, $\underline{Q}$, and $\underline{D}$ is closely tied to the controllability and observability structures of the plant. This will be treated in chapter 9. We content ourselves here with the observation that every assumption for $\underline{K}$, $\underline{H}$, and $\underline{D}$ which satisfies eq. (6.5.23) and for which $\underline{K}$, $\underline{H}$, and $\underline{Q}$ are realizable, leads to a control system whose transfer matrix has the denominator $\det(\underline{A}(z)+\underline{D}(z))$ and whose observer poles are the zeros of $\det \underline{Q}(z)$.

The general polynomial synthesis procedure, outlined in the preceeding remark, will only be elaborated here for the case of systems with one input (r=1) and s linearly independent outputs. Here A(z) is the least common multiple of all denominators of the s transfer functions from u to $\underline{y}$. $\underline{B}(z)$ is a column vector consisting of the s corresponding numerator polynomials. As previously, D(z) is the difference between the desired and the given characteristic polynomials. The controller order is fixed by the degree ρ of

$$Q(z) = q_0+q_1 z+\ldots+q_{\rho-1}z^{\rho-1}+z^{\rho} \qquad (6.5.24)$$

K(z) is as previously of degree $\rho-1$, i.e.

$$K(z) = k_0+k_1z+...+k_{\rho-1}z^{\rho-1} \tag{6.5.25}$$

$\underline{H}(z)$ is a row vector consisting of s polynomials of degree ρ, i.e.

$$\underline{H}(z) = [(h_{10}+...+h_{1\rho}z^{\rho})...(h_{so}+...+h_{s\rho}z^{\rho})] \tag{6.5.26}$$

The extension of the system of eq. (6.5.15) by the additional $\underline{B}$ and $\underline{H}$ elements leads to an underdetermined equation, i.e. there are arbitrarily many solutions. This fact may be exploited to reduce the order ρ of the controller. From section 6.4, it is known that a controller of order $\nu-1$ (ν = observability index) exists, so the synthesis eq. (6.5.23) must have at least one solution for $\rho = \nu-1$. A further reduction of the controller order no longer allows arbitrary placement of plant and observer poles.

If the knowledge of the observability structure is used in the assumed controller structure, then the calculation of its parameters is simplified by the polynomial synthesis of eq. (6.5.23). This is especially true if the plant is described by its transfer matrix $\underline{H}_z(z)$. Notice here exercise 6.5. For the calculation of pole assignment controllers of minimal order also see [69.6], [70.7].

6.6 Pole-Zero-Cancellations

The effect of pole-zero-cancellation in continuous systems was discussed in section 2.3.7. These results can immediately be transferred to discrete systems, but they must be extended if the output between the sampling instants is considered. The canonical decomposition of the plant into four subsystems (distinguished by their controllability and observability) can also be accomplished for sampled-data systems. In practice the sampling period is never chosen such that the controllability or observability is lost by discretization (see sections 4.1 and 5.1). Under this assumption the decomposition of eq. (2.3.47) also remains unchanged by discretization. Such a decomposition can also be performed for the closed loop. All poles vanishing in the transfer function because of the cancellation no longer belong to the controllable and observable subsystem. For example this holds for the cancellation of the observer pole in the transition form eq. (6.5.5) to (6.5.7). The zeros of Q are the eigenvalues of the subsystem, which is uncontrollable from w, but is observable from y. This cancellation is admissible since it occurs in a desirable region Γ of the z-plane due to the appropriate assignment of Q. Other cancellations of zeros of A or B are also possible. The given plant, however, dictates where these cancellations will occur. Thus, some distinctions between cases are necessary. We will work from the polynomial synthesis equation

$$KA + HB = QD \tag{6.6.1}$$

which can be written as follows by adding QA to both sides and inserting D + A = P:

$$(Q+K)A + HB = QP \tag{6.6.2}$$

In this form the right-hand side QP is the closed-loop characteristic polynomial. The various possible cancellations of zeros of A or B will now be discussed.

a) Cancellation of plant poles

This cancellation arises if one assigns an eigenvalue of the open loop, i.e. a zero of A, unchanged for the closed loop, i.e. as zero of Q or P. Because A and B are prime, it follows from eq. (6.6.2) that H must have this zero as well. The corresponding eigenvalues of the plant are unobservable from the feedback Hy in figure 6.7. Therefore they are also unobservable from the feedback signal $Q^{-1}(Ku+Hy)$ and cannot be changed by closing of the loop.

Now two cases can be distinguished, depending on whether the cancellation is made in P or Q. The cancellation in P corresponds to the assignment of an open-loop pole to the closed loop by state-vector feedback. From eq. (4.4.5)

$$u = -\underline{K}'\underline{x} = -\underline{e}'P(\underline{A})\underline{x} = -\underline{e}'(\underline{A}-z_1\underline{I})(\underline{A}-z_2\underline{I})\dots(\underline{A}-z_n\underline{I})\underline{x} \quad (6.6.3)$$

If an assigned pole z_i is also an eigenvalue of $\underline{A}$, i.e. of the open loop, then det $P(\underline{A}) = 0$ and there exists an initial state $\underline{x}_o \neq 0$ for which $u \equiv 0$. For q unchanged eigenvalues, there are q vectors $\underline{x}_{o1} \dots \underline{x}_{oq}$ spanning the unobservable subspace in which $u \equiv 0$ and rank $P(\underline{A}) = n-q$.

If the cancellation is made in Q, then eigenvalues of the plant are assigned unchanged to the observer. Figure 6.7 shows that these controller eigenvalues are then no longer controllable from y because of the cancellation in H and Q.

In the classical control loop configuration of figure 3.8 a cancellation of a plant pole by a controller zero is possible as well. See, for example, eq. (3.4.18). There this cancellation was used in order to remove the cancelled plant pole from the reference transfer function. Thus the deadbeat response required only two sampling intervals after a reference step input. The cancelled eigenvalue can still be stimulated by disturbances or initial states, however. See example (3.6.2). The result, that the order of the reference transfer function is equal to the order of the plant, is reached in another way for the control loop structure with observer and feedback of the constructed states or through the polynomial synthesis of eq. (6.5.6). The poles in Q are made uncontrollable from the reference input irrespective of whether they are identical to those of A or not. Therefore the system of figure 6.8 combines the advantages of both controllers (3.4.16) and (3.4.18) having a two step deadbeat response as in (3.4.18). However, for $q_o = 0$ all eigenvalues are placed at zero as with the controller of eq. (3.4.16), so that all transients are finite.

Cancellation of plant zeros

This cancellation arises if a plant zero, i.e. a zero of B, is assigned as a closed-loop eigenvalue, i.e. a zero of Q or P. In eq. (6.6.2) this common factor of B and QP cannot be contained in A because A and B are relatively prime. It must therefore occur in Q + K. This is illustrated by transforming the block diagram 6.7 into the classical control loop configuration 6.9.

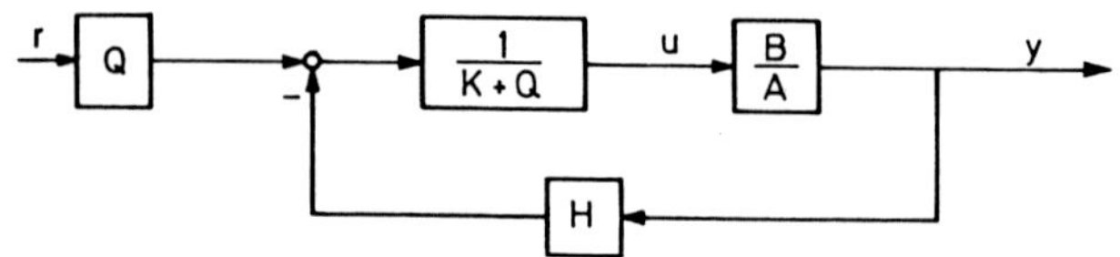

Figure 6.9 Alternative representation of block diagram 6.7

As in continuous systems the cancellation of a plant zero may result in large actuator inputs $|u|$. In addition there is another undesirable effect in sampled-data systems: the cancelled controller pole is not observable from $y(kT)$, but it can be observed from $u(kT)$ and $y(kT+\gamma T)$, $0 < \gamma < 1$. This leads to hidden oscillations between the sampling instants as illustrated in figure 3.18.

In summary it is recommended that pole-zero cancellations are made only in the region of nice stability according to figure 3.11. Here they have only minor influence on the system behavior. The cancellation simplifies the synthesis eqs. (6.6.1) and (6.6.2).

6.7 Closed-loop Transfer Function and Prefilter

The response of the system of figure 6.7 to a reference input w is described by $y_z(z) = f_z(z)w_z(z)$, where $f_z(z)$ is the closed-loop transfer function of eq. (6.5.9)

$$f_z(z) = B(z)[A(z)+D(z)]^{-1}L(z)^{-1}M(z) \tag{6.7.1}$$

First steady-state accuracy of a step response is required. By the final value theorem (B.8.1) this requires $f_z(1) = 1$. For exactly known A(1) and B(1) this could be achieved by a constant prefactor

$$\frac{M}{L} = \frac{A(1)+D(1)}{B(1)} \tag{6.7.2}$$

as in eq. (3.4.27). In most cases, however, robustness of steady-state accuracy with respect to plant uncertainty is required and the feedforward path of the loop must contain an integration. This can be an integral controller as in figure 3.9. In this figure the feedback $d_z(z)$ may also be designed using the structure of figure 6.7. The resulting structure is shown in figure 6.10 with $M_I(1)/L_I(1) = 1$.

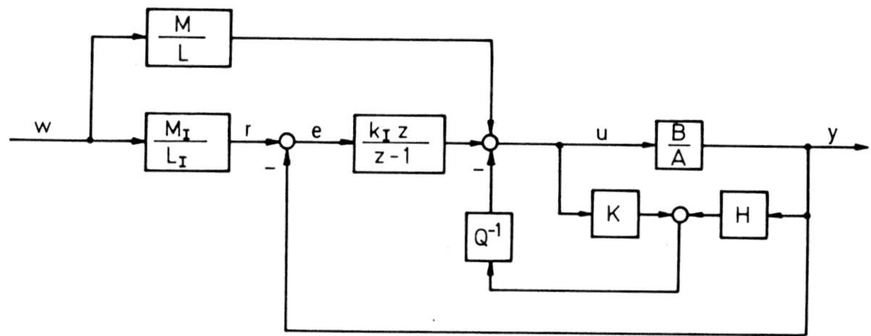

Figure 6.10 Feedback structure with integral controller. $M_I(1)/L_I(1) = 1$.

For a general study of this structure and its extension to internal models of reference and noise signal generators and to the multivariable case the reader is referred to [81.1]. Many plants contain an integration, for example all mechanical systems where u is a force and y a velocity or position. Then the integral part of the controller is not needed. We will pursue only this simpler case $k_I = 0$ described by eq. (6.7.1).

For the transient behavior a first essential step has been made by assigning the poles in $[A(z)+D(z)]^{-1}$ in the nice stability region in the z-plane. Thus y(kT) is composed of well damped, quickly decaying transients in addition to the desired terms entering by w(kT).

The reference response also depends on the zeros of the transfer function. In eq. (6.7.1) it is seen that the closed loop has the same zeros of B as the open loop unless they are cancelled. Suppose that the plant has m finite zeros outside the unit circle and that they constitute a factor $B_o(z)$ of $B(z) = B_o(z)B_i(z)$, $\deg B_o(z) = m_o$, $\deg B_i(z) = m_i$. Clearly $B_o(z)$ cannot be cancelled. Also the pole excess $p = \deg A - \deg B = n - m_o m_i$ cannot be reduced. In plants without time delay the pole excess is generally $p = 1$ (ignoring the pathological choice of the sampling interval as in figure 3.6). Then the "best" realizable closed-loop transfer function is $f_z(z) = B_o(z)/z^{m_o+p}$, e.g. $f_z(z) = 1/z$ for $m_o = 0$, $p = 1$ as in figure 3.18. This is still not a desirable transient because badly damped poles in L are cancelled but still observable from $u(kT)$ and $y(kT+\gamma T)$, $0 < \gamma < 1$. They have only been made unobservable from $y(kT)$. Thus it is more realistic to include all plant zeros outside the "nice stability region" (figure 3.11) into $B_o(z)$. This redefines m_o and m_i. This approach results in m_i well damped poles in L; their effect is visible in $y(t)$ between the sampling instants and may be tolerable. The study of open-loop deadbeat inputs u to the plant in section 4.2 has shown that for a single-input system n sampling intervals are necessary to transfer the system from the zero initial state to an arbitrarily specified final state, e.g. the steady state of the step response. This result must also hold for the special case that u is generated by feedback and prefilter. In a true deadbeat response not only $y(nT) = 1$ but $\dot{\underline{x}}(nT) = \underline{0}$ must be reached. In this case the closed loop transfer function is of the form $f_z(z) = B(z)/z^n$. This is achieved by assigning $P(z) = A(z) + D(z) = z^n$.

Frequently large actuator magnitudes $|u|$ arise in the n-step deadbeat solution. In this case it is possible to use a dynamic prefilter with $L = z^q$ and to give the system n+q sampling intervals for settling after a reference step. Such prefilters are designed via the open loop deadbeat control. From eq. (4.1.2) with $k = 0$

$$[\underline{A}^{N-1}\underline{b}, \underline{A}^{N-2}\underline{b}\ldots\underline{b}]\underline{u}_{N-1} = \underline{x}(NT) - \underline{A}^N\underline{x}(0) \qquad (6.7.3)$$

with $\underline{u}_{N-1} = [u(0), u(T)\ldots u(NT-T)]'$

Let $\underline{x}(0) = \underline{0}$ and suppose $\underline{x}$ must reach a steady state after N sampling intervals. This stationary state is characterized by

$$\underline{x}(NT+T) = \underline{x}(NT)$$
$$\underline{x}(NT) = \underline{A}\underline{x}(NT)+\underline{b}u_{stat} \tag{6.7.4}$$

If the plant does not contain an integration, i.e. if $\underline{A}$ does not have an eigenvalue at $z = 1$, then $\underline{x}(NT) = (\underline{I}-\underline{A})^{-1}\underline{b}u_{stat}$ holds and the stationary output is

$$y_{stat} = \underline{c}'\underline{x}(NT) = \underline{c}'(\underline{I}-\underline{A})^{-1}\underline{b}u_{stat} = h_z(1)u_{stat} \tag{6.7.5}$$

i.e.

$$u_{stat} = \frac{1}{h_z(1)} \times y_{stat} \tag{6.7.6}$$

If the plant contains an integration, then $u_{stat} = 0$. In this case $\underline{x}(NT)$ satisfies

$$(\underline{I}-\underline{A})\underline{x}(NT) = \underline{0}$$
$$y_{stat} = \underline{c}'\underline{x}(NT) \tag{6.7.7}$$

As is shown in the following examples a compromise between the required actuator signal magnitude and the duration of the dead-beat response can be made by the choice of N. If $u_z(z)$ has been determined, then the required prefilter is obtained from eq. (6.5.17)

$$u_z = A(A+D)^{-1}r_z = A(A+D)^{-1}L^{-1}Mw_z$$

and with $w_z = z/(z-1)$, $(A+D) = z^n$

$$\frac{M(z)}{L(z)} = \frac{z^{n-1}(z-1)u_z}{A(z)} \tag{6.7.8}$$

Example 1:

The plant $1/s(s+1)$ with sampler and hold has according to eq. (3.3.10) the z-transfer function

$$h_z(z) = \frac{B(z)}{A(z)} = \frac{(T-1+e^{-T})z+(1-Te^{-T}-e^{-T})}{(z-1)(z-e^{-T})} \qquad (6.7.9)$$

By partial fractions decomposition

$$h_z(z) = \frac{T}{z-1} - \frac{1-e^{-T}}{z-e^{-T}}$$

a state representation in diagonal form is obtained, see eq. (A.2.11)

$$\underline{x}(kT+T) = \begin{bmatrix} 1 & 0 \\ 0 & e^{-T} \end{bmatrix} \underline{x}(kT) + \begin{bmatrix} T \\ e^{-T}-1 \end{bmatrix} u(kT) \qquad (6.7.10)$$

$$y(kT) = [1 \quad 1]\underline{x}(kT)$$

The desired stationary state $y(t) \equiv 1$ is reached according to eq. (6.7.7) with

$$(\underline{I}-\underline{A})\underline{x}(NT) = \begin{bmatrix} 0 & 0 \\ 0 & 1-e^{-T} \end{bmatrix} \begin{bmatrix} x_1(NT) \\ x_2(NT) \end{bmatrix} = \begin{bmatrix} 0 \\ 0 \end{bmatrix}$$

$$\underline{c}'\underline{x}(NT) = [1 \quad 1] \begin{bmatrix} x_1(NT) \\ x_2(NT) \end{bmatrix} = 1$$

The stationary value of the state is $\underline{x}(NT) = [1 \quad 0]'$. The initial state is $\underline{x}(0) = \underline{0}$ and the corresponding control sequence satisfies eq. (6.7.3)

$$\begin{bmatrix} T & \cdots & T \\ e^{-(N-1)T}(e^{-T}-1) & \ldots e^{-T}(e^{-T}-1) & e^{-T}-1 \end{bmatrix} \begin{bmatrix} u(0) \\ \vdots \\ u(NT-T) \end{bmatrix} = \begin{bmatrix} 1 \\ 0 \end{bmatrix} \qquad (6.7.11)$$

i.e.

$$u(0)+u(T)+\ldots u(NT-T) = 1/T$$

$$e^{-(N-1)T}u(0)+e^{-(N-2)T}u(1)+\ldots e^{-T}u(NT-2T)+u(NT-T) = 0 \qquad (6.7.12)$$

For N = n = 2 the solution is unique and

$$u_0 = u(0) = \frac{1}{T(1-e^{-T})} , \quad u_1 = u(T) = -\frac{e^{-T}}{T(1-e^{-T})} \qquad (6.7.13)$$

As $T \to 0$, u_0 goes to infinity and u_1 to minus infinity. In the limiting case of the continuous system this solution is no longer reasonable. Here a δ-function and its derivative must be used as actuator input, in order to bring the system into the desired stationary state immediately. In sampled-data systems, however, this deadbeat response can be realized in practice with a finite transient process and finite input amplitudes u_0 and u_1, if the sampling period T is not chosen too small.

For T = 1 step responses in u and y are shown in figure 6.11. In figure 3.15 y is compared to other step responses.

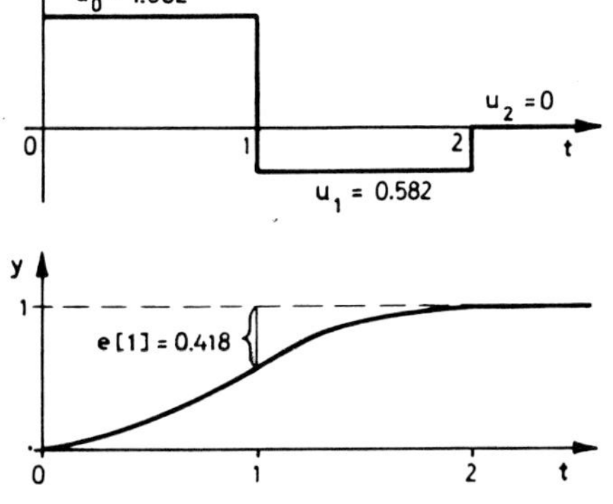

Figure 6.11

Deadbeat-step response y(t) and corresponding input variable u(t)

The numerical values are

$$u_z(z) = 1.582-0.582z^{-1} = 1.582(z-0.368)/z.$$

The prefilter of eq. (6.7.8) is

$$\frac{M}{L} = \frac{z(z-1)\times 1.582(z-0.368)}{(z-1)(z-0.368)z} = 1.582 \tag{6.7.14}$$

Example 2:

Assume that the maximal actuator input 1.582 in figure 6.10 is too large and should be reduced. Since $|u(0)|$ and $|u(T)|$ in eq. (6.7.13) decrease monotonically with growing T, one can consider enlarging the sampling period. However, a disadvantage is that a large T can produce a large time delay between 0 and T depending on the relative instants of reference step and sampling. Moreover this procedure can not be generalized to examples with complex eigenvalues, see section 4.2.

For comparison consider the time-optimal control to a desired steady state for $|u| \leq 1$. This is

$$u(t) = \begin{cases} 1 \text{ for } 0 < t < 1.585 \\ -1 \text{ for } 1.585 < t < 2.170 \end{cases} \tag{6.7.15}$$

This could suggest the application of unequal sampling intervals of 1.585 and 0.585 seconds. However, this would be unsuitable if the reference step was out of synchronism, i.e. if the 0.585-second interval came first after the reference step.

Finally, as a last resort the number of steps N could be increased to give the control system more time. In this case the solution of eq. (6.7.12) is no longer unique, i.e. additional demands can be placed on $\underline{u}_N$. For N = 3 this gives

$$\begin{aligned} u_0 + u_1 + u_2 &= 1/T \\ e^{-2T}u_0 + e^{-T}u_1 + u_2 &= 0 \end{aligned} \tag{6.7.16}$$

The time instant is here written as an index for simplification.

For $u_2 = 0$ the solution agrees with the previous one. Some alternative possibilities are as follows:

1. The requirement that the "input energy" $u_0^2 + u_1^2 + u_2^2$ should be minimal leads to

$$u_0 = \frac{e^T}{2T(e^T-1)}, \quad u_1 = \frac{1}{2T}, \quad u_2 = \frac{1}{2T(1-e^T)} \tag{6.7.17}$$

   and for T = 1: $u_0 = 0.791$, $u_1 = 0.5$, $u_2 = -0.292$

   The required prefilter L^{-1} is obtained according to eq. (6.7.8)

$$\frac{M}{L} = \frac{(A+D)u_z}{Aw_z} = \frac{z^2 \times u_z}{(z-1)(z-0.368)} \times \frac{z-1}{z}$$

$$= \frac{u_z \times z}{z-0.368} = \frac{0.791z^2+0.5z-0.292}{z(z-0.368)}$$

$$\frac{M(z)}{L(z)} = \frac{0.791(z+1)}{z} \tag{6.7.18}$$

2. For constrained $|u|$ a satisfactory solution is obtained with $u_0 = u_1$. It is

$$u_0 = u_1 = \frac{e^{2T}}{(2e^{2T}-e^T-1)T}, \quad u_2 = \frac{-(e^T+1)}{(2e^{2T}-e^T-1)T} \tag{6.7.19}$$

   and for T = 1: $u_0 = u_1 = 0.667$, $u_2 = -0.334$.

   The prefilter is

$$\frac{M(z)}{L(z)} = \frac{0.667(z+1.368)}{z} \tag{6.7.20}$$

3. The requirement that the minimum total time 3T should be reached for the transient process with $|u| \le 1$ leads to

$$T = 0.76 , \quad u_0 = u_1 = 1 , \quad u_2 = -0.684 \tag{6.7.21}$$

The total time of 2.28 time units lies close to the duration 2.17 time units of the time optimal solution from eq. (6.7.15). This solution cannot be significantly improved by increasing N. For example for N = 4 the result is T = 0.562, $u_0 = u_1 = u_2 = 0.927$, $u_3 = -1$; 4T = 2.25 time units.

6.8 Disturbance Compensation

Feedback and prefilter have been treated; now the third part of the control system will be introduced, namely the disturbance compensation. Assume first that a measurable disturbance $\underline{q}$ influences the plant as a disturbance input $\underline{s} = \underline{Mq}$ through a known matrix $\underline{M}$

$$x[k+1] = \underline{Ax}[k] + \underline{Bu}[k] + \underline{Mq}[k] \tag{6.8.1}$$

Form the input variable $\underline{u}$ as the sum of two parts

$$\underline{u} = \underline{u}_F + \underline{u}_D \tag{6.8.2}$$

The part $\underline{u}_F$ is determined by feedback as has already been shown. In contrast, the part $\underline{u}_D$ has the task of counteracting the disturbance input $\underline{q}$ [71.9]. Then

$$\underline{x}[k+1] = \underline{Ax}[k] + \underline{Bu}_F[k] + \underline{Bu}_D[k] + \underline{Mq}[k] \tag{6.8.3}$$

and $\underline{u}_D$ is determined such that the additional terms

$$\underline{e} = \underline{Bu}_D + \underline{Mq} \tag{6.8.4}$$

become zero or - if this is impossible - at least the quadratic additional term

$$V(\underline{u}_D) = \underline{e}'\underline{e} = (\underline{B}\underline{u}_D+\underline{M}\underline{q})'(\underline{B}\underline{u}_D+\underline{M}\underline{q}) \qquad (6.8.5)$$

is minimized. For the latter case the gradient of V with respect to $\underline{u}_D$ is set to zero:

$$\frac{dV}{d\underline{u}_D} = 2\underline{B}'(\underline{B}\underline{u}_D+\underline{M}\underline{q}) = \underline{0} \qquad (6.8.6)$$

Assume that the r inputs, r < n, are linearly independent, i.e. the n×r matrix B has rank r. Then the r×r matrix $\underline{B}'\underline{B}$ is regular and eq. (6.8.6) can be solved for $\underline{u}_D$:

$$\underline{u}_D[k] = -(\underline{B}'\underline{B})^{-1}\underline{B}'\underline{M}\underline{q}[k] \qquad (6.8.7)$$

This is actually a minimum because the second derivative 2 $\underline{B}'\underline{B}$ is positive definite.

Thus an additional part $\underline{u}_D[k]$ of the input is determined from the measured disturbance $\underline{q}[k]$ by eq. (6.8.7) which minimizes the influence of $\underline{q}[k]$ on $\underline{x}[k+1]$. (This approach may be modified to minimize the influence of $\underline{q}[k]$ on $\underline{x}[i]$, i > k [77.8].)

If $\underline{q}$ is unmeasurable but can be modelled by a disturbance generator, then a value correspondingly constructed by the disturbance observer of section 5.5 can be substituted for $\underline{q}$ in eq. (6.8.7).

Example:

The example from figure 5.5 is now worked through to give a complete system. The plant including a disturbance model is given by eq. (5.5.5), the observer by eq. (5.5.6). The variables x_1, $\hat{x}_2$, and $\hat{z}_s$ are available for feedback. The loop is closed by

$$u[k] = u_F[k] + u_D[k]$$

$$u_F[k] = -\underline{k}'\begin{bmatrix} x_1[k] \\ \hat{x}_2[k] \end{bmatrix} + Vw[k] \qquad (6.8.8)$$

$$u_D[k] = -k_D\hat{z}_s[k] \qquad (6.8.9)$$

$\underline{k}' = [k_1 \;\; k_2]$ is determined by state vector feedback, for example for a deadbeat solution $\underline{k}' = [1.582 \;\; 1.243]$ according to eq. (4.4.8). In the steady state $\hat{x}_2 = 0$, $u_F = -\underline{k}'\hat{\underline{x}} = 0$ and $x_1 = y = w$, i.e. from eq. (6.8.8) $u_F = 0 = -1.582w + Vw$, i.e. $V = 1.582$.

The second part of the input variable is set to $u_D = -z_s$ in order to compensate the step disturbance at the plant input. The control law is then

$$u[k] = -[1.582 \quad 1.243 \quad 1]\underline{y}^*[k] + 1.582w[k] \qquad (6.8.10)$$

$$\underline{y}^* = [x_1 \;\; \hat{x}_2 \;\; \hat{z}_s]'$$

The state equation of the closed loop is now calculated in order to check the result. These calculations are simplified if the observer errors $\tilde{x}_2 = x_2 - \hat{x}_2$ and $\tilde{z}_s = z_s - \hat{z}_s$ are used as state variables in place of $\hat{x}_2$ and $\hat{z}_s$. Eq. (6.3.10) is here

$$\underline{x}^*[k+1] = \begin{bmatrix} 1 & 0 & 0 & 0 & 0 \\ 0 & -0.418 & -1 & 0.632 & 0.785 \\ 0 & 0.175 & 0.418 & 0.368 & 0.457 \\ 0 & 0 & 0 & 0.418 & -1 \\ 0 & 0 & 0 & 0.175 & -0.418 \end{bmatrix} \underline{x}^*[k] + \begin{bmatrix} 0 \\ 1 \\ 0.582 \\ 0 \\ 0 \end{bmatrix} w[k]$$

$$y[k] = [0 \quad 0 \quad 1 \quad 0 \quad 0] \; \underline{x}^*[k]$$

$$(6.8.11)$$

with $\underline{x}^* = [z_s \; , \; x_2 \; , \; x_1 \; , \; \tilde{z}_s \; , \; \tilde{x}_2]'$

$$\underline{x}^*[k+2] = \begin{bmatrix} 1 & 0 & 0 & 0 & 0 \\ 0 & 0 & 0 & -0.231 & -1.745 \\ 0 & 0 & 0 & 0.501 & -0.231 \\ 0 & 0 & 0 & 0 & 0 \\ 0 & 0 & 0 & 0 & 0 \end{bmatrix} \underline{x}^*[k] + \begin{bmatrix} 0 \\ -1 \\ 0.418 \\ 0 \\ 0 \end{bmatrix} w[k] \begin{bmatrix} 0 \\ 1 \\ 0.582 \\ 0 \\ 0 \end{bmatrix} w[k+1]$$

$$\underline{x}^*[k+3] = \begin{bmatrix} 1 & 0 & 0 & 0 & 0 \\ 0 & 0 & 0 & 0 & 0 \\ 0 & 0 & 0 & 0 & 0 \\ 0 & 0 & 0 & 0 & 0 \\ 0 & 0 & 0 & 0 & 0 \end{bmatrix} \underline{x}^*[k] + \begin{bmatrix} 0 \\ -1 \\ 0.418 \\ 0 \\ 0 \end{bmatrix} w[k+1] + \begin{bmatrix} 0 \\ 1 \\ 0.582 \\ 0 \\ 0 \end{bmatrix} w[k+2]$$

Thus:

1. After two sampling steps the states x_1 and x_2 of the plant are no longer dependent upon their initial state, rather only upon the reference input w and the initial estimate errors $\tilde{z}_S[0]$ and $\tilde{x}_2[0]$.

2. If the initial estimate errors are zero, then the states x_1 and x_2 of the plant reach the stationary states $x_1 = 1$, $x_2 = 0$ in two sampling intervals after a reference step.

3. Two sampling intervals are required to remove the estimate errors after an initial estimate error or a disturbance step. After a further sampling interval the plant is in the zero state.

The minimal transition time is thus produced for all exitations.

The practical realization is simplified by representing the observer plus feedback of the observer states in the observer-canonical form. Eq. (6.8.10) is first written in the form

$$u[k] = 1.582\{w[k]-x_1[k]-[0.786 \quad 0.632][\hat{x}_2[k] \quad \hat{z}_S[k]]'\}$$

and with eq. (5.5.6)

$$u[k] = 1.582\{w[k]-2.979x_1[k]-[0.786 \quad 0.632]\hat{v}[k]\} \qquad (6.8.12)$$

The last term $u_V[k] = [0.786 \quad 0.632]\hat{v}[k]$ is formed by the observer eq. (5.5.6)

$$\underline{\hat{v}}[k+1] = \begin{bmatrix} -0.418 & 0.175 \\ -1 & 0.418 \end{bmatrix} \underline{\hat{v}}[k] - \begin{bmatrix} 1.488 \\ 2.165 \end{bmatrix} y[k] + \begin{bmatrix} 0.175 \\ 0.582 \end{bmatrix} u[k] \qquad (6.8.13)$$

and transformed into the observer-canonical form (A.3.22)

$$\hat{\underline{v}}_F[k+1] = \begin{bmatrix} 0 & 0 \\ 1 & 0 \end{bmatrix} \hat{\underline{v}}_F[k] + \begin{bmatrix} -2.539 \\ 0.560 \end{bmatrix} y[k] + \begin{bmatrix} -0.230 \\ -0.402 \end{bmatrix} u[k]$$

$$u_V[k] = [0 \quad 1] \hat{\underline{v}}_F[k]$$

$$u[k] = u_V[k] + 1.582(w[k] - 2.979y[k])$$

Figure 6.12 shows the complete control system.

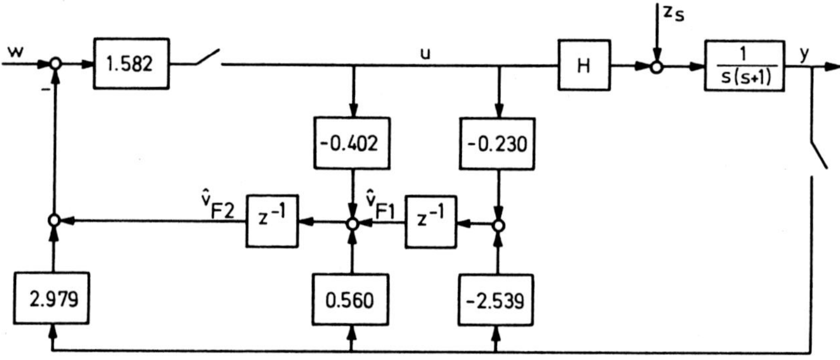

Figure 6.12 Deadbeat control system. Two sampling intervals after a step of w and three sampling intervals after a step of z_s are required for the transients.

6.9 Excercises

6.1 Calculate a sampled-data controller with a sampling period T = 1 for the plant of figure 6.13 such that y(t) becomes identically zero with a deadbeat transient after switching on the cosine disturbance at time t = 0.

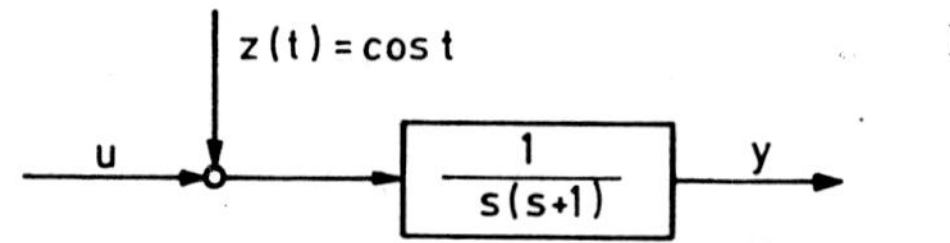

Figure 6.13 Exercise on disturbance compensation

6.2 For the following plant

$$\underline{x}[k+1] = \begin{bmatrix} 0 & -0.3 \\ 1 & 1.3 \end{bmatrix} \underline{x}[k] + \begin{bmatrix} -1 \\ 2 \end{bmatrix} u[k]$$

$$\underline{y}[k] = [0 \quad 1]\underline{x}[k]$$

a) Determine a state feedback such that the stable eigenvalue remains unchanged. How do the two feedback gains depend on the choice of the second real pole?

b) Calculate a first order observer with an eigenvalue at $z = 0.2$ and close the loop by the state vector feedback from a). Calculate $\underline{x}[k]$, $k = 1,2,3$ for $\underline{x}[0] = [1 \quad 1]$ and zero initial state of the observer.

c) Discuss the effect of the pole-zero-cancellation in a).

6.3 The continuous plant "triple integration" with a state representation

$$\dot{\underline{x}} = \begin{bmatrix} 0 & 1 & 0 \\ 0 & 0 & 1 \\ 0 & 0 & 0 \end{bmatrix} \underline{x} + \begin{bmatrix} 0 \\ 0 \\ 1 \end{bmatrix} u$$

$$y = [1 \quad 0 \quad 0] \; \underline{x}$$

is to be controlled by a second order observer with sampling period $T = 0.5$ and state feedback $u(kT) = -\underline{k}'\hat{\underline{x}}(kT) + w(kT)$ with a) $T = 0.5$, b) $T = 1$. All eigenvalues should be located at $z = 0$. Calculate the responses u and y to a step of the reference input w at initial state $\underline{x}[0] = [0 \quad -1 \quad -1]'$ and zero initial state of the observer.

6.4 The trolley position x_1 and the cable angle x_3 are measured for the loading bridge. Combine the two first order subobservers from exercise 5.2 with eigenvalues at $z = 0$ with the state feedback from exercise 4.6 for $z_3 = 0.75$, i.e.

$$\underline{k}' = [980 \quad 3614 \quad 1447. \quad 12342]$$

and thereby give a realization as a controller of first order.

6.5 A plant with two measured variables y_1 and y_2 has the z-transfer functions

$$\underline{y}_z(z) = \begin{bmatrix} \dfrac{z}{(z+1)(z-1)} \\ \dfrac{z}{(z+1)(z+1)} \end{bmatrix} u_z(z)$$

Calculate a controller of the lowest possible order such that all the eigenvalues of the closed loop lie at $z = 0$. Can this controller be chosen so that y_1 is fed back only proportionally, i.e. without dynamics?

6.6 By using the polynomial synthesis for the loading bridge, calculate a first order controller which places the poles at $z_{1,2} = 0.4876 \pm j0.3026$, $z_{3,4} = 0.75$ and has an observer pole at $z = 0$ which does not enter into the reference transfer function. Compare the result, solution method, and computational effort with the steps which led to the solution of exercise 6.4.

6.7 Give your opinion of the following sampled-data control loop:

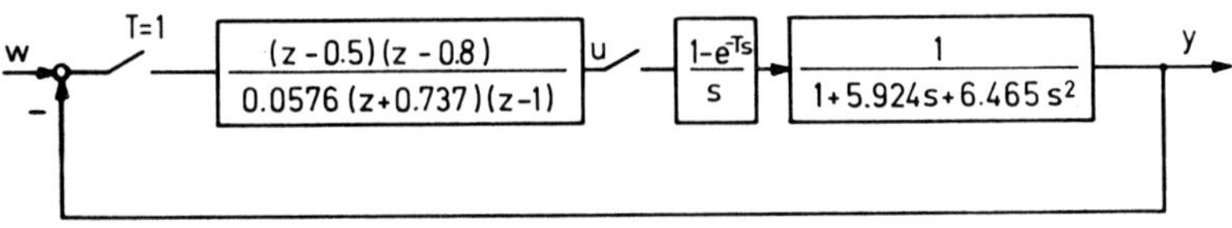

Figure 6.14 Sampled-data control loop

For a unit step $w(t) = 1(t)$ calculate a) $y(kT)$, b) $u(kT)$, c) $y(kT+0.5T)$. Does this controller have good properties?

6.8 For the plant of exercise 6.7 calculate by polynomial synthesis a sampled-data controller which assigns eigenvalues at $z = 0$. Determine a prefilter such that the step response of the closed loop does not exceed a maximal actuator signal magnitude of $|u| = 5$.

6.9. Reduce the effect of the wind disturbance described in exercise 5.4 on the loading bridge by a disturbance compensation.

7 Geometric Stability Investigation and Pole Region Assignment

7.1 Stability

The most important requirement on a control system is the stability of the closed loop. In section 3.2 it was indicated that a necessary and sufficient condition for the asymptotic stability of a discrete rational system is that all eigenvalues of its system matrix have an absolute value of less than one.

The control loop is therefore stable if and only if all zeros of the characteristic polynomial of the closed loop

$$P(z) = p_0 + p_1 + \dots + p_{n-1}z^{n-1} + z^n \qquad (7.1.1)$$

lie inside the unit circle. Stability criteria for testing this property are given in appendix C. A polynomial with all its zeros in the unit circle is called a Schur-Cohn polynomial [1917], [1922].

Stability tests played a central role in classical design methods. One had to take into consideration the available structurally simple analog controllers. Their design always included a stability test. The synthesis by state space methods has shown that every plant can be stabilized as long as its unstable eigenvalues are controllable and observable. Observer and pole assignment provide the controller structure by which the eigenvalues can be arbitrarily assigned. Since one always makes a stable assignment, a stability test of the closed loop is not necessary. For the exactly known, linear plant the stability problem is therefore of less importance. It retains its central importance, however, if plant parameter uncertainties or variations are considered.

For practical control problems an exact position of all eigenvalues is not specified. It suffices that they are located in a specified pole region Γ in the eigenvalue plane. By relaxing the eigenvalue specification we gain flexibility for trade-offs with other demands on the control system.

The set of all eigenvalue configurations in a given domain Γ corresponds to an admissible region K_Γ for the feedback vector $\underline{k}'$. The designer may select a solution from this admissible solution set based on additional requirements like small gains, gain reduction margins and robustness with respect to parameter variations, sensor failures or inaccurate controller implementation.

The representation of the stability region in a plane of two polynomial coefficients p_i, p_j or controller parameters κ_1, κ_2 goes back to Vishnegradsky [1876], Neimark [47.2], [48.2], Mitrovic [58.6] and Šiljak [64.10], [69.8]. In this so called D-decomposition method the characteristic polynomial

$$P(s,\kappa_1,\kappa_2) = p_0(\kappa_1,\kappa_2) + p_1(\kappa_1,\kappa_2)s + \ldots + p_{n-1}(\kappa_1,\kappa_2)s^{n-1} + s^n \tag{7.1.2}$$

is formed. The real and complex parts of this equation - taken for values of s on the boundary of Γ - are set to zero. This generates the equations for κ_1, κ_2 pairs which place the eigenvalues on the boundary. In the following a new derivation and further extensions of this approach are given which is particularly suited for computer graphics design of robust control systems.

In this chapter pole region assignment is treated in detail. It will be applied to the design of robust control systems in chapter eight.

7.2 Stability Region in P Space

The coefficients of the characteristic polynomial $P(z) = p_0 + p_1 z + \ldots + p_{n-1}z^{n-1} + z^n$ are interpreted as coordinates of an n-dimensional parameter space P. In other words: form the n-vector

$$\underline{p} := [p_o \quad p_1 \quad \dots \quad p_{n-1}]' \tag{7.2.1}$$

Then

$$P(z) = [\underline{p}' \quad 1]\underline{z}_n \text{ with } \underline{z}_n := [1 \quad z \quad z^2 \quad \dots \quad z^n]' \tag{7.2.2}$$

First the stability region is determined in the parameter space P in which the vector $\underline{p}$ lies. This stability region is bounded by surfaces corresponding to polynomials with zeros on the unit circle. By the continuity of the mapping

$$P(z) = [\underline{p}' \quad 1]\underline{z}_n = \prod_{i=1}^{n} (z-z_i) \tag{7.2.3}$$

from the zeros z_i to the $\underline{p}$-vector this boundary surface in P space is crossed whenever a real z_i or a complex conjugate pair z_i, $z_j = \bar{z}_i$ crosses the unit circle. There are three possible ways in which this crossing of the unit circle can occur:

1. At $z = 1$ for $P(1) = 0$ (7.2.4)

2. At $z = -1$ for $P(-1) = 0$ (7.2.5)

3. As a conjugate complex eigenvalue pair $z = \tau \pm j\eta$, with $\tau^2+\eta^2 = 1$. P(z) then contains a factor

   $(z-\tau-j\eta)(z-\tau+j\eta) = z^2-2\tau z+\tau^2+\eta^2 = z^2-2\tau z+1, \quad -1 \le \tau \le 1$

   i.e. P(z) can be written as

   $$P(z) = (z^2-2\tau z+1)R(z)\ , \ R(z) := z^{n-2} + r_{n-3}z^{n-3} + \dots + r_o \tag{7.2.6}$$

Let us first study the case n = 2 with R(z) = 1. The two real boundaries are the straight lines

$$P(1) = p_o+p_1+1 = 0$$

$$P(-1) = p_o-p_1+1 = 0$$

as shown in figure 7.1. The complex boundary is the straight line segment

$$\underline{p}' = [1 \quad -2\tau] \quad , \quad -1 \le \tau \le 1 \tag{7.2.7}$$

denoted by c in figure 7.1.

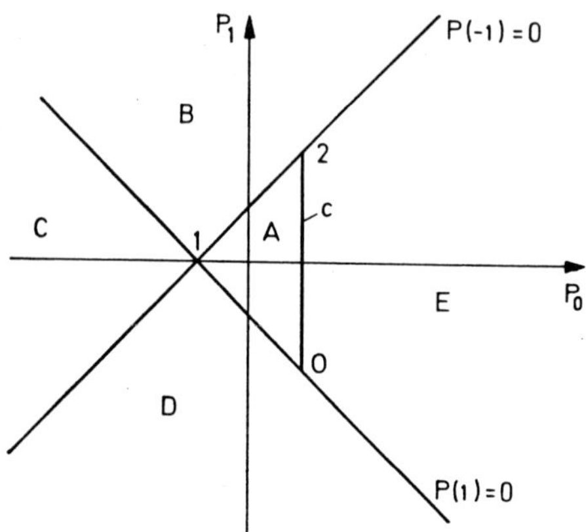

Figure 7.1 Partitions of the polynomial coefficient plane for n = 2

Consider the case $p_0 = p_1 = 0$, i.e. $P(z) = z^2$, so both roots are at z = 0. Now let the coefficients be varied continuously. If $\underline{p}$ crosses over the line P(1) = 0, then a root crosses the unit circle at z = 1, correspondingly for P(-1) = 0 and z = -1. If $\underline{p}$ crosses the complex boundary c, then a complex pole pair leaves the unit circle.

Point 0 in figure 7.1 is common to c and P(1) = 0. A double pole arises here for z = 1, i.e.

$$P_0(z) = (z-1)^2 = z^2-2z+1, \ \underline{p}_0' = [1 \quad -2]$$

correspondingly

$$\begin{aligned} P_1(z) &= (z-1)(z+1) = z^2-1, \ \underline{p}_1' = [-1 \quad 0] \\ P_2(z) &= (z+1)^2 = z^2+2z+1 \ , \ \underline{p}_2' = [\ 1 \quad 2] \end{aligned} \tag{7.2.8}$$

The two straight lines P(1) = 0 and P(-1) = 0 and the line segment c partition the P-plane into the following regions (EV = eigenvalue, UC = unit circle).

A: Both EVs in the UC

B: One EV in the UC, one to the left

C: One EV to the left, one to the right of the UC

D: One EV in the UC, one to the right

E: Both EVs outside the UC, either as a complex pair or both to the left, or both to the right. (Note that these cases can be transferred into each other by continuous motion of the eigenvalues across the branching points of a root locus without crossing the unit circle.)

Compare this with the algebraic stability criterion (C.2.14):

$P(1) > 0$ and $P(-1) > 0$ corresponds to the two real boundaries. $X - Y = 1 - p_0 > 0$ corresponds to a straight line. This extends the line segment c in figure 7.1 to an infinite line. Polynomials $P(z) = z^2-2\tau z+1$ for $|\tau| > 1$, however, have two real zeros z_1 and $z_2 = 1/z_1$. They are not located on the stability boundary.

We are mainly interested in the stability region A which is completely determined by the vertices of the triangle 012, see eq. (7.2.8).

Let us now return to eq. (7.2.6) for $n > 2$. If the polynomial factor $R(z)$ is fixed, i.e. $n - 2$ eigenvalues are fixed, then by eq. (7.2.6) $\underline{p}$ is linear in τ, which means that $\underline{p}$ moves along a straight line in P-space if the remaining two roots move as a conjugate complex pair along the unit circle. On the other hand for fixed τ, i.e. a pole pair fixed on the unit circle, $\underline{p}$ is linear in the $n - 2$ coefficients of the R-polynomial, yielding an $(n-2)$-dimensional hyperplane. If τ varies in the interval $[-1, +1]$, then this hyperplane moves with τ and forms the complex boundary surface c in the parameter space P.

The complex boundary c and the two hyperplanes $P(1) = 0$ and $P(-1) = 0$ partition the P-space. If $\underline{p}$ moves within one region, then the eigenvalues move in the z-plane without crossing over the unit circle. In order to test which eigenvalue position relative to the unit circle corresponds to which region, an arbitrary point in each region can be chosen and the roots of the corresponding polynomial determined. There relative location with respect to the unit circle applies to all points of the region.

Of special interest is the region for which all eigenvalues lie inside the unit circle. We will call this the "stability region in P-space". Since all stable roots are finite this region in P-space must also be finite. It is also continuously contractible to the point $\underline{p} = \underline{0}$, the "deadbeat point". This corresponds to a contraction of the unit circle to the point $z = 0$. In particular the stability region is connected and does not contain unstable enclaves. In every other region at least one root is located outside the unit circle. As this root moves to infinity, $\underline{p}$ also moves to infinity. Therefore the stability region is the only bounded region.

For $n = 3$ it is still possible to visualize the stability region. The two real boundaries are the planes

$$P(1) = p_0+p_1+p_2+1 = 0$$

$$P(-1) = p_0-p_1+p_2-1 = 0$$

Eq. (7.2.6) is here

$$\begin{aligned} P(z) &= (z^2-2\tau z+1)(z+r) = z^3+(r-2\tau)z^2+(1-2r\tau)z+r \\ \underline{p}' &= [r \quad 1-2r\tau \quad r-2\tau] \,, \quad -1 \le \tau \le 1 \end{aligned} \tag{7.2.9}$$

For constant r the following linear relation holds

$$\underline{p} = \begin{bmatrix} p_0 \\ p_1 \\ p_2 \end{bmatrix} = \begin{bmatrix} r \\ 1 \\ r \end{bmatrix} + \begin{bmatrix} 0 \\ -2r \\ -2 \end{bmatrix} \tau = \underline{r}_0+\underline{r}_1\tau \tag{7.2.10}$$

For constant τ a linear relation is also obtained

$$\underline{p} = \begin{bmatrix} p_0 \\ p_1 \\ p_2 \end{bmatrix} = \begin{bmatrix} 0 \\ 1 \\ -2\tau \end{bmatrix} + \begin{bmatrix} 1 \\ -2\tau \\ 1 \end{bmatrix} r = \underline{s}_0+\underline{s}_1 r \tag{7.2.11}$$

The complex boundary surface therefore contains these two families of straight lines.

The stability region is represented in figure 7.2. It is bounded by

a) the triangle 012, which lies in the $P(1) = 0$ plane,

b) the triangle 123, which lies in the $P(-1) = 0$ plane,

c) the complex boundary surface of eq. (7.2.9) with $|r| \leq 1$, $|\tau| \leq 1$.

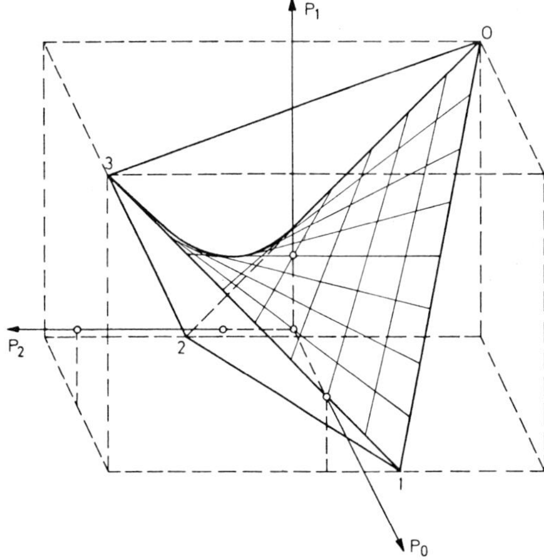

Figure 7.2 Stability region in the polynomial coefficient space for $n = 3$.

The corners of the stability region correspond to the four polynomials with zeros in $\{-1, +1\}$. They are again numbered such that $P_i(z) = (z+1)^i(z-1)^{n-i}$. Thus

$$\begin{aligned} \underline{p}_0' &= [-1 \quad 3 \quad -3] \\ \underline{p}_1' &= [\ 1 \quad -1 \quad -1] \\ \underline{p}_2' &= [-1 \quad -1 \quad 1] \\ \underline{p}_3' &= [\ 1 \quad 3 \quad 3] \end{aligned} \qquad (7.2.12)$$

If $\underline{p}$ moves along an edge from the corner i to the corner $i+1$, then a real root migrates from $z = +1$ to $z = -1$ and the other eigenvalues remain at $z = -1$ or $z = +1$. On the edge 12, for example, one eigenvalue lies at $z = -1$, one at $z = +1$, and one real eigenvalue in between.

If $\underline{p}$ moves along an edge from the corner i to the corner i+2, then a complex conjugate pole pair moves along the unit circle from $z = +1$ to $z = -1$. On the edge 13, for example, one eigenvalue lies at $z = -1$ and a conjugate complex pole pair on the unit circle. This edge also lies in the complex boundary surface of eq. (7.2.9). For $r = 1$ this yields $p_0 = 1$, $p_1 = 1-2\tau$, $p_2 = 1-2\tau$, $-1 \leq \tau \leq 1$.

Remark 7.1:

> It might obviously appear here that τ should be eliminated in order to obtain the line $p_0 = 1$, $p_2 = p_1$. Then, however, the constraint on the interval $-1 \leq \tau \leq 1$, i.e. on line segment 13, would be lost and the infinite line is described. For $|\tau| > 1$ the zeros of $z^2-2\tau z+1$ are real and reciprocal with each other, i.e. $z_1 = 1/z_2$ and do not lie on the stability boundary.

Eq. (7.2.10) produces a line segment from a point on 01 to a point on 23 for every constant value of r in the interval $-1 \leq r \leq 1$. The complex boundary surface can be thought of as being generated by the movement of this line segment. The graphic representation is constructed by dividing the edges 01 and 23 into the same number of pieces by markings and connecting corresponding markings by straight line segments. This surface is a "hyperbolic paraboloid". It also contains the second family of lines of eq. (7.2.11) which are produced for variable r with fixed τ. These straight lines connect corresponding points on the edges 02 and 13 and extend to infinity beyond these edges as a complex boundary surface. Only the portion $-1 \leq r \leq 1$ connecting the edges is plotted in figure 7.2, because only this bounds the stability region. The further partitioning of the unstable domain for $|r| > 1$ is not of practical interest.

Again we compare the geometric stability region with its algebraic representation by eq. (C.2.15): $P(1) > 0$ and $P(-1) < 0$ constitute the real boundaries. The complex boundary is

$$\det(\underline{X}-\underline{Y}) = 1 - p_1 + p_o p_2 - p_o^2 > 0 \tag{7.2.13}$$

It is easily tested, that this expression is identically zero for

$$P(z) = (z^2-2\tau z+1)(z+r) = z^3+z^2(r-2\tau) + z(1-2\tau r) + r$$

$$p_o = r,\ p_1 = (1-2\tau r),\ p_2 = (r-2\tau)$$

By eq. (C.2.8) $\det(\underline{X}-\underline{Y}) = (1-z_1z_2)(1-z_1z_3)(1-z_2z_3)$, i.e. the determinant also vanishes for real $z_1 = 1/z_2$, i.e. for cases when no root is on the unit circle.

The complex boundary is described by the saddle surface of figure 7.2 or by the nonlinear inequality $\det(\underline{X}-\underline{Y}) > 0$. It is useful to have simpler linear inequalities which are either necessary or sufficient stability conditions. For the third order stability region of figure 7.2 a necessary stability condition would be that $\underline{p}$ must be in the box $|p_o| < 1$, $-1 < p_1 < 3$, $|p_2| < 3$, See also eq. (C.3.3). A better linear necessary condition is that $\underline{p}$ must be in the tetrahedron 0123. This in fact is the convex hull of the stability region. (The convex hull of a set M is obtained by augmenting M by all line segments with endpoints in M or on the boundary of M.) The convex hull gives the "best" linear necessary conditions. In section 7.3 a test will be given which shows whether a given point $\underline{p}$ lies inside or outside of a tetrahedron with prescribed corners.

It is more difficult to find good linear sufficient stability conditions. A candidate is eq. (C.4.1) which, for $n = 3$, is $|p_o| + |p_1| + |p_2| \le 1$. This is a cube of edge length $\sqrt{2}$ with one corner at the saddle point $\underline{p} = [0\ \ 1\ \ 0]'$, one in the middle of the edge 12, i.e. at $\underline{p} = [0\ \ -1\ \ 0]$ and square faces in the triangles 123 and 012.

A better linear approximation of the stability region is a tetrahedron generated as follows: place a straight line parallel to the edge 03 passing through the saddle point. This line intersects the plane $P(1) = 0$ at a point A and the plane $P(-1) = 0$ at a point B. Thus the tetrahedron AB12 is completely contained in the stability region. Therefore a sufficient stability condition is that $\underline{p}$ is located inside this tetrahedron with vertices

$$\begin{aligned}
\underline{p}_A' &= [-0.5 \quad 1 \quad -1.5] \\
\underline{p}_1' &= [\ 1 \quad -1 \quad -1\ ] \\
\underline{p}_2' &= [-1 \quad -1 \quad 1\ ] \\
\underline{p}_B' &= [\ 0.5 \quad 1 \quad 1.5]
\end{aligned} \qquad (7.2.14)$$

For $n \geq 4$ the stability region is more complicated. For example, for two complex eigenvalue pairs on the unit circle the complex boundary intersects itself. The complex boundary gets a cusp if the two complex pairs merge into a double pair. We summarize the properties that generalize to arbitrary n:

1) The stability region is connected and contractible to $\underline{p} = 0$.
2) The stability region is finite. It is bounded by two hyperplanes $P(1) = 0$ and $P(-1) = 0$ and by the boundary describing all polynomials with a complex conjugate pair of roots on the unit circle. On this hypersurface

$$\det(\underline{X}-\underline{Y}) = \prod_{i>j}^{1\ldots n} (1-z_i z_j) = 0 \qquad (7.2.15)$$

where

$$\underline{X} = \begin{bmatrix} 1 & p_{n-1} & \cdots & p_2 \\ 0 & 1 & \ddots & \vdots \\ \vdots & & \ddots & p_{n-1} \\ 0 & \cdots & 0 & 1 \end{bmatrix}, \quad \underline{Y} = \begin{bmatrix} 0 & \cdots & 0 & p_0 \\ \vdots & & p_0 & p_1 \\ 0 & p_0 & & \vdots \\ p_0 & p_1 & \cdots & p_{n-2} \end{bmatrix}$$

This was shown by Jury and Pavlidis [63.4]. See also appendix C. Note that $\det(\underline{X}-\underline{Y})$ also vanishes for $z_1 = 1/z_2$ not on the unit circle.

3) The complex root boundary intersects P(1) for double roots at z = 1 and this intersecting hyperplane satisfies

$$P'(1) = \left.\frac{dP(z)}{dz}\right|_{z=1} = p_1 + 2p_2 + \ldots + (n-1)p_{n-1} + n = 0 \tag{7.2.16}$$

Similarly $P'(-1) = 0$ at the intersection with $P(-1)$.

4) The convex hull of the stability region is a polyhedron whose vertices correspond to the n+1 polynomials

$$P_i(z) = (z+1)^i(z-1)^{n-i}\ , \quad i = 0,1,2\ldots n \tag{7.2.17}$$

This necessary stability condition was derived by Fam and Meditch [78.1].

5. The edges with a vertex number difference of one or two are part of the stability boundary. Edges with a vertex number difference $m > 2$ are outside of the stability region. A motion along these edges corresponds to a motion of a complex conjugate root pair along the root locus of $(z-1)^m + K(z+1)^m = 0$, which consists of circles passing through $z = 1$ and $z = -1$ whose tangents intersect at $180^\circ/m$.

6. The hypercube

$$\sum_{k=0}^{n-1} |p_k| < 1 \tag{7.2.18}$$

is contained in the stability region. This sufficient stability condition was shown by Cohn [1922] and Thoma [62.3]. (The tightened condition (7.2.14) for n = 3 has not yet been generalized to $n \geq 4$.)

7.3 Barycentric Coordinates, Bilinear Transformation

In this section the condition that a point is located inside a polyhedron with n + 1 vertices is formulated in terms of barycentric coordinates. Applied to the convex hull of the stability region this is shown to be equivalent to a bilinear mapping of the unit circle onto the left half plane.

Consider a polyhedron with n + 1 vertices $\underline{p}_o$, $\underline{p}_1$ ... $\underline{p}_n$ and assume that it spans the n space, i.e. the vectors $\underline{p}_1-\underline{p}_o$, $\underline{p}_2-\underline{p}_o$ ... $\underline{p}_n-\underline{p}_o$ are linearly independent. The interior of this polyhedron is called a "simplex". Every vector $\underline{p}$ can be expressed uniquely as

$$\underline{p} = \sum_{i=0}^{n} \mu_i \underline{p}_i \qquad (7.3.1)$$

with

$$\sum_{i=0}^{n} \mu_i = 1 \qquad (7.3.2)$$

This may be interpreted as follows: given a unit mass 1 which can be split into n + 1 masses μ_i, eq. (7.3.2). By placing these masses μ_i at the vertices $\underline{p}_i$, any point $\underline{p}$ can be made the center of gravity, eq. (7.3.1). The μ_i are named "barycentric coordinates". $\underline{p}$ is located in the simplex if and only if all barycentric coordinates μ_i are positive [65.4].

Eqs. (7.3.1) and (7.3.2) can be written as

$$[\underline{p}' \quad 1] = \underline{\mu}'\underline{P}_n \qquad (7.3.3)$$

with

$$\underline{\mu} := [\mu_o \quad \mu_1 \ldots \mu_n] \quad , \quad \underline{P}_n := \begin{bmatrix} \underline{p}_o' & 1 \\ \underline{p}_1' & 1 \\ \vdots & \vdots \\ \underline{p}_n' & 1 \end{bmatrix} \qquad (7.3.4)$$

$\underline{p}$ lies in the simplex if and only if

$$\underline{\mu}' = [\underline{p}' \quad 1]\underline{P}_n^{-1} > 0 \qquad (7.3.5)$$

i.e. if all elements of the vector $\underline{\mu}$ are positive.

Example 1:

According to eq. (7.2.14) a sufficient stability condition for n = 3 is

$$\underline{\mu}' = [p_o \quad p_1 \quad p_2 \quad 1] \begin{bmatrix} -0.5 & 1 & -1.5 & 1 \\ 1 & -1 & -1 & 1 \\ -1 & -1 & 1 & 1 \\ 0.5 & 1 & 1.5 & 1 \end{bmatrix}^{-1}$$

$$\underline{\mu}' = \frac{1}{8}[p_o \quad p_1 \quad p_2 \quad 1] \begin{bmatrix} -2 & 3 & -3 & 2 \\ 2 & -2 & -2 & 2 \\ -2 & -1 & 1 & 2 \\ 2 & 2 & 2 & 2 \end{bmatrix}$$

1) $-p_o + p_1 - p_2 + 1 > 0$
2) $3p_o - 2p_1 - p_2 + 2 > 0$
3) $-3p_o + 2p_1 + p_2 + 2 > 0$
4) $p_o + p_1 + p_2 + 1 > 0$

1) and 4) are equivalent to $|p_o + p_2| - 1 < p_1$ and 2) and 3) to $p_1 < |1.5p_o - 0.5p_2| + 1$. The combined sufficient stability condition for n = 3 is

$$|p_o + p_2| - 1 < p_1 < 0.5|\,3p_o - p_2| + 1 \qquad (7.3.6)$$

Example 2:

According to eq. (7.2.12) a necessary stability condition for n = 3 is

$$\underline{\mu}' = [p_o \quad p_1 \quad p_2 \quad 1] \begin{bmatrix} -1 & 3 & -3 & 1 \\ 1 & -1 & -1 & 1 \\ -1 & -1 & 1 & 1 \\ 1 & 3 & 3 & 1 \end{bmatrix}^{-1}$$

$$= \frac{1}{8}[p_o \quad p_1 \quad p_2 \quad 1] \begin{bmatrix} -1 & 3 & -3 & 1 \\ 1 & -1 & -1 & 1 \\ -1 & -1 & 1 & 1 \\ 1 & 3 & 3 & 1 \end{bmatrix}$$

1) $-p_0 + p_1 - p_2 + 1 > 0$
2) $3p_0 - p_1 - p_2 + 3 > 0$
3) $-3p_0 - p_1 + p_2 + 3 > 0$
4) $p_0 + p_1 + p_2 + 1 > 0$

1) and 4) are equivalent to $|p_0 + p_2| - 1 < p_1$ and 2) and 3) to $p_1 < |3p_0 - p_2| + 3$. The combined necessary stability condition for n = 3 is

$$|p_0 + p_2| - 1 < p_1 < |3p_0 - p_2| + 3 \tag{7.3.7}$$

In the second example we saw that $\underline{P}_n^{-1} = \underline{P}_n/8$, i.e. $\underline{P}_n^2 = 8\underline{I}$. This is a general property of this particular matrix $\underline{P}_n$ with rows $[\underline{p}_i \quad 1]$, where

$[\underline{p}_i \quad 1]\underline{z}_n = (z+1)^i(z-1)^{n-i}$, i = 0,1,2...n. It is shown in [69.10] that

$$\underline{P}_n^2 = 2^n\underline{I}_{n+1}\ , \text{ i.e. } \underline{P}_n^{-1} = \frac{1}{2^n}\underline{P}_n \tag{7.3.8}$$

Thus the inversion in eq. (7.3.5) can be avoided for the necessary stability condition. It reads

$$\underline{\mu}' = \frac{1}{2^n}[\underline{p}' \quad 1]\underline{P}_n > 0 \tag{7.3.9}$$

where the factor $1/2^n$ may be omitted, i.e. all elements of $[\underline{p}' \quad 1]\underline{P}^n$ must be positive.

Example:

Necessary stability conditions for n = 4 are

$$[\underline{p}' \quad 1]\underline{P}_n = [p_0 \quad p_1 \quad p_2 \quad p_3 \quad 1]\begin{bmatrix} 1 & -4 & 6 & -4 & 1 \\ -1 & 2 & 0 & -2 & 1 \\ 1 & 0 & -2 & 0 & 1 \\ -1 & -2 & 0 & 2 & 1 \\ 1 & 4 & 6 & 4 & 1 \end{bmatrix}$$

$$
\begin{aligned}
p_0 - p_1 + p_2 - p_3 + 1 &> 0\\
-2p_0 + p_1 - p_3 + 2 &> 0\\
3p_0 - p_2 + 3 &> 0\\
-2p_0 - p_1 + p_3 + 2 &> 0\\
p_0 + p_1 + p_2 + p_3 + 1 &> 0
\end{aligned}
$$

These inequalities can be combined to

$$
\begin{aligned}
&|p_1 + p_3| - (1+p_0) < p_2 < 3(1+p_0)\\
&|p_1 - p_3| < 2(1-p_0)
\end{aligned}
\qquad (7.3.10)
$$

The transformation to barycentric coordinates is equivalent to a conformal mapping from the complex variable z to the complex variable $w = (z-1)/(z+1)$

This can be shown as follows:

Multiply eq. (7.3.3) by $\underline{z}_n = [1 \quad z \quad z^2 \; \ldots \; z^n]'$ to obtain the characteristic polynomial

$$
P(z) = [\underline{p}' \quad 1]\underline{z}_n = \underline{\mu}'\underline{P}_n\underline{z}_n \qquad (7.3.11)
$$

and for the particular $\underline{P}_n$ of the convex hull of the stability region, see eq. (7.2.17)

$$
\begin{aligned}
P(z) &= \sum_{i=0}^{n} \mu_i P_i(z) = \sum_{i=0}^{n} \mu_i (z+1)^i (z-1)^{n-i}\\
&= \mu_0(z-1)^n + \mu_1(z+1)(z-1)^{n-1} + \ldots + \mu_n(z+1)^n
\end{aligned}
\qquad (7.3.12)
$$

Dividing by $(z+1)^n$

$$
\frac{P(z)}{(z+1)^n} = \mu_0\left(\frac{z-1}{z+1}\right)^n + \mu_1\left(\frac{z-1}{z+1}\right)^{n-1} + \ldots + \mu_n \qquad (7.3.13)
$$

Now substitute the complex variable

$$w := \frac{z-1}{z+1} \quad , \quad z = \frac{1+w}{1-w} \tag{7.3.14}$$

$$M(w) := \frac{P(z)}{(z+1)^n} = \left(\frac{1-w}{2}\right)^n P\left(\frac{1+w}{1-w}\right) = \mu_o w^n + \mu_1 w^{n-1} + \ldots + \mu_n \tag{7.3.15}$$

The unit disk of the z plane is mapped onto the left half w-plane by the bilinear transformation (7.3.14). Thus the roots of P(z) are located inside the unit circle if and only if M(w) is a Hurwitz polynomial. The necessary stability condition that all barycentric coordinates μ_i must be positive is equivalent to the well-known necessary condition that all coefficients of a Hurwitz polynomial must be positive. There are no better linear necessary conditions because of the convex hull property.

The complex root boundary may be described in barycentric coordinates by Orlando's formula, eq. (C.1.10). However, the evaluation of $\Delta_{n-1} = 0$ is more complicated than that of $\det(\underline{X}-\underline{Y}) = 0$ in eq. (7.2.15) if the problem is given in terms of the polynomial P(z), see the example of eq. (C.2.10).

7.4 Γ-Stability

7.4.1 Circular Eigenvalue Regions

In the sections on root locus and pole assignment design, placement of the eigenvalues of a control system to achieve a short transient with sufficient damping was discussed. The result was the region Γ in figure 2.12 for continuous systems and figure 3.11 for discrete-time systems. In the continuous time case the straight lines for damping $1/\sqrt{2}$ and negative real part were approximated by a conic section, in particular a hyperbola. Similarly the image of the straight lines, mapped by $z = e^{Ts}$ into the z-plane may be approximated by a conic section. Figure 3.11 suggests that a circle Γ should be chosen, e.g. a circle with center z = 0.3 and radius 0.3. The property that all roots of a polynomial are in a specified region Γ in the z $(= \tau + j\eta)$-plane will be called "Γ-stability". Consider a circle with center $z = \tau_0$ and radius r, figure 7.3.

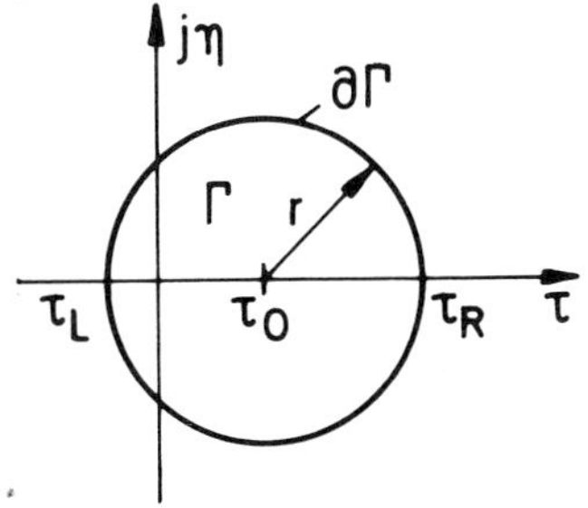

Figure 7.3

A circular region of nice stability

The geometric interpretation by barycentric coordinates follows closely to that of the unit circle case: corners of the Γ-stability region in P space are given by the n + 1 polynomials

$$P_i(z) = (z-\tau_L)^i (z-\tau_R)^{n-i} \quad , \ i = 0,1 \ldots n \qquad (7.4.1)$$

Example:

The zeros of $p_o + p_1 z + z^2$ should lie in the circle with $\tau_o = 0.45$, $r = 0.5$, i.e. $\tau_L = -0.05$, $\tau_R = 0.95$. From eq. (7.4.1) the corners of the triangle in the P-plane are obtained

$\underline{p}'_{\bar{o}} = [\ 0.9025 \quad -1.9]$

$\underline{p}'_{\bar{1}} = [-0.0475 \quad -0.9]$

$\underline{p}'_{\bar{2}} = [0.0025 \quad 0.1]$

This is represented together with the stability triangle in figure 7.4.

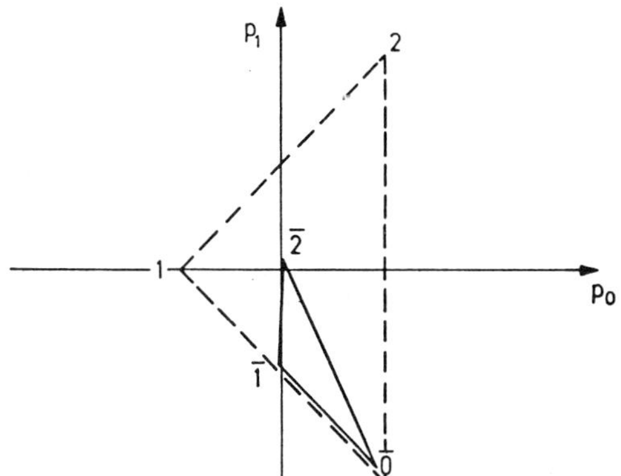

Figure 7.4

Triangle $\bar{0}\bar{1}\bar{2}$ for eigenvalues in the circle with center z = 0.45 and radius 0.5. Dotted lines: stability region of figure 7.1.

The representation of $\underline{p}$ in barycentric coordinates $\underline{\mu}$ is (corresponding to eq. (7.3.4))

$$[\underline{p}' \quad 1] = \underline{\mu}'\underline{P}_n \tag{7.4.2}$$

where now

$$\underline{P}_n \underline{z}_n = \begin{bmatrix} \underline{p}_o' & 1 \\ \underline{p}_1' & 1 \\ \vdots & \\ \underline{p}_n' & 1 \end{bmatrix} \begin{bmatrix} 1 \\ z \\ \vdots \\ z^n \end{bmatrix} = \begin{bmatrix} (z-\tau_R)^n \\ (z-\tau_R)^{n-1}(z-\tau_L) \\ \vdots \\ (z-\tau_L)^n \end{bmatrix} \tag{7.4.3}$$

Then

$$P(z) = [\underline{p}' \quad 1]\underline{z}_n = \underline{\mu P}_n \underline{z}_n = \sum_{i=1}^{n} \mu_i P_i(z)$$

$$= \mu_o(z-\tau_R)^n + \mu_1(z-\tau_R)^{n-1}(z-\tau_L) +\ldots+ \mu_n(z-\tau_L)^n$$

$$\frac{P(z)}{(z-\tau_L)^n} = \mu_o\left(\frac{z-\tau_R}{z-\tau_L}\right)^n + \mu_1\left(\frac{z-\tau_R}{z-\tau_L}\right)^{n-1} +\ldots+ \mu_n \tag{7.4.4}$$

Substitute the complex variable

$$w := \frac{z-\tau_R}{z-\tau_L}, \quad z = \frac{\tau_R - \tau_L w}{1-w} \tag{7.4.5}$$

$$M(w) := \frac{P(z)}{(z-\tau_L)^n} = \left(\frac{1-w}{\tau_R-\tau_L}\right)^n P\left(\frac{\tau_R-\tau_L w}{1-w}\right)$$

$$= \mu_o w^n + \mu_1 w^{n-1} + \ldots + \mu_n \tag{7.4.6}$$

Eq. (7.3.15) is the special case $\tau_L = -1$, $\tau_R = 1$.

The necessary and sufficient condition for P(z) to have all its roots inside the circle of figure 7.3 is that M(w) is a Hurwitz polynomial. Positivity of the barycentric coordinates μ_i is equivalent to the necessary condition that all coefficients of a Hurwitz polynomial must be positive .

For algebraic test of polynomials containing non-numerical parameters the form of eqs. (C.2.11-13) is most convenient. This form is obtained by mapping the circle of figure 7.3 first onto the unit circle in a $\tilde{z}$ plane, where

$$\tilde{z} := (z-\tau_o)/r \quad , \quad z = r\tilde{z}+\tau_o \qquad (7.4.7)$$

In other words: the zeros of P(z) are located inside the circle of figure 7.3 if and only if

$$N(\tilde{z}) = P(r\tilde{z}+\tau_o) \qquad (7.4.8)$$

is a Schur-Cohn polynomial. This form is most convenient for the condition (C.2.11) that $\underline{X}$-$\underline{Y}$ must be positive innerwise. The linear conditions (C.2.11 and 12) are obtained from the convex hull in the barycentric coordinates, i.e. by mapping the circle to the left half w-plane by

$$w = \frac{\tilde{z}-1}{\tilde{z}+1} = \frac{z-\tau_o-r}{z-\tau_o+r} = \frac{z-\tau_R}{z-\tau_L} \qquad (7.4.9)$$

as in eq. (7.4.5).

Example:

The zeros of $P(z) = p_o + p_1 z + p_2 z^2 + z^3$ should lie inside the circle with center $\tau_o = 0.4$ and radius $r = 0.4$, i.e. $\tau_L = 0$, $\tau_R = 0.8$.

$$\begin{aligned} N(\tilde{z}) &= P(0.4\tilde{z}+0.4) \\ &= (p_o+0.4p_1+0.4^2p_2+0.4^3) + (0.4p_1+2\times 0.4^2p_2+3\times 0.4^3)\tilde{z} + \\ &\quad + (0.4^2p_2+3\times 0.4^3)\tilde{z}^2 + 0.4^3\tilde{z}^3 \\ &= n_o + n_1\tilde{z} + n_2\tilde{z}^2 + n_3\tilde{z}^3 \end{aligned}$$

From eq. (C.2.15)

$$|\underline{X}-\underline{Y}| = n_3 - n_1 n_3 + n_0 n_2 - n_0^2$$

$$= 0.064p_0 - p_0^2 - 0.8p_0 p_1 - 0.16p_0 p_2 - 0.16p_1^2 - 0.064p_1 p_2$$

The linear conditions are derived from

$$0.8^3 M(w) = (1-w)^3 P\left(\frac{0.8}{1-w}\right)$$

$$= (p_0 + 0.8p_1 + 0.64p_2 + 0.512) + (-3p_0 - 1.6p_1 - 0.64p_2)w +$$

$$+ (3p_0 + 0.8p_1)w^2 - p_0 w^3$$

By eq. (C.2.15) they are

$$m_0 = P(0.8) = p_0 + 0.8p_1 + 0.64p_2 + 0.512 > 0$$

$$m_3 = -P(0) = -p_0 > 0$$

and one of the following two conditions must be satisfied

a) $m_1 = -3p_0 - 1.6p_1 - 0.64p_2 > 0$

or

b) $m_2 = 3p_0 + 0.8p_1 > 0$

7.4.2 Degree of Stability

It was shown that circular boundaries $\partial\Gamma$ with real center can be easily treated. Consider three circles of particular interest:

1) the unit circle, i.e. the stability boundary;

2) a circle approximating the desired eigenvalue region of figure 3.11; this guarantees fast and well-damped transients;

3) the limiting case $z \to 0$, i.e. the deadbeat solution.

For the formulation of robustness problems it is useful to introduce a measure "degree of stability". This may be done by defining a family of circles in the z-plane which contains the three above examples as special cases. The radius of these circles is then a feasible measure for the degree of stability. In reducing the radius from the unit circle, $r = 1$, we may first keep τ_R fixed at $z = 1$ until τ_L has reached $z = 0$ for $r = 0.5$ and then keep τ_L fixed until the radius becomes zero. Radius r and center τ_o then satisfy

$$\tau_o(1-\tau_o) = r(1-r) \quad , \quad \tau_o < 0.5 \qquad (7.4.10)$$

In order to have no common points of two circles and to have the deadbeat point $z = 0$ as an interior point for $r > 0$, the relation (7.4.10) may be slightly modified to

$$\tau_o(1-\tau_o) = 0.99\, r(1-r) \quad , \quad \tau_o < 0.45 \qquad (7.4.11)$$

For $r > 1$, this is augmented by concentric circles, $\tau_o = 0$. This family of circles is shown in figure 7.5.

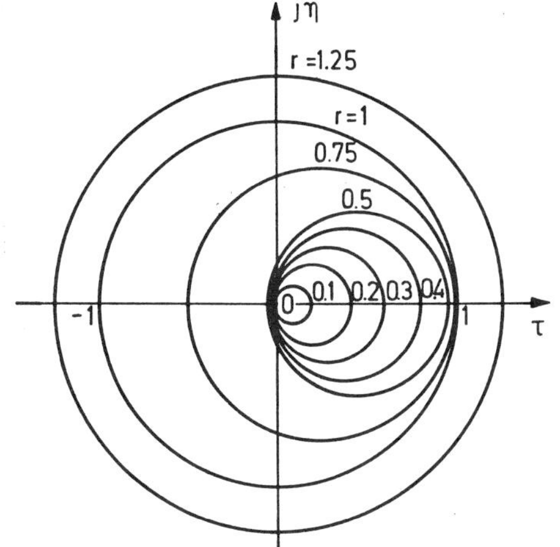

Figure 7.5

Definition of a degree of stability r for sampled-data systems.

7.4.3 Nice Stability Regions for Continuous Systems

We have seen how nice stability regions Γ in the form of circles and half planes can be represented algebraically as linear and nonlinear inequalities. In the following we will extend this approach to other nice stability regions.

In eq. (7.4.3) a circle was mapped onto the imaginary axis of a new complex w-plane, where $z = (\tau_R-\tau_L w)/(1-w)$. This expression was substituted into P(z) to obtain a polynomial which had to be tested for the Hurwitz property. Using higher order substitutions for z also other regions Γ can be treated in the same way [81.7], [82.1], [83.1].

Example:

For continuous time systems a hyperbola as shown in figure 2.12 is of particular interest. Let $s = \sigma + j\omega$ and consider the left branch of the hyperbola

$$\partial\Gamma: \quad \left(\frac{\sigma}{a}\right)^2 - \left(\frac{\omega}{b}\right)^2 = 1 \qquad a > 0 \quad , \quad b > 0 \qquad (7.4.12)$$

See figure 7.6 a. This is related to the formulation in terms of damping ζ in eq. (2.5.14) by $a = \sigma_o$, $b^2 = \sigma_o^2(1-\zeta^2)/\zeta^2$. In [83.1] we find the appropriate relation mapping the hyperbola onto the imaginary axis of a w-plane:

$$s =: \frac{-aw^2+2bw+a}{w^2+1} \qquad (7.4.13)$$

Substitute this into P(s) and take the numerator polynomial

$$M(w) = (w^2+1)^n \times P\left(\frac{-aw^2+2bw+a}{w^2+1}\right) \qquad (7.4.14)$$

P(s) is Γ-stable if and only if M(w) is a Hurwitz polynomial. M(w) is of degree 2n.

For example, let $n = 2$

$$P(s) = p_o+p_1s+s^2$$

$$M(w) = P(-a)w^4+2b(p_1-2a)w^3+2(p_o+2b^2-a^2)w^2+2b(p_1+2a)w+P(a) \qquad (7.4.15)$$

The complex boundary is obtained from the critical Hurwitz determinant

$$\Delta_3 = \begin{vmatrix} m_1 & m_3 & 0 \\ m_0 & m_2 & m_4 \\ 0 & m_1 & m_3 \end{vmatrix} = m_1 m_2 m_3 - m_0 m_3^2 - m_4 m_1^2$$

$$\Delta_3 = 16b^2 \times [p_1^2(a^2+b^2) - 4a^2 p_0 - 4a^2 b^2] = 0 \qquad (7.4.16)$$

This is a parabola in the p_0-p_1 plane, see figure 7.6 b. The real root boundary $P(-a) = 0$ is also shown. The region Γ to the left of the hyperbola in the s-plane is mapped into the region P_Γ in the upper right corner of figure 7.6 b. The conditions $P(-a) > 0$, $\Delta_3 > 0$ would also be satisfied below the lower branch of the parabola. This is excluded, however, by one of the Lienard-Chipart conditions of eqs. (C.1.7) or (C.1.8), e.g. $m_3 > 0$. Then also $m_0 > 0$, $m_1 > 0$, $m_2 > 0$ is satisfied in P_Γ.

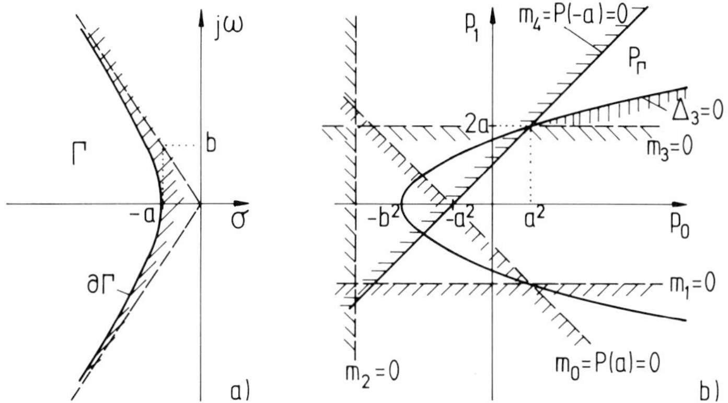

Figure 7.6 a) Hyperbolic boundary in s-plane
b) Boundaries and nice stability region P_Γ in P plane
(plotted for the numerical values $a = 1$, $b = 1.5$)

Other regions Γ may be found in [83.1], e.g. the interior of an ellipse in z plane

$$\left(\frac{\tau}{a}\right)^2 + \left(\frac{\eta}{b}\right)^2 = 1 \qquad (7.4.17)$$

with the mapping

$$z =: \frac{aw^2+2bw+a}{w^2-1} \tag{7.4.18}$$

For the ellipse see also [82.4]. Also mappings of higher order than two are possible. This, however, further increases the degree of the polynomial M(w).

7.4.4 General Eigenvalue Regions

In this book we will not pursue the formulation of Γ-stability problems in terms of nonlinear inequalities resulting from Hurwitz conditions. This would be the basis for numerical design techniques, e.g. by nonlinear programming. In view of the increasing availability of computer graphics we will pursue the representation of Γ-stability regions in two-dimensional cross sections in the sense of the classical D-decomposition. Here the choice of the cross section plane and fast computations allowing the designer to examine cross sections interactively will be important. First the computation of the real and complex boundaries of P_Γ will be treated for a quite general class of regions Γ.

Let Γ be a given region in the eigenvalue plane, see figure 7.7.

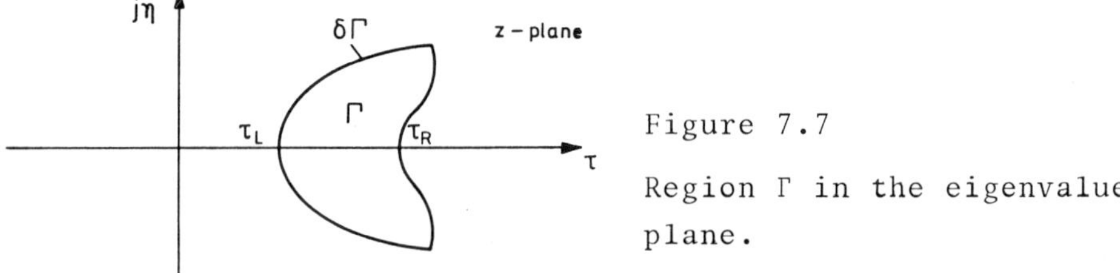

Figure 7.7
Region Γ in the eigenvalue plane.

Let the region be finite, symmetric with respect to the real axis and contractible. The boundary curve ∂Γ can be represented parametrically with the real parameter α, i.e. $\tau(\alpha)+j\eta(\alpha) \in \partial\Gamma$, $\alpha_0 \leq \alpha \leq \alpha_1$, or it is represented by a function $\eta^2 = f(\tau)$. ∂Γ may be defined piecewise.

Let P_Γ be the corresponding region in P space, i.e.

$$\underline{p} \in P_\Gamma \quad \text{if and only if} \quad z_i \in \Gamma \quad , \; i = 1,2\ldots n \tag{7.4.19}$$

where

$$P(z) = [\underline{p}' \quad 1]\underline{z}_n = \prod_{i=1}^{n}(z-z_i) \quad , \quad \underline{z}_n = [1 \quad z \ \dots \ z^n]' \tag{7.4.20}$$

Following the same argument as for the unit circle, P_Γ is bounded by three surfaces: two real root boundaries for the real axis intersections τ_L and τ_R of $\partial\Gamma$ and a complex root boundary. The two real root boundaries are the hyperplanes $P(\tau_L) = 0$ and $P(\tau_R) = 0$, which may be written

$$\begin{aligned} P(\tau_L) &= [\underline{p}' \quad 1]\underline{a}_L = 0 \quad , \quad \underline{a}_L := [1 \quad \tau_L \ \dots \ \tau_L^n]' \\ P(\tau_R) &= [\underline{p}' \quad 1]\underline{a}_R = 0 \quad , \quad \underline{a}_R := [1 \quad \tau_R \ \dots \ \tau_R^n]' \end{aligned} \tag{7.4.21}$$

The complex boundary describes all polynomials with at least one complex conjugate pair of roots $\tau \pm j\eta$ on $\partial\Gamma$. Such polynomials must contain the factor

$$Q(z) = (z-\tau-j\eta)(z-\tau+j\eta) = z^2-2\tau z+(\tau^2+\eta^2) = z^2+q_1z+q_0 \tag{7.4.22}$$

Depending on the parameterization of $\partial\Gamma$ we have either

$$q_1(\alpha) = -2\tau(\alpha) \quad , \quad q_0(\alpha) = \tau^2(\alpha)+\eta^2(\alpha) \tag{7.4.23}$$

or

$$q_1(\tau) = -2\tau \quad , \quad q_0(\tau) = \tau^2+\eta^2(\tau) \tag{7.4.24}$$

$P(z)$ is written

$$\begin{aligned} P(z) &= Q(z)R(z) \\ &= (q_0+q_1z+z^2)(r_0+r_1z+\dots+r_{n-3}z^{n-3}+z^{n-2}) \end{aligned} \tag{7.4.25}$$

Multiplication and equating polynomial coefficients with eq. (7.4.20) yields

$$[\underline{p}' \quad 1] = [0 \quad 0 \quad \underline{r}' \quad 1]\begin{bmatrix} 1 & 0 & \cdot & \cdot & \cdot & \cdot & \cdot & 0 \\ q_1 & 1 & & & & & & \cdot \\ q_o & q_1 & \cdot & & & & & \cdot \\ 0 & \cdot & \cdot & \cdot & & & & \cdot \\ \cdot & & \cdot & \cdot & \cdot & & & \cdot \\ \cdot & & & \cdot & \cdot & \cdot & & 0 \\ 0 & & & 0 & q_o & q_1 & & 1 \end{bmatrix} \tag{7.4.26}$$

where $\underline{r}' = [r_o \quad r_1 \; \dots \; r_{n-3}]$. Eq. (7.4.26) is inverted in order to eliminate the undetermined $\underline{r}$. A lower triangular matrix $\underline{D}$ with diagonal structure is obtained:

$$\begin{bmatrix} 1 & 0 & & & & 0 \\ q_1 & 1 & \cdot & & & \\ q_o & q_1 & \cdot & \cdot & & \\ 0 & \cdot & \cdot & \cdot & \cdot & \\ & & \cdot & \cdot & \cdot & 0 \\ 0 & & 0 & q_o & q_1 & 1 \end{bmatrix} \times \begin{bmatrix} d_o & 0 & & & & 0 \\ d_1 & d_o & & & & \\ d_2 & \cdot & \cdot & & & \\ & \cdot & \cdot & \cdot & & \\ & & \cdot & \cdot & \cdot & \\ d_n & & & & d_1 & d_o \end{bmatrix} = \begin{bmatrix} 1 & 0 & & & & 0 \\ 0 & 1 & & & & \\ & & \cdot & & & \\ & & & \cdot & & \\ & & & & \cdot & 0 \\ 0 & & & & 0 & 1 \end{bmatrix}$$

$$d_o = 1$$

$$q_1 d_o + d_1 = 0$$

$$q_o d_o + q_1 d_1 + d_2 = 0$$

$$\vdots$$

$$q_o d_{i-1} + q\, d_i + d_{i+1} = 0 \quad , \; i = 2,3\dots n-1$$

The following recursion formula is obtained for the coefficients d_i

$$d_{i+1} = -q_1 d_i - q_o d_{i-1} \; , \; i = 1,2\dots n-1 \tag{7.4.27}$$

with starting values $d_o = 1$ and $d_1 = -q_1$ [81.3]. In our low order examples the following first values are frequently used:

$$
\begin{aligned}
d_o &= 1 \\
d_1 &= -q_1 \\
d_2 &= q_1^2 - q_o \\
d_3 &= -q_1^3 + 2q_o q_1 \\
d_4 &= q_1^4 - 3q_o q_1^2 + q_o^2 \\
d_5 &= -q_1^5 + 4q_o q_1^3 - 3q_o^2 q_1
\end{aligned}
\qquad (7.4.28)
$$

From eq. (7.4.26)

$$
[\underline{p}' \quad 1]\begin{bmatrix} d_o & 0 & & 0 \\ d_1 & d_o & & \\ \vdots & & \ddots & \\ d_n & \cdots & d_1 & d_o \end{bmatrix} = [0 \quad 0 \quad \underline{r}' \quad 1] \qquad (7.4.29)
$$

The last column, $d_o = 1$, is trivial, the preceeding n - 2 columns have an undetermined right hand side, because no assumption was made on the remainder polynomial R(z). The following first two columns remain

$$
[\underline{p}' \quad 1][\underline{c}_1 \quad \underline{c}_2] = [0 \quad 0] \qquad (7.4.30)
$$

$$
\underline{c}_1 := \begin{bmatrix} d_o \\ d_1 \\ \vdots \\ d_n \end{bmatrix}, \quad \underline{c}_2 := \begin{bmatrix} 0 \\ d_o \\ \vdots \\ d_{n-1} \end{bmatrix}
$$

For each value of α on $\partial\Gamma$ two vectors $\underline{c}_1(\alpha)$ and $\underline{c}_2(a)$ are obtained and the polynomial coefficient vector $\underline{p}$ has to satisfy the two linear equations (7.4.30). Geometrically this is the intersection of two (n - 1) dimensional hyperplanes in P space, i.e. an (n-2) dimensional hyperplane. The two hyperplanes cannot be identical or parallel because $\underline{c}_1$ and $\underline{c}_2$ are linearly independent. If the complex pair of roots of Q(z) moves along $\partial\Gamma$, then the (n-2) dimensional hyperplane moves and forms the complex root boundary surface. Its shape depends on the particular form of $\partial\Gamma$.

Remark 7.2:

For hand calculations it is frequently convenient to replace the $\underline{c}_1$ and $\underline{c}_2$ by two other vectors in the same hyperplane, i.e. by two linear combinations of $\underline{c}_1$ and $\underline{c}_2$.

Substitute for example eq. (7.4.27) into $\underline{c}_1$

$$[\underline{p}' \quad 1]\begin{bmatrix} 1 & 0 \\ -q_1 & 1 \\ -q_1 d_1 - q_o d_o & d_1 \\ -q_1 d_2 - q_o d_1 & d_2 \\ \vdots & \\ -q_1 d_{n-1} - q_o d_{n-2} & d_{n-1} \end{bmatrix} = [0 \quad 0]$$

Multiply the second column by q_1 and add it to the first one

$$[\underline{p}' \quad 1]\begin{bmatrix} 1 & 0 \\ 0 & 1 \\ -q_o d_o & d_1 \\ -q_o d_1 & d_2 \\ \vdots & \vdots \\ -q_o d_{n-2} & d_{n-1} \end{bmatrix} = [0 \quad 0] \qquad (7.4.31)$$

This process may be repeated as follows

a) Multiply the second column by q_o and add the first one multiplied by $-q_1$. Substitute eq. (7.4.27);

b) Multiply the first column by q_o and add the second one multiplied by q_1. Substitute eq. (7.4.27).

Then a) again and so on. This replaces the d_i by q_o and q_1. Which is the most convenient form, depends on the shape of the boundary $\partial\Gamma$, i.e. on the particular form of $q_o(\alpha)$ and $q_1(\alpha)$. Eqs. (7.4.27) and (7.4.30) are better suited for numerical computer calculations.

Example 1:

For comparison with the example (7.4.12) take the hyperbola of figure 7.6 with a = 1, b = 1.5.

$$\partial\Gamma: \sigma^2 - \left(\frac{\omega}{1.5}\right)^2 = 1 \quad , \quad \sigma \leq -1$$

It may be parameterized by σ, i.e.

$$\tau = \sigma$$
$$\eta^2 = \omega^2 = 2.25(\sigma^2-1)$$

Then by eq. (7.4.24)

$$q_1 = -2\sigma \quad , \quad q_0 = \sigma^2 + \omega^2 = 3.25\sigma^2 - 2.25$$

For n = 2 there is no remainder polynomial to be eliminated and

$$p_1 = q_1 = -2\sigma \quad , \quad p_0 = q_0 = 3.25\sigma^2 - 2.25 \qquad (7.4.32)$$

Some values are listed in table 7.1

σ	$p_0 = 3.25\sigma^2 - 2.25$	$p_1 = -2\sigma$
-1	1	2
-1.1	1.6825	2.2
-1.2	2.43	2.4
-1.4	4.12	2.8
-1.6	6.07	3.2
-1.8	8.28	3.6
-2	10.75	4

Table 7.1 Complex boundary calculations

For $\sigma = -1$ the complex root boundary starts on the real root boundary

$$P(-1) = p_0 - p_1 + 1 = 0$$

Both boundaries are plotted in figure 7.8

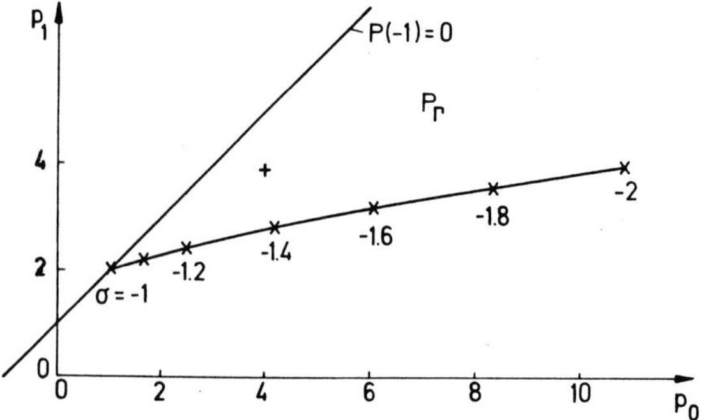

Figure 7.8 Γ-Stability region P_Γ for a hyperbola
$\partial\Gamma$: $\sigma^2 - (\omega/1.5)^2 = 1$

The boundaries partition the p_0-p_1-plane into three regions. Testing the point $p_0 = 4$, $p_1 = 4$, i.e. $P(s) = 4 + 4s + s^2 = (s+2)^2$, we find that P_Γ is the Γ-stability region. It is identical to P_Γ of figure 7.6 b.

Example 2:

Consider again a hyperbola

$$\left(\frac{\sigma}{a}\right)^2 - \left(\frac{\omega}{b}\right)^2 = 1 \quad , \quad \sigma \leq -a$$

and let n = 5. The real root boundary follows from eq. (7.4.21) with $\tau_R = -a$, $P(-a) = 0$. For the complex boundary

$$\begin{aligned} q_1 &= -2\sigma \\ q_0 &= \sigma^2+\omega^2 = \left[\left(\frac{b^2}{a^2}\right)+1\right]\sigma^2 - b^2 \end{aligned}$$

Eq. (7.4.28) becomes

$$\begin{aligned} d_0 &= 1 \\ d_1 &= 2\sigma \\ d_2 &= (3-b^2/a^2)\sigma^2 + b^2 \\ d_3 &= 4(1-b^2/a^2)\sigma^3 + 4b^2\sigma \\ d_4 &= (5-10b^2/a^2+b^4/a^4)\sigma^4 + 2b^2(5-b^2/a^2)\sigma^2 + b^4 \\ d_5 &= 32b^2\sigma^5/a^2 + 2(3b^2/a^2+16b^2+3)\sigma^3 - 6b^2\sigma \end{aligned} \qquad (7.4.33)$$

If the damping ζ is used as parameter instead of b, then $b^2 = a^2(1-\zeta^2)/\zeta^2$ must be substituted.

By eq. (7.4.30)

$$[p_0 \quad p_1 \quad p_2 \quad p_3 \quad p_4 \quad 1]\begin{bmatrix} d_0 & 0 \\ d_1 & d_0 \\ d_2 & d_1 \\ d_3 & d_2 \\ d_4 & d_3 \\ d_5 & d_4 \end{bmatrix} = [0 \quad 0] \tag{7.4.34}$$

For fixed numerical values of a, b and σ, the d_i are easily calculated. Eq. (7.4.34) describes the intersection of two four-dimensional hyperplanes in the five-dimensional P-space. This intersection is a three-dimensional hyperplane. We will see the use of eq. (7.4.34) if in a control system $\underline{p}$ depends on two free controller parameters or two physical parameters of the plant. Then the complex root boundary in the plane of two such parameters can easily be generated from eq. (7.4.34).

Example 3:

Consider the nice stability region bounded by the logarithmic spiral

$\partial\Gamma = e^{-\alpha} \times e^{\pm j\alpha}$

of eq. (3.5.3) and figure 3.11. This is produced by the irrational mapping $z = e^{Ts}$ from the sector in s plane for a damping of $1/\sqrt{2}$. It cannot be mapped rationally onto the imaginary axis, therefore a reduction to a Hurwitz problem is not possible. The real axis intersections are $\tau_R = 1$ for $\alpha = 0$ and $\tau_L = e^{-\pi} = -0.0432$ for $\alpha = \pi$. Let n = 3. The real boundary planes are

$$P(\tau_L) = [p_0 \quad p_1 \quad p_2 \quad 1]\begin{bmatrix} 1 \\ -0.0432 \\ 0.0432^2 \\ -0.0432^3 \end{bmatrix} = 0$$

$$P(\tau_R) = [p_o \quad p_1 \quad p_2 \quad 1]\begin{bmatrix}1\\1\\1\\1\end{bmatrix} = 0$$

On $\partial\Gamma$ we have $\tau = e^{-\alpha}\cos\alpha$, $\eta = \pm e^{-\alpha}\sin\alpha$, i.e. $q_o = \tau^2+\eta^2 = e^{-2\alpha}$, $q_1 = -2\tau = -2e^{-\alpha}\cos\alpha$. From eq. (7.4.28)

$$d_o = 1$$
$$d_1 = -q_1 = 2e^{-\alpha}\cos\alpha$$
$$d_2 = q_1^2 - q_o = e^{-2\alpha}(4\cos^2\alpha-1)$$

and eq. (7.4.31) becomes

$$[p_o \quad p_1 \quad p_2 \quad 1]\begin{bmatrix}1 & 0\\ 0 & 1\\ -e^{-2\alpha} & 2e^{-\alpha}\cos\alpha\\ -2e^{-3\alpha}\cos\alpha & e^{-2\alpha}(4\cos^2\alpha-1)\end{bmatrix} = [0 \quad 0]$$

A minor simplification is achieved as follows: Multiply the second column by $e^{-\alpha}$ and add the first one multiplied by $2\cos\alpha$

$$[p_o \quad p_1 \quad p_2 \quad 1]\begin{bmatrix}1 & 2\cos\alpha\\ 0 & e^{-\alpha}\\ -e^{-2\alpha} & 0\\ -2e^{-3\alpha}\cos\alpha & -e^{-3\alpha}\end{bmatrix} = [0 \quad 0] \qquad (7.4.35)$$

We will come back to this result later in section 7.5.

7.5 Pole-Region Assignment

7.5.1 Mapping from P space to K space

Chapter 7 has dealt so far with polynomials $P(z)$ and their roots $P(z) = (z-z_1)(z-z_2)\dots(z-z_n)$. We now relate this to discrete-time dynamical systems

$$\underline{x}[k+1] = \underline{A}\underline{x}[k] + \underline{b}u[k] \tag{7.5.1}$$

with state feedback

$$u[k] = -\underline{k}'\underline{x}[k] \tag{7.5.2}$$

Identify the closed loop characteristic polynomial with the previously studied polynomial

$$P(z) = \det(z\underline{I}-\underline{A}+\underline{b}\underline{k}') = [\underline{p}' \quad 1]\underline{z}_n \tag{7.5.3}$$

Then by eq. (4.4.4)

$$\underline{p}' = \underline{a}' + \underline{k}'\underline{W} \tag{7.5.4}$$

where

$$P_A(z) = \det(z\underline{I}-\underline{A}) = [\underline{a}' \quad 1]\underline{z}_n \tag{7.5.5}$$

is the open loop characteristic polynomial and $\underline{W}$ is given by eq. (2.6.12) as

$$\underline{W} = [\underline{D}_o\underline{b} \; , \; \underline{D}_1\underline{b} \; \ldots \; \underline{D}_{n-1}\underline{b}] \tag{7.5.6}$$

with $\underline{D}_i$ and a_i determined by Leverrier's algorithm, eq. (A.7.36).

In this control problem we are interested in nice stability regions in K space, that is the space in which $\underline{k}$ lies. For this purpose there are two approaches:

i) Study nice stability regions in P space first, see figures 7.1, 7.2, 7.4 and 7.6 b, then map them to K space by the inverse of eq. (7.5.4), which exists for controllable systems and is by eq. (4.4.6)

$$\underline{k}' = [\underline{p}' \quad 1]\underline{E} \tag{7.5.7}$$

In this approach significant points in P space like vertices of the convex hull of a nice stability region are mapped to

K space by the affine mapping of eq. (7.5.7). This affine mapping consists of a linear mapping $\underline{p}'\underline{W}^{-1}$, where $\underline{W}^{-1}$ is equal to the first n rows of $\underline{E}$, and a shift of the origin by $\underline{e}'\underline{A}^n$, the last row of $\underline{E}$, see eq. (4.4.7). This approach is conceptually simple for the design of robust control systems to be described in chapter 8.

ii) Use of computer graphics to represent nice stability regions in K space by two-dimensional cross sections. In this case points on the real and complex boundary must be calculated and connected to generate the computer graphic. For this purpose the relation $\underline{p}(\underline{k})$ for the assumed controller structure (for state feedback: $\underline{p}' = \underline{a}' + \underline{k}'\underline{W}$) is substituted into the equations for the real and complex boundaries. This approach has the advantage that other controller structures, like output feedback with $\underline{p}' = \underline{a}' + \underline{k}_y'\underline{CW}$ or dynamic output feedback can be used as well.

We will first map the simple nice stability regions studied previously by approach i).

7.5.2 Affine Mapping by Pole Assignment Matrix

For state-vector feedback, eq. (7.5.7) represents the affine mapping from P space to K space. Such a mapping shifts and distorts the stability region but preserves its basic shape. For example a triangle remains a triangle. The same is true for a hyperplane, a tetrahedron etc.

Example 1:

$$\underline{x}[k+1] = \begin{bmatrix} 0 & -4 \\ 1 & 4 \end{bmatrix} \underline{x}[k] + \frac{1}{16} \begin{bmatrix} 6 \\ -5 \end{bmatrix} u[k]$$

$$\underline{k}' = [p_0 \quad p_1 \quad 1] \begin{bmatrix} 5 & 6 \\ 6 & 4 \\ 4 & -8 \end{bmatrix} \qquad (7.5.8)$$

The corners $\underline{p}_o$, $\underline{p}_1$ and $\underline{p}_2$ of the stability triangle in figure 7.4 are mapped into

$$\begin{bmatrix} \underline{k}'_o \\ \underline{k}'_1 \\ \underline{k}'_2 \end{bmatrix} = \begin{bmatrix} \underline{p}'_o & 1 \\ \underline{p}'_1 & 1 \\ \underline{p}'_2 & 1 \end{bmatrix} \underline{E} = \begin{bmatrix} 1 & -2 & 1 \\ -1 & 0 & 1 \\ 1 & 2 & 1 \end{bmatrix} \begin{bmatrix} 5 & 6 \\ 6 & 4 \\ 4 & -8 \end{bmatrix} = \begin{bmatrix} -3 & -10 \\ -1 & -14 \\ 21 & 6 \end{bmatrix} \tag{7.5.9}$$

The stability triangle 012 is illustrated in the K plane of figure 7.9 by dotted lines. The triangle $\bar{0}\bar{1}\bar{2}$ for nice stability is mapped correspondingly by

$$\begin{bmatrix} \underline{k}'_{\bar{o}} \\ \underline{k}'_{\bar{1}} \\ \underline{k}'_{\bar{2}} \end{bmatrix} = \begin{bmatrix} 0.9025 & -1.9 & 1 \\ -0.0475 & -0.9 & 1 \\ 0.0025 & 0.1 & 1 \end{bmatrix} \begin{bmatrix} 5 & 6 \\ 6 & 4 \\ 4 & -8 \end{bmatrix} = \begin{bmatrix} -2.8875 & -10.185 \\ -1.6375 & -11.885 \\ 4.6125 & -7.585 \end{bmatrix}$$

This is shown in figure 7.9. In addition the deadbeat point $\underline{p} = \underline{0}$, i.e. $\underline{k}' = \underline{e}'\underline{A}^2 = [4 \quad -8]$ is indicated.

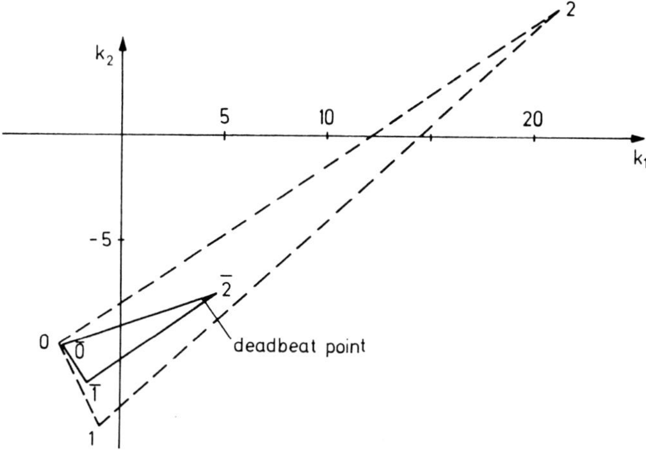

Figure 7.9 Triangles of figure 7.4 mapped into K plane for the system of eq. (7.5.8)

Example 2:

Modify eq. (7.5.8) to a continuous time example ($\underline{x}[k+1] \rightarrow \dot{\underline{x}}$) with the same pole assignment matrix $\underline{E}$. Map the nice stability region of figure 7.8 to the K plane.

The complex root boundary points of table 7.1 are substituted into

$$[k_1 \quad k_2] = [p_o \quad p_1 \quad 1]\begin{bmatrix} 5 & 6 \\ 6 & 4 \\ 4 & -8 \end{bmatrix} \qquad (7.5.10)$$

and the values of table 7.2 are obtained

σ	$k_1 = 5p_o + 6p_1 + 4$	$k_2 = 6p_o + 4p_1 - 8$
-1	21	6
-1.1	25.6125	10.895
-1.2	30.55	16.18
-1.4	41.4	27.29
-1.6	53.6	41.28
-1.8	67	56.08
-2	81.75	72.5

Table 7.2 Complex boundary points in K plane

The real root boundary remains a straight line under an affine mapping and is defined by two points on it. In addition to the above point for $\sigma = -1$ a second point on the line may be mapped, e.g. $p_o = 4$, $p_1 = 5$ resulting in $k_1 = 54$, $k_2 = 36$. The boundaries are plotted in figure 7.10.

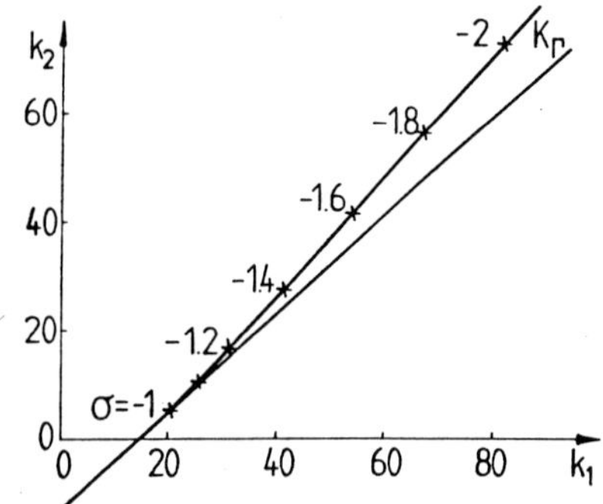

Figure 7.10

Γ-stability region K_Γ from figure 7.8 mapped for the continuous-time form of the system of eq. (7.5.8)

It is not necessary to calculate the boundaries of P_Γ in figure 7.8 first. Substitute eq. (7.4.32) into eq. (7.5.10) to obtain $k_1 = 16.25\sigma^2 - 12\sigma - 7.25$, $k_2 = 19.5\sigma^2 - 8\sigma - 21.5$ and table 7.2 can be calculated directly from σ.

For n = 3 and a circular pole region, figure 7.2 represents the Γ-stability region in P space. It is uniquely determined by the four vertices corresponding to the polynomials with zeros from the set $\{-1, 1\}$ or $\{\tau_L, \tau_R\}$. Only these vertices must be mapped into K space. By the convex hull property of eq. (7.2.17) this is true for arbitrary n. This is summarized in the following theorem:

> Determine n + 1 points $\underline{k}_i$, i = 0,1...n in K space by assigning $P(z) = (z-\tau_L)^i(z-\tau_R)^{n-i}$ by state feedback. Then the simplex with vertices k_i is the convex hull of the nice stability region K_Γ, where Γ is the circle in z plane with real center and real axis intersections at τ_L and τ_R.

7.5.3 K Space Boundaries for State Feedback

All nice stability conditions on $\underline{p}$ can be directly transformed into conditions on the state feedback vector $\underline{k}$ by substituting eq. (7.5.4), $\underline{p}' = \underline{a}' + \underline{k}'\underline{W}$.

The result of eq. (7.4.4) for circular regions may be rewritten as the following theorem:

> The eigenvalues of $\underline{A} - \underline{bk}'$ are located inside the circle with real center and real axis intersections τ_R and τ_L if and only if

$$M(w) = [\underline{a}'+\underline{k}'\underline{W} \quad 1]\begin{bmatrix} (1-w)^n \\ (1-w)^{n-1}(\tau_R-\tau_L w) \\ \vdots \\ (1-w)(\tau_R-\tau_L w)^{n-1} \\ (\tau_R-\tau_L w)^n \end{bmatrix} \tag{7.5.11}$$

> is a Hurwitz polynomial. $\underline{a}$ and $\underline{W}$ are uniquely determined by $\underline{A}$ and $\underline{b}$ via eqs. (7.5.5) and (7.5.6).

In eqs. (7.4.21) and (7.4.30) the real and complex boundaries were formulated for general regions Γ by

$$
\begin{aligned}
&[\underline{p}' \quad 1]\underline{a}_L = 0\\
&[\underline{p}' \quad 1]\underline{a}_R = 0\\
&[\underline{p}' \quad 1][\underline{c}_1 \quad \underline{c}_2] = [0 \quad 0]
\end{aligned} \tag{7.5.12}
$$

With $\underline{p}' = \underline{a}' + \underline{k}'\underline{W}$ this becomes

$$
\begin{aligned}
&[\underline{a}' + \underline{k}'\underline{W} \quad 1]\underline{a}_L = 0\\
&[\underline{a}' + \underline{k}'\underline{W} \quad 1]\underline{a}_R = 0\\
&[\underline{a}' + \underline{k}'\underline{W} \quad 1][\underline{c}_1 \quad \underline{c}_2] = [0 \quad 0]
\end{aligned} \tag{7.5.13}
$$

This is a relation for the complex and the two real boundaries of K_Γ in K space.

Example:

The system "triple integrator"

$$
\dot{\underline{x}} = \begin{bmatrix} 0 & 1 & 0 \\ 0 & 0 & 1 \\ 0 & 0 & 0 \end{bmatrix} \underline{x} + \begin{bmatrix} 0 \\ 0 \\ 1 \end{bmatrix} u \tag{7.5.14}
$$

sampled with T = 1 has by eq. (3.1.20) the discrete-time representation

$$
\underline{x}[k+1] = \begin{bmatrix} 1 & 1 & 0.5 \\ 0 & 1 & 1 \\ 0 & 0 & 1 \end{bmatrix} \underline{x}[k] + \frac{1}{6} \begin{bmatrix} 1 \\ 3 \\ 6 \end{bmatrix} u[k]
$$

The set of all state feedbacks $u = -\underline{k}'\underline{x}$ will now be represented which assign eigenvalues with damping greater than $1/\sqrt{2}$ to the closed loop, i.e. the discrete eigenvalues must be inside the logarithmic spiral $e^{-\alpha}e^{\pm j\alpha}$, $0 \le \alpha \le \pi$. From eq. (A.7.36)

$$
(z\underline{I}-\underline{A})^{-1} = \frac{\underline{D}_2 z^2 + \underline{D}_1 z + \underline{D}_0}{z^3 - 3z^2 + 3z - 1}
$$

$$
\underline{D}_2 = \begin{bmatrix} 1 & 0 & 0 \\ 0 & 1 & 0 \\ 0 & 0 & 1 \end{bmatrix}, \quad \underline{D}_1 = \begin{bmatrix} -2 & 1 & 0.5 \\ 0 & -2 & 1 \\ 0 & 0 & -2 \end{bmatrix}, \quad \underline{D}_0 = \begin{bmatrix} 1 & -1 & 0.5 \\ 0 & 1 & -1 \\ 0 & 0 & 1 \end{bmatrix}
$$

and with eq. (7.5.6)

$$\underline{W} = [\underline{D}_0\underline{b} \quad \underline{D}_1\underline{b} \quad \underline{D}_2\underline{b}] = \frac{1}{6}\begin{bmatrix} 1 & 4 & 1 \\ -3 & 0 & 3 \\ 6 & -12 & 6 \end{bmatrix}$$

Then

$$[\underline{p}' \quad 1] = [\underline{a}'+\underline{k}'\underline{W} \quad 1]$$

$$= [-1+(k_1-3k_2+6k_3)/6 \ , \ 3+(4k_1-12k_3)/6 \ , \ -3+(k_1+3k_2+6k_3)/6 \ , \ 1]$$

The boundary surfaces have already been determined in P space for the spiral $e^{-\alpha}e^{\pm j\alpha}$ in section 7.4. The vectors $\underline{a}_L$, $\underline{a}_R$ and $\underline{c}_1$, $\underline{c}_2$ (eq. (7.4.35)) in eq. (7.5.12) are

$$\underline{a}_L = \begin{bmatrix} 1 \\ -0.0432 \\ 0.0432^2 \\ -0.0432^3 \end{bmatrix}, \ \underline{a}_R = \begin{bmatrix} 1 \\ 1 \\ 1 \\ 1 \end{bmatrix}, \ \underline{c}_1 \begin{bmatrix} 1 \\ 0 \\ -e^{-2\alpha} \\ -2e^{-3\alpha}\cos\alpha \end{bmatrix}, \ \underline{c}_2 = \begin{bmatrix} 2\cos\alpha \\ e^{-\alpha} \\ 0 \\ -e^{-3\alpha} \end{bmatrix}$$

$$P(\tau_L) = [\underline{p}' \quad 1]\underline{a}_L = 0$$

The left real boundary becomes

$$P(\tau_L) = P(-0.0432) = [\underline{p}' \quad 1]\underline{a}_L = 0$$

$$= 0.1382k_1 - 0.4991k_2 + 1.0883k_3 - 1.135 = 0 \qquad (7.5.15)$$

and the right real boundary is

$$P(\tau_R) = P(1) = [\underline{p}' \quad 1]\underline{a}_R = k_1 = 0 \qquad (7.5.16)$$

For each value of α on $\partial\Gamma$ the complex boundary has to satisfy

$$[\underline{p}' \quad 1][\underline{c}_1 \ , \ \underline{c}_2] = [-1+3e^{-2\alpha}-2e^{-3\alpha}\cos\alpha, \ -2\cos\alpha+3e^{-\alpha}-e^{-3\alpha}] +$$

$$+ \frac{1}{6}[k_1 \quad k_2 \quad k_3]\begin{bmatrix} 1-e^{-2\alpha} & 2\cos\alpha+4e^{-\alpha} \\ -3-3e^{-2\alpha} & -6\cos\alpha \\ 6-6e^{-2\alpha} & 12\cos\alpha-12e^{-\alpha} \end{bmatrix}$$

$$= [0 \quad 0] \qquad (7.5.17)$$

$\pm\alpha$ is the angle between the complex conjugate pair of poles on $\partial\Gamma$ and the positive real axis. For each value $0 \le \alpha \le \pi$ a straight line in K space is obtained which is the intersection of the two planes represented by the above equation.

7.5.4 K Space Boundaries for Static Output Feedback

Assume now that only an output

$$\underline{y} = \underline{C}\underline{x} \qquad (7.5.18)$$

is available for feedback. The simplest controller structure is then static output feedback

$$u = -\underline{k}_y'\underline{y} = -\underline{k}_y'\underline{C}\underline{x} \qquad (7.5.19)$$

Comparing this with state feedback $u = -\underline{k}'\underline{x}$ it is seen that

$$\underline{k}' = \underline{k}_y'\underline{C} \qquad (7.5.20)$$

i.e. in K space only solutions which lie in the linear subspace spanned by the rows of the matrix $\underline{C}$ are admissible. Eq. (7.5.13) becomes

$$\begin{aligned}
&[\underline{a}' + \underline{k}_y'\underline{C}\underline{W} \quad 1]\underline{a}_L = 0\\
&[\underline{a}' + \underline{k}_y'\underline{C}\underline{W} \quad 1]\underline{a}_R = 0\\
&[\underline{a}' + \underline{k}_y'\underline{C}\underline{W} \quad 1][\underline{c}_1 \quad \underline{c}_2] = [0 \quad 0]
\end{aligned} \qquad (7.5.21)$$

An arbitrary $\underline{p}' = \underline{a}' + \underline{k}_y'\underline{C}\underline{W}$ can be generated only in the trivial case of rank $\underline{C} = n$ and $(\underline{A}, \underline{b})$ controllable, i.e. $\underline{W}$ regular. Now let rank $\underline{W} = n$ and rank $\underline{C} = s < n$. In this case the possible closed-loop eigenvalue locations are restricted. Some of them can be assigned arbitrarily; the location of the remaining eigenvalues is dictated, however, by this choice and may be unsuitable. It is easier in this case to satisfy the relaxed pole-region requirement.

Example:

$$\underline{x}[k+1] = \begin{bmatrix} 0 & 1 & 0 \\ 0 & 0 & 1 \\ -0.4 & -2 & -0.9 \end{bmatrix} \underline{x}[k] + \begin{bmatrix} 0 \\ 0 \\ 1 \end{bmatrix} u[k] \qquad (7.5.22)$$

$$\underline{y}[k] = \begin{bmatrix} 1 & 0 & 0 \\ 0 & 0 & 1 \end{bmatrix} \underline{x}[k]$$

This system is unstable ($|\underline{X}-\underline{Y}| = 1-a_1+a_0a_2-a_0^2 = -0.8 < 0$). Can it be stabilized by static output feedback $u = [k_1 \quad k_3]\underline{y}$?

The system is in control canonical form, therefore the transformation matrix $\underline{W}$ is the unit matrix and

$$\underline{p}' = \underline{a}' + \underline{k}'\underline{W} = [0.4 \quad 2 \quad 0.9] + [k_1 \quad 0 \quad k_3]$$

The real boundaries are

$$\begin{aligned} P(-1) &= [\underline{p}' \quad 1]\underline{a}_L \\ &= [0.4+k_1 \quad 2 \quad 0.9+k_3 \quad 1][1 \quad -1 \quad 1 \quad -1]' \\ &= k_1 + k_3 - 1.7 = 0 \end{aligned} \qquad (7.5.23)$$

$$\begin{aligned} P(1) &= [0.4+k_1 \quad 2 \quad 0.9+k_3 \quad 1][1 \quad 1 \quad 1 \quad 1]' \\ &= k_1 + k_3 + 4.3 = 0 \end{aligned} \qquad (7.5.24)$$

For the complex boundary $[\underline{p}' \quad 1][\underline{c}_1 \quad \underline{c}_2] = [0 \quad 0]$ by eqs. (7.4.30) and (7.4.28) with $q_0 = 1$, $q_1 = -2\tau$, $-1 \le \tau \le 1$,

$$\begin{bmatrix} d_0 & 0 \\ d_1 & d_0 \\ d_2 & d_1 \\ d_3 & d_2 \end{bmatrix} - \begin{bmatrix} 1 & 0 \\ 2\tau & 1 \\ 4\tau^2-1 & 2\tau \\ 8\tau^3-4\tau & 4\tau^2-1 \end{bmatrix}$$

The simplified eq. (7.4.31) is obtained by multiplying the second column by -2τ and adding it to the first one. The simplification may be continued, i.e. multiply the new first column by 2τ and add it to the second one.

1st step: $\begin{bmatrix} 1 & 0 \\ 0 & 1 \\ -1 & 2\tau \\ -2\tau & 4\tau^2-1 \end{bmatrix}$, 2nd step: $\begin{bmatrix} 1 & 2\tau \\ 0 & 1 \\ -1 & 0 \\ -2\tau & -1 \end{bmatrix} = [\underline{c}_1 \quad \underline{c}_2]$

Then

$$[0.4+k_1 \quad 2 \quad 0.9+k_3 \quad 1]\begin{bmatrix} 1 & 2\tau \\ 0 & 1 \\ -1 & 0 \\ -2\tau & -1 \end{bmatrix} = [0 \quad 0]$$

$$k_1 - k_3 - 2\tau - 0.5 = 0$$

$$(0.4+k_1)2\tau + 1 = 0$$

The intersection of these two straight lines is

$$\left.\begin{aligned} k_1(\tau) &= -0.4 - 1/2\tau \\ k_3(\tau) &= k_1(\tau) - 2\tau - 0.5 \end{aligned}\right\} \quad -1 \le \tau \le 1 \qquad (7.5.25)$$

Elimination of τ yields the hyperbola

$$k_3 = k_1 + \frac{1}{k_1+0.4} - 0.5$$

For complicated boundaries this elimination of τ is not possible. It is also generally undesirable because the limitation to the interval $-1 \le \tau \le 1$ is lost. Figure 7.11 shows the boundaries of eqs. (7.5.23), (7.5.24) and (7.5.25).

Figure 7.11 can be interpreted as follows: for a third order discrete system the stability region in P space is shown in figure 7.2. State feedback for the system (7.5.22) constitutes an affine mapping of this region into the three-dimensional K space. In the example the cross section in the plane $k_2 = 0$ was calculated and shown in figure 7.11. It intersects the stability region near the affine images of the vertices 0 and 3 of figure 7.2. Therefore the resulting region consists of two disjoint components near GH and EF. The cross section plane is parallel to the affine image of the edge 12 of figure 7.2, where P(z) contains the factor (z+1)(z-1). Thus the plane of figure 7.11 does not cut the intersection of the two real boundaries. In other words, with $k_2 = 0$ we can assign a pole at $z = 1$ or $z = -1$ but not both simultaneously. For $\tau = 0$ the complex boundary goes through infinity, i.e. it is impossible to assign a complex conjugate pair of poles at $z = \pm j$ with $k_2 = 0$.

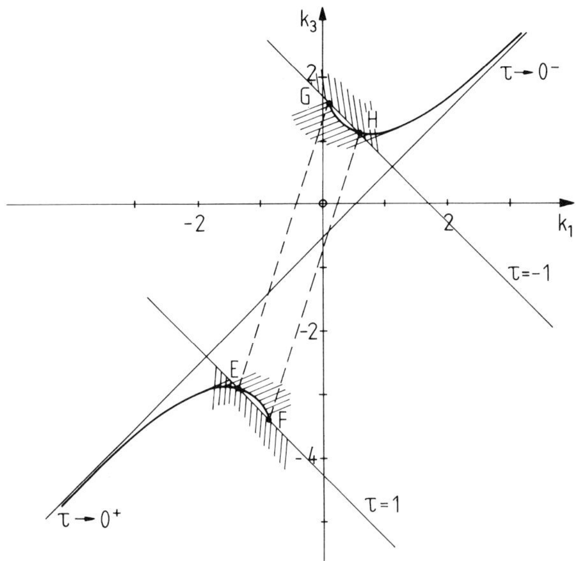

Figure 7.11 The stability region of the system (7.5.22) in the k_1-k_3 plane consists of two disjoint sets near GH and EF.

The two small stability regions near GH and EF are circum-shaded. The coordinates of the vertices and corresponding eigenvalues are given in the following table:

Vertex	k_1	k_3	Eigenvalues
E	-1.4	2.9	1, $\pm\exp j\pi/3$
F	-0.9	3.4	1, 1,05
G	0.1	1.6	-1, -1, -0.5
H	0.6	1.1	-1, $\pm\exp j2\pi/3$

An example of a stabilizing solution is $k_1 = -1.14$, $k_3 = -3.14$ with eigenvalues 0.959, $0.878e^{\pm j43^o}$.

Some general aspects of pole region assignment by output feedback are the following:

1. The Γ-stability region in P space and its affine image in K space are non-convex in most cases. The set of admissible solutions in a subspace is therefore not necessarily connected. For $n = 3$, Γ = unit circle and a two dimensional subspace the possible cases are apparent in figure 7.2:

   a) No intersection of the cross section plane with the stability region,

   b) One connected solution set,

   c) Two disjoint components of the solution set like in figure 7.11.

2. Such nonconvex constraints give rise to difficulties in numerical optimization algorithms, which search in the neighborhood of an initial guess. The resulting local optimum may depend on the initial guess. In the example of figure 7.2 a stabilizing value near GH (e.g. $k_1 = 0.3$, $k_2 = 1.3$) may be found, which results in a heavily oscillating actuator signal u, or a more satisfactory slowly decaying solution near EF may be found starting with a different initial guess.

3. An alternative numerical approach is a systematic search in a grid. Consider first the case of a circular pole region Γ. The convex hull of the stability region in feedback gain space is given by positivity of the coefficients of the polynomial M in eq. (7.5.11). This convex hull is first intersected with the admissible subspace of output feedback. In the example of figure 7.11 this is the quadrangle EFHG. In general the intersection is either empty (no Γ-stabilization possible) or it is a convex polyhedron. This polyhedron defines the search space for possible solutions. A crude grid would consist of the vertices of the polyhedron and one internal point, e.g. the center of gravity. If necessary it may be refined by introducing barycentric coordinates inside the triangles, tetrahedra etc. of neighboring points of the crude grid. For each

grid point it is tested whether $\underline{X}-\underline{Y}$ is positive innerwise, see eq. (C.2.7). There remains the set of Γ-stabilizing grid points in K space. If it is empty, the grid may be refined. However, from an engineering point of view, we are interested only in solutions for which also some neighborhood in K space is Γ-stabilizing. Thus it is not important if a local solution is not found because it is located between the points of a sufficiently fine grid.

The example of figure 7.11 suggests that promising candidates are located adjacent to the real boundaries. But for $n > 3$ this is not necessarily so.

For noncircular boundaries with real axis intersections at τ_L and τ_R the real boundaries are the same as for the circle passing through τ_L and τ_R. This includes the contour on the real root hyperplane formed by the intersection with the complex boundary. Along this contour there is a double eigenvalue at τ_L or τ_R, i.e. the derivative polynomial $P'(z) = dP(z)/dz$ also has a zero, i.e.

$$P'(\tau_L) = 0 \ , \quad P'(\tau_R) = 0 \tag{7.5.26}$$

Figure 7.2 indicates for third order systems that some deformation of ∂Γ from its circular form is tolerable before the complex boundary leaves the tetrahedron. If we want to be sure, the enclosing circle of ∂Γ may be taken to generate the search space.

4. The root locus method allows a search along a straight line in K space. The example in figure 7.11 shows that there is no stable solution for $k_1 = 0$ and arbitrary k_3 or $k_3 = 0$ and arbitrary k_1. A feasible straight line in figure 7.11 is $k_3 = 3k_1 + 0.3$. It passes to both parts of the stability region. It corresponds to the controller

$$u = -k(x_1+3x_3) - 0.3x_3 \tag{7.5.27}$$

The root locus shows two intervals of k for which all three eigenvalues are located inside the unit circle. The plot gives detailed information about the root location as a function of one controller parameter. The technique can be extended to two controller parameters, but the arising family of root locus plots is complicated [77.7]. In contrast the parameter-plane method only exhibits the information of number of poles in a desired region Γ (and their closeness to particular boundaries). However, this is easily achieved in a plane of two controller parameters. The essential computational difference is that the root locus implicitely factorizes the characteristic polynomial into $P(z) = (z-z_1)(z-z_2)\dots(z-z_n)$, whereas the parameter-space method takes $(z-z_1)$ on a real boundary or $(z-z_1)(z-\bar{z}_1)$ on a complex boundary and evaluates $P(z)$. This latter step requires less computational effort and allows one to "scan" through a higher dimensional parameter space by moving a two-dimensional cross-section plane and doing all calculations on-line.

5. Γ-stability regions may also be represented in a space of three controller parameters. In this case first the boundaries are calculated and restricted to the exact parameter ranges. A wire frame representation of the Γ-stability region can then be visualized by rotating, zooming etc. The complex boundary is described by quadrangles, see figure 7.2. Uniform color-shading of each quadrangle gives a realistic impression of a body. Putz and Wozny [84.7], [85.2] do this by three-dimensional interactive computer graphics. This approach allows the designer to look into the parameter space and to see the design specifications as geometric objects which can be manipulated in real time.

A flight control example with static output feedback will be studied in section 8.5. See eq. (8.5.3).

7.5.5 Pole-Region Assignment for Observer and Compensator Feedback

If a satisfactory solution with static output does not exist, then it becomes necessary to use a controller structure with feedback dynamics. This may be feedback of observer states or a filter or compensator in the controller.

In the observer structure of figure 5.1 the observer eigenvalues are assigned by the choice of $\underline{H}$. In the single output case, $y = \underline{c}'\underline{x}$ this is by eq. (5.2.15)

$$\underline{h} = \underline{F}\begin{bmatrix}\underline{q}\\1\end{bmatrix}, \quad \underline{F} = [\underline{f}, \quad \underline{Af} \ \dots \ \underline{A}^n\underline{f}] \tag{7.5.28}$$

$$\underline{f} = \begin{bmatrix}\underline{c}'\\ \underline{c}'\underline{A}\\ \vdots\\ \underline{c}'\underline{A}^{n-1}\end{bmatrix}^{-1}\begin{bmatrix}0\\ \vdots\\ 0\\ 1\end{bmatrix}$$

where Q(z) is the characteristic polynomial of the reconstruction error and $\underline{q}$ is its coefficient vector. Eq. (7.5.28) corresponds to eq. (7.5.7) as can be seen by transposing: $\underline{h}' = [\underline{q}' \quad 1]\underline{F}'$. It may be used to map a nice stability region from Q space to H space. Also for the reduced order observer a pole assignment problem must be solved, see eq. (5.3.15).

Example:

Consider the system of eq. (7.5.22)

$$\underline{x}[k+1] = \begin{bmatrix}0 & 1 & 0\\ 0 & 0 & 1\\ -0.4 & -2 & -0.9\end{bmatrix}\underline{x}[k] + \begin{bmatrix}0\\0\\1\end{bmatrix}u[k] \tag{7.5.29}$$

with a measured output

$$y = [1 \quad 0 \quad 0]\underline{x}[k]$$

Static output feedback cannot stabilize this system, see figure 7.11 with $k_3 = 0$. Find a reduced, i.e. second order observer and the region of gains that place the observer poles inside a circle passing through $\tau_L = 0$ and $\tau_R = 0.5$. Feedback of the observer states then allows arbitrary pole assignment or pole region assignment.

In eqs.(5.3.4) and (5.3.5) we have

$$\underline{z} = \begin{bmatrix} x_2 \\ x_3 \end{bmatrix}, \quad \underline{P} = \begin{bmatrix} 0 & 1 \\ -2 & -0.9 \end{bmatrix}, \quad \underline{Q} = \begin{bmatrix} 0 \\ -0.4 \end{bmatrix}, \quad \underline{D} = \begin{bmatrix} 0 \\ 1 \end{bmatrix}$$

$$y = x_1, \quad \underline{R} = [\,1 \quad 0\,], \quad S = 0, \quad E = 0$$

Then by eqs. (5.3.11) to (5.3.13) with $\underline{H} = \begin{bmatrix} h_1 \\ h_2 \end{bmatrix}$

$$\underline{L} = \underline{P} - \underline{HR} = \begin{bmatrix} -h_1 & 1 \\ -2-h_2 & -0.9 \end{bmatrix}$$

$$\underline{M} = \underline{Q} - \underline{HS} + \underline{LH} = \begin{bmatrix} -h_1^{\,2} + h_2 \\ -0.4 - 2h_1 - h_1 h_2 - 0.9h_2 \end{bmatrix}$$

$$\underline{N} = \underline{D} - \underline{HE} = \begin{bmatrix} 0 \\ 1 \end{bmatrix}$$

The observer of eq. (5.3.6) is

$$\underline{\hat{v}}[k+1] = \underline{L\hat{v}}[k] + \underline{M}y[k] + \underline{N}u[k]$$

$$\underline{\hat{z}}[k] = \underline{\hat{v}}[k] + \underline{H}y[k]$$

The error dynamics are specified by

$$Q(z) = \det(z\underline{I}-\underline{L}) = \det(z\underline{I}-\underline{P}+\underline{HR})$$

In eq. (7.5.28) $\underline{c}' = \underline{R}$, $\underline{A} = \underline{P}$

$$\underline{f} = \begin{bmatrix} 1 & 0 \\ 0 & 1 \end{bmatrix}^{-1} \begin{bmatrix} 0 \\ 1 \end{bmatrix} = \begin{bmatrix} 0 \\ 1 \end{bmatrix}$$

$$\underline{F} = \begin{bmatrix} 0 & 1 & -0.9 \\ 1 & -0.9 & -1.19 \end{bmatrix}$$

The corners of the nice stability triangle are

$$[\underline{h}_o,\ \underline{h}_1,\ \underline{h}_2] = \underline{F}\begin{bmatrix} \underline{q}_o & \underline{q}_1 & \underline{q}_2 \\ 1 & 1 & 1 \end{bmatrix}$$

where

$$[1 \quad z \quad z^2]\begin{bmatrix} \underline{q}_o & \underline{q}_1 & \underline{q}_2 \\ 1 & 1 & 1 \end{bmatrix} = [z^2 \quad z(z-0.5) \quad (z-0.5)^2]$$

$$[\underline{h}_o,\quad \underline{h}_1,\quad \underline{h}_2] = \begin{bmatrix} 0 & 1 & -0.9 \\ 1 & -0.9 & -1.19 \end{bmatrix}\begin{bmatrix} 0 & 0 & 0.25 \\ 0 & -0.5 & -1 \\ 1 & 1 & 1 \end{bmatrix}$$

$$= \begin{bmatrix} -0.9 & -1.4 & -1.9 \\ -1.19 & -0.74 & -0.04 \end{bmatrix}$$

The nice stability triangle is shown in figure 7.12. The corner 0 corresponds to a deadbeat observer.

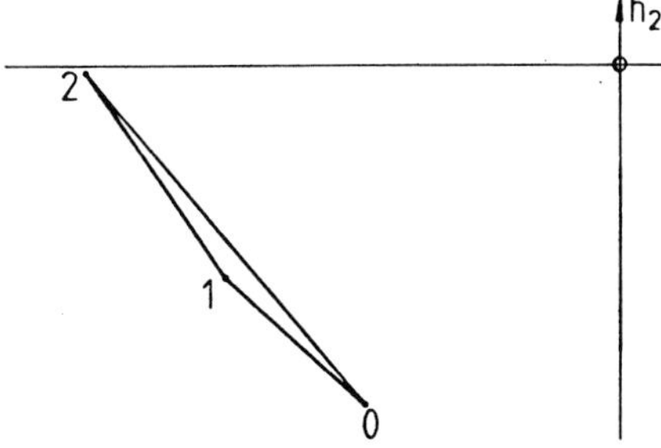

Figure 7.12

Observer gains h_1, h_2 in the triangle 012 place the observer poles into the circle between $\tau_L = 0$ and $\tau_R = 0.5$

For more complicated boundaries and higher order systems the observer equivalent of eq. (7.5.4) is used. By duality this is

$$\underline{q} = \underline{a} + \underline{Vh} \tag{7.5.30}$$

where

$$\underline{V} := \begin{bmatrix} \underline{c}'\underline{D}_o \\ \underline{c}'\underline{D}_1 \\ \vdots \\ \underline{c}'\underline{D}_{n-1} \end{bmatrix} \tag{7.5.31}$$

Then $\underline{q}(\underline{h})$ is substituted into the boundary eq. (7.5.12), where $\underline{q}$ replaces $\underline{p}$. Thus the boundaries in H space are obtained.

Also for transfer function descriptions of plant and controller it is easy to evaluate $\underline{p}(\underline{k})$, where the vector $\underline{k}$ contains the free controller parameters.

Example:

For the example of eq. (7.5.29) the z-transfer function is

$$h_z(z) = \frac{1}{0.4+2z+0.9z^2+z^3}$$

Closing the loop by a first order controller

$$d_z(z) = \frac{d_0+d_1 z}{c_0+z}$$

the closed-loop characteristic polynomial becomes

$$P(z) = d_0 + d_1 z + (c_0+z)(0.4+2z+0.9z^2+z^3)$$

$$\underline{p}' = [(d_0+0.4c_0) \quad (d_1+0.4+2c_0) \quad (2+0.9c_0) \quad (c_0+0.9)] \tag{7.5.32}$$

$\underline{p}'$ is linear in the controller parameters $\underline{k}' = [d_0 \quad d_1 \quad c_0]$

$p(\underline{k})$ can now be substituted into the boundary equations (7.4.21) and (7.4.30) for n = 4. This describes the nice stability boundaries in the three-dimensional controller parameter space.

7.6 Graphic Representation in Two-dimensional Cross Sections

7.6.1 Linear and Affine Subspaces

In the preceeding sections several graphical representations of Γ-stability regions were given. In these examples the choice of the two controller parameters was dictated by the problem. In many cases, however, we will have more than two controller parameters and cannot exhibit the Γ-stability region boundaries as a function of all of them simultaneously. For a third parameter a perspective view as shown in figure 7.2 is possible but it becomes difficult to read off the coordinates of a point and its distances from the boundaries. Therefore we prefer to use two-dimensional cross sections. The designer has to decide how to choose the particular section. A first choice may be to use a linear subspace, i.e. a plane which contains the point $\underline{k} = \underline{0}$. This occured naturally in the case of output feedback and suggests a definition of the cross section plane by introducing an artificial output feedback

$$u = -[\kappa_1 \quad \kappa_2]\begin{bmatrix}\underline{\gamma}'_1 \\ \underline{\gamma}'_2\end{bmatrix}\underline{x} \tag{7.6.1}$$

The row vectors $\underline{\gamma}'_1$ and $\underline{\gamma}'_2$ span the cross section plane in K space and κ_1 and κ_2 are the coordinates in this plane, the plane of the graphic terminal screen.

Also an affine subspace may be used which does not contain $\underline{k} = \underline{0}$. Instead the origin is shifted by $\underline{\gamma}'_3$ and eq. (7.6.1) is modified to

$$u = -[\kappa_1 \quad \kappa_2 \quad 1]\begin{bmatrix}\underline{\gamma}'_1 \\ \underline{\gamma}'_2 \\ \underline{\gamma}'_3\end{bmatrix}\underline{x} \tag{7.6.2}$$

Example 1:

The three-dimensional stability region was calculated for the discrete triple integrator and closed loop eigenvalues in the spiral $\partial\Gamma(\alpha) = e^{-\alpha}e^{\pm j\alpha}$, $0 \le \alpha \le \pi$. In eqs. (7.5.15), (7.5.16) and (7.5.17) the result is given in terms of k_1, k_2, k_3. It can, for example, be visualized in cross sections for fixed values of k_1. Then in eq. (7.6.2) $\kappa_1 = k_2$, $\kappa_2 = k_3$ and

$$\begin{bmatrix} \underline{\gamma}'_1 \\ \underline{\gamma}'_2 \\ \underline{\gamma}'_3 \end{bmatrix} = \begin{bmatrix} 0 & 1 & 0 \\ 0 & 0 & 1 \\ k_1 & 0 & 0 \end{bmatrix}$$

For $k_1 = 0$ there is a real eigenvalue at $z = 1$ in the entire plane, see eq. (7.5.16). This boundary will therefore not appear in parallel planes with $k_1 \neq 0$. There remains the left real root boundary of eq. (7.5.15)

$$-0.4991\,k_2 + 1.0883\,k_3 = 1.1353 - 0.1382\,k_1 \tag{7.6.3}$$

For fixed k_1 this is a straight line in the k_2-k_3-plane. By eq. (7.5.17) the complex root boundary is

$$[k_2 \quad k_3]\begin{bmatrix} -3(1+e^{-2\alpha}) & -6\cos\alpha \\ 6(1-e^{-2\alpha}) & 12(\cos\alpha-e^{-\alpha}) \end{bmatrix} =$$

$$= [6-18e^{-2\alpha}+12e^{-3\alpha}\cos\alpha-k_1(1-e^{-2\alpha})\ ,\ 12\cos\alpha-18e^{-\alpha}+6e^{-3\alpha}-k_1(2\cos\alpha+4e^{-\alpha})]$$

$$= [0 \quad 0].$$

Solve for k_3 with $\cos\alpha \neq 0$

$$k_3 = \frac{9+2k_1+(6+2k_1)e^{-2\alpha}-3e^{-4\alpha}+2e^{-\alpha}(k_1-12)\cos\alpha+12e^{-2\alpha}\cos^2\alpha}{6[1+e^{-2\alpha}-2e^{-\alpha}\cos\alpha]} \tag{7.6.4}$$

and k_2 can be determined by substituting k_3 into one of the previous equations.

The assignment of a polynomial

$$P(z) = (z-0.5)^3 = z^3 - 1.5z^2 + 0.75z - 0.125$$

whose zeros lie approximately in the middle of the region Γ, provides a clue for an initial choice of k_1. This requires

$$\underline{k}' = (\underline{p}'-\underline{a}')\underline{W}^{-1} = ([-0.125 \quad 0.75 \quad -1.5] - [-1 \quad 3 \quad -3])\underline{W}^{-1}$$

$$= \frac{1}{6}\,[0.875 \quad -2.25 \quad 1.5]\begin{bmatrix} 6 & -6 & 2 \\ 6 & 0 & -1 \\ 6 & 6 & 2 \end{bmatrix}$$

$$= [1/8 \quad 5/8 \quad 7/6]$$

$k_1 = 1/8 = 0.125$ is chosen. Then the real boundary is

$$k_3(\alpha) = \frac{9.25+6.25e^{-2\alpha}-3e^{-4\alpha}-23.75e^{-\alpha}\cos\alpha+12e^{-2\alpha}\cos^2\alpha}{6[1+e^{-2\alpha}-2e^{-\alpha}\cos\alpha]}$$

$$k_2(\alpha) = \frac{6(1-e^{-2\alpha})k_3(\alpha)-5.875+17.875e^{-2\alpha}-12e^{-3\alpha}\cos}{3(1+e^{-2\alpha})}$$

These boundaries are represented in figure 7.13. A double real eigenvalue at the left real boundary $z = \tau_L = -0.0432$ is produced by the point A ($k_2 = 1.203$, $k_3 = 1.579$). A simple eigenvalue lies at τ_L and a conjugate complex pole pair lies on the logarithmic spiral $\partial\Gamma = e^{-\alpha} \pm e^{\pm j\alpha}$ at $\alpha \approx 16^{\circ}$ for the point B ($k_2 = 0.569$, $k_3 = 1.288$). C is the starting point for $P(z) = (z-0.5)^3$.

We will now make a comparison of the calculation times. The calculation of a point $k_2(\alpha)$, $k_3(\alpha)$ of the complex boundary requires 8.8 seconds with a HP 67 pocket calculator. The factorization for any point in the plane would require 160 seconds. In this time 18 points can be determined which suffice to plot the entire figure 7.13. A significant point is B. Only a few iterations (of 8.8 seconds each) suffice to obtain for B the value $\alpha = 16.15^{\circ}$ and thereby the factorization $z_1' = -0.043$, $z_{2,3} = 0.754 \times e^{\pm j16.15^{\circ}}$.

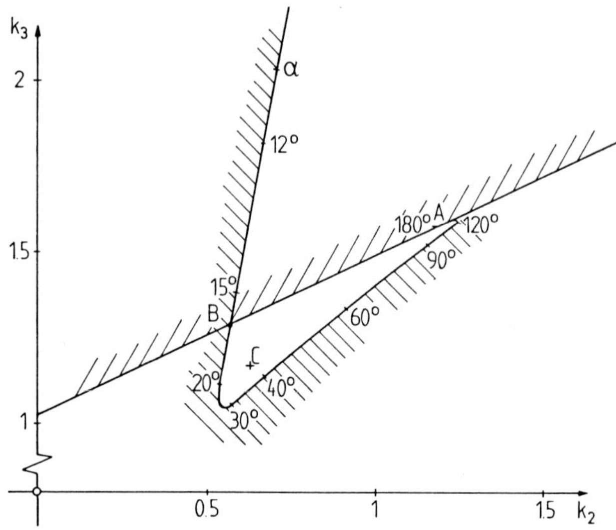

Figure 7.13 Nice stability region ($\partial\Gamma = e^{-\alpha} \times e^{\pm j\alpha}$) for the discrete triple integrator with $k_1 = 0.125$.

Example 2:

The loading bridge from eq. (2.1.15) with the physical parameter values $m_C = 1000$ kg, $m_L = 3000$ kg, $\ell = 10$ m, $g = 10$ m/sec^2 is to be controlled in continuous time so that all eigenvalues lie to the left of the left branch of the hyperbola

$$\left(\frac{\sigma}{0.25}\right)^2 - \left(\frac{\omega}{0.5}\right)^2 = 1\ , \quad \sigma < -0.25 \tag{7.6.5}$$

in the $s(=\sigma+j\omega)$-plane. Let the structure of the feedback be constrained as follows:

a) The cable angular velocity x_4 is difficult to measure. If possible, the control system should avoid reconstructing x_4 by an observer or approximating it by a differentiator, i.e. $k_4 = 0$ is chosen.

b) The position of the trolley x_1 must be fed back, because otherwise an eigenvalue at $s = 0$ is unobservable. The first choice of k_1 follows from the fact that $u(0) = -k_1$

for the typical transition from $\underline{x}(0) = [1 \quad 0 \quad 0 \quad 0]'$ to $\underline{x} = \underline{0}$. In order to limit the initial peak force to 500 Newton for a 1-meter transition, k_1 is fixed at $k_1 = 500$ (N/m). The feedback is consequently

$$u = -[500 \quad k_2 \quad k_3 \quad 0]\underline{x} \qquad (7.6.6)$$

Or in the notation of eq. (7.6.2): $\kappa_1 = k_2$, $\kappa_2 = k_3$,

$$\begin{bmatrix} \underline{\gamma}'_1 \\ \underline{\gamma}'_2 \\ \underline{\gamma}'_3 \end{bmatrix} = \begin{bmatrix} 0 & 1 & 0 & 0 \\ 0 & 0 & 1 & 0 \\ 500 & 0 & 0 & 0 \end{bmatrix}$$

The nice stability region of eq. (7.6.5) will be represented in this k_2-k_3-plane.

A real boundary arises at $\sigma = -0.25$, i.e.

$$[p_0 \quad p_1 \quad p_2 \quad p_3]\underline{\beta} = -0.25^4 \ , \ \underline{\beta}' = [1 \quad -0.25 \quad 0.25^2 \quad -0.25^3] \qquad (7.6.7)$$

The complex boundary for the hyperbola with $a = 0.25$, $b = 0.5$ follows from the first five rows of eq. (7.4.33) as

$$[p_0 \quad p_1 \quad p_2 \quad p_3 \quad 1] \begin{bmatrix} 1 & 0 \\ 2\sigma & 1 \\ -\sigma^2+0.5^2 & 2\sigma \\ -12\sigma^3+\sigma & -\sigma^2+0.5^2 \\ -19\sigma^4+0.5\sigma^2+0.5^4 & -12\sigma^3+\sigma \end{bmatrix} = [0 \quad 0]$$

and with the simplification of eq. (7.4.31)

$$[p_0 \quad p_1 \quad p_2 \quad p_3 \quad 1] \begin{bmatrix} 1 & 0 \\ 0 & 1 \\ -5\sigma^2+0.5^2 & 2\sigma \\ -10\sigma^3+0.5\sigma & -\sigma^2+0.5^2 \\ 5\sigma^4-1.5\sigma^2+0.5^4 & -12\sigma^3+\sigma \end{bmatrix} = [0 \quad 0] \qquad (7.6.8)$$

(Instead of this simplification the element σ^4 can also be eliminated.)

$\underline{p}'$ is now replaced by $\underline{a}' + \underline{k}'\underline{W}$ in eqs. (7.6.7) and (7.6.8). $\underline{W}$ and $\underline{a}'$ were already determined in eq. (2.3.10) as

$$\underline{W} = \frac{1}{\ell m_c}\begin{bmatrix} g & 0 & \ell & 0 \\ 0 & g & 0 & \ell \\ 0 & 0 & -1 & 0 \\ 0 & 0 & 0 & -1 \end{bmatrix} = 10^{-4}\begin{bmatrix} 10 & 0 & 10 & 0 \\ 0 & 10 & 0 & 10 \\ 0 & 0 & -1 & 0 \\ 0 & 0 & 0 & -1 \end{bmatrix}$$

$$\underline{a}' = [0 \quad 0 \quad \omega_L^2 \quad 0] \; , \quad \omega_L = \frac{m_L + m_c}{m_c} \times \frac{g}{\ell} = 4$$

Therefore

$$\underline{p}' = \underline{a}' + \underline{k}'\underline{W} = [0 \quad 0 \quad 4 \quad 0] + [500 \quad k_2 \quad k_3 \quad 0]\underline{W}$$

$$= [0.5k_1 \quad 0.001k_2 \quad 4.5-0.0001k_3 \quad 0.001k_2] \qquad (7.6.9)$$

The real boundary

$$k_3 = -42.5k_2 + 125625 \qquad (7.6.10)$$

results from eq. (7.6.7) and the complex boundary from eq. (7.6.8). The solution of these two linear equations is

$$k_2(\sigma) = \frac{\sigma(35+200\sigma^2-2000\sigma^4)}{(\sigma^2-0.05)(\sigma^2-0.25)}$$

$$k_3(\sigma) = \frac{(12.5-10\sigma^2)}{2\sigma}\, k_2(\sigma) + 5000(10-12\sigma^2) \qquad (7.6.11)$$

k_2 and k_3 go to infinity for

a) $\sigma = \pm\sqrt{0.05} = \pm 0.2236$ outside Γ

b) $\sigma = \pm\sqrt{0.25} = \pm 0.5$ only -0.5 in Γ

c) $\sigma \to \pm\infty$

The asymptote for $\sigma = -0.5$ is

$$k_3 = -10k_2 + 35\,000 \tag{7.6.12}$$

$k_2(-0.25) = 4233.333$ and $k_3(-0.25) = -54291.666$ for the intersection of the complex boundary with the real axis at $s = \sigma = -0.25$. This point A also lies on the straight line (7.6.10). The region of nice stability is represented in figure 7.14.

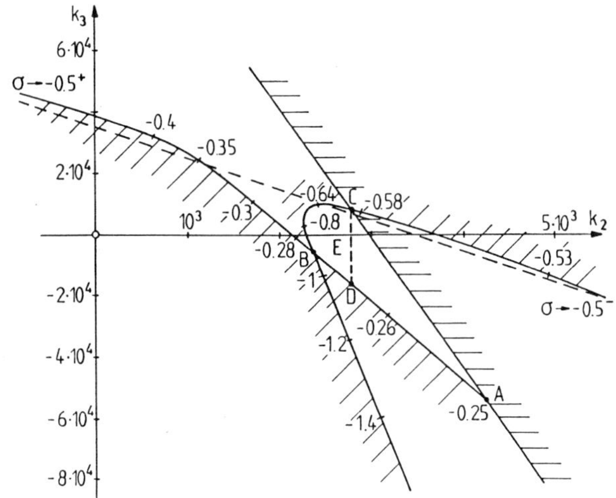

Figure 7.14 Nice stability region ABC for the loading bridge in the intersection plane $k_1 = 500$, $k_4 = 0$.

For $\sigma = -0.25$ the complex boundary begins at A (a double real pole) and goes to infinity along the asymptote for $\sigma \rightarrow -0.5$. It returns from the opposite direction and intersects the real boundary at point C for $\sigma = -0.591$. Here a real eigenvalue lies at $s = -0.25$ and a complex pair on the boundary $\partial\Gamma$ at $s = -0.591 \pm j1.071$. At point B ($k_2 = 2366$, $k_3 = -5000$) the complex boundary intersects with itself, i.e. two conjugate complex pole pairs lie on the boundary; for $\sigma = -0.275$ at $s = -0.275 \pm j0.231$ and for $\sigma = -0.908$ at $s = -0.908 \pm j1.745$. As $\sigma \rightarrow \infty$ the asymptotic behavior results finally from eq. (7.6.11).

Feedback gains inside the region ABC place all eigenvalues to the left of the hyperbola (7.6.5). As an example we choose the point E ($k_2 = 2667$, $k_3 = 0$), for which also the feedback of the cable angle x_3 can be dispensed with. We calculate $P(s) = 0.5 + 2.667s + 4.5s^2 + 2.667s^3 + s^4$ with roots $-0.420 \pm j0.057$ and $-0.914 \pm j1.397$ from eq. (7.6.9). This factorization requires 160 seconds with the HP 67 pocket calculator, while in contrast the determination of a point of the complex boundary from eq. (7.6.11) requires only 4.8 seconds. Therefore one could calculate 33 boundary points in the same time, but so many points are not at all necessary, particularly since we know the asymptote.

In the previous examples the cross section planes were chosen parallel to some of the coordinate axes, i.e. some of the k_i of state feedback were fixed at constant values. Eq. (7.6.2) would also allow arbitrary planes at an arbitrary angle with respect to the coordinate axes. The next section deals with suitable choices of such cross section planes.

7.6.2 Invariance Planes

In a practical design using two-dimensional cross sections we will proceed in iterative design steps to find an initial solution and improve it. A basic design step is to shift only one or two eigenvalues and to keep all other eigenvalues fixed. This idea was developed in section 2.6.3 and is applied here to the choice of the appropriate cross section plane. Let

$$P_A(z) = [\underline{a}' \quad 1]\underline{z}_n = \det(z\underline{I}-\underline{A}) = W(z)R(z) \tag{7.6.13}$$

with

$$R(z) = r_0 + r_1 z + \ldots + r_{n-3}z^{n-3} + z^{n-2} = [\underline{r}' \quad 1]\underline{z}_{n-2}$$
$$W(z) = w_0 + w_1 z + z^2 \tag{7.6.14}$$

and consider the polynomials $P(z)$ which have the common divisor $R(z)$ with $P_A(z)$:

$$P(z) = [\underline{p}' \quad 1]\underline{z}_n = V(z)\,R(z) \tag{7.6.15}$$

$$\text{with } V(z) = v_0 + v_1 z + z^2 \tag{7.6.16}$$

Then

$$P = VR = WR + (V-W)R = P_A + (V-W)R$$

$$P(z) = P_A(z) + (\kappa_0+\kappa_1 z)(r_0+r_1 z+\ldots+r_{n-3}z^{n-3}+z^{n-2})$$

$$\text{with } \kappa_0 = v_0-w_0\ ,\ \kappa_1 = v_1-w_1$$

$$[\underline{p}' \quad 1]\underline{z}_n = [\underline{a}' \quad 1]\underline{z}_n + [\kappa_0 \quad \kappa_1]\begin{bmatrix}\underline{r}' & 1 & 0\\ 0 & \underline{r} & 1\end{bmatrix}\underline{z}_{n-1} \tag{7.6.17}$$

Matching the coefficients of the powers of z on both sides yields

$$\underline{p}' = \underline{a}' + [\kappa_0 \quad \kappa_1]\begin{bmatrix}\underline{r}' & 1 & 0\\ 0 & \underline{r}' & 1\end{bmatrix} \tag{7.6.18}$$

This is a two-dimensional affine subspace of P space. For $V(z) = W(z)$, i.e. $\kappa_0 = \kappa_1 = 0$, it passes through $\underline{a}$, the open loop characteristic polynomial. The two vectors $[\underline{r}' \quad 1 \quad 0]$ and $[0 \quad \underline{r}' \quad 1]$ originating at $\underline{p} = \underline{a}$ span the "invariance plane" for $R(z)$ in P space. It is easy to calculate and display the intersection of this plane with higher dimensional Γ-stability regions. For circular boundaries ∂Γ this is simply the Γ-stability triangle for $V(z)$.

Example:

$$P_A(z) = (z^2+z+1.25)R(z) = [\underline{a}' \quad 1]\underline{z}_n \tag{7.6.19}$$

The roots of $R(z)$ are inside the unit circle. The set of stable polynomials $P(z) = (z^2+v_1 z+v_0)R(z)$ is given by eq. (7.6.18) with

$$\kappa_0 = v_0 - w_0 = v_0 - 1.25$$

$$\kappa_1 = v_1 - w_1 = v_1 - 1$$

The stability triangle of figure 7.1 with $p_0 = v_0$, $p_1 = v_1$ has the corners (v_{00}, v_{10}) for $v(z) = (z-1)^2$, (v_{01}, v_{11}) for $v(z) = (z-1)(z+1)$ and (v_{02}, v_{12}) for $v(z) = (z+1)^2$. In numerical values

$$\begin{bmatrix} v_{00} & v_{10} \\ v_{01} & v_{11} \\ v_{02} & v_{12} \end{bmatrix} = \begin{bmatrix} 1 & -2 \\ -1 & 0 \\ 1 & 2 \end{bmatrix}$$

in κ coordinates this is shifted to

$$\begin{bmatrix} \kappa_{00} & \kappa_{10} \\ \kappa_{01} & \kappa_{11} \\ \kappa_{02} & \kappa_{12} \end{bmatrix} = \begin{bmatrix} -0.25 & -3 \\ -2.25 & -1 \\ -0.25 & 1 \end{bmatrix}$$

All κ in this triangle, see figure 7.15, substituted into eq. (7.6.18) give a stable $\underline{p}$.

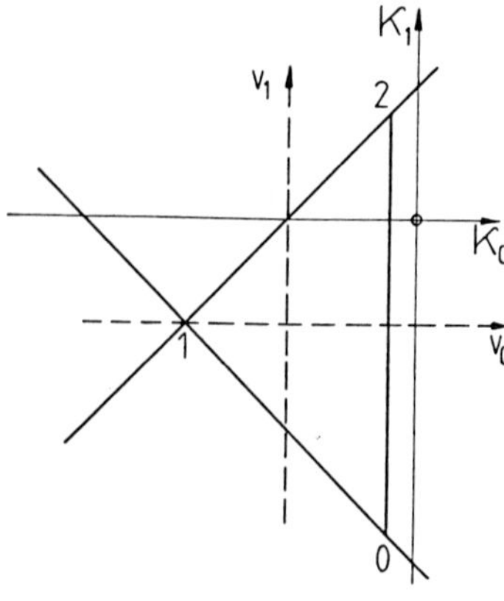

Figure 7.15

Stability triangle in an invariance plane.

If $\underline{R}(z)$ has unstable eigenvalues, then 012 is not the stability region but the region with the maximum number of eigenvalues inside the unit circle.

For general regions Γ in the complex plane we have to study the Γ-stability boundaries only for the second order polynomial $V(z)$. For a hyperbola, for example, the region in the V plane is P_Γ of figure 7.8. In the κ plane this region is shifted parallel to the coordinate axes.

The invariance plane can easily be mapped into the K space of state vector feedback. For this purpose $\underline{p}$ of eq. (7.6.18) is substituted into eq. (7.5.7)

$$\underline{k}' = [\underline{p}' \quad 1]\underline{E} \quad , \quad \underline{E} = \begin{bmatrix} \underline{e}' \\ \underline{e}'\underline{A} \\ \vdots \\ \underline{e}'\underline{A}^n \end{bmatrix} \tag{7.6.20}$$

Because of

$$[\underline{a}' \quad 1]\underline{E} = \underline{0} \tag{7.6.21}$$

eqs. (7.6.20) and (7.6.18) reduce to

$$\underline{k}' = [\kappa_0 \quad \kappa_1]\begin{bmatrix} \underline{r}' & 1 & 0 \\ 0 & \underline{r} & 1 \end{bmatrix}\begin{bmatrix} \underline{e}' \\ \underline{e}'\underline{A} \\ \vdots \\ \underline{e}'\underline{A}^{n-1} \end{bmatrix} \tag{7.6.22}$$

The $2\times n$ matrix

$$\underline{C}_R = \begin{bmatrix} \underline{r}' & 1 & 0 \\ 0 & \underline{r}' & 1 \end{bmatrix}\begin{bmatrix} \underline{e}' \\ \underline{e}'\underline{A} \\ \vdots \\ \underline{e}'\underline{A}^{n-1} \end{bmatrix} =: \begin{bmatrix} \underline{e}_R' \\ \underline{e}_R'\underline{A} \end{bmatrix} \tag{7.6.23}$$

must be a right factor of $\underline{k}' = [\kappa_0 \quad \kappa_1]\underline{C}_R$, i.e. it has the same meaning as an output matrix $\underline{y}_R = \underline{C}_R\underline{x}$ for two measured variables in $\underline{y}_R$. Here this output is artificially introduced and chosen such that the eigenvalues in $R(z)$ become unobservable and therefore unshiftable by output feedback $u = -[\kappa_1 \quad \kappa_2]\underline{y}_R$.

Eq. (7.6.23) is equivalent to eq. (2.6.53).

Essentially

$$\underline{e}_R' = [\underline{r}' \quad 1]\begin{bmatrix} \underline{e}' \\ \underline{e}'\underline{A} \\ \vdots \\ \underline{e}'\underline{A}^{n-2} \end{bmatrix} = \underline{e}'R(\underline{A}) \tag{7.6.24}$$

must be computed in this approach.

Example:

Consider the track-guided bus of appendix D.2 with velocity $v = 10\ ms^{-1}$, mass $m = 16\ 000\ kg$ and road adhesion coefficient $\mu = 0.5$. The characteristic polynomial of eq. (D.2.9) is

$$P_F(s) = s^2(s+4.7)(s^2+3.378s+3.192) \tag{7.6.25}$$

The last two eigenvalues at $s_{4,5} = -1.689 \pm j0.582$ are favorably located as well as the actuator pole at $s_3 = -4.7$. Thus we keep these three eigenvalues and shift only the eigenvalues $s_{1,2} = 0$. We want to shift them across the left branch of the hyperbola

$$\left(\frac{\sigma}{0.06}\right)^2 - \left(\frac{\omega}{0.12}\right)^2 = 1 \tag{7.6.26}$$

With $W(s) = s^2$, $w_0 = 0$, $w_1 = 0$, we have $\kappa_0 = v_0$, $\kappa_1 = v_1$ and the Γ-stability region is unshifted. Its boundaries were constructed in figure 7.6 b, where $a = 0.06$ and $b = 0.12$ must be substituted.

Choosing sensor coordinates, the state equations (D.2.13) are

$$\underline{F}_s = \begin{bmatrix} -1.452 & -0.313 & 1.589 & -1.589 & -19.165 \\ -0.426 & -1.926 & 2.118 & -2.118 & 4.394 \\ 1 & 0 & 0 & 0 & 0 \\ 0 & 1 & 0 & 0 & 0 \\ 0 & 0 & 0 & 0 & -4.7 \end{bmatrix}, \quad \underline{g}_s = \begin{bmatrix} 0 \\ 0 \\ 0 \\ 0 \\ 4.7 \end{bmatrix}$$

$$\underline{e}'[\underline{g}_s \quad \underline{F}_s\underline{g}_s \quad \underline{F}_s^2\underline{g}_s \quad \underline{F}_s^3\underline{g}_s \quad \underline{F}_s^4\underline{g}_s] = [0 \quad 0 \quad 0 \quad 0 \quad 1]$$

$$\underline{e}'\begin{bmatrix} 0 & -90.076 & -547.68 & -2930.13 & 14577.5 \\ 0 & 20.652 & -98.47 & 178.01 & 129.8 \\ 0 & 0 & -90.08 & 547.68 & -2930.1 \\ 0 & 0 & 20.65 & -98.47 & 178.0 \\ 4.7 & 0 & 0 & 0 & 1 \end{bmatrix} =$$

$$= [0 \quad 0 \quad 0 \quad 0 \quad 1]$$

$$\underline{e}' = 10^{-4}[-0.674 \quad -2.938 \quad -9.060 \quad -35.66 \quad 0] \qquad (7.6.27)$$

The fixed polynomial is

$$R(s) = (s+4.7)(s^2+3.378s+3.192)$$

$$= s^3+8.078s^2+19.069s+15.002$$

Then by eq. (7.6.24)

$$\underline{e}_R' = [\underline{r}' \quad 1]\begin{bmatrix} \underline{e}' \\ \underline{e}'\underline{F}_s \\ \underline{e}'\underline{F}_s{}^2 \\ \underline{e}'\underline{F}_s{}^3 \end{bmatrix}$$

$$\underline{e}_R' = 10^{-4}[15.002 \quad 19.069 \quad 8.078 \quad 1] \times$$

$$\times \begin{bmatrix} -0.674 & -2.938 & -9.060 & -35.664 & 0 \\ -6.831 & -29.794 & -7.293 & 7.293 & 0 \\ 15.318 & 66.815 & -73.959 & 73.959 & 0 \\ -124.664 & -59.521 & 165.855 & -165.855 & 0 \end{bmatrix}$$

$$\underline{C}_R = \begin{bmatrix} \underline{e}_R' \\ \underline{e}_R'\underline{F}_s \end{bmatrix} = 10^{-4}\begin{bmatrix} -141.28 & -132.02 & -706.57 & 35.622 & 0 \\ -445.19 & 334.11 & -504.11 & 504.11 & 2127.68 \end{bmatrix}$$

(7.6.28)

$\underline{C}_R$ is constructed such that the three eigenvalues in R(s) are unobservable from an artificial output $\underline{y}_R = \underline{C}_R\underline{x}_s$. The output feedback structure is

$$u = -[\kappa_1 \quad \kappa_1]\underline{y}_R = -[\kappa_1 \quad \kappa_2]\underline{C}_R\underline{x}_s \tag{7.6.29}$$

A different procedure for shifting only two eigenvalues was proposed by Türk [82.3]. By a transformation $\underline{T}$ the system is transformed to

$$\underline{A}^* = \underline{TAT}^{-1} = \begin{bmatrix} \underline{A}^*_{11} & \underline{A}^*_{12} \\ \underline{0} & \underline{A}^*_{22} \end{bmatrix}, \quad \underline{b}^* = \underline{Tb} \tag{7.6.30}$$

where $\underline{A}^*_{11}$ is an $(n-2)\times(n-2)$ matrix with the fixed eigenvalues. Only the two eigenvalues of $\underline{A}^*_{22}$ are to be shifted. This is achieved if feedback changes only the last two columns of $\underline{A}^*$, i.e. $\underline{A}^*_{12}$ and $\underline{A}^*_{22}$. A feasible feedback is

$$\underline{k}^{*\prime} = [\kappa^*_1 \quad \kappa^*_2]\left[\begin{array}{ccc:cc} 0 & \dots & 0 & 1 & 0 \\ 0 & \dots & 0 & 0 & 1 \end{array}\right]$$

Then

$$\underline{A}^* - \underline{b}^*\underline{k}^{*\prime} = \begin{bmatrix} \underline{A}^*_{11} & \underline{G} \\ \underline{0} & \underline{H} \end{bmatrix} \tag{7.6.31}$$

In the original coordinates $\underline{k}' = \underline{k}^{*\prime}\underline{T} = [\kappa_1 \quad \kappa_2]\underline{C}_R$ with

$$\underline{C}_R = \left[\begin{array}{ccc:cc} 0 & \dots & 0 & 1 & 0 \\ 0 & \dots & 0 & 0 & 1 \end{array}\right]\underline{T} \tag{7.6.32}$$

$\underline{C}_R$ is the fictitious output matrix such that $\underline{A}^*_{11}$ is not observable from $\underline{y}_R = \underline{C}_R\underline{x}$. $\underline{C}_R$ consists of the last two rows of the transformation matrix $\underline{T}$ that transforms $\underline{A}$ to $\underline{A}^*$ as in eq. (7.6.30). Feasible forms of $\underline{A}^*$ are Jordan form or the real Schur form [80.4], [81.16].

Note that $\underline{C}_R$ of eq. (7.6.32) is not identical to $\underline{C}_R$ of eq. (7.6.23), but the rows of the two matrices span the same plane. The Γ stability boundaries in κ_1-κ_2-plane are then obtained by eq. (7.5.21).

That is

$$[\underline{a}' + [\kappa_1 \quad \kappa_2]\underline{C}_R\underline{W} \quad 1]\underline{a}_L = 0$$

$$[\underline{a}' + [\kappa_1 \quad \kappa_2]\underline{C}_R\underline{W} \quad 1]\underline{a}_R = 0 \qquad (7.6.33)$$

$$[\underline{a}' + [\kappa_1 \quad \kappa_2]\underline{C}_R\underline{W} \quad 1][\underline{c}_1(\alpha) \quad \underline{c}_2(\alpha)] = [0 \quad 0]$$

7.7 Exercises

7.1 For a polynomial $P(z) = p_0 + p_1 z + z^2$, plot the region P_Γ in the p_0-p_1-plane such that all polynomials in P_Γ have complex conjugate roots inside the unit circle. (Hint: The boundary case between two real roots and a complex pair is a double real root).

7.2 For a polynomial $P(z) = p_0 + p_1 z + p_2 z^2 + z^3$ show that the critical complex root condition $\det(\underline{X}-\underline{Y}) > 0$ of the Schur-Cohn-Jury test is equivalent to the Hurwitz condition $\Delta_2 > 0$ for the polynomial transformed by $z = (1+w)/(1-w)$.

7.3 For $P(z) = p_0 + p_1 z + p_2 z^2 + p_3 z^3 + p_4 z^4 + z^5$ formulate the necessary conditions which result from the convex hull of the stability region.

7.4 For $P(z) = p_0 + p_1 z + z^2$ plot the Γ-stability triangles for the family of circles of eq. (7.4.10).

7.5 The roots of a polynomial $P(z) = p_0 + p_1 z + z^2$, $z = \tau + j\eta$, are to be located inside the ellipse $\tau^2 + (\eta/1.5)^2 = 1$. Plot the Γ stability region in p_0-p_1-plane

a) by transformation of the polynomial (see eq. (7.4.18)),

b) pointwise (see eq. (7.4.30)).

7.6 Plot the root locus for the example of eq. (7.5.27).

7.7 For the example "discrete triple integrator" of eq. (7.5.14) and $\partial\Gamma = e^{-\alpha} \times e^{\pm j\alpha}$ fix k_1 such that the resulting k_2-k_3 plane contains the deadbeat point. Plot the Γ-stability region in this plane. Compare the result with figure 7.13.

7.8 A dc motor with state vector $\underline{x} = [\alpha \quad \omega \quad i]'$ (α = rotation angle, ω = angular velocity, i = current) with a load torque proportional to ω, has a state representation

$$\dot{\underline{x}} = \begin{bmatrix} 0 & 1 & 0 \\ 0 & -2 & 1 \\ 0 & -1 & -1 \end{bmatrix} \underline{x} + \begin{bmatrix} 0 \\ 0 \\ 1 \end{bmatrix} u$$

Find the set of all state feedbacks $u = -x_1 - k_2 x_2 - k_3 x_3$ such that the closed loop has one negative real eigenvalue and two complex conjugate eigenvalues with damping $\geq 1/\sqrt{2}$.

7.9 The loading bridge of chapter 2 is controlled in discrete time with a sampling interval $T = \pi/8$. The same controller structure is assumed as in the continuous time case, eq. (7.6.6). Let m_C = 1000 kg, m_L = 3000 kg, ℓ = 10 m, g = 10 m/s². In the k_2-k_3 plane plot the region for which the closed-loop eigenvalues are located in a circle with radius 0.5 and center z = 0.45.

8 Design of Robust Control Systems

8.1 Robustness Problems

For most plants the mathematical model is not known exactly or it is too complicated for the controller design (for example it may be nonlinear). The usual procedure is to design the controller with a simplified model and nominal values of the plant parameters. An important design goal is therefore to reduce the influence of parameter uncertainty, neglected dynamics, and nonlinearity on the dynamics of the closed loop. A survey on the origin of uncertainty in modelling and model simplifications and on design of robust, adaptive and intelligent control systems is given in [85.1].

8.1.1 Sensitivity

Some basic design problems for a control loop with unit feedback and the transfer function $KG_0(s)$ in the forward path will now be explained. The closed-loop transfer function of the system is $KG_0(s)/[1+KG_0(s)]$. For large K this approximates the ideal value of unity and, further, the perturbations of $G_0(s)$ have little influence as long as the loop remains nicely stable. Stability for $K \to \infty$ can be achieved only if $G_0(s)$ has no zeros in the right half s-plane and if the pole excess of $G_0(s)$ is at most two, as can be seen from the asymptotes of the root locus. This high-gain approach can easily lead to instability due to model uncertainty at high frequencies, for example due to neglected actuator dynamics or structural vibrations. High loop gains are recommend only within the desired bandwidth of the control system. For high frequencies the loop gain should be small and decrease as $1/s^2$. The stability margin is determined in the transition between these frequency bands (see figure 3.24 for the discrete-time case) and acceptable margins can be secured by a variety of design methods. The following factors play a role in the determination of the bandwidth:

- Bandwidth of the uncontrolled plant
- Bandwidth of the reference and disturbance inputs
- Limitation of the bandwidth below the frequency of the structure vibrations in mechanical systems
- Limitation of the bandwidth by the actuator.

The model uncertainty may be of different types. If very little is known about the uncertainty, then one of the structures of figure 8.1 can be assumed, where P is a nominal plant and D is a deviation or perturbation from it. D may be a transfer function or a nonlinearity.

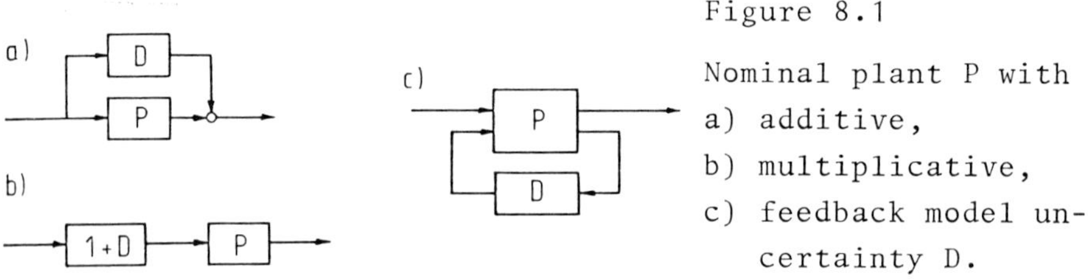

Figure 8.1

Nominal plant P with
a) additive,
b) multiplicative,
c) feedback model uncertainty D.

The usual assumption is that D is small in some sense. The controller is primarily designed for the nominal plant P. Typically also a desired nominal closed-loop behavior is specified (e.g. reference model, reference trajectory). An additional design goal is then to have a small influence of D on the deviations from the nominal behavior. The low sensitivity is a local property in the neighborhood of D = 0. In some publications this design for low sensitivity is also called robust control system design. We will use this expression in a different sense as will be explained in the next section.

8.1.2 Robustness

In a different class of problems the structure of the mathematical model is known and some physical parameters in it may vary. As an example consider the loading bridge of eq. (2.1.15). The rope length ℓ enters only into f_{43} and g_4 as a factor $1/\ell$. The rope length can vary between its minimum and maximum values. It is not sensible to imbed this problem into one where f_{43} and g_4 can vary independently between their minimum and maximum values. Similarly the load mass m_L enters only into f_{23} and f_{43} and varies independently of ℓ. In a plane of physical parameters we have the rectangle of figure 8.2.

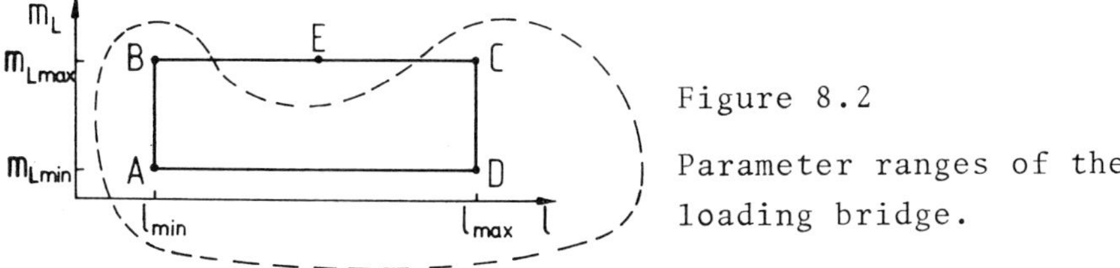

Figure 8.2

Parameter ranges of the loading bridge.

We can assume that the trolley mass m_C does not vary. Similarly in the track-guided bus example of section D.2 a rectangular parallelepiped is obtained. Its corners are determined by the minimum and maximum values of the speed, load and road adhesion coefficients.

In the aircraft example of section D.1 the ranges of the physical parameters speed and altitude are not independent because the aircraft cannot fly slow in thin air at high altitude and it cannot fly fast in thick air at low altitude. The possible combinations in stationary flight are given by the flight envelope of figure D.2.

A robust controller design tries to achieve acceptable closed-loop properties for the entire range of possible parameters. For this problem it is not necessary to select a nominal parameter set (operating condition). Also the requirement of a nominal behavior may be too stringent. The pilot expects that the aircraft shows a faster reaction in terrain following than in landing approach. This is also reflected in military specifications which give different ranges for the natural frequency of the short period mode for these two operating conditions. Figure 8.3 shows a result for the loading bridge which will be derived later.

Γ stability is defined here by the region to the left of the left branch of the hyperbola $\omega^2 = 4\sigma^2 - 0.25$. The loading bridge is Γ-stabilized for the range of load masses from 50 kg to 2395 kg. The rope length is fixed.

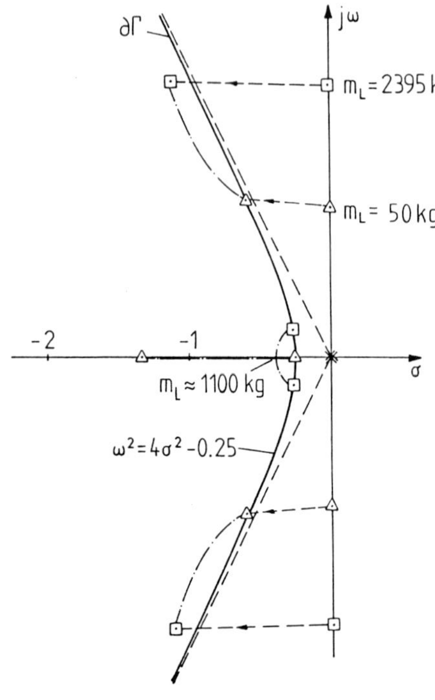

Figure 8.3

Loading bridge. The fast system remains fast, the slow system remains slow.

On the imaginary axis figure 8.3 shows the open-loop eigenvalues. The uncontrolled system is faster for larger load m_L, correspondingly also the controlled system is faster, i.e. it has eigenvalues further removed from the origin. In this design the complex eigenvalues are essentially moved left in order to give them sufficient damping. No attempt is made to give them also the same frequency. The other two eigenvalues move along the dotted root locus for varying load mass. At $m_L \approx 1100$ kg the root locus branches off the real axis. Around this value the eigenvalue location is very sensitive to changes in load mass, but this is not considered a disadvantage in robust design as long as all eigenvalues stay on the proper side of the boundary $\partial\Gamma$.

A definition of robustness of a control system has two essential ingredients:

a) a system property (e.g. Γ-stability, a stability margin in frequency domain, maximum overshoot or stationary accuracy of the step response), and

b) a class of perturbations against which the system property is robust (e.g. uncertain physical parameters, neglected actuator dynamics and nonlinearity, modelling uncertainty, non-ideal controller implementation, sensor or actuator failure).

The main emphasis of this chapter is the design of control systems for robustness of Γ-stability with respect to uncertain physical parameters and sensor failures.

8.1.3 Multi-Model Problem Formulation

Frequently a nonlinear plant is linearized for small deviations from a stationary state. Then a linear differential equation

$$\underline{\dot{x}} = \underline{A}(\underline{\Theta})\underline{x} + \underline{b}(\underline{\Theta})u$$

or difference equation

$$\underline{x}[k+1] = \underline{A}(\underline{\Theta})\underline{x}[k] + \underline{b}(\underline{\Theta})u[k] \tag{8.1.1}$$

is obtained in which the dynamics matrix $\underline{A}$ and the input vector $\underline{b}$ depend on the physical parameter vector $\underline{\Theta}$. An admissible region Θ is known for $\underline{\Theta}$. Now $\underline{W}$ and $\underline{a}$ in eqs. (7.5.4) and (7.5.5) also depend on $\underline{\Theta}$. The Leverrier algorithm of eq. (A.7.36) is suitable for such computations with nonnumerical variables $\underline{\Theta}$. With a state feedback controller the coefficient vector of the characteristic polynomial becomes

$$\underline{p}'(\underline{k},\underline{\Theta}) = \underline{a}'(\underline{\Theta}) + \underline{k}'\underline{W}(\underline{\Theta}) \tag{8.1.2}$$

The boundaries of nice stability regions described by eq. (7.5.13) are characterized by

$$\begin{aligned} [\underline{p}'(\underline{k},\underline{\Theta}) \quad 1]\, \underline{a}_L &= 0 \\ [\underline{p}'(\underline{k},\underline{\Theta}) \quad 1]\, \underline{a}_R &= 0 \\ [\underline{p}'(\underline{k},\underline{\Theta}) \quad 1][\underline{c}_1 \quad \underline{c}_2] &= [0 \quad 0] \end{aligned} \tag{8.1.3}$$

The general parameter space is now the combined space of $\underline{k}$ and $\underline{\Theta}$. Eq. (8.1.3) generally describes a nonlinear mapping and the nice properties of the affine mapping are lost. For example the first two real root hyperplanes in P space remain hyperplanes in K space only for constant $\underline{\Theta}$, but they do not remain hyperplanes in the general parameter space. A useful procedure is to discretize $\underline{\Theta}$, i.e. to represent Θ by some typical and extremal values $\underline{\Theta}_i \in \Theta$. In some cases - as in the aircraft example of appendix D.1 - the pair $\underline{A}$, $\underline{b}$ is known only for such discrete values of $\underline{\Theta}$:

$$\underline{A}_j = \underline{A}(\underline{\Theta}_j) \ , \ \underline{b}_j = \underline{b}(\underline{\Theta}_j) \ , \ \underline{\Theta}_j \in \Theta \ , \ j = 1,2 \ldots J \qquad (8.1.4)$$

We call this a "multi-model problem". $\underline{\Theta}$ may also be a variable, which admits only discrete values, e.g. for a nominal and for a failed component in the system.

The design is then carried out as before in the K space, but now Γ-stability must be achieved simultaneously for all members of the plant family (8.1.4). In the loading bridge example of figure 8.2 the four extremal points A, B, C and D may be used in this first step. Once a controller $\underline{k}'$ is fixed, we may also use eq. (8.1.3) to map the boundaries $\partial\Gamma$ into the $\underline{\Theta}$-space or a cross section of it. Assume for the loading bridge that this Γ-stability boundary is the dotted line in figure 8.2. By construction it must include the points A, B, C, D. But it does not necessarily contain the entire rectangle. In this example the model family should be augmented by the point E for a redesign.

It is advantageous to formulate the problem such that the output matrix $\underline{C}$ in $\underline{y} = \underline{Cx}$ is independent of $\underline{\Theta}$. The easiest way to achieve this is to use sensor coordinates, i.e. the measured variables are state variables. In the loading bridge example the measurable variables x_1 = trolley position and x_3 = cable angle were introduced as state variables. If one alternatively uses the position and velocity of the common center of gravity of the trolley and load as state variables (see exercise 2.1) then the $\underline{A}$-matrix is simplified to a block diagonal form. But this has the disadvantage that $\underline{C}$ is now dependent on $\underline{\Theta}$.

In robust control design we are interested especially in constant controllers whose structure and parameters are independent of $\underline{\Theta}$. In the following procedure first a structure of the controller is assumed and then the controller parameters in this structure are determined so that the controller simultaneously gives a satisfactory response to all the members of the family of system models $\underline{A}_j$, $\underline{b}_j$. Assume for example state feedback $u = -\underline{k}'\underline{x}$ and describe the required system property by an eigenvalue region Γ. The problem is then to find a $\underline{k}'$ such that all zeros of

$$P_J(z) = \prod_{j=1}^{J} \det(z\underline{I}-\underline{A}_j+\underline{b}_j\underline{k}') \quad , \quad j = 1,2...J \tag{8.1.5}$$

lie in Γ. In order to obtain a global picture, we are especially concerned to determine the set of all such $\underline{k}'$. The basic solution idea uses the pole region assignment introduced in chapter 7. Each pair $(\underline{A}_j, \underline{b}_j)$ generates a corresponding pole assignment matrix $\underline{E}_j$. This represents a mapping $\underline{k}' = [\underline{p}' \quad 1]\underline{E}_j$ from P space to K space. Figure 8.4 shows this mapping schematically for two plant models with the pole assignment matrices $\underline{E}_1$ and $\underline{E}_2$.

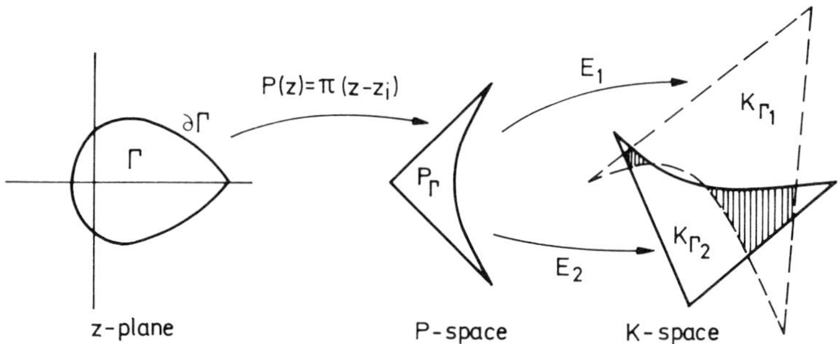

Figure 8.4 Pole region assignment for two system models with the pole assignment matrices $\underline{E}_1$ and $\underline{E}_2$

The boundaries of P_Γ are two hyperplanes for the real eigenvalues and the complex root boundary surface. P_Γ is mapped into K_{Γ_1} by $\underline{E}_1$ and into K_{Γ_2} by $\underline{E}_2$. The set of all feedback gains which shift the eigenvalues of both systems into the region Γ is the cross-hatched intersection set. Since P_Γ and therefore $K_{\Gamma i}$ need not be convex, the intersection set of K_{Γ_1} and K_{Γ_2} can consist of several disjoint components, as in the represented case. Similarly further plant models and corresponding pole-assignment matrices $\underline{E}_3$, $\underline{E}_4$ ... may be included. In sections three to five of this chapter the simultaneous pole-region assignment for a family of models is executed further by parameter-space methods for the design of robust control systems. In section 8.6 the problem is solved by optimization of a vector performance criterion. In the next section some questions of structure and existence of controllers for simultaneous pole-region assignment are discussed. The engineer, who is mainly interested in the practical application, may proceed directly to section 8.3.

8.2 Structural Assumptions and Existence of Robust Controllers

The structure of controllers for one individual controllable plant model ($\underline{A}$, $\underline{b}$) has already been discussed in chapter 6. For a controllable and observable system all eigenvalues can be placed arbitrarily. Therefore, the problem of pole-region assignment always has solutions. This no longer holds for the simultaneous pole region assignment for a family of models. It is an open question for which assumptions the problem can be solved by dynamic state vector feedback. For the practical cases studied so far good results were obtained by proportional state feedback, and feedback dynamics were needed only in order to reconstruct non-measured states. The following examples should illustrate some non-trivial effects and difficulties in the simultaneous stabilization problem for two plant models. Constructive procedures giving the order and parameters of a simultaneous Γ-stabilizer for a given family of plant models are not yet available.

8.2.1 Examples for Simultaneous Stabilization

Example 1: $n = 1$, Γ = left half plane

$$x_s = \frac{b_j}{s-a_j} u_s \ , \quad u_s = k_s x_s \ , \quad j = 1,2 \tag{8.2.1}$$

$$k_s = \frac{C(s)}{D(s)} = \frac{c_o + c_1 s + \ldots\ldots + c_m s^m}{d_o + d_1 s + \ldots + d_{m-1} s^{m-1} + s^m}$$

The closed-loop characteristic polynomial is

$$P_j(s) = b_j C(s) + (s-a_j) D(s) = p_{jo} + p_{j1} s + \ldots + p_{jm} s^m + s^{m+1}$$

$$\begin{aligned}
p_{jo} &= b_j c_o - a_j d_o \\
p_{j1} &= b_j c_1 - a_j d_1 + d_o \\
&\vdots \\
p_{jm-1} &= b_j c_{m-1} - a_j d_{m-1} + d_{m-2} \\
p_{jm} &= b_j c_m - a_j + d_{m-1}
\end{aligned}$$

Static state feedback ($m = 0$, $k_s = k = c_o, d_o = 1$) stabilizes for $p_{jo} = b_j k - a_j > 0$ for all j. Now assume

a) there does not exist such k,

b) there does exist a stabilizing controller of order $m > 0$, i.e. $p_{ji} > 0$ for all j and i.

It will be shown that this leads to a contradiction, i.e. feedback dynamics do not help in this case.

$p_{jm} > 0 \Rightarrow b_j c_m - a_j > - d_{m-1}$

By assumption a) there does not exist $k = c_m$ such that $b_j c_m - a_j > 0$, thus $d_{m-1} > 0$.

$p_{jm-1} > 0 \Rightarrow b_j \dfrac{c_{m-1}}{d_{m-1}} - a_j > - \dfrac{d_{m-2}}{d_{m-1}}$

By the same conclusion $d_{m-2} > 0$ and so on until $d_o > 0$

$p_o > 0 \Rightarrow b_j \dfrac{c_o}{d_o} - a_j > 0$ contradicts assumption a).

Example 2: $n = 1$, Γ = unit circle

The result of example 1 does not carry over to the discrete-time case. This is shown by the following example (Kraus, Mansour, unpublished):

In example 1 let $s = z$, $a_1 = a_2 = 1.5$, $b_1 = 1$, $b_2 = 0.2$.

a) $m = 0$ $\quad P_1(z) = z - 1.5 + k$, stable for $0.5 < k < 2.5$

$\quad P_2(z) = z - 1.5 + 0.2k$, stable for $2.5 < k = 12.5$

No simultaneous stabilization possible.

b) $m = 1$ $\quad k_z = \dfrac{1.55 + 2z}{0.4 + z}$ (8.2.3)

stabilizes both plants. (Exercise: Plot the root loci for a) and b) with parameter b_j).

(This corrects an error in the example of the German edition).

For the simple case of first-order systems it is possible to solve a more general problem than the two-model stabilization. Consider

$$
\begin{aligned}
x[k+1] &= ax[k] + bu[k] \\
u[k] &= -kx[k]
\end{aligned}
\tag{8.2.4}
$$

where a and b belong to a given set Ω which may be described as a region in the a-b-plane. This set of plants can be simultaneously stabilized if there exists a feedback gain k such that $|a-bk| < 1$ for all $(a, b) \in \Omega$. The conditions

$$
-1+bk < a < 1+bk \tag{8.2.5}
$$

are satisfied between to parallel lines through a = -1, b = 0, and a = 1, b = 0 with slope 1/k. A simultaneously stabilizing k exists for the set Ω shown in figure 8.5 a. This result for discrete-time systems [84.9] corresponds to a result of Saeks and Murray [82.2] for continuous-time systems.

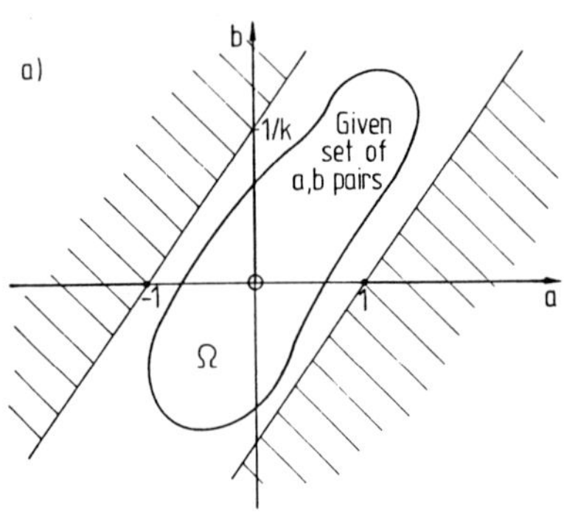

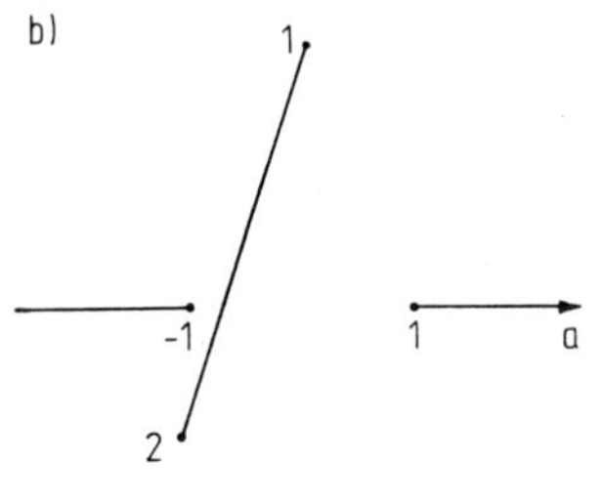

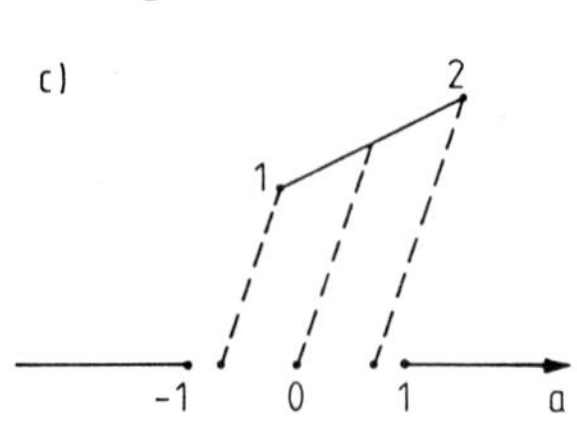

Figure 8.5

a) A given set of a, b pairs can be simultaneously stabilized if there exists a finite k such that the set is enclosed between the two straight lines through a = -1 and a = 1 with slope 1/k.

b) Straight line through 1 and 2 intersects b = 0, $|a| < 1$.

c) Projection of 12 is in b = 0, $|a| < 1$.

Figure 8.5 a not only describes an existence result, but it illustrates a graphical procedure for finding a feasible k if it exists. The solution to the two-model problem of eq. (8.2.1) is contained in the above result as a special case. Figure 8.5 b shows the case of different signs of b_1 and b_2. For simultaneous stabilization the straight line through 1 and 2 must intersect $b = 0$, $|a| < 1$. This is true for $|\alpha_-| < 1$, where

$$\alpha_- = \frac{1}{b_2 - b_1} \det \begin{bmatrix} a_1 & b_1 \\ a_2 & b_2 \end{bmatrix} \tag{8.2.6}$$

Figure 8.5 c shows the case of identical signs of b_1 and b_2. Here the projection of the line segment 12 must be in $b = 0$, $|a| < 1$, where the direction of projection is given by the connecting line between $a = b = 0$ and the center of 12. This is true for $|\alpha_+| < 1$ where

$$\alpha_+ = \frac{1}{b_2 + b_1} \det \begin{bmatrix} a_1 & b_1 \\ a_2 & b_2 \end{bmatrix} \tag{8.2.7}$$

The result is: a necessary and sufficient condition for simultaneous stabilizability of (a_1, b_1) and (a_2, b_2) is that

$$\frac{1}{|b_2| + |b_1|} \left| \det \begin{bmatrix} a_1 & b_1 \\ a_2 & b_2 \end{bmatrix} \right| < 1 \tag{8.2.8}$$

Now consider second-order systems. A generalization of the solution for a set of plants would be complicated. The "diaphragm" $b = 0$, $|a| > 1$ of non-stabilizable plants generalizes to non-stabilizable surfaces in the space of the six parameters of $(\underline{A}, \underline{b})$, [84.9]. It is easy, however, to represent stability regions in the plane of the two characteristic-polynomial coefficients or state-feedback gains.

For second-order systems it does not suffice to consider proportional state feedback as in the first-order case. The following example shows, that the introduction of dynamics in state feedback can allow simultaneous stabilization where it is impossible by proportional state feedback.

Example 3:

$$\underline{x}[k+1] = \underline{A}_j\underline{x}[k] + \underline{b}_j u[k] \ , \ j = 1,2$$

$$\underline{A}_1 = \begin{bmatrix} 0 & 1 \\ -1.5 & 0 \end{bmatrix}, \quad \underline{b}_1 = \begin{bmatrix} 0 \\ 1 \end{bmatrix} \tag{8.2.9}$$

$$\underline{A}_2 = \underline{A}_1 \ , \quad \underline{b}_2 = -\underline{b}_1$$

The pole assignment matrices are

$$\underline{E}_1 = \begin{bmatrix} 1 & 0 \\ 0 & 1 \\ -1.5 & 0 \end{bmatrix}, \quad \underline{E}_2 = \begin{bmatrix} -1 & 0 \\ 0 & -1 \\ 1.5 & 0 \end{bmatrix}$$

They map the stability triangle of figure 7.1 into the two triangles of figure 8.6 for state feedback $u = -[k_1 \quad k_2]\underline{x}$.

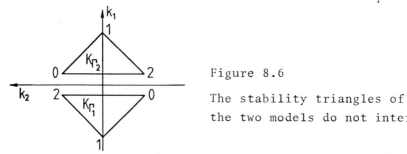

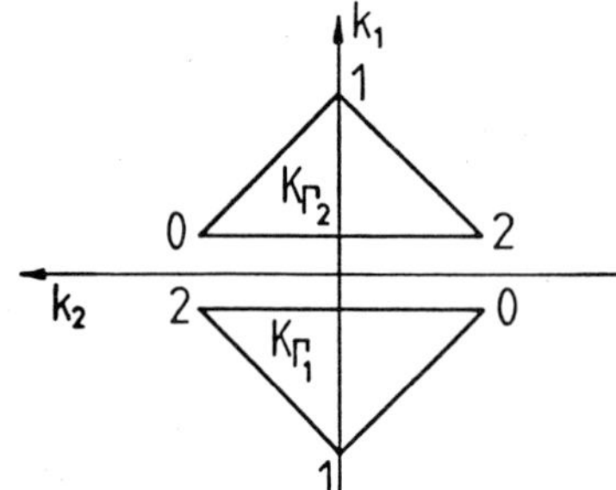

Figure 8.6

The stability triangles of the two models do not intersect

Now introduce a controller state $x_0[k+1] = x_1[k]$ and augment the feedback law to $u = -[k_0 \quad k_1 \quad k_2]\underline{x}$. The additional feedback parameter k_0 admits a simultaneous stabilization of the two systems

$$\underline{A}_1 = \begin{bmatrix} 0 & 1 & 0 \\ 0 & 0 & 1 \\ 0 & -1.5 & 0 \end{bmatrix}, \quad \underline{b}_1 = \begin{bmatrix} 0 \\ 0 \\ 1 \end{bmatrix} \tag{8.2.10}$$

$$\underline{A}_2 = \underline{A}_1 \ , \quad \underline{b}_2 = -\underline{b}_1$$

Stability region is now the tetrahedron with saddle surface

of figure 7.2. This is mapped into K space by the pole assignment matrices

$$\underline{E}_1 = \begin{bmatrix} 1 & 0 & 0 \\ 0 & 1 & 0 \\ 0 & 0 & 1 \\ 0 & -1.5 & 0 \end{bmatrix}, \quad \underline{E}_2 = -\underline{E}_1$$

The upper part of figure 8.7 shows the two stability regions.

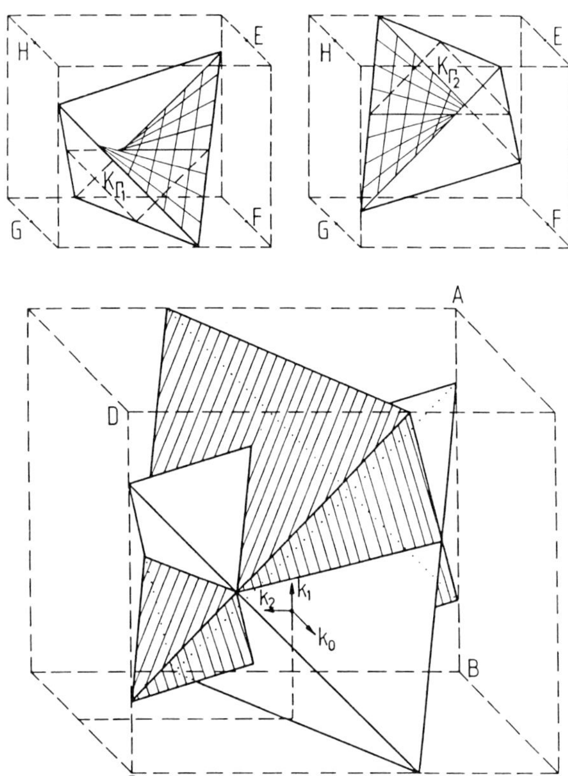

Figure 8.7 Upper part: Two stability regions K_{Γ_1} and K_{Γ_2}
Lower part: The intersection of the convex hulls of K_{Γ_1} and K_{Γ_2}.

K_{Γ_1} is below the saddle surface of the left figure, K_{Γ_2} is above the saddle surface of the right figure. The lower part shows the intersection of the convex hulls, i.e. of the two tetrahedra. $\underline{E}_1$ produces the white tetrahedron with the edge for (z+1)(z-1) in the bottom of the dotted box (i.e. same orientation as in figure 7.2). $\underline{E}_2$ produces the shaded tetrahedron which is upside down, i.e. the edge for (z+1)(z-1) is in the top of the dotted box. The two triangles of figure 8.6 lie in the cross section EFGH for $k_o = 0$ as indicated in the upper part of figure 8.7. Using k_o in the feedback law the cross section plane can be tilted. The cross section $k_2 = 3k_o$ which passes diagonally through the dotted enclosing box (i.e. through the corners ABCD), is promising. This is obtained by substituting

$$\underline{k}' = [k_o \quad k_1 \quad 3k_o] = [k_o \quad k_1]\begin{bmatrix}1 & 0 & 3\\ 0 & 1 & 0\end{bmatrix}$$

into $\underline{p}' = \underline{a}' + \underline{k}'\underline{W}$ with $\underline{a}' = [0 \quad 1.5 \quad 0]$ and $\underline{W}_1 = \underline{I}$, $\underline{W}_2 = -\underline{I}$, i.e.

$$\underline{p}_1' = [\;k_o \quad 1.5+k_1 \quad 3k_o]$$
$$\underline{p}_2' = [-k_o \quad 1.5-k_1 \quad -3k_o]$$

This in turn is substituted into the eqs. (7.4.21) for the real boundaries and (7.4.30) for the complex boundary.

$$[\underline{p}' \quad 1]\begin{bmatrix}1\\-1\\1\\-1\end{bmatrix} = 0 \;, \quad [\underline{p}' \quad 1]\begin{bmatrix}1\\1\\1\\1\end{bmatrix} = 0$$

As in example (7.5.22) we may use

$$[\underline{p}' \quad 1]\begin{bmatrix}1 & 2\tau\\ 0 & 1\\ -1 & 0\\ -2\tau & -1\end{bmatrix} = [0 \quad 0]$$

For $\underline{A}_1$, $\underline{b}_1$ and the resulting $\underline{p}_1'$ the boundaries are:

real: $-2.5 + 4k_o - k_1 = 0$

$\quad\quad 2.5 + 4k_o + k_1 = 0$

complex: $-2k_o - 2\tau = 0$, $0.5 + 2\tau k_o + k_1 = 0$

i.e. $k_o = -\tau$, $k_1 = -0.5 + 2\tau^2$

and for $\underline{A}_2$, $\underline{b}_2$ and $\underline{p}_2'$

real: $-2.5 - 4k_o + k_1 = 0$
$2.5 - 4k_o - k_1 = 0$

complex: $2k_o - 2\tau = 0$, $0.5 - 2\tau k_o - k_1 = 0$

i.e. $k_o = \tau$, $k_1 = 0.5 - 2\tau^2$

The boundaries are plotted in figure 8.8. A simultaneous stabilization is obtained in the two regions framed by shading.

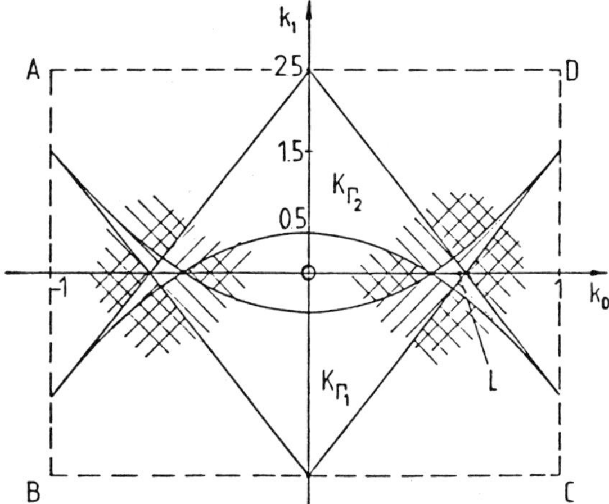

Figure 8.8 Two intersections of K_{Γ_1} and K_{Γ_2} around the intervals $-0.625 < k_o < -0.5$ and $0.5 < k_o < 0.625$.

For example with $\underline{k}' = [0.6 \quad 0 \quad 1.8]$ (point L in figure 8.8) the characteristic polynomials are

$$P_1(z) = 0.6 + 1.5z + 1.8z^2 + z^3$$
$$= (z+0.868)(z^2+0.932z+0.691)$$

$$P_2(z) = -0.6 + 1.5z - 1.8z^2 + z^3$$
$$= (z-0.868)(z^2-0.932z+0.691)$$

Both systems of eq. (8.2.4) are stabilized by the same first order dynamic feedback $u[k] = -1.8x_2[k]-0.6x_1[k-1]$.

The last example shows that there are cases where simultaneous stabilization is impossible by proportional state feedback, but is possible by dynamic state feedback. This poses several questions: when does simultaneous stabilization become possible with dynamics? What is the minimum order of the controller? How do we systematically find its parameters? Complete answers to these questions are not yet available, but we can make some further useful observations in the next example.

Example 4:

Imbed example 3 into

$$\underline{A}_1 = \begin{bmatrix} 0 & 1 \\ -a & 0 \end{bmatrix} , \quad \underline{b}_1 = \begin{bmatrix} 0 \\ 1 \end{bmatrix} \tag{8.2.11}$$

$$\underline{A}_2 = \underline{A}_1 \quad , \quad \underline{b}_2 = -\underline{b}_1$$

Figures 8.6, 8.7 and 8.8 were constructed for $a = 1.5$. Increasing parameter value a moves K_{Γ_1} down and K_{Γ_2} up. For $a > 3$ the convex hulls of K_{Γ_1} and K_{Γ_2} do not intersect and the first order controller can no longer stabilize both plants. However, a second order controller can still work. Assume the controller has the structure of figure 8.9.

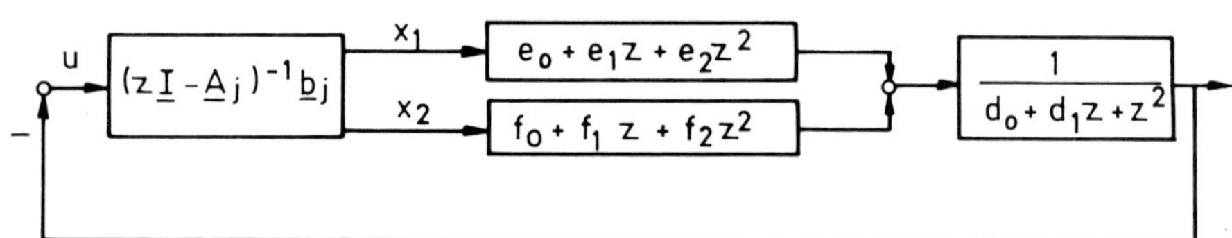

Figure 8.9 Second order controller for second order plant.

The plant is described in frequency domain by

$$\underline{x}_z = (z\underline{I}-\underline{A}_j)^{-1}\underline{b}_j u_z \quad , \quad j = 1,2$$

The transfer functions are

$$(z\underline{I}-\underline{A}_1)^{-1}\underline{b}_1 = \frac{1}{z^2+a}\begin{bmatrix}1\\ z\end{bmatrix} \tag{8.2.12}$$

The closed loop characteristic polynomial is then

$$P(z) = (z^2+a)(d_o+d_1z+z^2) \pm [(e_o+e_1z+e_2z^2) + z(f_o+f_1z+f_2z^2)]$$

where the "+" applies to $p_1(z)$ and the "-" to $p_2(z)$. Equating coefficients with the following general polynomials

$$P_1(z) = p_{10} + p_{11}z + p_{12}z^2 + p_{13}z^3 + z^4$$

$$P_2(z) = p_{20} + p_{21}z + p_{22}z^2 + p_{23}z^3 + z^4$$

yields a set of linear equations $\underline{Mv} = \underline{q}$ with

$$\underline{v} = [f_o \quad f_1 \quad f_2 \quad e_o \quad e_1 \quad e_2 \quad d_o \quad d_1]'$$

$$\underline{M} = \begin{bmatrix} 0 & 0 & 0 & 1 & 0 & 0 & a & 0\\ 1 & 0 & 0 & 0 & 1 & 0 & 0 & a\\ 0 & 1 & 0 & 0 & 0 & 1 & 1 & 0\\ 0 & 0 & 1 & 0 & 0 & 0 & 0 & 1\\ 0 & 0 & 0 & -1 & 0 & 0 & a & 0\\ -1 & 0 & 0 & 0 & -1 & 0 & 0 & a\\ 0 & -1 & 0 & 0 & 0 & -1 & 1 & 0\\ 0 & 0 & -1 & 0 & 0 & 0 & 0 & 1 \end{bmatrix}, \quad \underline{q} = \begin{bmatrix} p_{10}\\ p_{11}\\ p_{12}-a\\ p_{13}\\ p_{20}\\ p_{21}\\ p_{22}-a\\ p_{23} \end{bmatrix} \tag{8.2.13}$$

The first and fifth columns of $\underline{M}$ are identical and the same is true for the second and sixth columns. Thus we cannot assign arbitrary polynomials but we can choose $\underline{q}$ such that the linearly dependent equations are consistent and the two polynomials are stable. For $a = 0.99^2(3+\sqrt{8}) = 5.712$ this is for example achieved by

$$\underline{q} = \begin{bmatrix} 0.960596 \\ 3.881196 \\ 0.168159 \\ 3.96 \\ 0.960596 \\ -3.881196 \\ 0.168159 \\ -3.96 \end{bmatrix} \quad \rightarrow \quad \underline{v} = \begin{bmatrix} 3.881196 \\ 0 \\ 3.96 \\ 0 \\ 0 \\ 0 \\ 0.168159 \\ 0 \end{bmatrix} \qquad (8.2.14)$$

This controller assigns the characteristic polynomials $P_1(z) = (z+0.99)^4$ and $P_2(z) = (z-0.99)^4$ and thus stabilizes both plants. In the above solution $e_0 = e_1 = e_2 = 0$, i.e. x_1 is not fed back. The necessary stability condition $p_2 < 6$, see eq. (C.3.3) shows that this example with $p_{12} = p_{22} = 6\times0.99^2$ is very close to the maximum a, that can be admitted with n = 4, i.e. with a second order controller. For a ≥ 6 a third order controller may be used and so on. A sequence of such examples with increasing a and increasing controller order m may be constructed by assigning $(z\pm0.99)^{2+m}$. Thus there is no general upper limit on the required controller order m. Now shift the convex hulls in figure 8.7 in the opposite direction. For a = -1, K_{Γ_1} has been moved up and K_{Γ_2} has been moved down so far that now the two convex hulls just touch each other along the intersection of the real boundaries. For a < -1 the tetrahedra do not intersect. Is it possible, to have simultaneous stabilization with a second order controller? The answer is no. This can be shown from eq. (8.2.13) and some simple necessary stability conditions: The sum of rows 1 and 5 yields

$$2ad_0 = p_{10} + p_{20}$$

Now by eq. (C.3.2) a necessary stability condition is $|p_{10}| < 1$, $|p_{20}| < 1$. This implies

$$|ad_0| < 1 \qquad (8.2.15)$$

The sum of rows 1, 3, 5 and 7 of eq. (8.2.13) may be written as

$$P_1(1) + P_2(1) + P_1(-1) + P_2(-1) = 4(1+a)(d_o+1) \qquad (8.2.16)$$

A necessary stability condition is that the expression of eq. (8.2.16) is positive. This can be satisfied in two ways:

a) $a < -1 \rightarrow d_o < -1$
b) $a > -1 \rightarrow d_o > -1$

a) contradicts eq. (8.2.15) and thus b) is the only stabilizable case. Thus $a > -1$ is a necessary condition for simultaneous stabilizability by proportional, first and second order state-feedback controller.

8.2.2 Conflicting Real Root Conditions

The result of the previous example has a more general background. This is formulated as the following theorem:

Let $P_j(z) = \det(z\underline{I}-\underline{A}_j+\underline{b}_j\underline{k}')$. If there does not exist a $\underline{k}'$ that satisfies

$$P_j(1) > 0 \text{ for all } j \qquad (8.2.17)$$

then the plant family $(\underline{A}_j, \underline{b}_j)$ cannot be stabilized simultaneously by a stable constant linear controller of any order m.

The corresponding result also holds for $P_j(-1)$.

For the proof of the theorem assume a stable, linear, time-invariant dynamic state-vector feedback as

$$u_z(z) = -\frac{\underline{h}'(z)}{D(z)}\underline{x}_z(z) \;,\; \underline{h}'(z) = [H_1(z) \ldots H_n(z)] \qquad (8.2.18)$$

By stability $D(1) > 0$. Assume that this controller simultaneously stabilzes a plant family

$$\underline{x}_z(z) = (z\underline{I}-\underline{A}_j)^{-1}\underline{b}_j u_z(z) = \frac{\operatorname{adj}(z\underline{I}-\underline{A}_j)\underline{b}_j}{\det(z\underline{I}-\underline{A}_j)} u_z(z) \;.$$

With $Q_j(z) = \det(z\underline{I}-\underline{A}_j)$ the closed-loop characteristic polynomial is

$$P_j(z) = Q_j(z)D(z) + \underline{h}'(z)\mathrm{adj}(z\underline{I}-\underline{A}_j)\underline{b}_j \tag{8.2.19}$$

This is stable by assumption, thus $P_j(1) > 0$ for $j = 1,2...J$. Now compare this with proportional state-vector feedback $u = -\underline{k}'\underline{x}$ with

$$\underline{k}' = \frac{\underline{h}'(1)}{D(1)} \tag{8.2.20}$$

and characteristic polynomial

$$\begin{aligned}\bar{P}_j(z) &= \det(z\underline{I}-\underline{A}_j+\underline{b}_j\underline{k}') \\ &= Q_j(z) + \underline{k}'\mathrm{adj}(z\underline{I}-\underline{A}_j)\underline{b}_j \\ &= \frac{1}{D(1)} \times [Q_j(z)D(1) + \underline{h}'(1)\mathrm{adj}(z\underline{I}-\underline{A}_j)\underline{b}_j]\end{aligned} \tag{8.2.21}$$

Inview of $D(1) > 0$ it satisfies

$$\bar{P}_j(1) = P_j(1)/D(1) > 0 \quad \text{for all } j \tag{8.2.22}$$

This contradicts the assumption that there does not exist a $\underline{k}'$ such that all $\bar{P}_j(1) > 0$.

8.2.3 Remarks and Recommendations

For more than two plant models many cases must be distinguished. As an example figure 8.10 shows the case of three plants being stabilizable only pairwise. The three conflicting boundaries are indicated by dotted extensions. If these boundaries result from $P_j(1) < 1$, then stable dynamic state feedback cannot simultaneously stabilize all three plants.

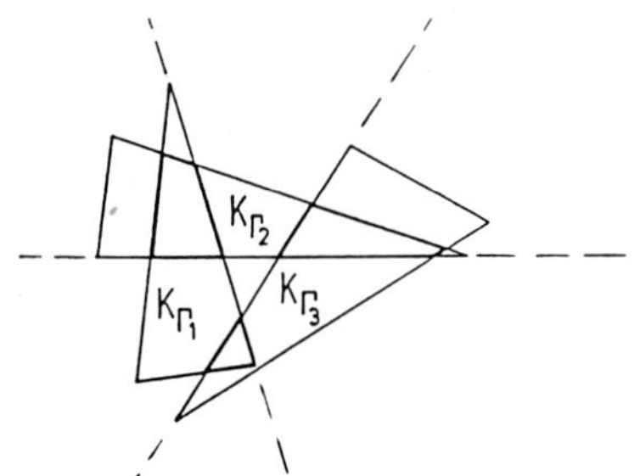

Figure 8.10

Three second order plants which can be stabilized only pairwise but not altogether.

A problem related to that of simultaneous stabilization is the design of a stable compensator for the stabilization of one plant [74.5]. This may be viewed as the problem of simultaneous stabilization of the plant and a second "plant with transfer function zero" [82.5].

General conditions for simultaneous stabilization were given in [82.2], [82.5] for the linear case and in [84.6] for the nonlinear case. An interesting procedure for a time-varying controller was given in [83.3]. For each pair $(\underline{A}_j, \underline{b}_j)$ i = 1,2...J calculate the deadbeat controller $\underline{k}'_{Dj}$, see eq. (4.4.8). Then apply

$$\underline{k}' = \begin{cases} \underline{k}'_{D1} & \text{for} \quad 0 < t < (n-1)T \\ \underline{k}'_{D2} & \text{for} \quad nT < t < (2n-1)T \\ \vdots & \qquad\qquad \vdots \\ \underline{k}'_{DJ} & \text{for} \quad (J-1)nT < t < (Jn-1)T \end{cases} \tag{8.2.23}$$

and repeat periodically, i.e.

$$\underline{k}' = \underline{k}'_{D1} \quad \text{for} \quad JnT < t < [(J+1)n-1]T \quad \text{etc.}$$

Then for the actual pair $(\underline{A}_j, \underline{b}_j)$ in one of the intervals the appropriate deadbeat control is applied and the state goes to zero. This approach assumes that only the models assumed in the given family can occur and no intermediate values of the parameters are allowed.

Presently these results do not provide a recipe for the designer, and the selection of a controller structure is more an art than a science. In this sense the parameter space techniques for finding parameter regions in an assumed controller structure must be regarded as a design tool rather than a fool-proof synthesis method. So far, several application examples have shown that surprisingly large variations of physical parameters can be accommodated by a constant controller in situations where gain scheduling or parameter adaptation were believed to be unavoidable. The limited experience with practical examples showed also that they were not nearly as nasty as the constructed examples of this section.

A recommended practical approach is to try first to find a common Γ-stabilizer for the plant family in the form of proportional state feedback and then to replace unmeasured states by filtered outputs. This will be illustrated in detail for the flight control example in section 8.5. Here the idea is only sketched for the track-guided bus example of appendix 2. Feedback of all five states is first assumed and the design performed in the invariance plane of eq. (7.6.28). It is possible to accommodate all variations of the three physical parameters including a speed variation from 1 m/s to 20 m/s. Now only the states δ_R, δ_F and β are measured. The missing states $\dot{\delta}_R$, $\dot{\delta}_F$ are reconstructed within the required bandwidth by a differentiator with low pass filter, see figure 8.11.

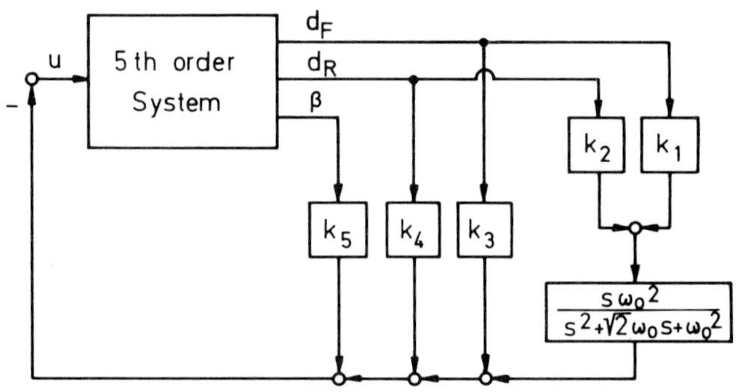

Figure 8.11 Reconstruction of missing state variables

Although the stability of a seventh order system must now be studied only five feedback parameters are required. This is possible by fixing the two filter parameters (damping = $1/\sqrt{2}$ and natural frequency ω_0 = desired bandwidth) to satisfy other requirements than simultaneous stabilization. Essentially the seventh order system shows a similar behavior to the fifth order system with state feedback.

An observer would cause more difficulties than the filter, because it can be matched to the plant only for a particular parameter set.

8.3 Simultaneous Pole Region Assignment

8.3.1 Graphic Solution

The fundamental idea of the design of a robust control loop has been introduced in section 8.1. See figure 8.4. The practical application of this design tool will now be illustrated for some detailed examples.

Circular eigenvalue regions Γ, for example those introduced by the family of circles of figure 7.5, are comparitively easy to deal with. In the first example let the region of nice stability be a circle with intersection points τ_L and τ_R with the real axis. We begin with systems of second order with state-vector feedback $u = -[k_1 \quad k_2]\underline{x}$. For a given pair $(\underline{A}_j, \underline{b}_j)$ with the pole assignment matrix $\underline{E}_j$, the region of nice stability in the k_1-k_2-plane is the triangle with the corners

$$\begin{bmatrix} \underline{k}'_{j0} \\ \underline{k}'_{j1} \\ \underline{k}'_{j2} \end{bmatrix} = \begin{bmatrix} \underline{p}'_0 & 1 \\ \underline{p}'_1 & 1 \\ \underline{p}'_2 & 1 \end{bmatrix} \underline{E}_j \tag{8.3.1}$$

Here $\underline{p}_i$, $i = 0,1,2$ are the coefficient vectors of the polynomial $P_i(z) = (z-\tau_L)^i(z-\tau_R)^{2-i}$, i.e. $\underline{p}'_0 = [\tau_R^2 \quad -2\tau_R]$,
$\underline{p}'_1 = [\tau_R\tau_L \quad -(\tau_R+\tau_L)]$, $\underline{p}'_2 = [\tau_L^2 \quad -2\tau_L]$.

If two pairs $(\underline{A}_j, \underline{b}_j)$, $j = 1,2$ are given, then the intersection of the two triangles in the k_1-k_2-plane must be determined.

Example 1:

The example from eq. (7.5.8) with

$$\underline{A}_1 = \begin{bmatrix} 0 & -4 \\ 1 & 4 \end{bmatrix}, \quad \underline{b}_1 = \begin{bmatrix} 0.375 \\ -0.3125 \end{bmatrix}, \quad \underline{E}_1 = \begin{bmatrix} 5 & 6 \\ 6 & 4 \\ 4 & -8 \end{bmatrix}$$

is augmented by a second plant model

$$\underline{A}_2 = \begin{bmatrix} 0.32 & -2.5 \\ 0.32 & 0.6 \end{bmatrix}, \quad \underline{b}_2 = \begin{bmatrix} 0.25 \\ 0 \end{bmatrix}, \quad \underline{E}_2 = \begin{bmatrix} 0 & 12.5 \\ 4 & 7.5 \\ 3.68 & -5.5 \end{bmatrix} \tag{8.3.2}$$

As in the previous example, nice stability with respect to a circle with center 0.45 and radius 0.5 is required, i.e. $\tau_L = -0.05$, $\tau_R = 0.95$. The corners of the second nice stability triangle are

$$\begin{bmatrix} \underline{k}_{20} \\ \underline{k}_{21} \\ \underline{k}_{22} \end{bmatrix} = \begin{bmatrix} 0.9025 & -1.9 & 1 \\ -0.0475 & -0.9 & 1 \\ 0.0025 & 0.1 & 1 \end{bmatrix} \begin{bmatrix} 0 & 12.5 \\ 4 & 7.5 \\ 3.68 & -5.5 \end{bmatrix} = \begin{bmatrix} -3.92 & -8.47 \\ 0.08 & -12.84 \\ 4.08 & -4.72 \end{bmatrix}$$

Figure 8.12 shows this triangle together with the first nice stability triangle which was taken from figure 7.9

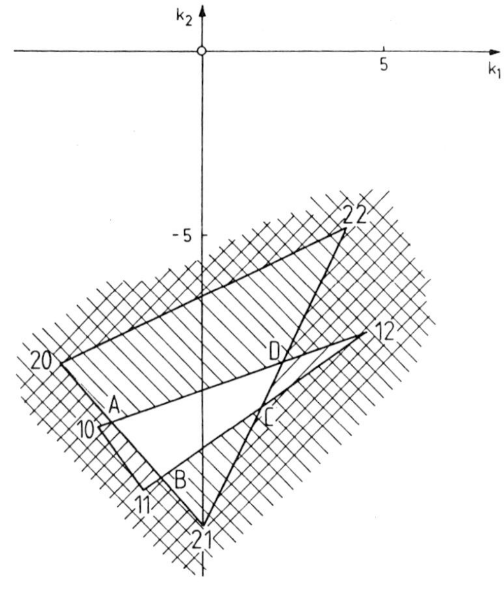

Figure 8.12

The quadrilateral ABCD is the intersection set of triangles 1-012 and 2-012

The intersection set of both triangles is the quadrilateral ABCD. All points in this quadrilateral simultaneously satisfy the condition for nice stabilization for the systems $(\underline{A}_1, \underline{b}_1)$ and $(\underline{A}_2, \underline{b}_2)$.

Further pairs $(\underline{A}_j, \underline{b}_j)$, $j = 3,4 \ldots J$ can be handled in a corresponding way. The task of finding the intersection of the triangles is generally tedious on a computer. The polygon corners would have to be calculated numerically and this is a much more involved path than would be taken by a human. For two triangles the possible cases are represented in figure 8.13.

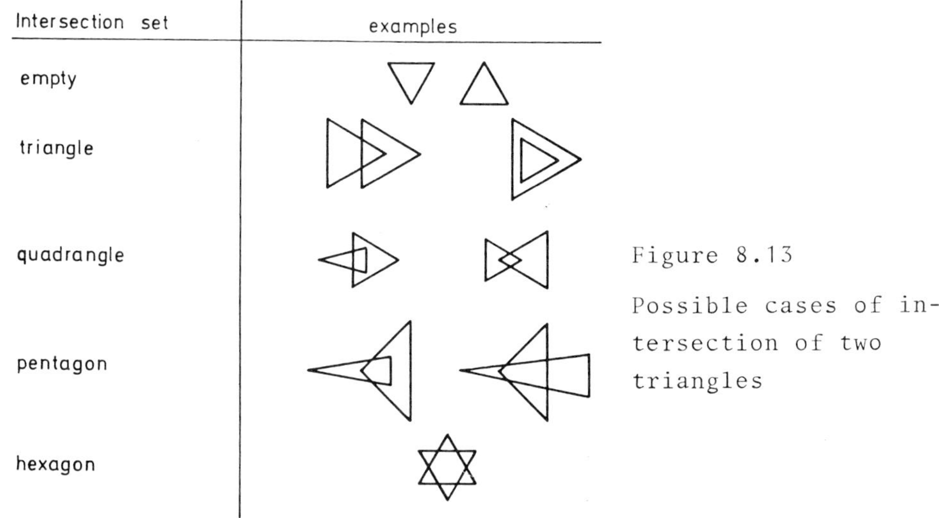

Figure 8.13

Possible cases of intersection of two triangles

When arranging the division of labor between computer graphics and the operator, this problem should be handled by man. For several triangles it is not important for the designer to know, for example, that the intersection is an polygon with 86 exactly calculated vertices. What is much more important is the approximate size and location of the intersection set as we can see it at a glance from a picture. This difference between pattern recognition and exact calculation of the intersection set becomes still more extreme if it is not a question of triangles but nonconvex regions which are bounded by complicated contours. For two-dimensional intersections, the use of computer graphics is preferable to the numerical determination of the intersection set. In the next three examples a plant parameter varies continuously in a given interval.

Example 2:

$$\underline{x}[k+1] = \begin{bmatrix} 0 & 1-\alpha \\ -\alpha & 0 \end{bmatrix} \underline{x}[k] + \begin{bmatrix} \alpha \\ 1-\alpha \end{bmatrix} u[k] \qquad (8.3.3)$$

Determine the set of all state-vector feedbacks which stabilize the system in the interval $0 \le \alpha \le 1$. The corners of the stability triangle are

$$\begin{bmatrix} \underline{k}_0'(\alpha) \\ \underline{k}_1'(\alpha) \\ \underline{k}_2'(\alpha) \end{bmatrix} = \begin{bmatrix} 1 & -2 & 1 \\ -1 & 0 & 1 \\ 1 & 2 & 1 \end{bmatrix} \begin{bmatrix} \alpha-1 & \alpha \\ -\alpha^2 & -1+2\alpha-\alpha^2 \\ \alpha-2\alpha^2+\alpha^3 & -\alpha^2+\alpha^3 \end{bmatrix} \times \frac{1}{N(\alpha)}$$

$$N(\alpha) = -1+3\alpha-3\alpha^2$$

$$\begin{bmatrix} \underline{k}_0'(\alpha) \\ \underline{k}_1'(\alpha) \\ \underline{k}_2'(\alpha) \end{bmatrix} = \frac{1}{N(\alpha)} \times \begin{bmatrix} -1+2\alpha+\alpha^3 & 2-3\alpha+\alpha^2+\alpha^3 \\ 1-2\alpha^2+\alpha^3 & -\alpha-\alpha^2+\alpha^3 \\ -1+2\alpha-4\alpha^2+\alpha^3 & -2+5\alpha-3\alpha^2+\alpha^3 \end{bmatrix}$$

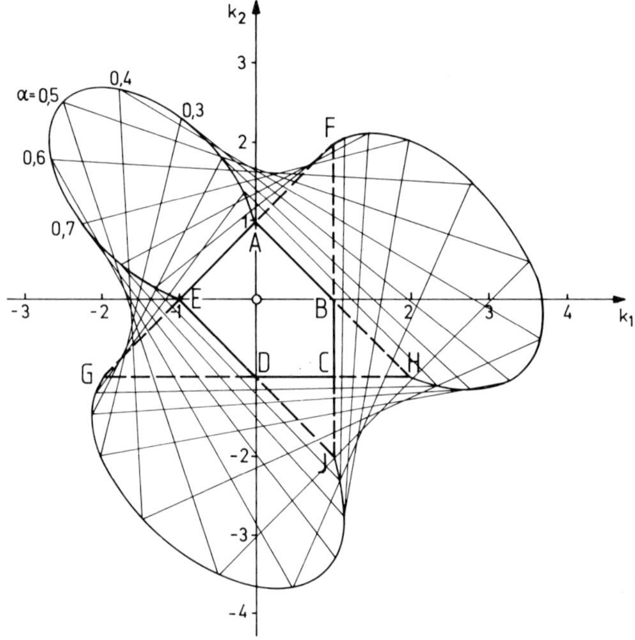

Figure 8.14 The pentagon ABCDE is the intersection set of all stability triangles for the system (8.3.3) with $0 \le \alpha \le 1$

Figure 8.14 shows the familiy of stability triangles, beginning with EFJ for $\alpha = 0$ and moving forward in α steps of 0.1 until AHG for $\alpha = 1$. The corners of the stability triangle move along the enveloping curve of the figure from E to A, from F to H, and from J to G. The intersection of all triangles is the pentagon ABCDE.

For the visualization by computer graphics it is useful to color-shade the forbidden region, e.g. outside of each of the triangles in figure 8.14. Then the intersection set is the region which is not color-shaded. This also holds if one (as in figure 7.1) maps not only the stability region, but also all the boundary lines of the D-decomposition which lie outside.

Example 3:

In order to show an interesting singularity, example 2 is modified by changing the sign of the element a_{21}, i.e.

$$\underline{x}[k+1] = \begin{bmatrix} 0 & 1-\alpha \\ \alpha & 0 \end{bmatrix} \underline{x}[k] + \begin{bmatrix} \alpha \\ 1-\alpha \end{bmatrix} u[k] \tag{8.3.4}$$

The corners of the stability triangle are

$$\begin{bmatrix} \underline{k}'_0(\alpha) \\ \underline{k}'_1(\alpha) \\ \underline{k}'_2(\alpha) \end{bmatrix} = \frac{1}{N(\alpha)} \times \begin{bmatrix} -1-\alpha^3 & 2-3\alpha+3\alpha^2-\alpha^3 \\ 1-2\alpha+2\alpha^2-\alpha^3 & -\alpha+\alpha^2-\alpha^3 \\ -1+4\alpha^2-\alpha^3 & -2+5\alpha-\alpha^2-\alpha^3 \end{bmatrix}$$

$$N(\alpha) = -1+3\alpha-3\alpha^2+2\alpha^3$$

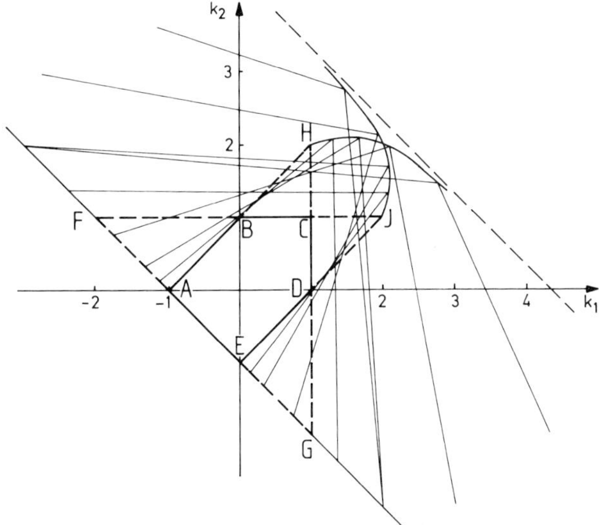

Figure 8.15 The pentagon ABCDE is the intersection set of all stability triangles for the system (8.3.4) with $0 \leq \alpha \leq 1$.

The stability triangles are represented in figure 8.15 for $0 \le \alpha \le 1$. $N(\alpha)$ has a zero in this interval for $\alpha = 0.5$. Here the system looses its controllability. For $\alpha = 0.5$, $\underline{A}$ has eigenvalues at $z_1 = -0.5$ and $z_2 = 0.5$. By the Hautus-criterion eq. (4.1.9) the eigenvalue $z_2 = 0.5$ is uncontrollable. In the neighborhood of $\alpha = 0.5$, z_2 is badly controllable, very high feedback gains are thus required to assign eigenvalues at +1 or -1. This means that the corners of the stability triangle go to infinity. In figure 8.15 A ($\alpha = 0$) moves on the line $k_2 = -1 - k_1$ to the left with growing α and goes to infinity for $\alpha = 0.5$. This corner returns from the right along the same line and reaches E for $\alpha = 1$. Corner H proceeds to the right towards the asymptote $k_2 = 4.333 - k_1$ and returns along this asymptote from the left, finally arriving at J. G moves on the line $k_2 = -1 - k_1$ to the right, returns from the left and ends in F. The intersection of all the stability triangles is the pentagon ABCDE.

It has already been shown in the discussion of the D-decomposition in section 7.6 that the two-dimensional graphic representation of stability regions is not restricted to systems of second order. The essential restriction is that only two free parameters can be examined in a design step. We now consider an example of order four.

Example 4:

Loading bridge in continuous time. The nice stability region Γ is described by the hyperbola (7.6.5). As in eq. (7.6.6), let the controller structure be $u = -[500 \quad k_2 \quad k_3 \quad 0]$.

The nice stability region has been determined in the k_2-k_3-plane and represented in figure 7.14 for a load mass $m_L = 3000$ kg. This design should now be modified so that the loading bridge is stable in a domain 50 kg $< m_L < m_{Lmax}$, where the load domain m_{Lmax} should be as large as possible.

The general procedure now would be, to calculate further regions for $m_L = 50$ kg and larger values corresponding to the nice stability region ABC in figure 7.14, and to intersect them. Here we find it convenient that the load mass only occurs in k_3. From eq. (2.6.35) $k_3 = k_{30} + m_L g$. The region of nice stability in figure 7.14 consequently does not alter its form, rather it is only shifted upwards by $10(m_L - 3000)$ if the load mass is changed from 3000 kg to m_L. For the empty load hook of $m_L = 50$ kg it is shifted down by 29500.

The largest allowable load domain is obtained if k_2 is chosen so that K_Γ has the maximal range in the k_3-direction. This is obviously the straight line CD in figure 7.14 with the corresponding value $k_2 = 2769$. The allowable load variation is then a tenth of the interval CD, i.e. $(7943 + 15503)/10 = 2345$ kg. Now k_3 must be chosen such that the allowable load domain begins at $m_L = 50$ kg and consequently extends up to $m_L = 2395$ kg. For $m_L = 50$ kg the point C is shifted to $k_3 = 7943 - 29500 = -21557$. This same point then corresponds to D for 2395 kg, see figure 8.16.

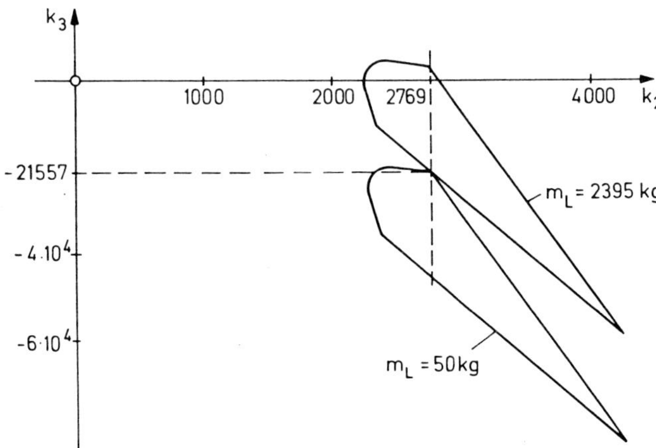

Figure 8.16 For $k_2 = 2769$, $k_3 = -21557$ the maximal load variation $50 \text{ kg} < m_L < 2395$ kg is obtained.

The result of this design is the following: With the controller

$$u = -\begin{bmatrix} 500 & 2769 & -21557 & 0 \end{bmatrix} \underline{x} \tag{8.3.5}$$

the loading bridge exhibits the following properties:

- for a transition from $\underline{x}(0) = [1 \quad 0 \quad 0 \quad 0]'$ to $\underline{x} = \underline{0}$ the initial peak force is 500 Newton. The simulation actually shows that $|u(t)| \leq 500$ for all t,
- no measurement or reconstruction of the cable angle velocity x_4 is required,
- the maximal load variation is reached under these constraints. The eigenvalues lie to the left of the hyperbola $\omega^2 = 4\sigma^2 - 0.25$ if 50 kg < m_L < 2395 kg. For m_L = 50 kg and m_L = 2395 kg some eigenvalues lie on the hyperbola, see figure 8.3.

8.3.2 Γ-Contraction

In the example of the loading bridge the desired system property Γ was specified in the form of a hyperbola and the allowable disturbance was maximized, i.e. the load variation, so that the given system property is robust with respect to large disturbances in m_L. Also the opposite problem formulation may be useful, namely to fix the parameter variation and to find the best system property, which can be made robust against this class of perturbations. The family of circles of figure 7.5 may be used to define a measure r for the system property.

One possible procedure is, first to examine the intersection of the convex hulls of the stability regions. These convex hulls no longer intersect, once a certain reduced radius r of Γ is reached. Thus there is no common nice stabilizer for such a value of r, because the necessary condition that the convex hulls intersect, is not satisfied. One then increases r once again until an intersection of the convex hulls occurs, and tests also the complex eigenvalue condition. This procedure will be illustrated by the following example performed by D. Kaesbauer:

Example 5:

In appendix D the problem of the stabilization of the short period mode of an F4-E airplane with additional canards is formulated. The linearized equations of motion have the form

$$\dot{\underline{x}} = \begin{bmatrix} f_{11} & f_{12} & f_{13} \\ f_{21} & f_{22} & f_{23} \\ 0 & 0 & -14 \end{bmatrix} \underline{x} + \begin{bmatrix} g_1 \\ 0 \\ 14 \end{bmatrix} u \qquad (8.3.6)$$

For the given seven general coefficients, the numerical values for four flight conditions are given. They vary with altitude and velocity of the aircraft. The system has been discretized with sampling period T = 0.1 seconds. The state variable x_3 is the deviation of the rudder deflection from its trim position. This third state variable is not fed back in order to avoid estimating the trim position. For proportional feedback of the normal acceleration x_1, measured by the accelerometer, and the pitch rate x_2, measured with a gyroscope, the structure of the control law is

$$u = -[k_1 \quad k_2 \quad 0]\underline{x} \qquad (8.3.7)$$

Each of the four flight conditions produces a pattern in k_1-k_2-k_3-space affine to figure 7.2 for circular eigenvalue regions. The intersection is the admissible solution set for state feedback. Because of the controller structure (8.3.7) we are interested in the two-dimensional cross section in the plane $k_3 = 0$. First we examine a single flight condition (here no. 2, see appendix D). Instead of the stability region from figure 7.2 first of all the enveloping tetrahedron is examined. Its intersection with a plane can be empty, a triangle, or a quadrilateral. In special boundary cases it can also be a point or a line. Here it is a quadrilateral in the $k_3 = 0$ plane, and for the unit circle, it is the outermost quadrilateral (r = 1) in figure 8.17.

The corners are named q_{ij} where ij is the intersected edge of the tetrahedron. For example, q_{01} is the intersection of the $k_3 = 0$ plane with the affine map of the edge 01, see figure 7.2. On this edge 01 a real eigenvalue moves along the real axis from right to left, while two eigenvalues remain at the right real axis intersection of Γ, i.e. two eigenvalues lie at z = 1 and one in the interval $-1 < z < 1$ at the point q_{01}. Figure 8.17 also shows the corresponding quadrilaterals for smaller pole regions Γ_r from figure 7.5.

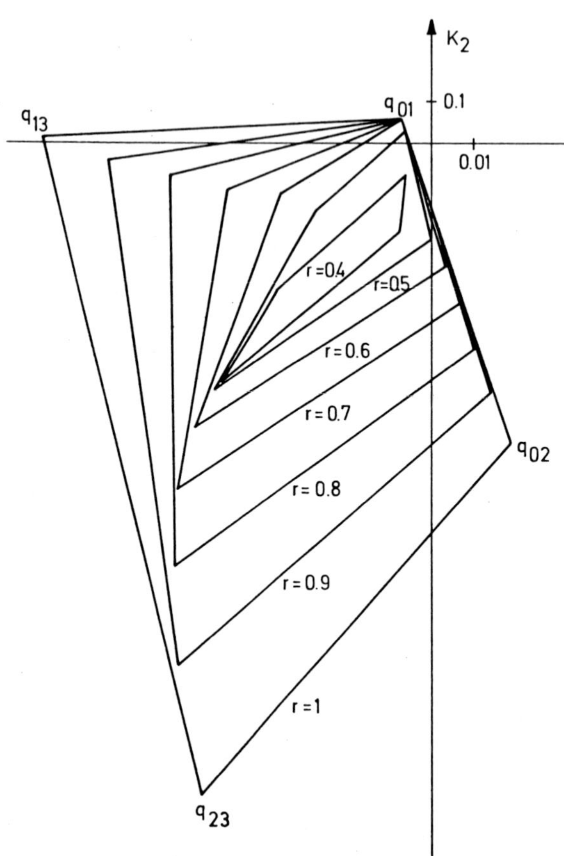

Figure 8.17

Intersection of the enveloping tetrahedrons of the stability regions with the plane $k_3 = 0$ for different circular pole regions from figure 7.5

If one forms the corresponding intersections for all four flight conditions, this yields figure 8.18 for r = 0.4.

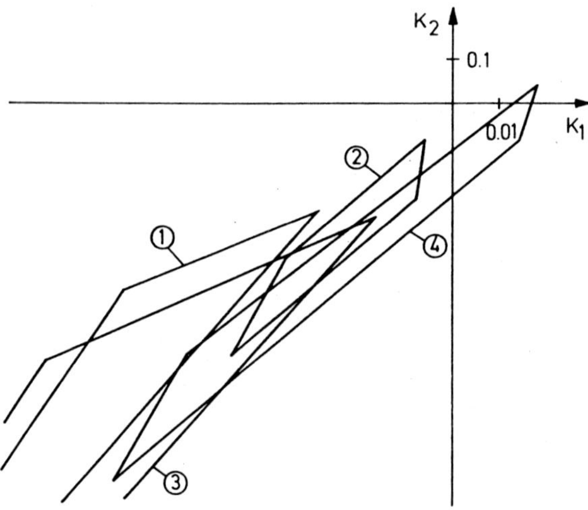

Figure 8.18

Convex hulls of the stability regions no longer intersect if the radius of the pole regions is reduced to r = 0.4

One recognizes that the flight conditions 1 and 2 cannot be simultaneously Γ-stabilized for r = 0.4. The radius is therefore enlarged to r = 0.5. The intersection of the convex hulls is traced in figure 8.19 by dark lines.

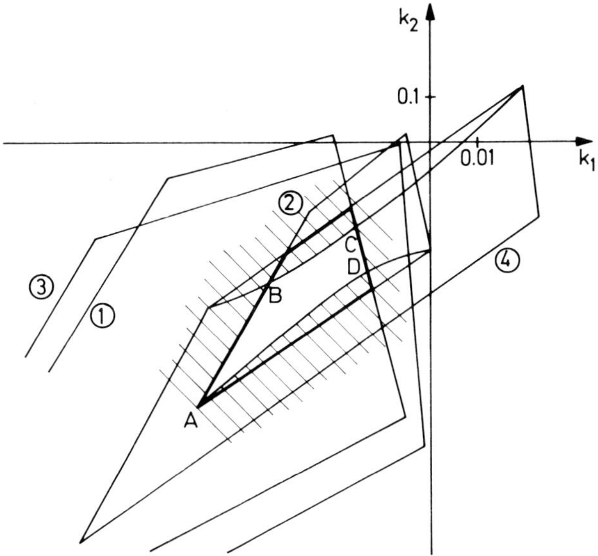

Figure 8.19 The region in which the linear conditions for simultaneous Γ-stabilization (with r = 0.5) are satisfied is traced by dark lines. By the nonlinear conditions the region is reduced to ABCD in which all necessary and sufficient conditions are satisfied.

The complex boundary surfaces are also represented for $k_3 = 0$ and r = 0.5. The admissible region is reduced from above by the complex root boundary of flight condition 4 and from below by that of flight condition 2. The remaining conditions for complex eigenvalues are satisfied in the resulting region ABCD.

The result allows the corner points A, B, C and D to be interpreted respectively for the four flight conditions (FC):

A double real eigenvalue at $\tau_L = -0.05$ in FC 2

B simple real eigenvalue at $\tau_L = -0.05$ in FC 2 and complex eigenvalue pair on $\partial\Gamma_{0.5}$ in FC 4

C complex eigenvalue pair on $\partial\Gamma_{0.5}$ in FC 4 and a real eigenvalue at $\tau_R = 0.95$ in FC 1

D real eigenvalue at $\tau_R = 0.95$ in FC 1 and complex pole pair on $\partial\Gamma_{0.5}$ in FC 2.

All other eigenvalues lie in $\Gamma_{0.5}$.

8.3.3 Summary of Graphical Approaches

The examples have shown that the examination of admissible feedback gains in parameter space is not a well-defined design procedure, but rather an approach to the problem of robustness which forms the basis for different problem formulations and design strategies. For three free controller parameters 3D-graphics may be used [84.7]. Here we discuss 2D-graphics first. If a system has more than two free controller parameters, then 2D-graphic methods must be applied iteratively. The choice of the intersection plane in K space is an essential element of the design strategy in every design step. In the airplane example the choice of $k_3 = 0$ was dictated by practical considerations. This example will be taken up once again with an extended controller structure in section 8.5. Two feedback gains k_1 and k_4 were also fixed for the loading bridge. It can also be reasonable to lay the intersection plane in K space oblique to the coordinate system, i.e. to consider two linear combinations of feedback gains as free parameters. For example it can be advantageous to choose the plane so that n-2 eigenvalues remain unchanged. This is done by constructing an artificial output $\underline{y}_R = \underline{C}_R\underline{x}$ from which the n-2 eigenvalues are unobservable. For the track-guided bus example this was done in eq. (7.6.28). In a detailed study [82.3] a controller with five free parameters was designed for the bus.

A good solution for the whole range of speed, load and road adhesion (see appendix D.2) was found in the above mentioned cross section plane, which was chosen for the worst cases in load (max) and road adhesion (min) and an average speed. In speed both limiting cases are critical: the slow bus (1m/s) is too sluggish, the fast bus (20 m/s) is not sufficiently damped for good passenger comfort. A good compromise is possible by design in the invariance plane for a medium speed (10 m/s). (Readers who want to repeat this example should note that in [82.3] the moment of inertia of Θ_o = 10570 kgm^2 was too low by a factor of ten. The correct value is Θ_o = 105700 kgm^2. However, an equally good solution exists for the true value. Interestingly the solution obtained with the wrong Θ_o was considered good in road tests. The parameters mass, moment of inertia and road adhesion are less critical than the wide variation of speeds. See exercise 8.14.)

The graphical plane intersection procedure has the disadvantage that only two free parameters or two linear combinations of more parameters can be used in each design step. This requires the development of a suitable design strategy for the problem.

On the other hand one can exploit essential advantages of the graphical procedure:

1. Computer graphics are becoming cheaper and increasingly available at the engineer's work place.
2. A human can grasp and interpret information much faster if it is presented in graphical form rather than as a column of numbers.
3. A human can easily recognize the intersection of several curves and the intersection sets of regions. Figure 8.20 illustrates this point.

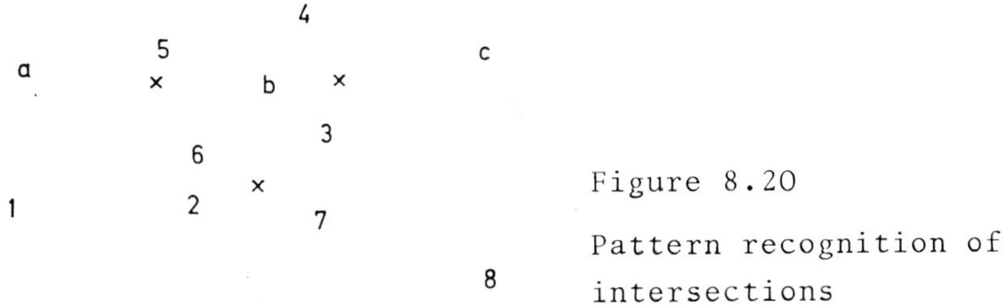

Figure 8.20
Pattern recognition of intersections

A number curve (1,2...8) and a letter curve (a, b, c) are given point-by-point, representing the boundary curves of our problem. Let the sign conditions be satisfied within the loop of the number curve and below the letter curve. The two curves can be easily drawn, the three intersection points being designated by crosses and, for example, an ellipse plotted between the points b, 6 and 3. Thus the allowable region is approximated well enough for our design purpose. For given coordinates

1,2...8, a, b, c the purely numerical calculation of such an ellipse would be much more involved. Of course the graphics computer connects the calculated points with a curve and controls the step distance for the calculation of these points.

4. At a graphics terminal with input capability the designer can communicate the relevant curve portions so that all other curve portions can be erased from the screen.
5. The computation of a nice stability region in a two-dimensional cross section is very fast. This allows the cross section plane to be moved on line in the direction of other free controller parameters. The designer can immediately decide which cross section gives the most favorable intersection set and can choose other free controller parameters from this starting point.
6. 3D graphics with wireframe or solids representations of Γ-stability regions become clear if the penetration lines are calculated. Then the representation can be restricted to the intersection set. The precalculated stability regions and their intersections can also be studied by rotating and zooming in an interactive 3D computer graphic system [84.7]. Different colors can be used for stability regions of different operating conditions.
7. A versatile program allows (for selected test points in K space) the eigenvalue locations, step responses, Nyquist diagrams etc. to be displayed. Also a combination of 3D graphics for visualization with moving 2D cross sections for reading off gain values is recommended.

8.3.4 Computational Solution

An alternative to the graphical solution of the problem of simultaneous pole assignment is to reduce it to a numerically tractable problem. This is not necessarily an optimization procedure which finds a point of the admissible solution set. Rather, methods are sought which - just as the graphical ones - give a global overview of the admissible solution set. Since the solution set is an intersection of nonconvex figures, it can have several disjoint components. Optimization algorithms can then lead to a local optimum in one component without finding the global optimum in another one.

Consider the problem of simultaneous Γ-stabilization for a family of plants

$$\begin{aligned} \underline{x}[k+1] &= \underline{A}_j\underline{x}[k] + \underline{b}_j u[k] \\ \underline{y}[k] &= \underline{C}_j\underline{x}[k] \end{aligned} \qquad j = 1,2,\ldots J \tag{8.3.8}$$

by output feedback $u = -\underline{k}_y'\underline{y}$.

A first possibility is to put a grid into K space, calculate the n×J eigenvalues of $\underline{A}_j - \underline{b}_j\underline{k}_y'\underline{C}_j$ for every grid point $\underline{k}_y'$ and test their position relative to Γ.

A second possibility is to calculate the J characteristic polynomials $P_j(z) = \det(z\underline{I}-\underline{A}_j+\underline{b}_j\underline{k}_y'\underline{C}_j)$ with the polynomial coefficients

$$p_j' = \underline{a}_j' + \underline{k}_y' C_j W_j \tag{8.3.9}$$

and test algebraically for Γ-stability for every grid point $\underline{k}_y'$. As Sondergeld [83.1] has shown, the regions Γ whose boundaries ∂Γ are mapped by a rational mapping of degree m onto the imaginary axis, can be algebraically handled. Here the Γ-stability problem is reduced to a Hurwitz problem for a polynomial of degree mn. In practice the cases of interest are

m = 1 for half planes and circular pole regions
m = 2 for all other conic section regions.

In the following we limit the discussion to circular pole regions, which are of special interest in sampled-data control systems. The two possibilities mentioned earlier require first a decision on how far the grid must be extended. Such a search region can be obtained from the convex hull property. This follows directly from eq. (7.5.11) applied to the model family $\underline{A}_j$, $\underline{b}_j$:

> The eigenvalues of $\underline{A}_j - \underline{b}_j\underline{k}_y'\underline{C}_j$, j = 1,2...J are located inside the circle with real center and real axis intersections τ_L and τ_R if and only if the J polynomials

$$M_j(w) = [\underline{a}_j' + \underline{k}_y'\underline{C}_j\underline{W}_j \quad 1] \begin{bmatrix} (w-1)^n \\ (w-1)^{n-1}(w\tau_R-\tau_L) \\ \vdots \\ (w\tau_R-\tau_L)^n \end{bmatrix} \qquad (8.3.10)$$

are Hurwitz polynomials. $\underline{W}_j$ and $\underline{a}_j$ are determined from $(\underline{A}_j,\ \underline{b}_j)$ via eqs. (7.5.5) and (7.5.6).

The convex hull property is expressed by the necessary condition that all coefficients of the $M_j(w)$ must be positive. It is relatively easy to deal with these linear inequalities. Thus we can first determine the intersection of the convex hulls. There are two possible results

a) The intersection of the convex hulls is empty. Then there does not exist a robust controller of the assumed structure. The nonlinear Hurwitz conditions need not be tested.

b) The intersection of the convex hulls is a polyhedron. This region gives all possible solution candidates. A grid may be defined in this region by splitting it into nonoverlapping simplices and discretizing their interior by positive barycentric coordinates. For each grid point the "critical" Hurwitz condition $\Delta_{n-1} > 0$ is tested, which is zero on the complex root boundary. If Δ_{n-1} gets small, then the other nonlinear Hurwitz conditions should be tested also, because Δ_{n-1} does not necessarily change its sign when crossing the complex root boundary.

Also for computational design techniques different strategies may be used, e.g. reduction of r in the boundary family Γ_r as in the aircraft example. By such numerical techniques we are no longer constrained to two controller parameters in one design step. Also methods of geometric programming [80.12] or computation of intersections of nonconvex objects [80.11] may be applied.

In developing numerical techniques for this problem two aspects are important:

1. In the relation

$$P(z) = [\underline{p}' \quad 1]\underline{z}_n = [\underline{a}'+\underline{k}'\underline{W} \quad 1]\underline{z}_n = \prod_{i=1}^{n} (z-z_i) \qquad (8.3.11)$$

   it is essentially simpler to assume polynomials with zeros z_i on $\partial\Gamma$ and to calculate the corresponding point or surface in K space than to assume $\underline{k}'$ and to test the relative position of the eigenvalues z_i with respect to Γ.

2. Only the approximate form and extent of the admissible region needs to be found for the design of robust control systems. Consequently, very accurate descriptions of the boundary surfaces, their intersections, and the resulting regions should not be the goal of the investigation. In any case, a robust solution is good only if a neighborhood of the point $\underline{k}'$ finally chosen also lies in the admissible region.

8.4 Selection of a Controller from the Admissible Solution Set

So far we have tried to obtain a global picture of an admissible solution set. Of course we finally have to select a point with desirable properties from this set. A common procedure is to define design as an optimization problem. Here, however, design is understood as a tradeoff between several goals formulated by "soft" inequalities, i.e. performance bounds which may be shifted during the design process. A design method should offer some insights into which requirements are in conflict and which are not. Also the designer would like to learn easily which additional or tightened specifications could be satisfied, and for which specifications a high price must be paid.

In this section several typical design requirements are discussed which can be easily interpreted in K space. The next section shows the application to a flight control problem. In the final section of this chapter an optimization technique is used to refine the solution such that it is improved as far as possible in specific aspects without worsening other properties already achieved.

8.4.1 Simulation with Nonlinear Plant

The multi-model problem frequently arises from a nonlinear plant which is linearized in different operation conditions. In the airplane example the linear description by eq. (8.3.6) holds only for small deviations from flight conditions at constant altitude and with constant velocity. Also the slow phygoid mode and the coupling with the lateral motion is neglected. For changing flight conditions a nonlinear simulation is required to test and refine the controller. The solution obtained by solving the multi-model problem satisfies only a necessary condition for the nice stability of the nonlinear plant, not a sufficient one. A controller structure and an admissible parameter region are obtained which can be used as promising controller candidates for the nonlinear simulation. In other words: many controllers are excluded from the beginning from the expensive simulation because they do not satisfy the necessary condition that the locally linearized airplane is nicely stabilized.

8.4.2 Solutions with Small Loop Gains

In control systems only a constrained input amplitude $|u| \leq U$ is frequently available. For state feedback $u = -\underline{k}'\underline{x}$ the magnitude $|u|$ is bounded by

$$|u| = |\underline{k}'\underline{x}| \leq ||\underline{k}'|| \times ||\underline{x}|| \tag{8.4.1}$$

Usually very little is known about the probability distribution of the states $\underline{x}$, and a uniform distribution over the unit ball is the simplest assumption, i.e. $|\underline{x}'\underline{x}| = 1$. Here it is assumed that the individual state variables x_i are always scaled with respect to their maximal possible or expected values. The worst case of equality arises in eq. (8.4.1) if $\underline{k} = c\underline{x}$ for some real scalar c. Thus the absolute value $|\underline{k}|$ is immediately suitable as a measure for the maximal necessary input amplitude $|u|$.

There is a second practical reason why the loop gain should not be chosen too high. Practically

$$u = -\underline{k}'(\underline{x}+\Delta\underline{x}) \qquad (8.4.2)$$

is formed instead of the ideal control law $u = -\underline{k}'\underline{x}$. $\Delta\underline{x}$ can be a measurement noise or a quantization error in the analog-digital conversion, for example.

As the gain $|\underline{k}|$ decreases, the effect of such an error $\underline{k}'\Delta\underline{x}$ is reduced as well. According to these considerations it is reasonable to select the point from the allowable solution set which is closest to the origin in K space. In the example from figure 8.12 this is the corner D of the admissible quadrilateral.

8.4.3 Safety Margin from the Boundary Surfaces

If a corner point is chosen in the allowable solution set, as D in figure 8.12, then small gain changes in k_1 or k_2 bring the system out of the admissible region. This can be caused by a quantized storage of the control coefficients as $\underline{k}' + \Delta\underline{k}'$ or can arise from the fact that the two system models 1 and 2 were not exact. Consequently, in practice a safety margin from the boundary of the nice stability region is maintained which, for example, ensures that the hypercube $k_i \pm \Delta k$, $i = 1,2...n$ for a given maximum quantization error Δk is contained in the admissible region. In the example of figure 8.12 this is a square of the edge lengths $2\Delta k$ with edges parallel to the axes. (A hypersphere can also be used, in the two-gain example a circle around $\underline{k}'$).

Figure 8.12 is repeated once again as figure 8.21 where point $E(\underline{k}' = [-0.95 \quad 10.35])$ produces the maximal safety margin $\Delta k = 0.65$ and point F $(\underline{k}' = [1.45 \quad -9.2])$ represents a suitable compromise with the demands of small loop gain $|\underline{k}'|$ for given $\Delta k = 0.3$.

Also for solutions found by optimization procedures, it is helpful to examine the neighborhood of the solution - especially its distance from the boundary - in different two-dimensional intersections (margin analysis).

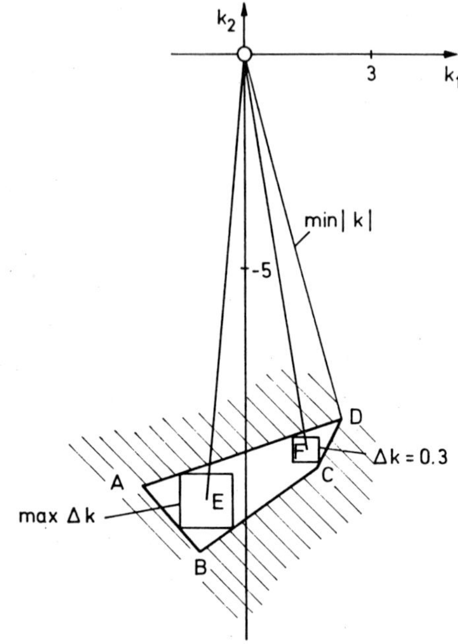

Figure 8.21

D: Minimal loop gain $|k| = 8.8$

E: Maximal safety margin
$\Delta k = 0.65$, $|k| = 10.4$

F: Compromise for given
$\Delta k = 0.3$, $|k| = 9.3$

8.4.4 Gain Reduction Margins

In some design problems we are interested in stability or Γ-stability margins for gain reductions. This is for example true if one of several parallel components may fail or if a saturation nonlinearity as in figure 8.22 occurs in the actuator.

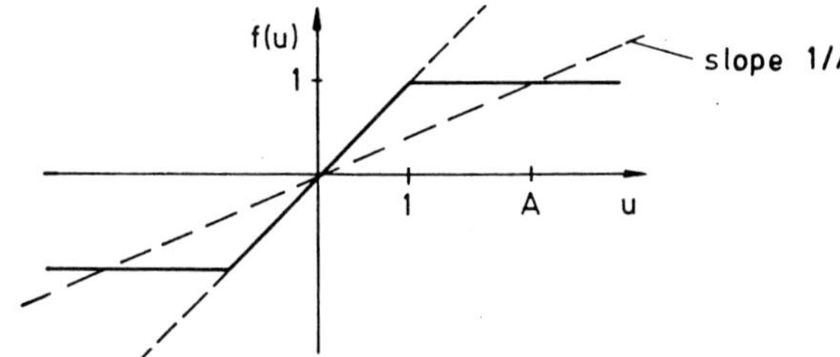

Figure 8.22

Saturation and its gain reduction effect

For a maximum input amplitude A a reduction of the linearized gain to 1/A occurs. Here we have two design possibilities:

a) Use a low loop gain in order to keep the actuator in its linear operating range. Within this range the pole region assignment guarantees Γ-stability.

b) Let the actuator go into saturation to make use of its available maximum output magnitude. Then the stability of the nonlinear loop must be tested. The Tsypkin or circle criterion, as described in section 3.9, is suited for this purpose. In this case absolute stability in the sector $1/A < K < 1$ must be assured. This requires a linear gain reduction margin from 1 to 1/A. The position in the loop where the gain reduction may occur is now important. Consider the three candidates a, b and c in figure 8.23.

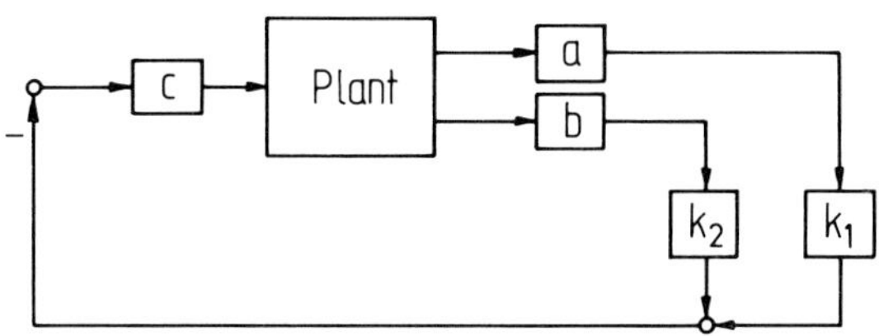

Figure 8.23

Gain reductions may occur at a, b or c.

Assume now that the plant is second order and the Γ-stability region in k_1-k_2-plane is the triangle of figure 8.24.

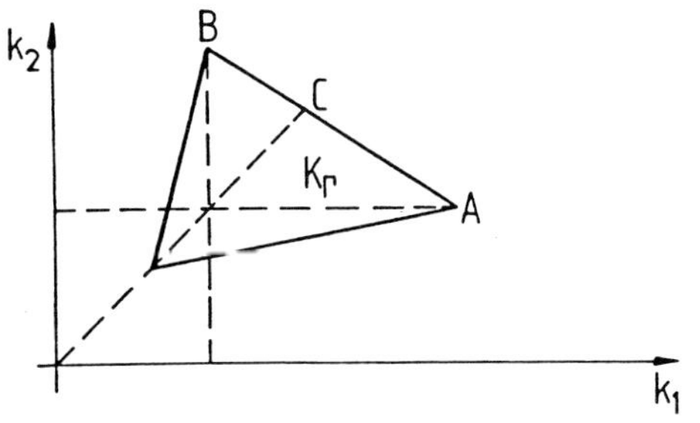

Figure 8.24

Maximal gain reduction margin.

A has the largest gain reduction margin in the direction k_1, i.e. for case a in figure 8.23. Correspondingly B belongs to case b and C to case c.

8.4.5 Robustness against Sensor Failures [84.2]

Usually control systems are designed under the assumption that sensors do not fail. Redundancy management has to provide then the required measurements with only very short interruptions by failures of individual sensors. If the plant is an unstable aircraft, for example, this means that failure detection is vital for stabilization. Detection must operate quickly and this is in conflict with the requirement of low probability of false alarms.

An alternative is the use of a hierarchical concept. Its basic level is a fixed gain control system which is designed such that pole region requirements are robust with respect to component failures and uncertain parameters. All the more sophisticated tasks, like failure detection and redundancy management, plant parameter identification and controller parameter adaptation or gain scheduling are assigned to higher levels, if they are required for best performance. The higher levels process more information and are operating in a slower time scale than the basic level. Since the higher levels are not vital for stabilization they can make their decisions without panic.

Assume that a sensor failure has two effects as illustrated by figure 8.25.

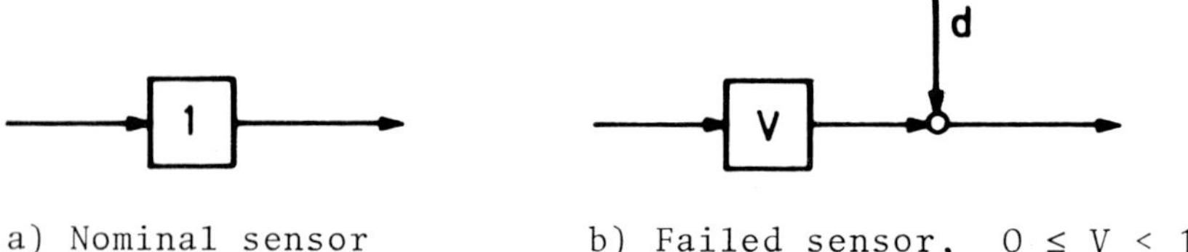

a) Nominal sensor b) Failed sensor, $0 \leq V < 1$

Figure 8.25 Model of sensor failures

1) The multiplicative effect reduces the gain V from its nominal value 1 to zero or some value in between;

2) The additive effect introduces a bias or noise d at the output.

As far as the eigenvalue location is concerned, only the multiplicative effect is important. The additive effect may require that the failed sensor signal is switched off. This decision of a failure detecting system may be slow, e.g. if the plant operates in a steady state and one of the sensors sticks at one value.

If the measured variables are used as state variables ('sensor coordinates'), then the sensor failure is equivalent to reducing the corresponding feedback gain. For $V = 0$ a solution, for which Γ-stability is robust against failure of the sensor for the state variable x_i, is characterized in gain space by the fact that the projection of $\underline{k}$ on the subspace $k_i = 0$ is contained in K_Γ. A two-dimensional cross-section through gain space is shown in figure 8.26. Assume that the admissible region K_Γ is the triangle ABC. Robustness against failure of sensor i is achieved if the projection of the appropriate gain is on GE. This property holds for DEFG. Similarly HCJKL is robust against failure of sensor j and KLMN is robust against failure of either i or j.

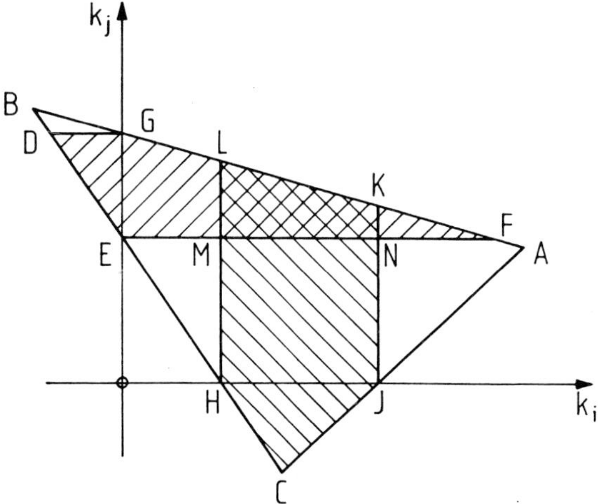

Figure 8.26

In KLMN Γ-stability is robust against failures of the type $k_i = 0$ or $k_j = 0$

If no such intersection exists, then the designer has two choices:

1) Ask for the best possible system property that can still be achieved under sensor failure. Here the degree r of stability as shown in figure 7.5 may be used.

2) Use redundant sensors in parallel.

The first possibility is illustrated by figure 8.27.

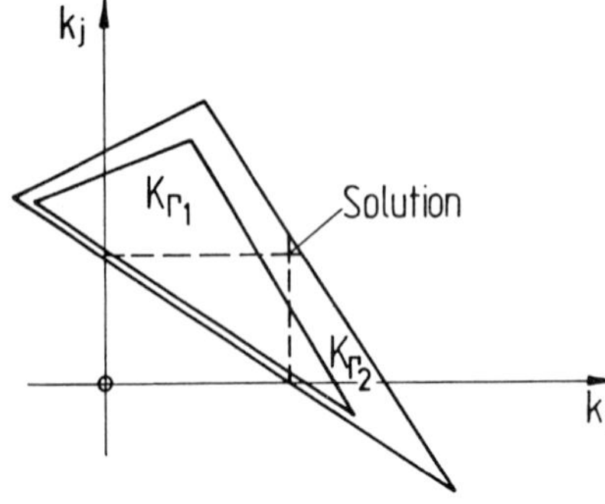

Figure 8.27

Sensor failure robustness cannot be achieved for Γ_1-stability but for Γ_2-stability

For the second approach the simplest possibility is to use two parallel sensors. Their outputs are multiplied by a factor 1/2 and added to produce x_i in the unfailed case. If one of the two sensors fails, then k_i is reduced by 50 %. In order to achieve robustness, k has to be selected such that it has a 50 % gain reduction margin in k_i for Γ-stability. If K_Γ is the triangle ABC in figure 8.28, then DEF is the region for which Γ-stability is achieved after 50 % reduction of k_i. Thus the triangle EGH contains the admissible points which are Γ-stable and also remain Γ-stable after 50 % gain reduction.

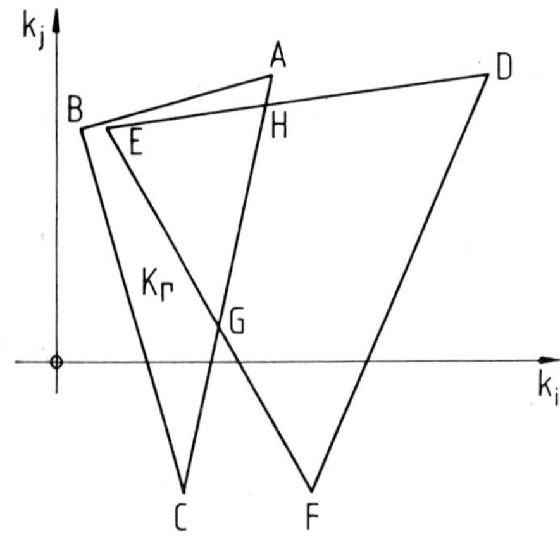

Figure 8.28

In EGH Γ-stability is preserved under 50 % gain reduction in k_i.

If an intersection still does not exist for two parallel sensors, then three sensors in parallel may be considered. An advantage of this choice is that it can be combined with a failure detection procedure on the next higher hierarchical level. Its structure is shown in figure 8.29.

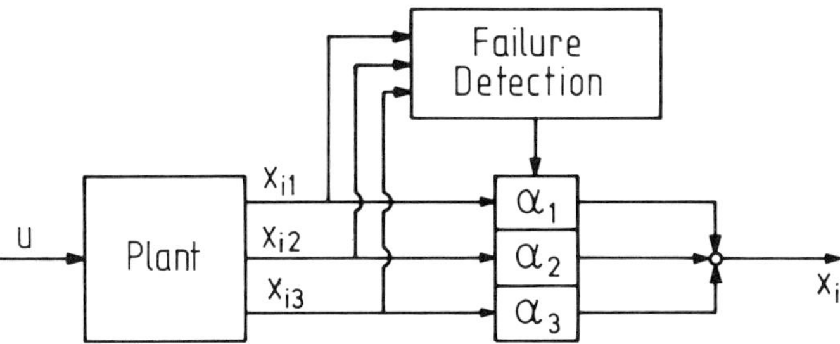

Figure 8.29 A triplex system

The failure detection system forms the three decision functions

$$\begin{aligned} d_1(t) &= [x_{i_1}(t) - x_{i_2}(t)][x_{i_1}(t) - x_{i_3}(t)] \\ d_2(t) &= [x_{i_2}(t) - x_{i_3}(t)][x_{i_2}(t) - x_{i_1}(t)] \\ d_3(t) &= [x_{i_3}(t) - x_{i_1}(t)][x_{i_3}(t) - x_{i_2}(t)] \end{aligned} \tag{8.4.3}$$

The d_k's are nominally zero; $|d_k| > \varepsilon$ indicates a failure of sensor k. In order to avoid false alarms from short impulses, $d_k(t)$ is low-pass filtered first and then compared to a threshold value.

$$\begin{aligned} \dot{f}_1(t) &= af_1(t) + d_1(t) \\ \dot{f}_2(t) &= af_2(t) + d_2(t) \\ \dot{f}_3(t) &= af_3(t) + d_3(t) \end{aligned} \tag{8.4.4}$$

The decision logic is then

Nominal state:

$$\begin{aligned} &|f_1(t)| < \varepsilon \,, \quad |f_2(t)| < \varepsilon \,, \quad |f_3(t)| < \varepsilon \\ &\alpha_1 = \alpha_2 = \alpha_3 = 1/3 \end{aligned} \tag{8.4.5}$$

Failure of sensor k:

$$\begin{aligned} &|f_k(t)| > \varepsilon \,, \quad |f_j(t)| < \varepsilon \quad \text{for } j \neq k \\ &\alpha_k = 0, \quad \alpha_j = 1/2 \quad \text{for } j \neq k \end{aligned} \tag{8.4.6}$$

No changes in α's after second and third failure.

The parameters a and ε are chosen in view of safety against false alarms, i.e. both not too small. This means that the decision may take some time. Between failure and decision times the gain is reduced to 2/3. If a second failure occurs after the first decision, then the gain is reduced to 1/2. In the unlikely case that a second failure occurs before the first one is detected, the gain is only 1/3. Thus the basic robust control system should be designed for this gain reduction margin of 50 % or 67 % for Γ-stability.

In applications where the sensors are expensive, it is desirable to substitute some measurements by feedback variables generated by feedback dynamics. The observer-state-feedback structure is feasible for this purpose, but other structures may be more convenient in the design process. This is particularly true for the design for robustness with respect to sensor failures. An observer with reduced sensor gain still has the full gain for the control input u. Thus its behavior is very different from the nominal observer. The filter feedback structure of figure 8.30 is more convenient.

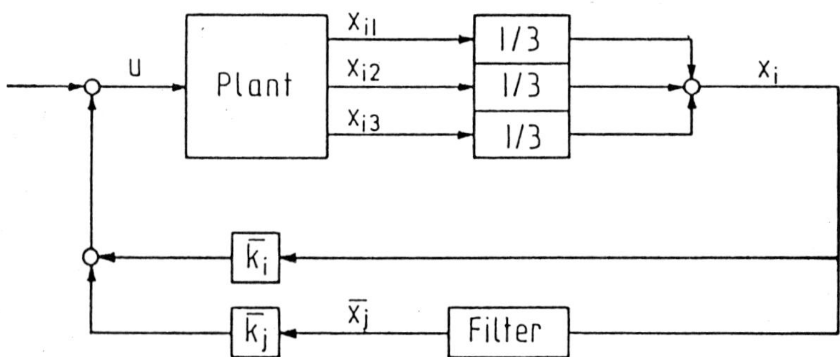

Figure 8.30 Substituting the measurement of x_j by a filter producing $\bar{x}_j$.

Here a filter produces a substitute feedback variable $\bar{x}_j$ for the true state variable x_j. The filter transfer function is chosen such that the transfer function from u to $\bar{x}_j$ at least crudely approximates the transfer function from u to x_j within the desired closed loop bandwidth and over the range of plant parameter variations.

The controller structure of figure 8.30 is especially useful if the transfer function from u to x_i is minimum-phase, because then cancellations or near-cancellations by filter poles can be made. Instability of the transfer function is no disadvantage if the same instability also occurs - as is usual - in the transfer function from u to x_j. In section 8.5.3 this concept will be illustrated by the example of an unstable aircraft.

In the filter structure a failure effects both feedback channels simultaneously, i.e. both gains $\bar{k}_i$ and $\bar{k}_j$ are reduced. For example let a failure of one sensor occur, such that the sensor and feedback gain is reduced by a factor 2/3. If ABC in figure 8.31 is the Γ-stability region, then DEF is the region for which Γ-stability is maintained after one third gain reduction in $\bar{k}_i$ and $\bar{k}_j$. Thus AGH is the region in which Γ-stability is achieved both nominally and after the sensor failure.

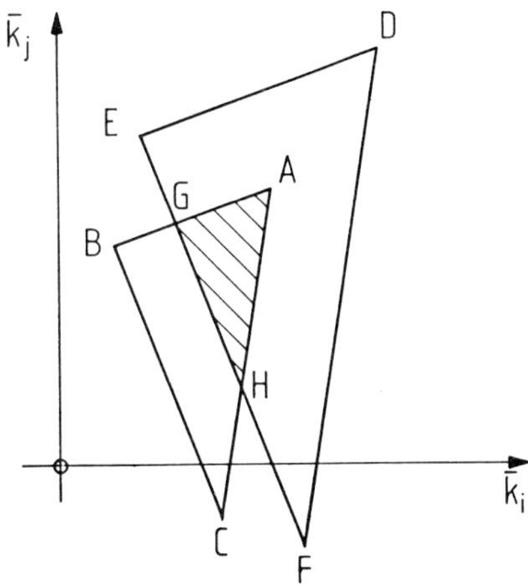

Figure 8.31

In AGH Γ-stability is preserved under 1/3 gain reduction in both channels.

The configuration of figure 8.30 with one measured state variable immediately gives the same gain reduction margins for the failure of one of several parallel actuators.

In summary, a useful design approach is:

1. Assume a state or output feedback structure and find the set of controllers which gives Γ-stability for all plant models. If no such set exists, then either the specification for Γ

must be relaxed, or the controller order can be increased, or a gain scheduling or adaptive structure may be assumed.

2. Depending on the shape of the admissible region in K space and on the cost of the sensors, a dynamic output feedback structure with filters indicated in figure 8.30 is assumed.

3. In the space of the new feedback gains, the exact common Γ-stability region for all plant models is determined.

4. From this region a design point is selected considering all other design requirements mentioned in this section.

8.5 Stabilization of the Short-period Longitudinal Mode of an F4-E with Canards

8.5.1 Design Specifications

In this section several results for the design of robust control systems are applied to a practical example in flight control. The design is for the continuous-time system.

An F4-E aircraft is fitted with supplementary canards in order to improve its maneuverability. This has the result that it is longitudinally unstable at subsonic velocities. The short period longitudinal mode then has two real poles of which one lies in the right half s-plane. A weakly damped complex pole pair arises in supersonic flight which also requires nice stabilization by the control. Data for four typical flight conditions are given in appendix D.

The example has already been formulated and examined in eq. (8.3.6). Here, however, the actual specifications for the eigenvalues of the closed loop are based on investigations of flying qualities of piloted aircraft. Pilots' eigenvalue requirements have been specified [69.7] which must be verified in the qualification tests of a new aircraft model. For the characteristic polynomial of the short-period longitudinal mode

$$s^2 + p_1 s + p_o = s^2 + 2\zeta\omega s + \omega^2 = 0 \tag{8.5.1}$$

the following boundaries are prescribed

$$\begin{aligned} 0.35 \le \zeta \le 1.3 \\ \omega_a \le \omega \le \omega_b \end{aligned} \tag{8.5.2}$$

The boundaries ω_a and ω_b for the natural frequency depend on the flight condition. Here a basic rule of robust control with constrained input amplitudes is built in. That is, a quickly reacting system (for example an airplane in high-speed low-level flight), though controlled, should remain quick. Similarly, a slowly reacting system (for example an airplane during the landing approach) should also remain slow. For the flight conditions examined here, the following limiting values ω_a and ω_b are specified [69.7]:

Flight condition	velocity (mach)	altitude (feet)	ω_a (radian / second)	ω_b (radian / second)
1	0.5	5000	2.02	7.23
2	0.85	5000	3.50	12.6
3	0.9	35000	2.19	7.86
4	1.5	35000	3.29	11.8

The region of nice stability is defined in the p_o-p_1-plane by eq. (8.5.2). It is bounded by the lines $p_o = \omega_a^2$ and $p_o = \omega_b^2$ as well as the two parabolas $p_o = \omega^2$, $p_1 = 2\zeta_{min}\omega = 0.7\omega$ and $p_o = \omega^2$, $p_1 = 2\zeta_{max}\omega = 2.6\omega$. The region for flight condition 2 is represented in figure 8.32.

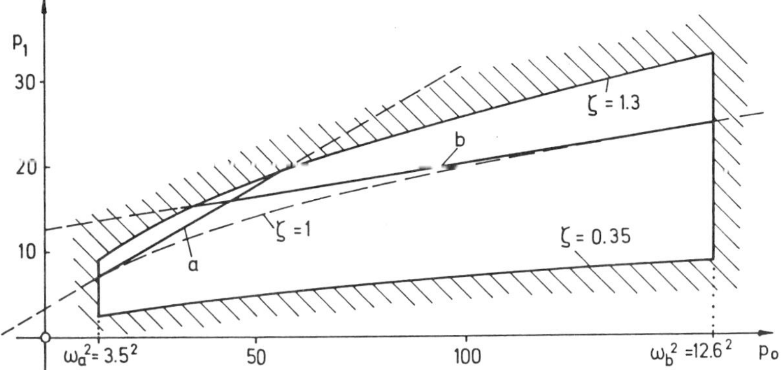

Figure 8.32 Nice stability region for flight condition 2 from eq. (8.5.2) and replacement of the boundary $\zeta = 1.3$ by the two real boundaries ω_a and ω_b.

If ζ is increased beyond one, the complex pole pair unites at a branching point for $\zeta = 1$ and then separates into two real poles of which one moves to the left and the other to the right as ζ increases. The pole moving to the right often leads to an undesired decrease in the bandwidth of the closed loop; this corresponds in the time domain to a sluggish response. These less desirable domains lie above the parabola for $\zeta = 1$, $p_1 = 2\omega$ in figure 8.32.

The demands for the additional eigenvalues, which come from the actuator or controller dynamics, can be formulated better in the s-plane. The specification $\omega_a < \omega < \omega_b$ describes an annulus. The requirement ζ greater than 0.35 cuts out a segment which resembles a pineapple segment Γ, see figure 8.33.

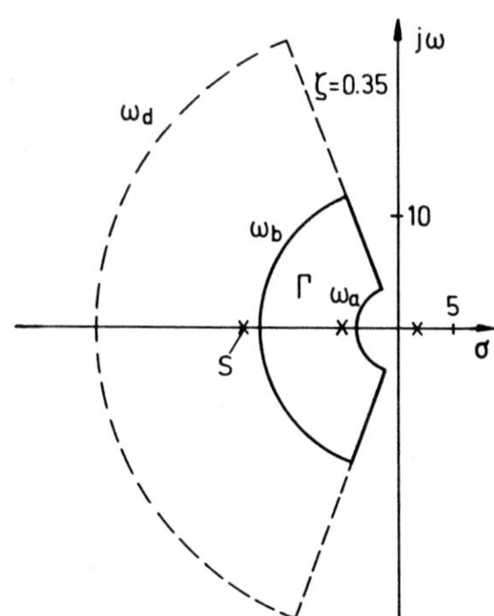

Figure 8.33

Nice stability in the s-plane

For $\zeta_{max} < 1$ the real axis is excluded. For $\zeta_{max} = 1.3$ real pole pairs can arise from which one pole may also lie a little to the left or right of the pineapple segment. Since all real pole pairs between $s = -\omega_a$ and $s = -\omega_b$ satisfy the condition $\zeta \le 1.3$, these two real boundaries can be employed instead of $\zeta \le 1.3$; this corresponds to the line a for ω_a and the line b for ω_b in figure 8.32. These are the tangents to the curve $\zeta = 1$ since a polynomial with a double pole at $s = -\omega_a$ or $s = -\omega_b$ lies both on the parabola and on the straight lines in P space. The pineapple segment is consequently fixed as the region Γ of nice stability. This case lies between the dampings 1 and 1.3 according to figure 8.32.

The region Γ could also be specified for the additional eigenvalues. Then no distinction need be made in the characteristic equation of the closed loop as to the origin of the eigenvalues. In practice this would not be a good design if the actuator pole at $s = -14$, denoted by S in figure 8.23, would have to be shifted to the right. It would be preferable to leave it to the left of the pineapple segment. In this way the eigenvalues remain distinct.

In general it must be observed that for mechanical systems the state variables are input variables for further systems which are usually left unmodelled and describe the elastic degrees of freedom. The rigid body control should be designed such that the high frequency structural oscillations are not stimulated. The transfer functions from the actuator to all individual state variables must therefore be constrained in bandwidth below the first structural-oscillation frequency. This bandwidth constraint for ω_d is satisfied by restricting the additional eigenvalues to the bandwidth circle with radius ω_d (represented in figure 8.33 by dotted lines). To simplify matters, the requirement $\zeta \leq 0.35$ is adopted for the outer region.

8.5.2 Robustness with Respect to Flight Condition

As already discussed in eq. (8.3.6), the controller structure which is easiest to implement is

$$u = -[k_{Nz} \quad k_q \quad 0]\underline{x} = -\underline{k}'\underline{x} \tag{8.5.3}$$

where x_1 is measured by an accelerometer and x_2 with a gyro. The resulting two-dimensional cutting plane through the three-dimensional region of nice stability is shown in figure 8.34 for $\omega_d = 70$ radians/second and flight condition 2.

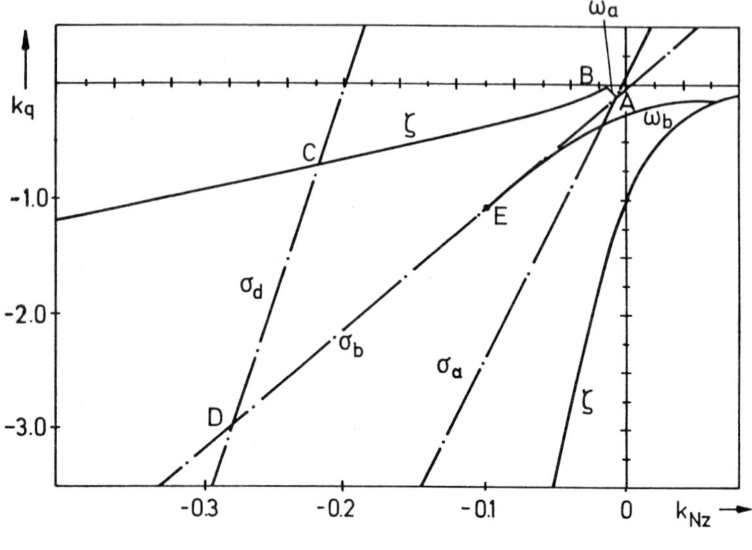

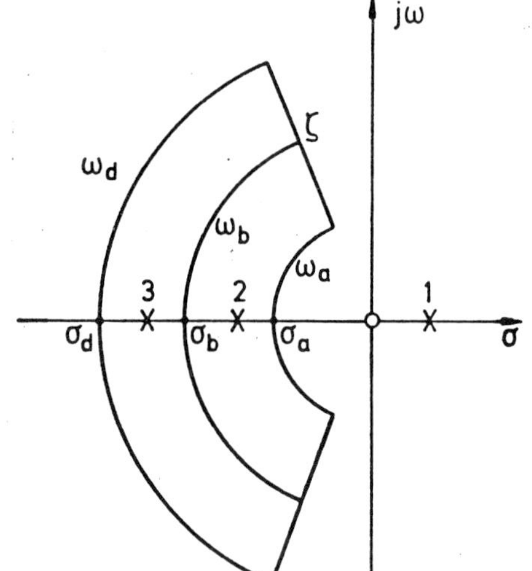

Figure 8.34

Two-dimensional cross section of the nice stability region in K space and the corresponding nice stability region in the s-plane

For $\underline{k} = \underline{0}$ the eigenvalues are those of the open loop, see 1, 2 and 3 in the s-plane figure. If the boundary σ_a in K space is crossed, then eigenvalue 2 moves across the point σ_a to the right. The eigenvalue pair (1; 2) which describes the short period longitudinal mode, now moves as a conjugate complex pole pair into the desired region if the boundaries ω_a or ζ are crossed. The desired region is thus ABCDE. On AB the eigenvalues 1 and 2 lie on the circle ω_a, on BC they lie on ζ, and on CD eigenvalue 3 lies at σ_d. On DE one of the eigenvalues (1; 2) lies at σ_b, on EA eigenvalue 3 lies at σ_b.

Remark 8.1:

A one-to-one correspondence of the eigenvalues of the closed loop to those of the open loop is still possible on the described path in the k_q-k_{Nz}-plane. Such a relation can be lost, for example, if eigenvalues 2 and 3 combine to form a complex pair, then a branching point is crossed. This would be the case, for example, if after leaving $\underline{k} = \underline{0}$, first σ_b and subsequently ω_b were crossed. Pole 3 then moves to the right first and the pair (2; 3) moves over the boundary ω_b in the s-plane.

Correspondingly the Γ-stability regions are determined for the other three flight conditions. All four regions intersect in the region represented in figure 8.35. Thus the assumed controller structure leads to an admissible solution set.

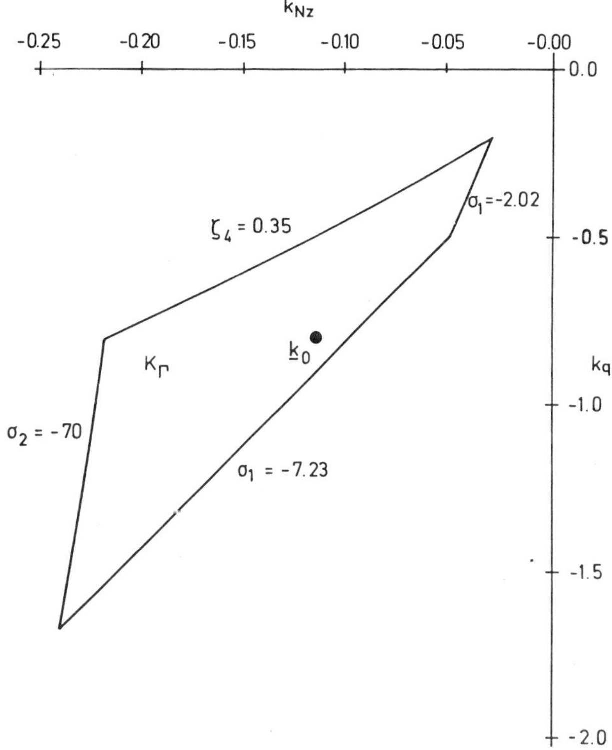

Figure 8.35 Intersection of the regions of nice stability for all four flight conditions

Figure 8.35 provides the design engineer with essential information as to which demands are critical in which flight cases. The flight condition numbers are used as indices at the boundary line names. The stimulation of structural oscillations is especially critical near the boundary $\sigma_2 = -70$, i.e. for flight condition 2 (high velocity, low-level). Insufficient damping is critical near the boundary ζ_4 for flight condition 4 (fast flight at high altitude). The other two boundaries originate from flight condition 1 (landing approach). Near the boundary $\sigma_1 = -2.02$ a real pole moves towards the origin. A real pole would cross over to the left at the boundary $\sigma_1 = -7.23$. Flight condition 3 is not critical in this controller structure.

Information on the possible demands that can be met by a control system and the possibilities for compromise is usually more helpful for the engineer than an optimization process, which requires that the trade-offs for all conceivable conflicts are decided beforehand by the choice of weightings in a performance criterion.

Figure 8.35 suggests different possibilities for coming closer to the selection of a point from the solution set. One possibility is to reduce the eigenvalue region Γ of figure 8.33. Figure 8.36 shows the tightened region obtained by a reduction of ω_d from 70 to 50 radians/second, an increase of the minimum damping from 0.35 to 0.5 and an increase of all minimum natural frequencies by 50 %. There still exists an intersection for the four flight conditions and the designer may decide which of these specifications he wants to tighten even more and thereby narrow the admissible region further.

However, other requirements can also be incorporated into the solution choice which have not been taken into consideration so far. As an example, suppose the limitation of the elevator deflection x_3 and its derivative $\dot{x}_3 = -14x_3 + 14\underline{k}'\underline{x}$ is essential. Then a solution with a smaller loop gain must be chosen as explained in section 8.4.2. In order to illustrate this effect, the C^*-response to a unit step input given by the pilot has been calculated, as well as the required elevator deflection x_3 for the points $\underline{g}_1$,

$\underline{g}_2$ and $\underline{g}_3$ in figure 8.36, see figure 8.37. (The output variable C* is introduced in appendix D.)

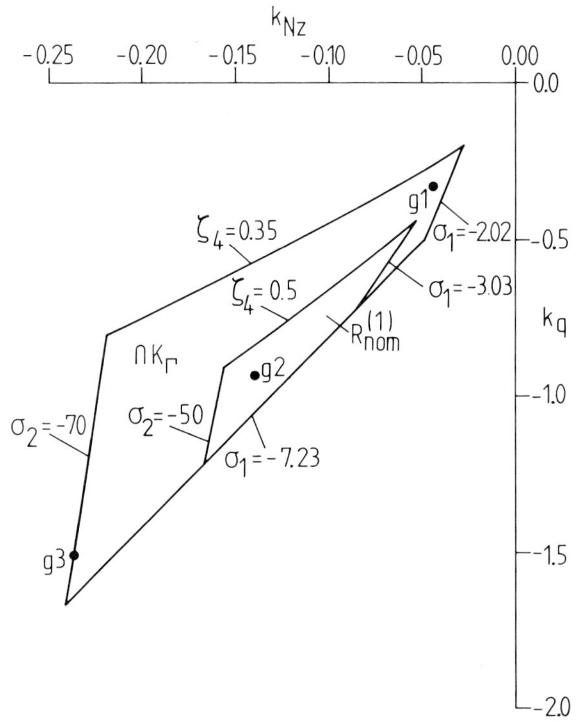

Figure 8.36

Tightened specifications reduce the admissible solution set

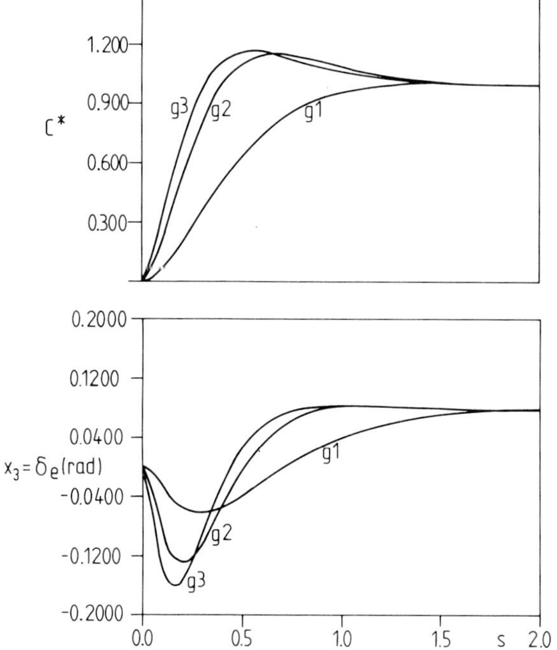

Figure 8.37

C*-response and elevator deflection δ_e for low (g_1), medium (g_2), and high (g_3) loop gain

The elevator deflection is decreased significantly for a small loop gain g_1 and the step response is correspondingly slow. A high loop gain g_3 is unsatisfactory due to too much overshoot of the step response, greater input amplitude, and extension of the bandwidth to the vicinity of the structural-oscillation frequency. The medium loop gain $\underline{g}_2$ is the best in this comparison, but it produces a solution which is closer to $\underline{g}_3$ than to $\underline{g}_1$ and a point $\underline{k}_o$ slightly closer to $\underline{g}_1$ was finally selected. The controller is

$$u = -\underline{k}_o\underline{x} = -[-0.115 \quad -0.8 \quad 0]\underline{x} \tag{8.5.4}$$

$\underline{k}_o'$ is indicated in figure 8.35. Calculating and testing the eigenvalues shows, that they lie in their respective prescribed regions for all flight conditions.

8.5.3 Robustness against Sensor Failures

In a further refinement of the design, the requirement of robustness with respect to gyro or accelerometer failure is introduced. Figure 8.35 shows that the nice stability region does not intersect the axes k_{Nz} and k_q. Therefore one of the two feedbacks alone is not sufficient for nice stabilization.

A first attempt would be to use two parallel gyros and two parallel accelerometers. From figure 8.35, it is seen that there are no points in the nice stability region which admit 50 % gain reduction both in k_{Nz} and k_q. Only emergency requirements (level 3 of the flying qualities [69.7]) can be achieved [81.6]. Interestingly the shape of K_Γ in figure 8.35 with its lengthy extension away from the origin is such that the flying qualities are improved if a second failure in the other channel occurs, i.e. for a simultaneous 50 % reduction of gyro and accelerometer gain. This indicates that a controller structure as in figure 8.30 with only gyros or with only accelerometers is advantageous. Here any gain reduction occurs in both feedback gains simultaneously.

Of course the unmeasured variable cannot be really reconstructed for all four flight conditions simultaneously by a constant filter. Also the system order is increased by the filter such that additional eigenvalues occur. Therefore the new stability region in the plane of the two feedback gains must be established. A similar form of the stability region can be expected, at least if the filter is chosen so that it yields an approximation for the unmeasured state variable.

In deciding whether to use only gyros or only accelerometers we consider the zeros of the two transfer functions. In the four flight conditions the zeros in the accelerometer transfer function vary between $-0.4\pm j5.7$ and $-0.9\pm j9.1$. They are minimum phase but outside the nice stability region. Also they vary widely such that an approximate cancellation by the polynomial of averaged zeros $s^2 + 1.172\,s + 49.9$ is not advisable. This would generate closed-loop poles which are not Γ-stable, possibly not even stable.

The gyro transfer function has one real zero varying between -0.64 and -1.57. Here an approximate cancellation by the polynomial of averaged zeros $s + 0.98$ is no problem. By the almost cancellation, this filter pole will be only weakly controllable from u and will not be shifted much by closing the loop. Also it is weakly observable from the pitch rate and can be exempted from the pole region requirement.

Both transfer functions from u to $N_Z = x_1$ and $q = x_2$ have the same poles; one pole is unstable in subsonic flight. The filter must replace the numerator of the gyro transfer function by the numerator of the accelerometer transfer function. Taking the averaged zeros of the two transfer functions the filter is

$$f(s) = A \times \frac{s^2+1.172s+49.9}{s+0.98} \times \frac{10}{s+10} \tag{8.5.5}$$

The term $10/(s+10)$ was introduced to make the filter realizable. The gain ratio A varies between 0.527 and 0.577 in the four flight conditions and was chosen at the average value $A = 0.543$.

By introducing the filter, the system order is raised to $n = 5$, i.e. the region of nice stability in P space is five dimensional.

The chosen controller structure of figure 8.30 defines a plane of the free controller parameters $\bar{k}_{Nz}$ and $\bar{k}_q$. In this plane the intersection of nice stability regions for the four flight conditions must be analyzed. It is represented in figure 8.38.

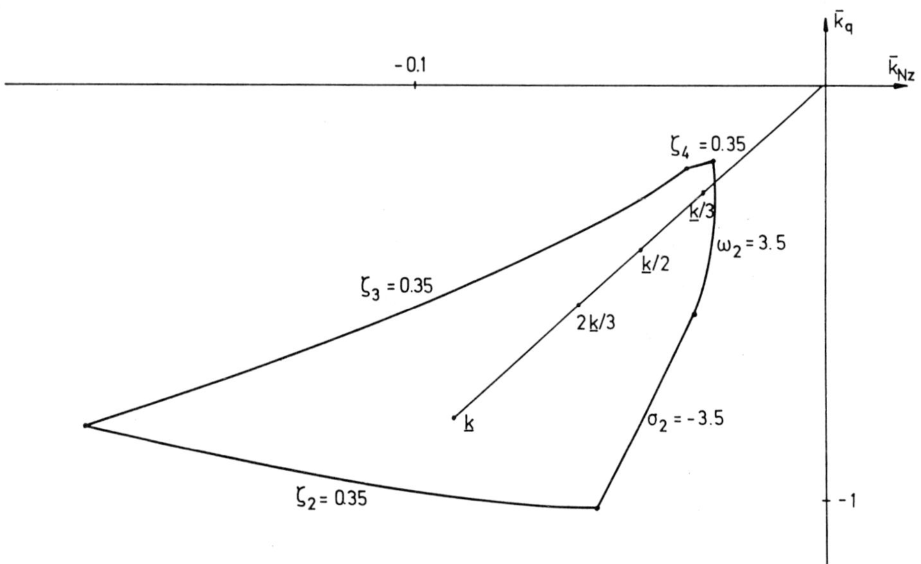

Figure 8.38 Intersection of nice stability regions for four flight conditions for the controller structure using only gyros and a feedback filter

The allowable region is clearly different from the region in figure 8.35. Also flight condition 3 instead of flight condition 1 now contributes to the boundary. Nevertheless, it confirms our expectation that a large simultaneous gain-reduction margin is now achievable, in the extreme case 80 %. If one chooses the indicated point $\underline{k} = [-0.09 \quad -0.8 \quad 0]'$, then the points $2\underline{k}/3$, $\underline{k}/2$ and $\underline{k}/3$ lie well in the admissible region. Therefore the system can handle the failure of two of the three parallel gyros, even if the second failure occurs before the first one is detected. For the C*-step responses in figure 8.39 it is also worth remarking how little the step response is changed in all four flight conditions if $\underline{k}$ (steeply climbing curve) is reduced to $\underline{k}/2$.

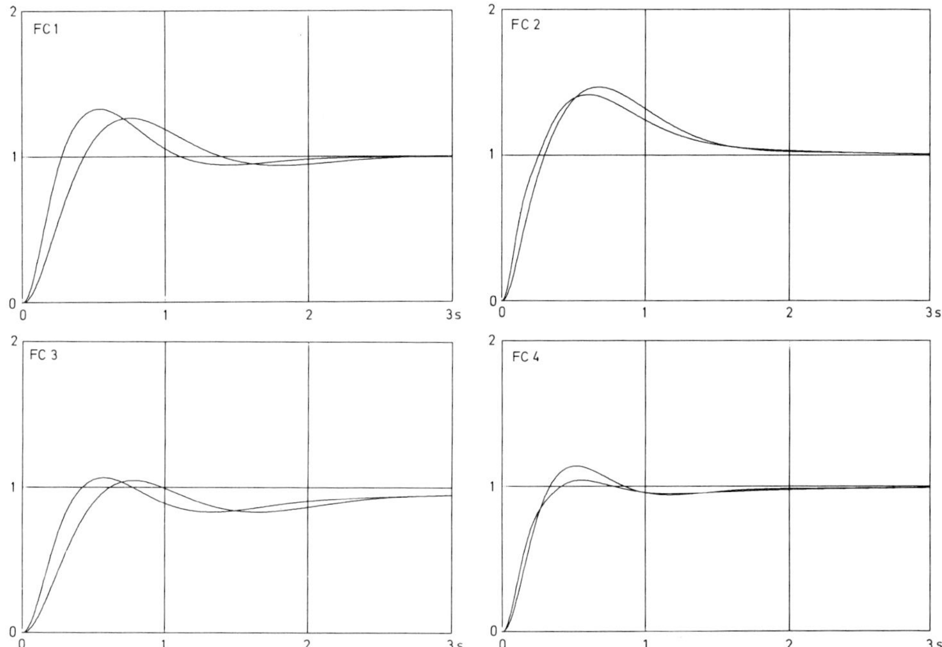

Figure 8.39 C*-step response in four flight conditions. Each case shows nominal and 50 % reduced loop gains. The nominal gain results in the steeper initial rise.

If the two feedback paths via $\bar{k}_q$ and $\bar{k}_{Nz}$ are combined, then the following controller is obtained

$$\frac{-u_S(s)}{x_{2S}(s)} = -0.8 - 0.09 \times 0.543 \times \frac{s^2+1.172s+49.9}{(s+0.98)(s+10)}$$

$$= -\frac{0.8489s^2+8.8413s+10.2786}{s^2+10.98s+9.8} \qquad (8.5.6)$$

$$= -0.8489 \frac{(s+1.333)(s+9.082)}{(s+0.98)(s+10)}$$

The minus sign is explained by the common flight mechanics definition of the sign of the elevator deflection x_3 and the pitch angle q_1 which causes a minus sign of the plant transfer function.

The procedure for the design of robust stabilization for the F4-E is certainly not a general design recipe; but similar results can be expected for other aircraft. In the robust design of a back-up controller for the new Swedish fighter JAS 39 it was assumed from the beginning that only the gyro is used for feedback. The controller in this case was assumed as a proportional channel plus two channels with first order delay. A robust level 1 controller for 10 flight conditions was designed using both the parameter space technique and the vector performance optimization described in section 8.6 [84.4].

This extensive F4-E example shows how the design tool of two-dimensional cross sections through regions of nice stability is applied and combined with other design considerations. The controller structure was developed in steps. The design would be less transparent if the second order controller of eq. (8.5.6) with five free controller parameters had been assumed from the beginning.

8.6 Design by Optimization of a Vector Performance Criterion

A compromise between very different requirements must often be made in the design of control systems. A typical situation occurs when a solution has been found, by previous design steps, that already achieves some essential specifications, but is unsatisfactory in regard to other requirements. For example, in the initial design process it would be quite time consuming to calculate a larger number of step responses or other trajectories which are finally used to judge the quality of the design. Once we are close to a solution we may wish to reduce the overshoot of the step response for one of the operating points, or to improve a particular disturbance response, or to reduce an initial peak in $|u|$ etc. Here it is helpful to optimize the control parameters in subsequent design steps. The goal of this procedure is to improve certain criteria as far as possible while preserving the performance that has already been obtained.

In the optimization of a vector performance criterion according to Kreisselmeier and Steinhauser [79.4], all requirements on the control system are formulated as individual performance criteria g_i, $i = 1,2...N$, which are always greater than or equal to zero and which should be minimized or made smaller than an assigned value. Such criteria may include properties of typical trajectories or their deviations from reference trajectories, frequency domain stability margins, the absolute values of feedback gains or maximum actuator signals, or the position of eigenvalues. The N performance criteria are combined into a performance vector

$$\underline{g}(\underline{k}) = [g_1[\underline{k}] \quad g_2[\underline{k}] \quad \dots \quad g_N[\underline{k}]]' \tag{8.6.1}$$

The criteria depend on the parameters of the plant and the controller. For a family of plant models the number N of the individual criteria is raised correspondingly.

A Pareto-optimal value of $\underline{g}$ is wanted. The definition of Pareto-optimality is illustrated by figure 8.40 for the case of two indices g_1 and g_2.

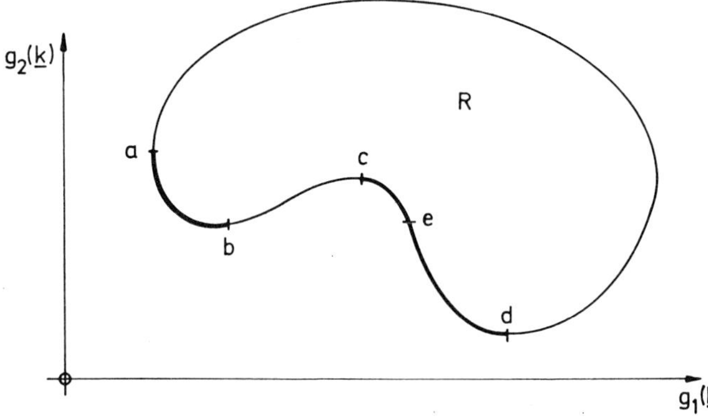

Figure 8.40
On the definition of Pareto-optimal solutions

The closed region R describes all possible values of the performance vector for admissible values of $\underline{k}$ (e.g. $\underline{k} \in K_\Gamma$ from pole region assignment, or all real $\underline{k}$). Now all $\underline{g}$-vectors are excluded for which, locally, both criteria g_1 and g_2 can be improved simultaneously. There remain some parts of the boundary, in the example of figure 8.40 the parts ab and cd. In these portions one performance index can only be improved at the expense of the other one. These solutions are called Pareto-optimal. The portions ab and ed are globally Pareto-optimal, the portion ce is only locally Pareto-optimal. Ideally we try to find a globally optimal solution. But a procedure, which finds a locally optimal solution, is also a useful tool.

Suppose the designer knows the set of Pareto-optimal solutions. Then he has information on which tradeoffs are possible for a given plant and he can select the most desired solution from the Pareto-optimal set. Practically it is difficult to compute and visualize the set of Pareto-optimal solutions.

The design strategy of Kreisselmeier and Steinhauser [79.4] allows a systematic search for a desirable Pareto-optimal solution. During this interactive search the designer learns about the conflicts in design objectives and the possible tradeoffs. The idea is illustrated by figure 8.41.

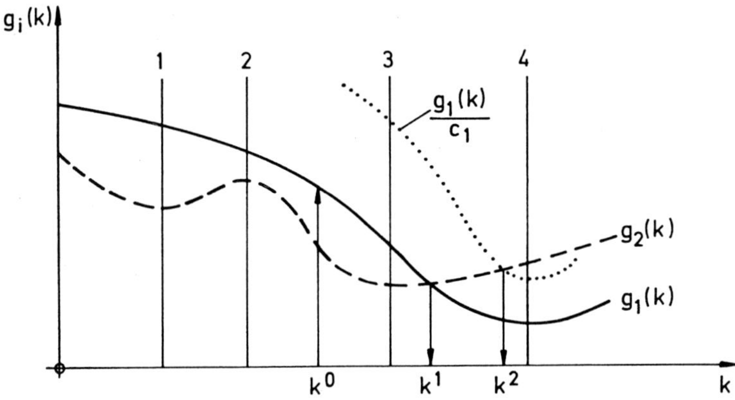

Figure 8.41 Two performance indices g_1 and g_2 depending on only one controller parameter k.

Pareto-optimal solutions are located between the vertical lines 1 and 2 (local solutions) and between 3 and 4 (global solutions). Starting from an initial guess k^0 a Pareto-optimal solution k^1 is found by minimizing the maximal component of $\underline{g}$

$$\underline{g}^1 = \underline{g}(k^1) = \min_k \max\{g_1(k), \ g_2(k)\} \tag{8.6.2}$$

Now assume the designer does not like this solution. He wants to reduce g_1. He can achieve this by dividing $g_1(k)$ by a factor c_1, $0 < c_1 < 1$, i.e. he computes

$$\underline{g}^2 = \underline{g}(k^2) = \min_k \max \left\{\frac{g_1(k)}{c_1}, \quad g_2(k)\right\} \tag{8.6.3}$$

He obtains a different Pareto-optimal solution k^2. Each Pareto-optimal solution can be found by appropriate choice of c_1.

In general the design can be steered in a desired direction in a systematic way by the following procedure:

i) Choose $\underline{c}^1$ such that

$$\underline{g}(\underline{k}^o) < \underline{c}^1 \quad (8.6.4)$$

(The notation $\underline{a} < \underline{b}$ means $a_i \leq b_i$ for $i = 1,2...M$ and $\underline{a} \neq \underline{b}$).

ii) In the ν-th design step a vector of design parameters $\underline{c}^\nu$ is chosen such that

$$\underline{g}(\underline{k}^{\nu-1}) < \underline{c}^\nu < \underline{c}^{\nu-1} \quad (8.6.5)$$

$\underline{k}^{\nu-1}$ denotes the result of the previous design step. If some components of $\underline{g}$ have already reached satisfactory values, then the corresponding components of $\underline{c}$ may be kept constant: $c_i^\nu = c_i^{\nu-1}$. For components of $\underline{g}$ which should be reduced the best achieved value is chosen: $c_j^\nu = g_j(\underline{k}^{\nu-1})$.

iii) The solution of the min-max problem for finding $\underline{k}^\nu$ may be approximated by the scalar optimization problem [79.4], [83.8]

$$\min_{\underline{k}} \{\frac{1}{\rho} \ln \sum_{i=1}^{L} \exp [\rho g_i(\underline{k})/c_i^\nu]\} \quad (8.6.6)$$

with sufficiently large ρ. Here an unconstrained optimization is used as a design tool instead of more complicated nonlinear programming techniques.

iv) The iteration terminates when

$$\underline{g}(\underline{k}^\nu) \approx \underline{c}^\nu \quad (8.6.7)$$

Then by construction the sequence of design vectors is monotonically decreasing, i.e.

$$\underline{g}(\underline{k}^{\nu}) < \underline{c}^{\nu} < \underline{c}^{\nu-1} < \ldots < \underline{c}^{1} \tag{8.6.8}$$

This design method is implemented in the computer program REMVG by Steinhauser [80.7]. He also worked out the following example with REMVG.

Example: Loading bridge

The starting point is the result of eq. (8.3.5)

$$u = -[500 \quad 2769 \quad -21557 \quad 0]\underline{x} = -\underline{k}^{O}{}'\underline{x} \tag{8.6.9}$$

For loads in the interval 50 kg < m_L < 2395 kg this controller shifts the eigenvalues to the left side of the left branch of the hyperbola $\omega^2-4\sigma^2+0.25 = 0$. The maximal force is $\max|u(t)| = 500$ Newton for a transition from $\underline{x}_A = \underline{x}(0) = [1 \quad 0 \quad 0 \quad 0]'$ to $\underline{x}_E = \underline{0}$. These requirements are expressed by the following criteria in the performance vector $\underline{g} = [g_1 \quad g_2 \quad g_3]'$

$$g_1 = \max_i \exp\{10[\omega_i^2-4\sigma_i^2+0.25]\}$$

ω_i and σ_i are determined from the eigenvalues of the closed loop for $m_L = 2395$ kg

$$g_2 = \max|u(t)| \text{ for } 0 < t < 5 \text{ seconds} \tag{8.6.10}$$

for the transition from $\underline{x}_A$ to $\underline{x}_E$ with $m_L = 2395$ kg

g_3 as g_1 but for $m_L = 50$ kg.

$g_1 \le 1$, $g_3 \le 1$, and $g_2 \le 500$ are required. $\underline{c}^1 = [1 \quad 500 \quad 1]'$ is given as the first assignment vector. The solution of the minimax problem yields

$$\underline{g}^1 = [0.96 \quad 516 \quad 0.88] \tag{8.6.11}$$

It is impossible to reduce all criteria simultaneously because the starting value was already optimal. Therefore an improvement can result only from an altered controller structure.

The actuator signal u(t) is represented in figure 8.42 a. This shows that the maximal value u = - 500 Newton arises only as an initial peak and only much smaller forces are needed later.

This suggests searching for solutions with a smaller g_2. The initial peak in u is caused by the proportional feedback of x_1 with k_1 = 500. A reduction of k_1 only would slow down the solution and still not make efficient use of the actuator power.

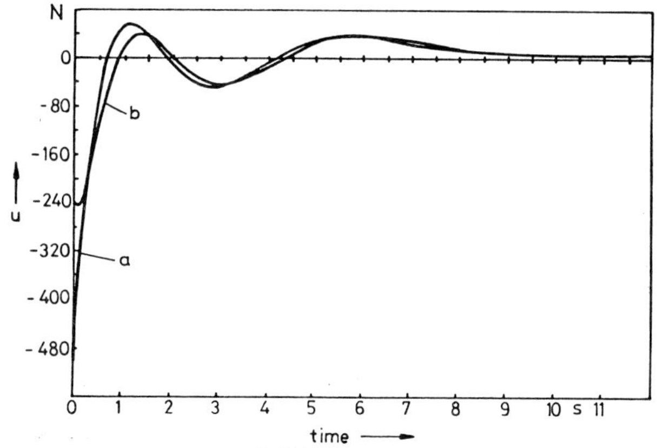

Figure 8.42 Input signal for m_L = 2395 kg

a) only proportional feedback of the trolley position x_1

b) proportional and delayed feedback

But if the reduced k_1 is paralleled by a feedback with a small time constant (see figure 8.43) then after a short transient the sum of the two feedbacks of x_1 is effective, but the short initial peak in u is reduced. (An alternative would be a low pass prefilter at w.)

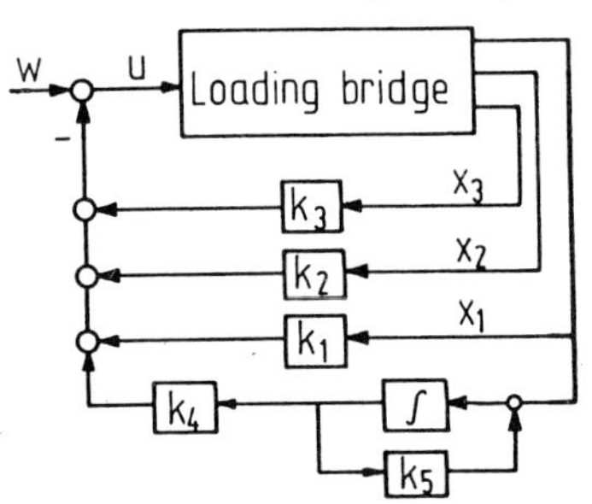

Figure 8.43

Extended controller structure with delayed feedback of the trolley position x_1

k_1, k_2 and k_3 from eq. (8.6.9) and $k_4 = k_5 = 0$ are used as starting values, i.e.

$$\underline{k}^{o} = [500 \quad 2769 \quad -21557 \quad 0 \quad 0]' \tag{8.6.12}$$

The averaged quadratic deviation of the trolley position from its end position at zero for times after eight seconds (since then the fast parts of the oscillation process have faded. See figure 8.42) is now introduced as a criterion. The performance vector (8.6.4) therefore is extended by

$$g_4 = \frac{1}{t_2 - t_1} \int_{t_1}^{t_2} x_1^2 dt \;, \; t_1 = 8s, \; t_2 = 30s, \; m_L = 2395 \text{ kg}$$

$$g_5 \text{ as } g_4 \text{ but for } m_L = 50 \text{ kg} \tag{8.6.13}$$

In design steps of the kind (8.6.8) c_2 was considerably diminished in order to reduce the input amplitude without worsening the other quality criteria. The result was

$$\underline{k}' = [234 \quad 3048 \quad -22701 \quad 984 \quad -2.96] \tag{8.6.14}$$

$$\underline{g}' = [0.92 \quad 234 \quad 0.80 \quad 0.66 \quad 0.67] \tag{8.6.15}$$

The maximal input amplitude g_2 was reduced from 500 N to 234 N, see figure 8.42 b. It occurs after a short time $t > 0$, not at $t = 0$. The pole region requirements $g_1 < 1$, $g_3 < 1$ are also improved to $g_1 < 0.92$, $g_3 < 0.80$ for the two extreme cases $m_L = 50$ kg and $m_L = 2395$ kg. (Whether the requirements are met for the entire interval of m_L remains to be checked.)

In this chapter two tools for the design of robust control systems were introduced, namely parameter space design and the optimization of a vector performance criterion. Both procedures can be effectively combined.

Parameter space studies are especially suitable in order to satisfy the requirement of simultaneous nice stabilization, using the most important controller parameters. It can provide a global overview of the allowable solution set, for example the shape and size of its separate components. It shows the conflicts between different specifications and possible compromises. The numerical parameter optimization is especially suitable for the systematic improvement of a satisfactory starting value in order that additional demands on the control system are achieved. Using a large computer, a great variety of criteria can be employed, even criteria which require the frequent calculation of trajectories. These can be determined in principle even by simulation with a nonlinear plant model. In the loading bridge example the repeated calculation of eigenvalues was required for the determination of g_1 and g_3 and the repeated calculation of trajectories for g_2, g_4 and g_5. By comparison, the parameter space solution of figure 8.16 was found with a pocket calculator, paper and pencil. If a solution was calculated by parameter optimization, then it is helpful to examine the neighborhood of the solution in different two-dimensional cross-section planes through the space of the control parameters or of the plant parameters. Figures 8.35 and 8.38 illustrate for example that a minimization of $||k||$ leads to a solution in a corner point of the admissible solution set and it is advisable to modify the solution slightly such that some safety margins from the boundaries are maintained.

We have not yet pursued the question of mapping nice stability regions for a given controller into the space of physical parameters $\underline{\theta}$, see figure 8.2 and the discussion after eq. (8.1.4). This, however, can be developed from eq. (8.1.3) in the same manner as the mapping of boundaries into K space. It must be considered that $\underline{p}(\underline{k}, \underline{\theta})$ is generally a nonlinear function of $\underline{\theta}$ but a linear function of $\underline{k}$.

8.7 Exercises

8.1 For a first order system $x[k+1] = ax[k] + bu[k]$ sketch some typical sets in the a-b-plane for which a simultaneous stabilization is not possible by a constant, finite dimensional, linear controller.

8.2 For the pair of discrete-time plants

$$\underline{A}_1 = \begin{bmatrix} 0 & 1 \\ -a & 0 \end{bmatrix}, \quad \underline{b}_1 = \begin{bmatrix} 0 \\ 1 \end{bmatrix}.$$

$$\underline{A}_2 = \underline{A}_1 \quad , \quad \underline{b}_2 = -\underline{b}_1$$

choose a convenient value a in the interval $6 < a < 10$ and find a constant third order controller for simultaneous stabilization. (Hint: Start with the assigned polynomials, e.g. $P_1(z) = (z+0.99)^5$ and $(z-0.99)^5$. Then determine a, $\underline{q}$ and $\underline{v}$ as in eq. (8.2.14).)

8.3 Construct an example of three second order discrete-time plants, with the property that each pair of two of them can be simultaneously stabilized by static or dynamic state feedback, but not all three simultaneously.

8.4 Given

$$\dot{\underline{x}} = \begin{bmatrix} 0 & 1 & 0 \\ 0 & 0 & 1 \\ 0 & 0 & 0 \end{bmatrix} \underline{x} + \begin{bmatrix} 0 \\ 0 \\ 1 \end{bmatrix} u$$

Assume the feedback structure

$$u = - [0.125 \quad k_2 \quad k_3]\underline{x}$$

a) Determine the region in k_2-k_3-plane in which the closed loop has damping $\geq 1/\sqrt{2}$.

b) Find the feedback vector with minimum norm $\sqrt{k_1^2+k_2^2}$ which assigns damping $= 1/\sqrt{2}$, and determine the corresponding closed-loop eigenvalues.

8.5 Given:

$$\underline{x}[k+1] = \begin{bmatrix} 2 & 0 \\ \frac{1}{1+\alpha} & 1 \end{bmatrix} \underline{x}[k] + \begin{bmatrix} 1 \\ 0 \end{bmatrix} u[k]$$

Find

a) The set of all state feedbacks $\underline{k}' = [k_1 \quad k_2]$ such that $u = -\underline{k}'\underline{x}$ stabilizes the system for all $\alpha > 0$.

b) From this set the point which allows the maximal error Δk in $u = -[\underline{k}_1 \pm \Delta k \quad k_2 \pm \Delta k]\underline{x}$ without loss of stability.

8.6 Given:

$$\underline{x}[k+1] = \begin{bmatrix} 2 & 0 \\ 2+\frac{1}{\alpha} & 0 \end{bmatrix} \underline{x}[k] + \begin{bmatrix} 1 \\ 1 \end{bmatrix} u[k]$$

Find

The set of all state feedbacks $u = -\underline{k}'\underline{x}$ such that

1. The system is stabilized for all α in the interval $1 \leq \alpha \leq 2$, and moreover
2. this stabilization is still preserved for the two sensor failure situations $\underline{k}_a' = [0 \quad k_2]$ and $\underline{k}_b' = [k_1 \quad 0]$.

8.7 Let the two values $\alpha_1 = 0.45$ and $\alpha_2 = 0.55$ be given for α in example (8.3.4). Examine the simultaneous Γ-stabilization for

a) Γ = unit circle
b) Γ = circle with center $z = 0$ and radius $r = 0.4$.

Discuss what happens for $\alpha_3 = 0.5$.

8.8 The loading bridge was examined in continuous time in figure 8.16. Suppose the input variable is sampled and held with a sampling period $T = \pi/8$, and the load mass lies in the interval 50 kg < m_L < 3000 kg. Let the controller structure be $u = -[500 \quad k_2 \quad k_3 \quad 0]$. Determine the region in this k_2-k_3-plane for which all eigenvalues lie in a circle with radius 0.5 around $z = 0.45$. Select a central point $\underline{k}'$ in the admissible region. Calculate the actuator input u and the load position y_L for the three load cases a) m_L = 50 kg, b) m_L = 1000 kg, c) m_L = 3000 kg in the transition from $\underline{x}_A = [1 \quad 0 \quad 0 \quad 0]'$ to $\underline{x}_E = \underline{0}$.

8.9 Let $x_2 = \dot{x}_1$ be unmeasurable for the system from exercise 8.8. It is replaced by a variable $\hat{x}_2$ which is produced from x_1 by an approximately differentiating filter $f_z(z)=2(z-1)/T(z+1)$ according to eq. (3.4.32). Solve exercise 8.8 for the controller structure $u = -500x_1 - k_2\hat{x}_2 - k_3x_3$.

8.10 Extend the controller structure from problem 8.9 so that it also produces $x_4 = \dot{x}_3$, approximated as $\hat{x}_4$, by a differentiating filter. Treat exercise 8.8 in the k_3-k_4-plane which is fixed by the controller structure $u = -500x_1 - 3000\hat{x}_2 - k_3x_3 - k_4\hat{x}_4$.

8.11 Extend the problem from exercise 8.10 to robustness with respect to variation of the cable length ℓ. Examine the four cases 1) $m_L = 50$ kg, $\ell = 5$ m, 2) $m_L = 3000$ kg, $\ell = 5$ m, 3) $m_L = 3000$ kg, $\ell = 20$ m, 4) $m_L = 50$ kg, $\ell = 20$ m.

8.12 (Project) Choose an admissible controller for problem 8.11 and map the nice stability boundary (center $z = 0.45$, radius 0.5) into the ℓ-m_L-plane.

8.13 For the direct current motor of exercise 7.8 start with the values $k_2 = -0.4$, $k_3 = 1$. Examine Γ-stability for the reduction of the gain k_3. Determine the set of controller parameters k_2, k_3 for which k_2 as well as k_3 can be continuously reduced from their nominal value to zero without leaving the admissible pole region.

8.14 (Project) For the track-guided bus of appendix D.2 determine a controller in the structure of figure 8.11 such that for all eight extremal combinations of speed, mass and road adhesion, the closed-loop eigenvalues are left of the left branch of the hyperbola

$$\left(\frac{\sigma}{0.06}\right)^2 - \left(\frac{\omega}{0.12}\right)^2 = 1$$

Use the invariance plane as discussed in section 8.3.3.

9 Multivariable Systems

In the last two decades many papers and books on linear systems with several inputs and several outputs, i.e. multivariable systems, have been published. More and more abstract notations and definitions have been introduced in both frequency domain and state space descriptions and other newly developed formalisms. It is not the aim of this last chapter to review this vast amount of literature. The main motivation for this chapter is the observation that some practically important concepts can be made clear and applicable in a simple language if we restrict ourselves to the discrete-time case. This applies in particular to the theory of finite effect sequences (FES) which is developed and applied in this chapter. Therefore this topic seems to be an appropriate conclusion for a book on sampled-data systems.

9.1 Controllability and Observability Structure

9.1.1 Feedforward Control

Consider a plant with r actuator inputs $u_1, u_2 \dots u_r$, combined into an input vector $\underline{u} = [u_1, u_2 \dots u_r]'$. Its continuous-time state-space model is

$$\dot{\underline{x}} = \underline{F}\underline{x} + \underline{G}\underline{u} \tag{9.1.1}$$

The discretization with synchronously sampled and held inputs yields

$$\underline{x}[k+1] = \underline{A}\underline{x}[k] + \underline{B}\underline{u}[k] \tag{9.1.2}$$

$\underline{A}$ and $\underline{B}$ are calculated as described in section 3.1 by

$$\underline{R} = \int_0^T e^{\underline{F}v} dv = T \sum_{m=0}^{\infty} \frac{1}{(m+1)!} \underline{F}^m T^m$$

$$\underline{A} = e^{\underline{F}T} = \underline{I} + \underline{FR}$$

$$\underline{B} = \underline{RG} \tag{9.1.3}$$

Let rank $\underline{B} = r$, that is, redundant inputs have been eliminated. The solution of the difference eq. (9.1.2) for N sampling intervals is generated iteratively as follows:

$$\begin{aligned} \underline{x}[1] &= \underline{A}\,\underline{x}[0] + \underline{Bu}[0] \\ \underline{x}[2] &= \underline{A}^2\underline{x}[0] + \underline{ABu}[0] + \underline{Bu}[1] \\ &\vdots \\ \underline{x}[N] &= \underline{A}^N\underline{x}[0] + \underline{A}^{N-1}\underline{Bu}[0] + \underline{A}^{N-2}\underline{Bu}[1] + \ldots + \underline{Bu}[N-1] \end{aligned} \tag{9.1.4}$$

With the input sequence vector

$$\underline{u}_{N-1} := \begin{bmatrix} \underline{u}[0] \\ \underline{u}[1] \\ \vdots \\ \underline{u}[N-1] \end{bmatrix} \tag{9.1.5}$$

eq. (9.1.4) can be written as

$$\underline{x}[N] = \underline{A}^N\underline{x}[0] + [\underline{A}^{N-1}\underline{B}\ ,\ \underline{A}^{N-2}\underline{B}\ \ldots\ \underline{B}]\underline{u}_{N-1} \tag{9.1.6}$$

The system can be transferred from any given initial state $\underline{x}[0]$ into any desired final state $\underline{x}[N] = \underline{x}_E$ if and only if

$$\text{rank}\ [\underline{A}^{N-1}\underline{B}\ ,\ \underline{A}^{N-2}\underline{B}\ \ldots\ \underline{B}] = n \tag{9.1.7}$$

where n is the order of the system. This relation has already been examined as eq. (4.1.4) for the single-input case with a column vector $\underline{b}$ instead of the n×r matrix $\underline{B}$. The rank relation shown in eq. (4.1.6) now holds for the submatrices: if the columns of $\underline{A}^i\underline{B}$ are linearly dependent on the columns of $\underline{B}$, $\underline{AB}$ ... $\underline{A}^{i-1}\underline{B}$, then the same holds for $\underline{A}^{i+1}\underline{B}$. This implies that the rank of the matrix in eq. (9.1.7) must have reached its maximum for N = n.

Therefore this condition is equivalent to

$$\text{rank}\ [\underline{A}^{n-1}\underline{B}\ ,\ \underline{A}^{n-2}\underline{B}\ \dots\ \underline{B}] = n \tag{9.1.8}$$

This is the necessary and sufficient condition for controllability and reachability of the system (9.1.2).

Remark 9.1:

> As was discussed in section 4.1, eq. (9.1.8) is strictly speaking the necessary and sufficient condition for reachability. Controllablity follows from reachability. But the inverse implication holds only if all zero-eigenvalues are reachable, i.e. by the Hautus criterion if
>
> $$\text{rank}\ [\underline{A}\ ,\ \underline{B}] = n \tag{9.1.9}$$
>
> If $\underline{A}$ is generated by discretization of a continuous system as in eq. (9.1.3) then this condition is always satisfied. This remains true if the zero eigenvalues originate from additional time-delay, see section 3.8. Thus for all practical purposes the two properties are equivalent and we will speak only about controllability in the following.

In the controllable single-input case the choice of N = n leads to a unique input sequence. In the controllable multi-input case, however, the equation

$$\underline{x}[n] - \underline{A}^n\underline{x}[0] = [\underline{A}^{n-1}\underline{B}\ ,\ \underline{A}^{n-2}\underline{B}\ \dots\ \underline{B}]\underline{u}_{n-1} \tag{9.1.10}$$

admits many solutions $\underline{u}_{n-1}$. The following examination of these solutions will lead to interesting insights into the controllability structure of multi-variable systems.

Consider first only the input variable which has been designated as u_1 and the corresponding column $\underline{b}_1$ from $\underline{B}$. If all elements of $\underline{A}$ and $\underline{b}_1$ are produced by a random number generator then $(\underline{A}, \underline{b}_1)$ will be controllable with probability one. The same holds for the other input variables $u_2 \dots u_r$, i.e. the random system is controllable from each individual input. Any linear dependencies between the columns $\underline{b}_i$, $\underline{Ab}_i$ ... $\underline{A}^{n-1}\underline{b}_i$ are singular cases.

Our control systems in practice are not produced by random number generators, however. They frequently have a structure in which certain connections do not exist. In the open loop for example, actuator dynamics are controllable only from the corresponding input. By neglecting small coupling effects, a system can sometimes be divided into separately controllable subsystems, examples are the longitudinal and lateral motion of an airplane or the three axis stabilization of a satellite with orthogonal gas jet pairs.

Mechanical systems with forces as input variables $\underline{u}$ are described by a vector differential equation of second order:

$$\underline{M\ddot{z}} + \underline{D\dot{z}} + \underline{Sz} = \underline{Eu}$$

Here $\underline{M}$ = mass matrix, $\underline{D}$ = damping matrix, $\underline{S}$ = stiffness matrix. $\underline{x} = \begin{bmatrix} \underline{z} \\ \underline{\dot{z}} \end{bmatrix}$ is introduced as a state vector and the first order vector differential equation becomes

$$\underline{\dot{x}} = \begin{bmatrix} \underline{0} & \underline{I} \\ -\underline{M}^{-1}\underline{S} & -\underline{M}^{-1}\underline{D} \end{bmatrix}\underline{x} + \begin{bmatrix} \underline{0} \\ \underline{M}^{-1}\underline{E} \end{bmatrix}\underline{u} = \underline{Fx} + \underline{Gu} \qquad (9.1.11)$$

Now rank $[\underline{G}, \underline{FG}] = n$ only if rank $\underline{M}^{-1}\underline{E} = n/2$.

Integer zeros and ones in $(\underline{F}, \underline{G})$ indicate a "controllability structure". It is physically not meaningful to substitute a real number $\pm\varepsilon$ or $1\pm\varepsilon$ for such an integer zero or one. These effects may be hidden, however, for example by discretization, eq. (9.1.3), or by a basis transformation $\underline{x}^* = \underline{Tx}$. After such operations only real numbers remain. It is therefore recommended that physical state variables be used in modelling and the controllability structure be analyzed for the continuous-time system (9.1.1), i.e. for the controllability matrix $[\underline{G}, \underline{FG} \ldots \underline{F}^{n-1}\underline{G}]$. If the rules of section 4.3 on the choice of the sampling interval T are obeyed, then no eigenvalue looses its controllability from the inputs connected with it. Thus the controllability structure is completely preserved under sampling. In the following discussion the discrete-time version is discussed, because this is easier.

Remark 9.2:

In addition to the effect of structural relationships the controllability can also get lost by specific numerical values of parameters. For example in eq. (4.1.17) the loading bridge becomes uncontrollable if a load

$$m_L = \frac{q^2\pi^2 \ell m_c}{T^2 g} - m_c \qquad (9.1.12)$$

is attached to it. It must be guaranteed by the choice of the sampling interval T, that these values of the load m_L are beyond the maximum load of the bridge. Generally the neighborhood of such "numerical uncontrollability" must be avoided as well, because large input magnitudes are required in this region of weak controllability. The practical approach is to go back to the physical reasons if such numerical uncontrollability occurs and to try to remove it by a change of construction or parameters.

How can the orders of the subsystems, controllable from one input or a selected group of inputs, now be recognized for a given pair $(\underline{F}, \underline{G})$ or $(\underline{A}, \underline{B})$? This is shown first of all by the following

Example:

$$\underline{A} = \begin{bmatrix} 1 & 1 & 0 & 0 & 0 \\ 0 & 1 & 1 & 0 & 1 \\ -1 & 1 & 1 & 0 & -1 \\ 0 & 0 & 1 & 1 & 1 \\ 1 & -1 & 0 & 0 & 2 \end{bmatrix}, \quad \underline{B} = \begin{bmatrix} 1 & -1 & -2 \\ 2 & -1 & -1 \\ 1 & -1 & -1 \\ 1 & 0 & -1 \\ 1 & 1 & 1 \end{bmatrix} \qquad (9.1.13)$$

First we examine only the controllability from u_1 by the controllability matrix

$$[\underline{b}_1, \underline{Ab}_1, \underline{A}^2\underline{b}_1, \underline{A}^3\underline{b}_1, \underline{A}^4\underline{b}_1] = \begin{bmatrix} 1 & 3 & 7 & 13 & 21 \\ 2 & 4 & 6 & 8 & 10 \\ 1 & 1 & 1 & -1 & -9 \\ 1 & 3 & 5 & 7 & 9 \\ 1 & 1 & 1 & 3 & 11 \end{bmatrix}$$
$$\qquad\qquad\qquad\qquad\qquad\qquad * \quad * \quad * \quad * \quad 0$$

The linearly independent columns are designated by a star (*). These are produced if the selection starts from the left. The zero (0) designates linearly dependent columns, here

$$[\underline{b}_1, \underline{Ab}_1, \underline{A}^2\underline{b}_1, \underline{A}^3\underline{b}_1, \underline{A}^4\underline{b}_1] \begin{bmatrix} 2 \\ -7 \\ 9 \\ -5 \\ 1 \end{bmatrix} = \underline{0}$$

The rank of the controllability matrix is four, therefore a fourth order subsystem is controllable from u_1. By eq. (2.3.36) the characteristic polynomial of this subsystem can be written down immediately as

$$P_1(z) = 2 - 7z + 9z^2 - 5z^3 + z^4$$

Correspondingly for the second input

$$[\underline{b}_2, \underline{Ab}_2, \underline{A}^2\underline{b}_2, \underline{A}^3\underline{b}_2] \begin{bmatrix} \underline{p}_2 \\ 1 \end{bmatrix} = \begin{bmatrix} -1 & -2 & -3 & -4 \\ -1 & -1 & -1 & -1 \\ -1 & -2 & -3 & -4 \\ 0 & 0 & 0 & 0 \\ 1 & 2 & 3 & 4 \end{bmatrix} \begin{bmatrix} -1 \\ 3 \\ -3 \\ 1 \end{bmatrix} = \underline{0}$$
$$\qquad\qquad\qquad\qquad\qquad\qquad * \quad * \quad * \quad 0$$

u_2 controls a third order subsystem with the characteristic polynomial $P_2(z) = -1 + 3z - 3z^2 + z^3$. The corresponding relations for the third input are

$$[\underline{b}_3, \underline{Ab}_3, \underline{A}^2\underline{b}_3, \underline{A}^3\underline{b}_3]\begin{bmatrix}\underline{p}_3\\1\end{bmatrix} = \begin{bmatrix}-2 & -3 & -4 & -5\\-1 & -1 & -1 & -1\\-1 & -1 & 0 & 3\\-1 & -1 & -1 & -1\\1 & 1 & 0 & -3\end{bmatrix}\begin{bmatrix}-2\\5\\-4\\1\end{bmatrix} = \underline{0}$$

$$\begin{matrix}\because & \because & \because & 0\end{matrix}$$

$$P_3(z) = -2 + 5z - 4z^2 + z^3$$

We now attempt to control the system with the input variables u_1 and u_2. Since the individual inputs can control only sub-systems with a maximal order of four, the control - if it is at all possible - can be completed in four sampling steps. Eq. (9.1.6) is therefore written for N = 4

$$\underline{x}[4]-\underline{A}^4\underline{x}[0] = [\underline{A}^3\underline{b}_1, \underline{A}^3\underline{b}_2, \underline{A}^2\underline{b}_1, \underline{A}^2\underline{b}_2, \underline{Ab}_1, \underline{Ab}_2, \underline{b}_1, \underline{b}_2]\underline{u}_3$$

$$\underline{u}_3 = [u_1[0], u_2[0], u_1[1], u_2[1], u_1[2], u_2[2], u_1[3], u_2[3]]' \qquad (9.1.14)$$

Solutions then exist if and only if the matrix contains five linearly independent vectors. There are now several possible ways to choose these, depending on the sequence in which the vectors are tested for linear dependency. For example, it is easy to employ the four vectors, which have already been found for the first input and to include the first vector $\underline{A}^3\underline{b}_2$ of the second input variable. Setting those input amplitudes to zero, which belong to the unused vector, i.e. $u_2[1] = u_2[2] = u_2[3] = 0$, eq. (9.1.14) reduces to

$$\underline{x}[4] - \underline{A}^4\underline{x}[0] = [\underline{A}^3\underline{b}_1, \underline{A}^3\underline{b}_2, \underline{A}^2\underline{b}_1, \underline{Ab}_1, \underline{b}_1]\underline{w}$$

$$\underline{w} = [u_1[0], u_2[0], u_1[1], u_1[2], u_1[3]]'$$

Since the quadratic matrix which arises here is invertible, $\underline{w}$ can be calculated.

Now consider the question of whether the control task can be shared by the two inputs and if, as a result, control can be achieved more rapidly. In fact

$$\text{rank}\,[\underline{A}^3\underline{b}_1,\ \underline{A}^3\underline{b}_2,\ \underline{A}^2\underline{b}_1,\ \underline{A}^2\underline{b}_2,\ \underline{A}\underline{b}_1] = 5$$

so that the solution $u_2[2] = u_1[3] = u_2[3] = 0$ is also possible. Eq. (9.1.6) can be written for N = 3 sampling steps and solved for $\underline{w}$.

$$\underline{x}[3] - \underline{A}^3\underline{x}[0] = [\underline{A}^2\underline{b}_1,\ \underline{A}^2\underline{b}_2,\ \underline{A}\underline{b}_1,\ \underline{A}\underline{b}_2,\ \underline{b}_1]\underline{w}$$

$$\underline{w} = [u_1[0],\ u_2[0],\ u_1[1],\ u_2[1],\ u_1[2]]'$$

Finally we ask: can the transient be shortened even further if the third input is also used? The rank test yields

$$\text{rank}\,[\underline{A}^2\underline{b}_1,\ \underline{A}^2\underline{b}_2,\ \underline{A}^2\underline{b}_3,\ \underline{A}\underline{b}_1,\ \underline{A}\underline{b}_3] = 5$$

Since only two sampling intervals are required now, eq. (9.1.6) can be written as

$$\underline{x}[2] - \underline{A}^2\underline{x}[0] = [\underline{A}\underline{b}_1,\ \underline{A}\underline{b}_2,\ \underline{A}\underline{b}_3,\ \underline{b}_1,\ \underline{b}_3]\underline{w}$$

$$\underline{w} = [u_1[0],\ u_2[0],\ u_3[0],\ u_1[1],\ u_3[1]]' \tag{9.1.15}$$

Thus the system is transferred from the initial state $\underline{x}[0]$ to the zero state $\underline{x}[2]$ with the minimum-time deadbeat control sequence

$$\begin{bmatrix} u_1[0] \\ u_2[0] \\ u_3[0] \\ u_1[1] \\ u_3[1] \end{bmatrix} = \begin{bmatrix} -1.5 & 0 & 0.5 & 0.5 & -1.5 \\ 0 & 1 & -0.5 & -1 & -0.5 \\ 4 & 0 & -2.5 & -1 & 1.5 \\ 1.5 & 0 & -1 & -0.5 & 1 \\ -7 & 0 & 5 & 3 & -3 \end{bmatrix} \underline{x}[0] \tag{9.1.16}$$

9.1.2 Controllability Indices

The example of the minimum-time deadbeat control sequences motivates the following definition:

The maximal controllability index μ of a controllable pair $(\underline{A}, \underline{B})$ is the smallest integer N such that

$$\operatorname{rank} [\underline{B}, \underline{AB} \ldots \underline{A}^{N-1}\underline{B}] = n \tag{9.1.17}$$

Obviously the time-optimal solution of the control problem (9.1.6) with arbitrary initial and final states requires exactly μ sampling steps. It satisfies

$$\underline{x}[\mu] - \underline{A}^{\mu}\underline{x}[0] = [\underline{A}^{\mu-1}\underline{B}, \underline{A}^{\mu-2}\underline{B} \ldots \underline{B}][\underline{u}'[0], \underline{u}'[1] \ldots \underline{u}'[\mu-1]]' \tag{9.1.18}$$

The dimension of the solution space is $\mu r-n$. In the previous example $\mu r-n = 2\times 3-5 = 1$ and

$$[\underline{AB}, \underline{B}] \times [1 \quad -2 \quad 0 \quad -1 \quad 2 \quad 2]' = \underline{0}$$

Thus a sequence

$$\Delta\underline{u}[0] = a\begin{bmatrix} 1 \\ -2 \\ 0 \end{bmatrix}, \quad \Delta\underline{u}[1] = a\begin{bmatrix} -1 \\ 2 \\ 2 \end{bmatrix}$$

with arbitrary a can be added to the solution of eq. (9.1.16) and still $\underline{x}[2] = \underline{0}$ is reached. The solution can be shortened only in one of the input sequences, e.g. by the requirement $u_2[1] = 0$. This results if, beginning from the left, the first n linearly independent vectors of $[A^{\mu-1}\underline{B} \ldots \underline{B}]$ are chosen in eq. (9.1.18).

Remark 9.3:

Our presentation of structural properties of multi-variable systems assumes in the rank determination, that a newly introduced vector is tested for its linear dependency from the previously selected linearly independent vectors. A possible procedure would be the Gram-Schmidt orthonormalization [59.4]. It should be noted, however, that numerically efficient methods proceed differently, for example via the calculation of singular values [80.9], [81.4], or via transformation to Hessenberg form, see appendix A.5.

A selection of linearly independent columns is taken from the controllability matrix for the calculation of time-optimal controls. The vectors are formed and tested successively in the column sequence of $\underline{B}$, $\underline{AB}$ ... $\underline{A}^{\mu-1}\underline{B}$. The notation "reg $\underline{X}$" is introduced for this selection process. Reg $\underline{X}$ denotes the matrix, in which all linearly dependent vectors of $\underline{X}$ are eliminated, beginning from the left. The matrix $\underline{R}_\mu = \text{reg}\ [\underline{B}, \underline{AB} \ldots \underline{A}^{\mu-1}\underline{B}]$ contains the columns

$$\begin{array}{llll} \underline{b}_1 & \underline{b}_2 & \cdots & \underline{b}_r \\ \underline{Ab}_1 & \cdot & & \cdot \\ \cdot & \cdot & & \cdot \\ \cdot & \cdot & & \cdot \\ \underline{A}^{\mu_1-1}\underline{b}_1 & \underline{A}^{\mu_2-1}\underline{b}_2 & \cdots & \underline{A}^{\mu_r-1}\underline{b}_r \end{array} \qquad (9.1.19)$$

We call these the "regular columns". The individual vector chains $\underline{b}_i$, $\underline{Ab}_i$ ... $\underline{A}^{\mu_i-1}\underline{b}_i$ have no gaps. This can be shown as follows: Assume that $\underline{A}^k\underline{b}_i$ is linearly dependent on its predecessors, i.e. there exists $\underline{q}$ such that

$$\underline{A}^k\underline{b}_i = [\underline{B}, \underline{AB} \ldots \underline{A}^{k-1}\underline{B}, \underline{A}^k\underline{b}_1 \ldots \underline{A}^k\underline{b}_{i-1}]\underline{q}$$

Multiplication of this equation by $\underline{A}$ yields

$$\underline{A}^{k+1}\underline{b}_i = [\underline{AB}, \underline{A}^2\underline{B} \ldots \underline{A}^k\underline{B}, \underline{A}^{k+1}\underline{b}_1 \ldots \underline{A}^{k+1}\underline{b}_{i-1}]\underline{q} \qquad (9.1.20)$$

Therefore $\underline{A}^{k+1}\underline{b}_i$ is also linearly dependent on its predecessors.

The lengths μ_i of the vector chains resulting from the selection procedure (9.1.19) are now defined as follows:

> The <u>controllability index μ_i</u> is the smallest integer such that $\underline{A}^{\mu_i}\underline{b}_i$ is linearly dependent on its predecessors in $[\underline{B}, \underline{AB} \ldots]$.

The ordered controllability indices are also called "Kronecker indices".

Example:

For the example (9.1.13)

$\underline{R}_\mu = \text{reg}\,[\underline{B}, \underline{AB}] = [\underline{b}_1, \underline{b}_2, \underline{b}_3, \underline{Ab}_1, \underline{Ab}_3]$

i.e. $\mu_1 = 2, \mu_2 = 1, \mu_3 = 2$.

In the following the assumption of controllability will always be made. Then obviously

$$\mu_1 + \mu_2 + \ldots + \mu_r = n$$
$$\mu = \max_i \mu_i \qquad (9.1.21)$$
$$\frac{n}{r} \leq \mu \leq n-r+1$$

For the last inequality the assumption rank $\underline{B} = r$ was used.

It was shown by Brunovsky [70.1] that the μ_i constitute a complete set of independent invariants under all transformations

$$(\underline{A}, \underline{B}) \rightarrow (\underline{T}(\underline{A}-\underline{BK})\underline{T}^{-1}, \underline{TBM}), \ \det \underline{M} \neq 0 \qquad (9.1.22)$$

9.1.3 α and β Parameters, Input Normalization

In eq. (9.1.19) for each input a vector chain $\underline{b}_i, \underline{Ab}_i \ldots$ is obtained. $\underline{A}^{\mu_i}\underline{b}_i$ is the first vector in this chain, which is linearly dependent on its predecessors in $[\underline{B}, \underline{AB} \ldots]$. Therefore it can be expressed as

$$-\underline{A}^{\mu_i}\underline{b}_i = [\underline{B}, \underline{AB} \ldots \underline{A}^{\mu_i-1}\underline{B} \vdots \underline{A}^{\mu_i}\underline{b}_1 \ldots \underline{A}^{\mu_i}\underline{b}_{i-1}]\begin{bmatrix}\underline{a}_i\\ \underline{\beta}_i\end{bmatrix}$$

or

$$[\underline{B}, \underline{AB} \ldots \underline{A}^{\mu_i-1}\underline{B}]\underline{a}_i + \underline{A}^{\mu_i}[\underline{b}_1 \ldots \underline{b}_i]\begin{bmatrix}\underline{\beta}_i\\ 1\end{bmatrix} = \underline{0} \qquad (9.1.23)$$

Here $\underline{a}_i$ and $\underline{\beta}_i$ can be fixed uniquely by using only the regular vectors in eq. (9.1.23) and setting the remaining elements in $\underline{a}_i$ and $\underline{\beta}_i$ to zero. Then in particular $\beta_{ih} = 0$ if $\underline{A}^{\mu_i}\underline{b}_h$ is not a regular column, i.e.

$$\beta_{ih} = 0 \text{ for } \mu_i \geq \mu_h \tag{9.1.24}$$

Example:

For the system of eq. (9.1.13)

$$\begin{array}{lllllllll}
\underline{b}_1 & = [\ 1 & 2 & 1 & 1 & 1]' & & * \\
\underline{b}_2 & = [-1 & -1 & -1 & 0 & 1]' & & * \\
\underline{b}_3 & = [-2 & -1 & -1 & -1 & 1]' & & * \\
\underline{A}\,\underline{b}_1 & = [\ 3 & 4 & 1 & 3 & 1]' & & * \\
\underline{A}\,\underline{b}_2 & = [-2 & -1 & -2 & 0 & 2]' & = -0.5\underline{b}_1 + \underline{b}_2 + \underline{b}_3 + 0.5\underline{Ab}_1 & 0 \\
\underline{A}\,\underline{b}_3 & = [-3 & -1 & -1 & -1 & 1]' & & * \\
\underline{A}^2\underline{b}_1 & = [\ 7 & 6 & 1 & 5 & 1]' & = -\underline{b}_1 + 2\underline{b}_3 + 2\underline{Ab}_1 - 2\underline{Ab}_3 & 0 \\
\underline{A}^2\underline{b}_2 & = \underline{AAb}_2 & & & & & & 0 \\
\underline{A}^2\underline{b}_3 & = [-4 & -1 & 0 & -1 & 0]' & = 0.5\underline{b}_1 - 3\underline{b}_3 - 0.5\underline{Ab}_1 + 3\underline{Ab}_3 & 0
\end{array}$$

Again the star designates a regular column and 0 designates a linearly dependent column. For i = 1,2,3 eq. (9.1.23) is

$$\begin{aligned}
&[\underline{b}_1, \underline{b}_2, \underline{b}_3, \underline{Ab}_1, \underline{Ab}_3][1 \quad 0 \quad -2 \quad -2 \quad 2]' + \underline{A}^2\underline{b}_1 = \underline{0} \\
&[\underline{b}_1, \underline{b}_2, \underline{b}_3][0.5 \quad -1 \quad -1]' + [\underline{Ab}_1, \underline{Ab}_2][-0.5 \quad 1]' = \underline{0} \\
&[\underline{b}_1, \underline{b}_2, \underline{b}_3, \underline{Ab}_1, \underline{Ab}_3][-0.5 \quad 0 \quad 3 \quad 0.5 \quad -3]' + \underline{A}^2\underline{b}_3 = \underline{0}
\end{aligned} \tag{9.1.25}$$

The selection of vectors necessary for the time optimal control is obtained by multiplying the vector chain $\underline{b}_i$, $\underline{Ab}_i \ldots \underline{A}^{\mu_i-1}\underline{b}_i$ by $\underline{A}^{\mu-\mu_i}$. Then eq. (9.1.23) yields

$$[\underline{A}^{\mu-\mu_i}\underline{B}, \underline{A}^{\mu-\mu_i+1}\underline{B} \ldots \underline{A}^{\mu-1}\underline{B}]\underline{a}_i + \underline{A}^{\mu}[\underline{b}_1 \ldots \underline{b}_i]\begin{bmatrix}\underline{\beta}_i \\ 1\end{bmatrix} = \underline{0}$$

This is now written for i = 1,2, ... r and combined into a matrix equation

$$[\underline{B},\ \underline{AB}\ \ldots\ \underline{A}^{\mu-1}\underline{B}]\begin{bmatrix} \underline{0} & \underline{0} & \cdots & \boxed{\underline{a}_m} & \cdots & \underline{0} \\ \boxed{\underline{a}_1} & \boxed{\underline{a}_2} & & & & \boxed{\underline{a}_r} \end{bmatrix} + \underline{A}^{\mu}\underline{B}\ \underline{M} = \underline{0} \qquad \mu_m = \mu \tag{9.1.26}$$

Here the following matrix arises as a factor of $\underline{A}^{\mu}\underline{B}$

$$\underline{M} := \begin{bmatrix} 1 & \boxed{\beta_2} & \boxed{\underline{\beta}_3} & \cdots & \boxed{\underline{\beta}_r} \\ 0 & 1 & & & \\ \vdots & & & & \\ 0 & \cdots & & & 1 \end{bmatrix} \qquad \underline{\beta}_i = \begin{bmatrix} \beta_{i1} \\ \vdots \\ \beta_{ii-1} \end{bmatrix} \tag{9.1.27}$$

Example:

The second equation in (9.1.25) is multiplied by $\underline{A}$ and combined with the first and third equation into

$$[\underline{B},\ \underline{AB}]\begin{bmatrix} 1 & 0 & -0.5 \\ 0 & 0 & 0 \\ -2 & 0 & 3 \\ -2 & 0.5 & 0.5 \\ 0 & -1 & 0 \\ 2 & -1 & -3 \end{bmatrix} + \underline{A}^2\underline{B}\underbrace{\begin{bmatrix} 1 & -0.5 & 0 \\ 0 & 1 & 0 \\ 0 & 0 & 1 \end{bmatrix}}_{\underline{M}} = \underline{0} \tag{9.1.28}$$

The controllability indices are $\mu_1 = 2$, $\mu_2 = 1$, $\mu_3 = 2$. From eq. (9.1.24) $\beta_{31} = 0$, because $\mu_3 \geq \mu_1$, and $\beta_{32} = 0$, because $\mu_3 \geq \mu_2$.

As the example illustrates, β-coefficients arise only if a value μ_i is smaller than its predecessors in the sequence of controllability indices $\mu_1, \mu_2 \ldots \mu_r$. If the μ_i form a monotonically nondecreasing sequence, then $\underline{M}$ is the unit matrix. The triangular matrix $\underline{M}$ in eq. (9.1.27) is fully occupied only in the special case that the μ_i form a monotonically decreasing sequence.

It was shown by Popov [72.4] that the μ_i and the β-parameters constitute a complete set of independent invariants under all transformations

$$(\underline{A}, \underline{B}) \rightarrow (\underline{T}(\underline{A}-\underline{BK})\underline{T}^{-1}, \underline{TB}) \tag{9.1.29}$$

Thus only the a-parameters can be changed by state feedback. However, in the design procedure this is obscured by the existence of β-parameters. The later derivation of feedback laws can be simplified if we introduce an "input normalization" here in the discussion of feedforward control. We make use of the fact that the β-parameters can be easily eliminated by an input transformation

$$\underline{v} := \underline{M}^{-1}\underline{u} \quad , \quad \underline{u} = \underline{Mv} \tag{9.1.30}$$

This is illustrated by figure 9.1.

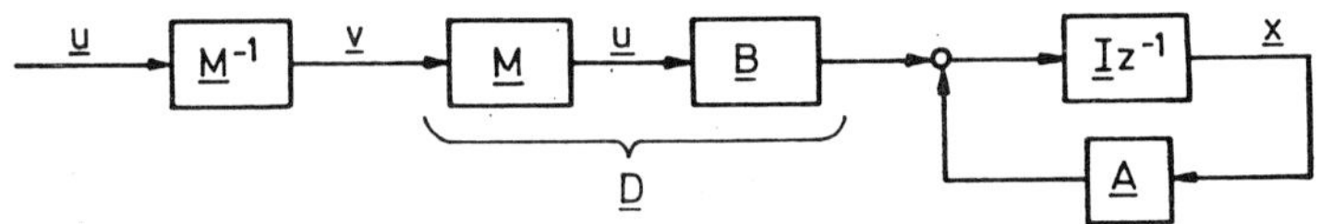

Figure 9.1 Input normalization

The normalized input matrix is

$$\underline{D} := \underline{BM} = [\underline{d}_1 \; \ldots \; \underline{d}_r] \tag{9.1.31}$$

During the design of feedforward and feedback controls, we now refer to the artificial input vector $\underline{v}$ and only return to the true input vector $\underline{u}$ at the end.

Eq. (9.1.26) is then transformed into

$$[\underline{D}, \underline{AD} \ldots \underline{A}^{\mu-1}\underline{D}] \begin{bmatrix} \underline{0} & \underline{0} & & \boxed{} & & \underline{0} \\ & & \ldots & \boxed{\underline{\alpha}_m} & \ldots & \boxed{} \\ \boxed{\underline{\alpha}_1} & \boxed{\underline{\alpha}_2} & & \boxed{} & & \boxed{\underline{\alpha}_r} \end{bmatrix} + \underline{A}^{\mu}\underline{D} = \underline{0}$$

$$\mu_m = \mu \tag{9.1.32}$$

where the elements of

$$\underline{\alpha}_i := \begin{bmatrix} \underline{M}^{-1} & & \underline{0} \\ & \ddots & \\ \underline{0} & & \underline{M}^{-1} \end{bmatrix} \underline{a}_i \quad , \; i = 1,2 \ldots r \qquad (9.1.33)$$

are defined as "α-parameters".

Remark 9.4:

Note that in some publications the a-parameters of eq. (9.1.23) are called α-parameters. The α's of eq. (9.1.33) are then called $\tilde{\alpha}$. In the application to design only the latter parameters will be used, thus they are more fundamental and more frequently used. Therefore they are introduced here without the tilda as α-parameters.

The ith column of eq. (9.1.32) can be written corresponding to eq. (9.1.23) as

$$[\underline{D}, \underline{AD} \ldots \underline{A}^{\mu_i - 1} \underline{D}]\underline{\alpha}_i + \underline{A}^{\mu_i}\underline{d}_i = \underline{0} \qquad (9.1.34)$$

The dependency on $\underline{A}^{\mu_i}\underline{d}_1 \ldots \underline{A}^{\mu_i}\underline{d}_{i-1}$ is eliminated by the input transformation. The α-parameters are designated with a triple index as α_{ijk}, where

$$\underline{A}^{\mu_i}\underline{d}_i + \sum_{j=1}^{r} \sum_{k=0}^{\mu_i - 1} \alpha_{ijk} \underline{A}^k \underline{d}_j = \underline{0} \qquad (9.1.35)$$

Due to the selection of linearly independent columns, according to eq. (9.1.19)

$$\alpha_{ijk} = 0 \quad \text{for} \quad k \geq \mu_j \qquad (9.1.36)$$

Example:

For the example of eq. (9.1.13) and substituting eq. (9.1.28)

$$\underline{D} = \underline{BM} = \begin{bmatrix} 1 & -1 & -2 \\ 2 & -1 & -1 \\ 1 & -1 & -1 \\ 1 & 0 & -1 \\ 1 & 1 & 1 \end{bmatrix} \begin{bmatrix} 1 & -0.5 & 0 \\ 0 & 1 & 0 \\ 0 & 0 & 1 \end{bmatrix} = \begin{bmatrix} 1 & -1.5 & -2 \\ 2 & -2 & -1 \\ 1 & -1.5 & -1 \\ 1 & -0.5 & -1 \\ 1 & 0.5 & 1 \end{bmatrix}$$

$\underline{d}_1 = [\,1 \quad 2 \quad 1 \quad 1 \quad 1\,]'$ ×

$\underline{d}_2 = [-1.5 \quad -2 \quad -1.5 \quad -0.5 \quad 0.5]'$ ×

$\underline{d}_3 = [-2 \quad -1 \quad -1 \quad -1 \quad 1\,]'$ ×

$\underline{A}\,\underline{d}_1 = [\,3 \quad 4 \quad 1 \quad 3 \quad 1\,]'$ ×

$\underline{A}\,\underline{d}_2 = [-3.5 \quad -3 \quad -2.5 \quad -1.5 \quad 1.5]'\,, \quad -\underline{d}_2 - \underline{d}_3 + \underline{Ad}_2 = \underline{0}$ 0

$\underline{A}\,\underline{d}_3 = [-3 \quad -1 \quad -1 \quad -1 \quad 1\,]'$ ×

$\underline{A}^2\underline{d}_1 = [\,7 \quad 6 \quad 1 \quad 5 \quad 1\,]'\,, \quad \underline{d}_1 - 2\underline{d}_3 - 2\underline{Ad}_1 + 2\underline{Ad}_3 + \underline{A}^2\underline{d}_1 = \underline{0}$ 0

$\underline{A}^2\underline{d}_2 = \underline{A} \times \underline{A}\,\underline{d}_2\,, \quad -\underline{A}\,\underline{d}_2 - \underline{A}\,\underline{d}_3 + \underline{A}^2\underline{d}_2 = \underline{0}$ 0

$\underline{A}^2\underline{d}_3 = [-4 \quad -1 \quad 0 \quad -1 \quad 0\,]'\,, \quad -0.5\underline{d}_1 + 3\underline{d}_3 + 0.5\underline{Ad}_1 - 3\underline{Ad}_3 + \underline{A}^2\underline{d}_3 = \underline{0}$ 0

For systems with $\mu_1 = 2$, $\mu_2 = 1$, $\mu_3 = 2$ eq. (9.1.32) reads

$$[\underline{d}_1, \underline{d}_2, \underline{d}_3, \underline{Ad}_1, \underline{Ad}_3] \begin{bmatrix} \alpha_{110} \\ \alpha_{120} \\ \alpha_{130} \\ \alpha_{111} \\ \alpha_{131} \end{bmatrix} + \underline{A}^2\underline{d}_1 = \underline{0}$$

$$[\underline{d}_1, \underline{d}_2, \underline{d}_3] \begin{bmatrix} \alpha_{210} \\ \alpha_{220} \\ \alpha_{230} \end{bmatrix} + \underline{Ad}_2 = \underline{0} \qquad (9.1.37)$$

$$[\underline{d}_1, \underline{d}_2, \underline{d}_3, \underline{Ad}_1, \underline{Ad}_3] \begin{bmatrix} \alpha_{310} \\ \alpha_{320} \\ \alpha_{330} \\ \alpha_{311} \\ \alpha_{331} \end{bmatrix} + \underline{A}^2\underline{d}_3 = \underline{0}$$

and for the given numerical values

$\alpha_{110} = 1 \qquad \alpha_{210} = 0 \qquad \alpha_{310} = -0.5$

$\alpha_{120} = 0 \qquad \alpha_{220} = -1 \qquad \alpha_{320} = 0$

$\alpha_{130} = -2 \qquad \alpha_{230} = -1 \qquad \alpha_{330} = 3$

$\alpha_{111} = -2 \qquad \alpha_{311} = 0.5$

$\alpha_{131} = 2 \qquad \alpha_{331} = -3$

The α-parameters play a central role in the finite effect sequences introduced in section 9.2.

9.1.4 Observability Structure

All definitions of this section can be easily dualized for the observability of a pair $(\underline{A}, \underline{C})$. Its "oberservability index" ν_i is defined as the smallest integer such that $\underline{C}_i\underline{A}^{\nu_i}$ is linear independent of its predecessors in $\begin{bmatrix} \underline{C} \\ \underline{CA} \\ \vdots \end{bmatrix}$.

Similarly the parameters a, β and α can be defined exactly as before by studying the controllability structure of a pair

$\bar{\underline{A}} = \underline{A}' \quad , \quad \bar{\underline{B}} = \underline{C}'$

Then

$$[\bar{\underline{B}}, \; \bar{\underline{A}}\bar{\underline{B}} \; \ldots] = \begin{bmatrix} \underline{C} \\ \underline{CA} \\ \vdots \end{bmatrix}' \tag{9.1.38}$$

Observers for multi-output systems have already been discussed in chapter 5. (See eq. (5.2.8) for the nth order observer and eqs. (5.3.6) and (5.3.7) for the reduced order observer.) The observer design problem is the exact dual of state feedback design. (See eqs. (5.2.11) and (5.2.12)). Therefore the theory and assignment of properties additional to pole locations will be developed only for the controllability and state feedback problem. It is recommended that the reader does this dualization as an exercise.

9.2 Finite Effect Sequences (FESs)

9.2.1 Introduction to FESs

Consider again the solution (9.1.6) of the difference equation written for initial state $\underline{x}[0] = \underline{0}$ and N+1 instead of N.

$$\underline{x}[N+1] = [\underline{A}^N\underline{B},\ \underline{A}^{N-1}\underline{B}\ \ldots\ \underline{B}]\underline{u}_N[0] \qquad (9.2.1)$$

$$\underline{u}_N[0] = \begin{bmatrix} \underline{u}[0] \\ \underline{u}[1] \\ \vdots \\ \underline{u}[N] \end{bmatrix}$$

In order to have the controllability matrix in the usual ordering rearrange eq. (9.2.1) to

$$\underline{x}[N+1] = [\underline{B},\ \underline{AB}\ \ldots\ \underline{A}^N\underline{B}]\underline{s}_N[0] \qquad (9.2.2)$$

$$\underline{s}_N[0] = \begin{bmatrix} \underline{u}[N] \\ \underline{u}[N-1] \\ \vdots \\ \underline{u}[0] \end{bmatrix}$$

Definition:

> Sequences $\underline{s}_N[0] = \underline{\hat{s}}_N[0]$ which bring the system from the zero state $\underline{x}[0] = \underline{0}$ to the zero state $\underline{x}[N+1] = \underline{0}$ for some N, are called "finite effect sequences" (FESs).

By eq. (9.2.2) FESs satisfy

$$[\underline{B},\ \underline{AB}\ \ldots\ \underline{A}^N\underline{B}]\underline{\hat{s}}_N[0] = \underline{0} \qquad (9.2.3)$$

Consider first a FES of a single-input system

$$[\underline{b},\ \underline{Ab}\ \ldots\ \underline{A}^N\underline{b}]\underline{\hat{s}}_N[0] = \underline{0} \qquad (9.2.4)$$

Now corresponding to eq. (2.3.35)

$$[\underline{b},\ \underline{Ab}\ \ldots\ \underline{A}^r\underline{b}]\begin{bmatrix}\underline{a}\\ 1\end{bmatrix} = \underline{0} \tag{9.2.5}$$

where $\underline{a} = [a_o\ \ a_1\ \ldots\ a_r]'$ consists of the characteristic polynomial coefficients of the controllable subsystem. By comparison

$$\underline{\hat{s}}_r[0] = \begin{bmatrix}\underline{a}\\ 1\end{bmatrix} \tag{9.2.6}$$

is a FES. If $(\underline{A},\ \underline{b})$ is controllable and $P_A(z) = \det(z\underline{I}-\underline{A}) = a_o+a_1z+\ldots+a_{n-1}z^{n-1}+z^n$, then the input sequence $u[0] = 1$, $u[1] = a_{n-1}$, ... $u[n] = a_o$ brings the system from $\underline{x}[0] = \underline{0}$ to $\underline{x}[n+1] = \underline{0}$.

For further illustration consider one system output $y = \underline{c}'\underline{x}$, and let the z-transfer function be

$$\frac{y_z(z)}{u_z(z)} = \underline{c}'(z\underline{I}-\underline{A})^{-1}\underline{b} = \frac{b_{n-1}z^{n-1}+\ldots+b_1+b_o}{z^n+a_{n-1}z^{n-1}+\ldots+a_o} \tag{9.2.7}$$

For the above solution we have

$$u_z(z) = 1+a_{n-1}z^{-1}+\ldots+a_oz^{-n} = z^{-n}(z^n+a_{n-1}z^{n-1}+\ldots+a_o) = z^{-n}P_A(z) \tag{9.2.8}$$

and therefore

$$y_z(z) = z^{-n}(b_{n-1}z^{n-1}+\ldots+b_1z+b_o) = b_{n-1}z^{-1}+b_{n-2}z^{n-2}+\ldots+b_oz^{-n} \tag{9.2.9}$$

The system and its input and output sequences are shown in figure 9.2.

Note that this FES and its response constitute a complete input-output description of the plant. State feedback will change the required closed-loop FES input, but not the FES at the actuator u and the corresponding response y.

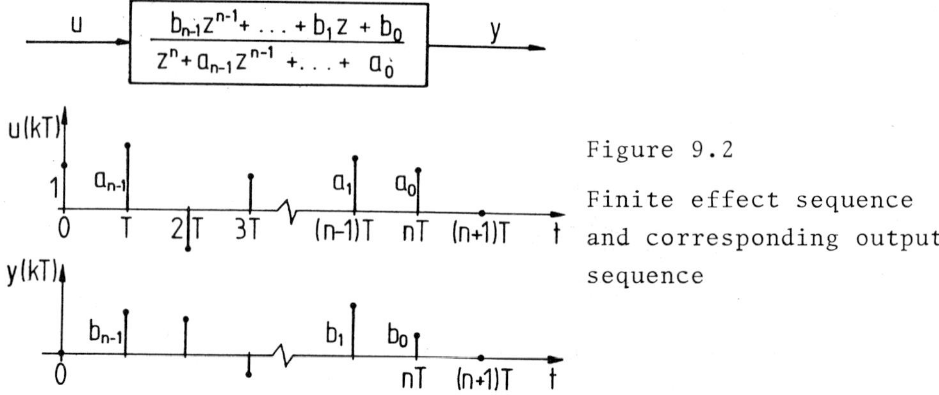

Figure 9.2

Finite effect sequence and corresponding output sequence

Assume now that the system is not completely observable from y(kT). Then the unoberservable common factor of the numerator and denominator can be cancelled in eq. (9.2.7) and $u_z(z)$ and $y_z(z)$ in eqs. (9.2.8) and (9.2.9) can be divided by this factor. Thus there exists a shorter input sequence that brings the system from $y[0] = 0$ and $\underline{x}[0] = \underline{0}$ to $y[m] = 0$, $m < n$. Then, however, $\underline{x}[m] \neq \underline{0}$ and the unobservable modes remain observable between the sampling instants, i.e. from $y(kT+\gamma T)$, $0 < \gamma < 1$.

In practice this situation is not likely to occur in the plant itself, but it can be generated in the closed loop by design techniques which make use of cancellations. We exclude this case here and study only FESs which bring not only the output but the complete state to zero.

9.2.2 Use of FESs

Before studying the properties of FESs further, their introduction is motivated by some examples.

1) Model testing
   To compare the real plant and its model, a very sensitive test is to compute FESs from the model, apply them as inputs to the real plant, and compare the responses. Techniques can be developed to use the deviation for model improvement. Model errors, defined for this case, may also be used in model reduction.

2) Failure detection
Precalculate FESs for all possible failures in the system. Apply them one after the other to the real plant to find out which of them brings the system to zero, i.e. what is the actual failure.

3) Reduction of u
Assume a feedforward control has been found for a desired transition from A to B, see fig. 9.3. Assume further that u[m] is too large and must be reduced. Then the indicated FES may be added to u such that the new u[m] is at the admissible limit. This either solves the problem or postpones it to a later time.

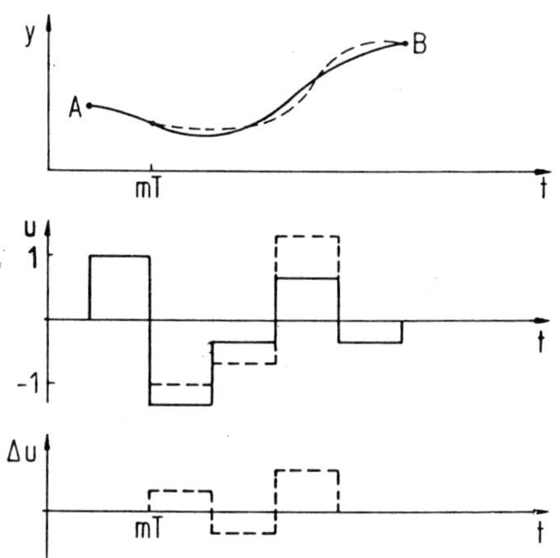

Figure 9.3

Reduction of u[m] by adding a FES

4) Error correction
Similarly, as in figure 9.3, assume that u[m] is not properly executed by the actuator, e.g. a gas jet for satellite attitude control. If the executed value can be determined during the interval from mT to (m+1)T, then the deviation from the command value is considered as the starting term of a FES. The further terms of this FES must now be added to the original control sequence, in order to cancel the effect of the error.

5) FES assignment
In the single-input example, the FES of figure 9.2, consisting of the coefficients of the characteristic polynomial, is exactly what can be changed arbitrarily by state feedback. This

concept can be generalized to systems with r inputs. It will be shown, that r basic FESs can be defined for this case. They generalize the characteristic polynomial and describe exactly what can be assigned to the system by state feedback.

9.2.3 Basic FESs

We will now study some elementary properties of FESs.

1. Time shift
   If $\underline{\hat{s}}_N[0]$ is a FES, then the shifted sequence $\underline{\hat{s}}_N[p]$ is also a FES. This is obvious by the time invariance of the plant.

2. Multiplication by a constant
   If $\underline{\hat{s}}_N[0]$ is a FES, then $b\underline{\hat{s}}_N[0]$ is also a FES. This follows from the linearity of the plant.

3. Addition
   If $\underline{\hat{s}}_{N1}[0]$ and $\underline{\hat{s}}_{N2}[0]$ are FESs, then $\underline{\hat{s}}_{N1}[0] + \underline{\hat{s}}_{N2}[0]$ is also a FES. This follows also from the linearity of the plant.

In z-transform notation a time shift is equivalent to multiplication with some power of z. Multiplication and addition apply directly to the z-transforms. Thus the three properties are equivalent to the statement that a polynomial representation of a FES can be multiplied by an arbitrary polynomial in z and a factor z^{-p} for some p and the result is a FES again.

Example:

For the system of figure 9.2 choose any polynomial $Q(z) = q_m z^m + q_{m-1} z^{m-1} + \ldots + q_o$ and any integer p. Form

$$z^{-n}(z^n + a_{n-1} z^{n-1} + \ldots + a_o)(q_m z^m + q_{m-1} z^{m-1} + \ldots + q_o) z^{-p} =$$

$$= q_m z^{m-p} + (a_{n-1} q_m + q_{m-1}) z^{m-p-1} + \ldots + a_o q_o z^{-n-p} \qquad (9.2.10)$$

Then the following input sequence of n+m+1 elements is a FES:

$$\begin{aligned} u[p-m] &= q_m \\ u[p-m+1] &= a_{n-1}q_m+q_{m-1} \\ &\vdots \\ u[p+n] &= a_o q_o \end{aligned}$$

Note, that in the present use of z-transform notation, we do not assume $f(kT) = 0$ for $k < 0$, as was done in appendix B. At the moment we are only talking about sequences with finite start and end times. Thus any questions of convergence of a z-transform are irrelevant here.

Remark 9.5:

> From a mathematical standpoint we are really talking about a module [69.1]. The set of all FESs may be called the "module of return to zero" [83.2]. It was introduced and its bases were studied in [78.6], [79.7] in the context of an algebraic theory of feedback. The importance of minimal bases for output feedback is further discussed in [81.11]. These references are "motivated by a desire to gain insight into the general nature of linear feedback" and are not specifically aimed at solving design problems.

In the controllable single-input case eq. (9.2.10), the characteristic polynomial $P_A(z)$, or equivalently the sequence consisting of its coefficients, was used as a basis from which all FESs for this system can be derived. This basis is also minimal, i.e. no lower degree polynomial could be used and no shorter FES exists. This is true because to satisfy eq. (9.2.5) with $r < n$ would contradict the assumption of controllability.

We now come back to the system with r inputs and study the minimal basis of their FESs. Comparing eqs. (9.2.3) and (9.1.23) it is seen that the r sequences

$$\underline{\hat{s}}_{\mu_i}[0] = \begin{bmatrix} \underline{a}_i \\ \underline{\beta}_i \\ 1 \\ 0 \\ \vdots \\ 0 \end{bmatrix} \left.\begin{matrix} \\ \\ \\ \\ \\ \end{matrix}\right\} r-i \qquad i = 1,2\ldots r \tag{9.2.11}$$

are FESs. Popov [72.4] has shown that the quantities μ_i, $\underline{\beta}_i$, $\underline{a}_i$, for i = 1,2...r constitute a complete set of independent invariants under all tranformations

$$(\underline{A}, \underline{B}) \rightarrow (\underline{TAT}^{-1}, \underline{TB}) \tag{9.2.12}$$

Therefore the basic FESs of eq. (9.2.11) cannot be further reduced and form a minimal basis for all FESs. All FESs $\underline{\hat{s}}_N$ can be generated from the r basic sequences by

. time shift

. multiplication by a constant

. addition.

This follows directly from the fact that each column of $[\underline{B}, \underline{AB}\ldots\underline{A}^N\underline{B}]$ in eq. (9.2.3) can be expressed uniquely in terms of the columns of $\underline{R}_\mu = \text{reg}[\underline{B}, \underline{AB} \ldots \underline{A}^\mu\underline{B}]$. Therefore any FES has a representation

$$\underline{\hat{s}}_N[0] = \sum_{i=1}^{r} \sum_{k=0}^{N-\mu_i} c_{ik}\, \underline{\hat{s}}_{\mu_i}[k] \tag{9.2.13}$$

A disadvantage of the form (9.2.11) of the FESs is, that the ith one begins with

$$\underline{u}[0] = \begin{bmatrix} \underline{\beta}_i \\ 1 \\ 0 \\ \vdots \\ 0 \end{bmatrix} \left.\begin{matrix} \\ \\ \\ \end{matrix}\right\} r-i \tag{9.2.14}$$

i.e. several inputs must be excited simultaneously at time 0.

We can, however, generate the basic FESs of eq. (9.2.11) from even simpler basic sequences, if we allow a linear transformation not only in the state space, but also in the input space. In eqs. (9.1.27) and (9.1.30), this was done such, that all β parameters are made zero. For the normalized input $\underline{v} = \underline{M}^{-1}\underline{u}$ basic FESs are obtained by comparison of eq. (9.1.34) and the equivalent of eq. (9.2.3) for the pair $(\underline{A}, \underline{D})$, i.e.

$$[\underline{D}, \underline{AD} \dots \underline{A}^N\underline{D}]\underline{\sigma}_N[0] = \underline{0} \tag{9.2.15}$$

The result is

$$\underline{\sigma}_{\mu_i}[0] = \begin{bmatrix} \underline{\alpha}_i \\ \underline{i}_i \end{bmatrix}, \quad i = 1,2 \dots r \tag{9.2.16}$$

where $\underline{i}_i$ is the ith column of the r×r unit matrix. Now the ith basic FES is started at the ith normalized input only by

$$\underline{v}_i[0] = \underline{i}_i \tag{9.2.17}$$

As in eq. (9.2.13), any FES for the normalized input $\underline{v}$ has a representation

$$\underline{\sigma}_N[0] = \sum_{i=1}^{r} \sum_{k=0}^{N-\mu_i} c_{ik}\, \underline{\sigma}_{\mu_i}[k] \tag{9.2.18}$$

Example:

For the example of eq. (9.1.13) with the α parameters of eq. (9.1.37) the basic FESs for the normalized input are

$$\underline{v}_1[0] = \begin{bmatrix} 1 \\ 0 \\ 0 \end{bmatrix}, \quad \underline{v}_1[1] = \begin{bmatrix} \alpha_{111} \\ 0 \\ \alpha_{131} \end{bmatrix} = \begin{bmatrix} -2 \\ 0 \\ 2 \end{bmatrix}, \quad \underline{v}_1[2] = \begin{bmatrix} \alpha_{110} \\ \alpha_{120} \\ \alpha_{130} \end{bmatrix} = \begin{bmatrix} 1 \\ 0 \\ -2 \end{bmatrix}$$

$$\underline{v}_2[0] = \begin{bmatrix} 0 \\ 1 \\ 0 \end{bmatrix}, \quad \underline{v}_2[1] = \begin{bmatrix} \alpha_{210} \\ \alpha_{220} \\ \alpha_{230} \end{bmatrix} = \begin{bmatrix} 0 \\ -1 \\ -1 \end{bmatrix} \tag{9.2.19}$$

$$\underline{v}_3[0] = \begin{bmatrix} 0 \\ 0 \\ 1 \end{bmatrix}, \quad \underline{v}_3[1] = \begin{bmatrix} \alpha_{311} \\ 0 \\ \alpha_{331} \end{bmatrix} = \begin{bmatrix} 0.5 \\ 0 \\ -3 \end{bmatrix}, \quad \underline{v}_3[2] = \begin{bmatrix} \alpha_{310} \\ \alpha_{320} \\ \alpha_{330} \end{bmatrix} = \begin{bmatrix} -0.5 \\ 0 \\ 3 \end{bmatrix}$$

By Popov's and Brunovsky's invariance results (eqs. (9.1.29) and (9.1.22)) only the α-parameters can be changed by state feedback, but not their structure as given by the μ_i. In the above example the open loop value $\alpha_{320} = 0$ can be changed to a nonzero value by closing the loop. But a parameter α_{321} does not exist by the controllability structure, in particular $\mu_3 = 2 > \mu_2 = 1$, see eq. (9.1.37). Thus the second element of $\underline{v}_3[1]$ cannot be made nonzero by feedback. Zeros in place of elements like α_{321} are called "structural zeros". This is in contrast to a "numerical zero" like $\alpha_{320} = 0$.

9.2.4 Matrix and Polynomial Notation of FESs

In the single-input case we have frequently used a vector of the coefficients of the characteristic polynomial, e.g. $[\underline{p}' \;\; 1]$ or $[\underline{a}' \;\; 1]$. Its generalization to the multi-input case can now be defined by the α parameters as

$$\underline{Q} := \left[\underbrace{\begin{matrix} \alpha_{110} & \cdots & \alpha_{11\mu_1-1} & 1 \\ \alpha_{120} & \cdots & & 0 \\ \vdots & & & \vdots \\ \alpha_{1r0} & \cdots & & 0 \end{matrix}}_{\mu_1+1} \left| \underbrace{\begin{matrix} \alpha_{210} & \cdots & 0 \\ \alpha_{220} & \cdots & 1 \\ & & 0 \\ & & \vdots \\ \alpha_{2r0} & & 0 \end{matrix}}_{\mu_2+1} \right| \cdots \left| \underbrace{\begin{matrix} \cdots\alpha_{r1\mu_r-1} & 0 \\ & \vdots \\ & 0 \\ \cdots\alpha_{rr\mu_r-1} & 1 \end{matrix}}_{\mu_r+1} \right.\right] \tag{9.2.20}$$

The block rows in the main diagonal of $\underline{Q}$ are fully occupied with α parameters, their last element is one. The off-diagonal row vector in the jth row and ith block column is by eq. (9.1.36)

$$\begin{aligned} &[\alpha_{ij0}\,,\ \alpha_{ij1} \cdots \qquad\qquad \alpha_{ij\mu_i-1}\,,\ 0] && \text{if } \mu_j \geq \mu_i \\ &[\alpha_{ij0}\,,\ \alpha_{ij1} \cdots \alpha_{ij\mu_i-1}\,,\ 0 \ \cdots \qquad 0] && \text{if } \mu_j < \mu_i \end{aligned} \tag{9.2.21}$$

Example:

For the example of eq. (9.1.13) with the basic FESs of eq. (9.2.19)

$$\underline{Q} = \left[\begin{array}{ccc|cc|ccc} \alpha_{110} & \alpha_{111} & 1 & \alpha_{210} & 0 & \alpha_{310} & \alpha_{311} & 0 \\ \alpha_{120} & 0 & 0 & \alpha_{220} & 1 & \alpha_{320} & 0 & 0 \\ \alpha_{130} & \alpha_{131} & 0 & \alpha_{230} & 0 & \alpha_{330} & \alpha_{331} & 1 \end{array}\right] \qquad (9.2.22)$$

$$= \left[\begin{array}{ccc|cc|ccc} 1 & -2 & 1 & 0 & 0 & -0.5 & 0.5 & 0 \\ 0 & 0 & 0 & -1 & 1 & 0 & 0 & 0 \\ -2 & 2 & 0 & -1 & 0 & 3 & -3 & 1 \end{array}\right]$$

The elements of the first basic FES appear in the first column block, where the first row contains the sequence at the first normalized input. The sequence at the second normalized input is written in the second row etc. All elements in one column refer to the same time of occurence of this value in the FES. In the single-input case $\underline{Q} = [\underline{a} \quad 1]$.
The matrix $\underline{Q}$ plays an important role in FES assignment for the closed loop, as will be shown in the next section.

An alternative notation is the polynomial form of eq. (9.2.20), which is simply obtained as

$$\underline{S}(z) := \underline{QZ} \qquad (9.2.23)$$

with

$$\underline{Z}(z) := \begin{bmatrix} \underline{z}_{\mu_1} & & \\ & \ddots & \\ & & \underline{z}_{\mu_r} \end{bmatrix}, \quad \underline{z}_{\mu_i} = \begin{bmatrix} 1 \\ z \\ \vdots \\ z^{\mu_i} \end{bmatrix} \qquad (9.2.24)$$

This has the effect that α_{ijk} is multiplied by z^k and the rows of each block column in eq. (9.2.20) are added. In detail:

$$\underline{QZ} = \begin{bmatrix} q_{11}(z) & \dots & q_{1r}(z) \\ \vdots & & \\ q_{r1}(z) & \dots & q_{rr}(z) \end{bmatrix} \qquad (9.2.25)$$

$$q_{ii}(z) = \alpha_{iio} + \alpha_{ii1}z + \dots + \alpha_{ii\mu_i-1}z^{\mu_i-1} + z^{\mu_i}$$

$$q_{ji}(z) = \begin{cases} \alpha_{ijo} + \alpha_{ij1}z + \dots + \alpha_{ij\mu_i-1}z^{\mu_i-1} & \text{for } \mu_j \geq \mu_i \\ \alpha_{ijo} + \alpha_{ij1}z + \dots + \alpha_{ij\mu_j-1}z^{\mu_j-1} & \text{for } \mu_j < \mu_i \end{cases}$$

Example:

The polynomial notation of eq. (9.2.22) is

$$\underline{QZ} = \begin{bmatrix} \alpha_{110}+\alpha_{111}z+z^2 & \alpha_{210} & \alpha_{310}+\alpha_{311}z \\ \alpha_{120} & \alpha_{220}+z & \alpha_{320} \\ \alpha_{130}+\alpha_{131}z & \alpha_{230} & \alpha_{330}+\alpha_{331}z+z^2 \end{bmatrix}$$

and with the numerical values

$$\underline{QZ} = \begin{bmatrix} 1-2z+z^2 & 0 & -0.5+0.5z \\ 0 & -1+z & 0 \\ -2+2z & -1 & 3-3z+z^2 \end{bmatrix} \qquad (9.2.26)$$

An advantage of the polynomial notation is that there is a very simple correspondence with the characteristic polynomial:

$$P_A(z) = \det(z\underline{I}-\underline{A}) = \det \underline{QZ} \qquad (9.2.27)$$

Note that $(z\underline{I}-\underline{A})$ is an n×n matrix, but $\underline{QZ}$ is only of dimension r×r. One possible proof of eq. (9.2.27) is via transformation to a Luenberger canonical form of $\underline{A}$. This is, however, quite tedious and is omitted here [77.5]. We call $\underline{QZ}$ the "characteristic polynomial matrix". In the single input case $\underline{QZ} = P_A(z)$.

Example:

For the $\underline{QZ}$ of eq. (9.2.26)

$$P_A(z) = \det \underline{QZ} = (z-1)^4(z-2) \qquad (9.2.28)$$

Here a 3×3 determinant is evaluated instead of a 5×5 determinant in $\det(z\underline{I}-\underline{A})$.

9.3 FES Assignment

We are ready now to close the loop by

$$\underline{u} = -\underline{Kx} + \underline{Vw} \qquad (9.3.1)$$

The design of $\underline{K}$ is simplified if we refer to the normalized input $\underline{v} = \underline{M}^{-1}\underline{u}$, i.e.

$$\underline{v} = -\underline{M}^{-1}\underline{Kx} + \underline{M}^{-1}\underline{Vw} \qquad (9.3.2)$$

First $\underline{F} = \underline{M}^{-1}\underline{K}$ is designed and finally

$$\underline{K} = \underline{MF} \qquad (9.3.3)$$

is calculated. Figure 9.4 shows the control system structure

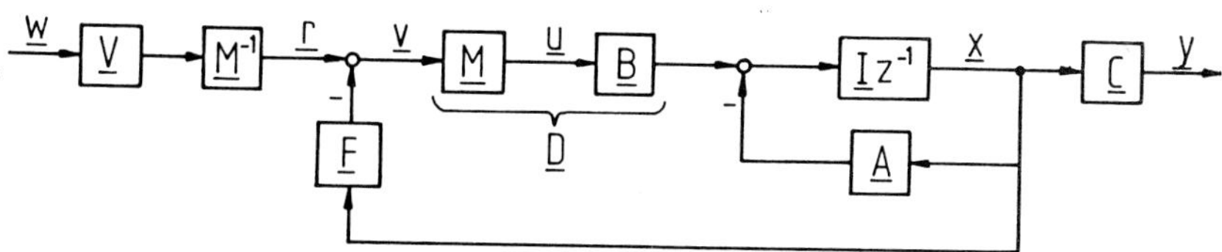

Figure 9.4 State feedback $\underline{F}$ to the normalized input $\underline{v}$ is equivalent to state feedback $\underline{K} = \underline{MF}$ to the plant input $\underline{u}$.

In many cases $\underline{V} = \underline{M}$ is an obvious choice for the prefilter. Then all β-parameters of the closed loop with input $\underline{w}$ are zero. For $\underline{V} = \underline{I}$ the β-parameters remain unchanged. In the following discussion we refer to $\underline{r} = \underline{M}^{-1}\underline{Vw}$ as the closed-loop input.

9.3.1 Assignment Formula

The normalized problem is then: given

$$\underline{x}[k+1] = \underline{A}\underline{x}[k] + \underline{D}\underline{v}[k] \tag{9.3.4}$$

find a state-feedback control law

$$\underline{v}[k] = -\underline{F}\underline{x}[k] + \underline{r}[k] \tag{9.3.5}$$

such that the closed loop with input $\underline{r}$ has FESs specified by

$$\underline{Q}_c = \left[\begin{array}{ccc|ccc|ccc} \pi_{110} & \cdots & \pi_{11\mu_i-1} & 1 & \pi_{210} \cdots & 0 & \cdots\pi_{r1\mu_r-1} & & 0 \\ \cdot & & & 0 & & 1 & & & \cdot \\ \cdot & & & \cdot & & 0 & & & \cdot \\ \cdot & & & \cdot & & \cdot & & & \cdot \\ \cdot & & & \cdot & & \cdot & & & 0 \\ \pi_{1r0} & & & 0 & \pi_{2r0} \cdots & 0 & \cdots\pi_{rr\mu_r-1} & & 1 \end{array}\right] \tag{9.3.6}$$

The index c stands for closed-loop. The π_{ijk} are the α parameters of the closed loop. Thus $\underline{Q}_c$ must have the same structural zeros as $\underline{Q}$. In polynomial notation $\underline{Q}_c$ becomes $\underline{Q}_c\underline{Z}$. In the single-input case $\underline{Q}_c = [\underline{p}' \quad 1]$ and $\underline{Q}_c\underline{Z} = P(z)$.

In the single-input case a form of the pole assignment formula is

$$\underline{k}' = [\underline{p}' \quad 1]\underline{E} \tag{9.3.7}$$

Its multivariable generalization is

$$\underline{F} = \underline{Q}_c\underline{E} \tag{9.3.8}$$

where the pole assignment matrix $\underline{E}$ is generalized to

$$\underline{E} := \begin{bmatrix} \underline{E}_1 \\ \vdots \\ \underline{E}_r \end{bmatrix}, \quad \underline{E}_i := \begin{bmatrix} \underline{e}_i' \\ \underline{e}_i'\underline{A} \\ \vdots \\ \underline{e}_i'\underline{A}^{\mu_i} \end{bmatrix} \tag{9.3.9}$$

and $\underline{e}_i'$ is formed as follows: arrange the columns of $\underline{R}_o = \text{reg}[\underline{B}, \underline{AB} \ldots]$, see eq. (9.1.19), as

$$\underline{R} := [\underline{b}_1, \underline{Ab}_1 \ldots \underline{A}^{\mu_i - 1}\underline{b}_1 \vdots \underline{b}_2 \ldots \vdots \vdots \ldots \underline{A}^{\mu_r - 1}\underline{b}_r] \tag{9.3.10}$$

Then in the same partition

$$\underline{R}^{-1} = \begin{bmatrix} \underline{S}_1 \\ \vdots \\ \underline{S}_r \end{bmatrix} \tag{9.3.11}$$

and $\underline{e}_i'$ is the last row of the $\mu_i \times n$ matrix $\underline{S}_i$. In the single-input case $\underline{E}$ specializes to the $\underline{E}$ used in eq. (9.3.7). Correspondingly $\underline{E}$ is called "FES-assignment matrix".

The following proof of eq. (9.3.8), see [77.5], uses the fact that both the given open-loop system $(\underline{A}, \underline{D})$ and the resulting closed-loop system $(\underline{A}-\underline{DF}, \underline{D})$ must have the same controllability indices μ_i, $i = 1,2\ldots r$. This follows directly from Brunovsky's invariance result, eq. (9.1.22), where the input normalization by $\underline{M}$ has been performed already in our problem formulation. Therefore both systems can be transformed to the same Brunovsky canonical form, eq. (A.4.15) by a change of basis in the state space and state feedback.

The state-space transformation to the Luenberger feedback-canonical form is given by eq. (A.4.13)

$$\underline{T} = \begin{bmatrix} \underline{T}_1 \\ \vdots \\ \underline{T}_r \end{bmatrix}, \quad \underline{T}_i = \begin{bmatrix} \underline{e}_i' \\ \underline{e}_i'\underline{A} \\ \vdots \\ \underline{e}_i'\underline{A}^{\mu_i - 1} \end{bmatrix} \tag{9.3.12}$$

This transformation yields

$$\underline{TAT}^{-1} = \underline{A}_L = \underline{A}_B - \underline{B}_B\underline{H}$$
$$\underline{TB} = \underline{B}_L = \underline{B}_B\underline{M}^{-1}$$

or for the normalized input

$$\underline{TD} = \underline{D}_L = \underline{B}_B$$

where $(\underline{A}_B, \underline{B}_B)$ is the Brunovsky form of eq. (A.4.15).

$$\underline{A}_B = \left[\begin{array}{cccc:c:cccc} 0 & 1 & & & & & & & \\ & & & 1 & \cdots & & \underline{0} & & \\ 0 & & & 0 & & & & & \\ \hdashline & \vdots & & & & & & & \\ \hdashline & & & & & 0 & 1 & & \\ & \underline{0} & & & & & & & \\ & & & & & & & 1 & \\ & & & & & 0 & & & 0 \end{array}\right], \quad \underline{B}_B = \left[\begin{array}{ccccc} 0 & & & & \\ \vdots & & & & \\ 0 & & & & \\ 1 & 0 & \cdots & & 0 \\ \hdashline & & \vdots & & \\ \hdashline 0 & & & & 0 \\ & & & & \\ & & & & 0 \\ 0 & & & 0 & 1 \end{array}\right]$$

$$\underbrace{\qquad\qquad}_{\mu_1} \quad \cdots \quad \underbrace{\qquad\qquad}_{\mu_r}$$

and $\underline{H}$ contains the open-loop α-parameters as in eqs. (A.4.1) and (A.4.14)

$$\underline{H} = \begin{bmatrix} \underline{h}'_{11} & \cdots & \underline{h}'_{1r} \\ \vdots & & \\ \underline{h}'_{r1} & \cdots & \underline{h}'_{rr} \end{bmatrix} \tag{9.3.13}$$

$$\underline{h}'_{ii} = [\alpha_{ii0} \quad \alpha_{ii1} \cdots \alpha_{ii\mu_i-1}]$$

$$\underline{h}'_{ij} = \begin{cases} [\alpha_{ji0} \quad \alpha_{ji1} \cdots \alpha_{ji\mu_j-1}] & \text{for } \mu_i \geq \mu_j \\ [\alpha_{ji0} \cdots \alpha_{ji\mu_i-1} \quad 0 \ldots 0] & \text{for } \mu_i < \mu_j \end{cases}$$

The plant is now described by

$$\underline{x}_L[k+1] = (\underline{A}_B-\underline{B}_B\underline{H})\underline{x}_L[k] + \underline{B}_B\underline{v}[k] \tag{9.3.14}$$

$$\underline{x}_L = \underline{Tx}\ ,\ \underline{v} = \underline{M}^{-1}\underline{u}$$

In the next step, feedback

$$\underline{v}[k] = \underline{Hx}_L[k] \tag{9.3.15}$$

may be used to bring the system to Brunovsky form ($\underline{A}_B$, $\underline{B}_B$). This skeleton of the system can now be revived by new closed-loop α-parameters, denoted by π in eq. (9.3.6).

Assemble them into $\underline{H}_C$, which has the same structural zeros as $\underline{H}$ and all α replaced by π. This is effected by the feedback

$$\underline{v}[k] = -(\underline{H}_c-\underline{H})\underline{x}_L[k] \tag{9.3.16}$$

and - going back to the original state variables - with eq. (9.3.5) and $\underline{r}[k] \equiv \underline{0}$.

$$\underline{v}[k] = -(\underline{H}_c-\underline{H})\underline{Tx}[k] = -\underline{Fx}[k]$$

Thus

$$\underline{F} = (\underline{H}_c-\underline{H})\underline{T} \tag{9.3.17}$$

Now $-\underline{HT}$ consists of the last block rows of

$$\underline{A}_L\underline{T} = \underline{TA}$$

and with $\underline{T}$ of eq. (9.3.12)

$$-\underline{HT} = \begin{bmatrix} \underline{e}_1'\underline{A}^{\mu_1} \\ \vdots \\ \underline{e}_r'\underline{A}^{\mu_r} \end{bmatrix}$$

Thus

$$\underline{F} = \underline{H}_c\underline{T} - \underline{H}\underline{T} = \underline{H}_c \begin{bmatrix} \underline{T}_1 \\ \vdots \\ \underline{T}_r \end{bmatrix} + \begin{bmatrix} \underline{e}_1'\underline{A}^{\mu_1} \\ \vdots \\ \underline{e}_r'\underline{A}^{\mu_r} \end{bmatrix}$$

$$= \left[\begin{array}{cc|cc|c|cc} \underline{h}_{c11}' & 1 & \underline{h}_{c12}' & 0 & & \underline{h}_{c1r}' & 0 \\ \underline{h}_{c21}' & 0 & \underline{h}_{c22}' & 1 & & \vdots & \vdots \\ \vdots & & \vdots & & & & 0 \\ \underline{h}_{cr1}' & 0 & \underline{h}_{cr2}' & 0 & & \underline{h}_{crr}' & 1 \end{array}\right] \left[\begin{array}{c} \underline{T}_1 \\ \underline{e}_1'\underline{A}^{\mu_1} \\ \hline \underline{T}_2 \\ \underline{e}_2'\underline{A}^{\mu_2} \\ \hline \vdots \\ \hline \underline{T}_r \\ \underline{e}_r'\underline{A}^{\mu_r} \end{array}\right]$$

$$= \underline{Q}_c\underline{E}$$

This completes the proof of eq. (9.3.8). Note that the transformation to Luenberger or Brunovsky form was only used for the proof. It should not be executed in the application of eq. (9.3.8). A numerically efficient method for FES assignment by transformation to Hessenberg form is given in appendix A.5.

Example:

For the example (9.1.13) by eq. (9.1.25)

$$\text{reg}[\underline{B}, \underline{AB}] = [\underline{b}_1, \underline{b}_2, \underline{b}_3, \underline{Ab}_1, \underline{Ab}_3]$$

$$\underline{R} = [\underline{b}_1, \underline{Ab}_1 \vdots \underline{b}_2 \vdots \underline{b}_3, \underline{Ab}_3] = \begin{bmatrix} 1 & 3 & -1 & -2 & -3 \\ 2 & 4 & -1 & -1 & -1 \\ 1 & 1 & -1 & -1 & -1 \\ 1 & 3 & 0 & -1 & -1 \\ 1 & 1 & 1 & 1 & 1 \end{bmatrix}$$

$$\begin{bmatrix} \underline{e}'_1 \\ \underline{e}'_2 \\ \underline{e}'_3 \end{bmatrix} = \begin{bmatrix} 0 & 1 & 0 & 0 & 0 \\ 0 & 0 & 1 & 0 & 0 \\ 0 & 0 & 0 & 0 & 1 \end{bmatrix} \underline{R}^{-1} = \begin{bmatrix} 0 & 0.5 & -0.75 & 0 & -0.25 \\ 0 & -1 & 0.5 & 1 & 0.5 \\ -1 & 0 & 0.5 & 1 & -0.5 \end{bmatrix}$$

$$\underline{E} = \begin{bmatrix} \underline{e}'_1 \\ \underline{e}'_1\underline{A} \\ \underline{e}'_1\underline{A}^2 \\ \hline \underline{e}'_2 \\ \underline{e}'_2\underline{A} \\ \hline \underline{e}'_3 \\ \underline{e}'_3\underline{A} \\ \underline{e}'_3\underline{A}^2 \end{bmatrix} = \begin{bmatrix} 0 & 0.5 & -0.75 & 0 & -0.25 \\ 0.5 & 0 & -0.25 & 0 & 0.75 \\ 1.5 & -0.5 & -0.25 & 0 & 1.75 \\ \hline 0 & -1 & 0.5 & 1 & 0.5 \\ 0 & -1 & 0.5 & 1 & 0.5 \\ \hline -1 & 0 & 0.5 & 1 & -0.5 \\ -2 & 0 & 1.5 & 1 & -0.5 \\ -4 & 0 & 2.5 & 1 & -1.5 \end{bmatrix} \qquad (9.3.18)$$

$$\underline{F} = \underline{Q}_c\underline{E} = \left[\begin{array}{ccc|cc|ccc} \pi_{110} & \pi_{111} & 1 & \pi_{210} & 0 & \pi_{310} & \pi_{311} & 0 \\ \pi_{120} & 0 & 0 & \pi_{220} & 1 & \pi_{320} & 0 & 0 \\ \pi_{130} & \pi_{131} & 0 & \pi_{230} & 0 & \pi_{330} & \pi_{331} & 1 \end{array}\right] \underline{E}$$

(9.3.19)

A test for $\underline{E}$ is given by the fact, that zero feedback $\underline{F} = \underline{0}$ cannot change $\underline{Q}$, therefore in eq. (9.3.8)

$$\underline{Q}\;\underline{E} = \underline{0} \qquad (9.3.20)$$

for the open-loop $\underline{Q}$ of eq. (9.2.20).

Example:

In the above example

$$\underline{QE} = \left[\begin{array}{ccc|cc|ccc} 1 & -2 & 1 & 0 & 0 & -0.5 & 0.5 & 0 \\ 0 & 0 & 0 & -1 & 1 & 0 & 0 & 0 \\ -2 & 2 & 0 & -1 & 0 & 3 & -3 & 1 \end{array}\right] \underline{E} = \underline{0}$$

There also exists a multi-variable generalization of formula (4.4.4).

$\underline{p}' = \underline{a}' + \underline{k}'\underline{W}$

This follows from eq. (9.3.17) by inversion as

$$\underline{H}_c = \underline{H} + \underline{FT}^{-1} \tag{9.3.21}$$

It shows what feedback $\underline{F}$ can do to the open-loop FESs in $\underline{H}$ in order to produce the closed-loop FESs in $\underline{H}_c$.

With $\underline{F} = \underline{F}_y\underline{C}$ for given $\underline{C}$, rank $\underline{C} = s < n$, eq. (9.3.21) directly applies to output feedback:

$$\underline{H}_c = \underline{H} + \underline{F}_y\underline{CT}^{-1} \tag{9.3.22}$$

Here $\underline{F}_y$ is an r×s matrix, therefore only r×s requirements on the closed loop can be satisfied. If r×s < n, then not even pole assignment is possible. Stabilizability by output feedback is studied in [84.5].

For the inversion of $\underline{T}$ note that the rows $\underline{e}_i'$ of $\underline{T}$ were computed by the inversion of $\underline{R}$, eq. (9.3.10).

Let

$\underline{V} := \underline{TR}$

then $\underline{V}$ contains r rows of the unit matrix and is easier to invert than $\underline{T}$. Thus

$$\underline{T}^{-1} = \underline{RV}^{-1} \tag{9.3.23}$$

Example:

For the example of eq. (9.1.13) $\underline{T}$ is extracted from eq. (9.3.18)

$$\underline{T} = \begin{bmatrix} \underline{e}_1' \\ \underline{e}_1'\underline{A} \\ \underline{e}_2' \\ \underline{e}_3' \\ \underline{e}_3'\underline{A} \end{bmatrix} = \begin{bmatrix} 0 & 0.5 & -0.75 & 0 & -0.25 \\ 0.5 & 0 & -0.25 & 0 & 0.75 \\ 0 & -1 & 0.5 & 1 & 0.5 \\ -1 & 0 & 0.5 & 1 & -0.5 \\ -2 & 0 & 1.5 & 1 & -0.5 \end{bmatrix} \tag{9.3.24}$$

$$\underline{V} = \underline{TR} = \begin{bmatrix} 0 & 1 & 0 & 0 & 0 \\ 1 & 2 & 0.5 & 0 & -0.5 \\ 0 & 0 & 1 & 0 & 0 \\ 0 & 0 & 0 & 0 & 1 \\ 0 & -2 & 0 & 1 & 3 \end{bmatrix}$$

$$\underline{T}^{-1} = \underline{RV}^{-1} = \begin{bmatrix} 1 & 3 & -1 & -2 & -3 \\ 2 & 4 & -1 & -1 & -1 \\ 1 & 1 & -1 & -1 & -1 \\ 1 & 3 & 0 & -1 & -1 \\ 1 & 1 & 1 & 1 & 1 \end{bmatrix} \times \begin{bmatrix} -2 & 1 & -0.5 & 0.5 & 0 \\ 1 & 0 & 0 & 0 & 0 \\ 0 & 0 & 1 & 0 & 0 \\ 2 & 0 & 0 & -3 & 1 \\ 0 & 0 & 0 & 1 & 0 \end{bmatrix}$$

$$= \begin{bmatrix} -3 & 1 & -1.5 & 3.5 & -2 \\ -2 & 2 & -2 & 3 & -1 \\ -3 & 1 & -1.5 & 2.5 & -1 \\ -1 & 1 & -0.5 & 2.5 & -1 \\ 1 & 1 & 0.5 & -1.5 & 1 \end{bmatrix} \qquad (9.3.25)$$

We will now relate the FES-assignment formula (9.3.8) to the closed-loop characteristic polynomial. As in eq. (9.2.27) this is

$$P(z) = \det(z\underline{I}-\underline{A}+\underline{DF}) = \det \underline{Q}_c\underline{Z} \qquad (9.3.26)$$

where the first determinant is n×n and the second one is only r×r and directly related to the closed-loop FESs, assigned by $\underline{Q}_c$, via eq. (9.3.8).

Also eq. (9.3.21) can be written in a polynomial version and thus be related to the characteristic polynomial. First write for the open and closed loops

$$\underline{Q}\,\underline{Z}_\mu = \mathrm{diag}\{z^{\mu_i}\} + \underline{H}\,\underline{Z}_{\mu-1}$$

$$\underline{Q}_c\underline{Z}_\mu = \mathrm{diag}\{z^{\mu_i}\} + \underline{H}_c\underline{Z}_{\mu-1}$$

where $\underline{Z}_\mu = \underline{Z}$ as defined in eq. (9.2.24) and correspondingly

$$\underline{Z}_{\mu-1} := \begin{bmatrix} \underline{z}_{\mu_1-1} & & \\ & \ddots & \\ & & \underline{z}_{\mu_r-1} \end{bmatrix} , \quad \underline{z}_{\mu_i-1} = \begin{bmatrix} 1 \\ z \\ \vdots \\ z^{\mu_i-1} \end{bmatrix} \qquad (9.3.27)$$

Then

$$\underline{Q}_c \underline{Z}_\mu = \underline{QZ}_\mu + (\underline{H}_c - \underline{H}) \underline{Z}_{\mu-1}$$

and with eq. (9.3.17)

$$\underline{Q}_c \underline{Z}_\mu = \underline{QZ}_\mu + \underline{FT}^{-1} \underline{Z}_{\mu-1}$$

Thus the closed-loop characteristic polynomial is

$$P(z) = \det \underline{Q}_c \underline{Z}_\mu = \det(\underline{QZ}_\mu + \underline{FT}^{-1} \underline{Z}_{\mu-1}) \qquad (9.3.28)$$

Example:

For the example of eq. (9.1.13) with $\underline{QZ}_\mu$ of eq. (9.2.26) and $\underline{T}^{-1}$ of eq. (9.3.25)

$$P(z) = \det \left\{ \begin{bmatrix} 1-2z+z^2 & 0 & -0.5+0.5z \\ 0 & -1+z & 0 \\ -2+2z & -1 & 3-3z+z^2 \end{bmatrix} + \underline{F} \begin{bmatrix} -3+z & -1.5 & 3.5-2z \\ -2+2z & -2 & 3-z \\ -3+z & -1.5 & 2.5-z \\ -1+z & -0.5 & 2.5-z \\ 1+z & 0.5 & -1.5+z \end{bmatrix} \right\} \qquad (9.3.29)$$

9.3.2 Design Choices

This section discusses some possibilities of specifying closed-loop FESs in $\underline{Q}_c$ for eq. (9.3.8), $\underline{F} = \underline{Q}_c\underline{E}$, and corresponding closed-loop characteristic polynomials $P(z) = \det \underline{Q}_c\underline{Z}$. Alternatively the inverse relation (9.3.21), $\underline{H}_c = \underline{H} + \underline{FT}^{-1}$, can be used in the design, which shows the influence of the feedback matrix $\underline{F}$ on the closed-loop FESs contained in $\underline{H}_c$. This last equation is particularly useful for the case of structural constraints on $\underline{F}$ like output feedback. In such a case an arbitrary $\underline{H}_c$ cannot be achieved, but an acceptable closed-loop characteristic polynomial P(z) may be obtainable.

1) Minimum-time deadbeat control

Every control law assigning $P(z) = z^n$ is called deadbeat, because $P(z) = \det(z\underline{I}-\underline{A}+\underline{DF}) = z^n$ implies by the Cayley-Hamilton theorem

$$(\underline{A}-\underline{DF})^n = \underline{0} \tag{9.3.30}$$

and therefore

$$\underline{x}[n] = (\underline{A}-\underline{DF})^n\underline{x}[0] = \underline{0} \text{ for all } \underline{x}[0] \tag{9.3.31}$$

Faster deadbeat responses are obtained by assignment of a minimal polynomial z^μ. This results in $(\underline{A}-\underline{DF})^\mu = \underline{0}$ and the zero state is reached in μ sampling intervals. The controllability index μ is the minimum number for which this is possible. This can be seen from eq. (9.1.18). For less than μ steps there does not exist a deadbeat feedforward control sequence for arbitrary initial states $\underline{x}[0]$. If a feedforward control sequence does not exist, then feedback cannot generate a solution either. For μ steps the solution is still not unique as eq. (9.1.18) shows. For feedforward control we can introduce the additional requirement, that the individual input sequences $u_i[k]$ are also as short as possible. A unique solution is obtained, if each input sequence $u_i[k]$ is of duration μ_i. Thus, only the longest one takes μ sampling intervals. Only this solution will be called minimum-time deadbeat.

For feedforward control, this sequence is achieved by selecting the first n linearly independent columns of

$$[\underline{A}^{\mu-1}\underline{B}\ ,\ \underline{A}^{\mu-2}\underline{B}\ \ldots\ \underline{B}]$$

The same solution is obtained in feedback form by selecting the first n linearly independent columns of

$[\underline{B}, \underline{AB} \dots \underline{A}^{\mu-1}\underline{B}]$
as in eq. (9.1.19) and the subsequent derivation of control laws. In terms of FESs, this is achieved by setting all closed-loop α parameters to zero, i.e. all $\pi_{ijk} = 0$ in eq. (9.3.6). After a unit impulse at closed-loop input r_i, see figure 9.4, nothing further must be done at $\underline{r}$. The system generates the minimum-time deadbeat sequences at $\underline{v}$ by feedback.

By eq. (9.3.6) with all $\pi_{ijk} = 0$ and eqs. (9.3.8) and (9.3.9) the feedback matrix $\underline{F}$ for minimum-time deadbeat control is

$$\underline{F}_D = \begin{bmatrix} \underline{e}_1'\underline{A}^{\mu_1} \\ \vdots \\ \underline{e}_r'\underline{A}^{\mu_r} \end{bmatrix} \tag{9.3.32}$$

This solution was derived in [68.2]. By eq. (9.3.17) with $\underline{H}_c = \underline{0}$, it may also be written

$$\underline{F} = -\underline{HT} \tag{9.3.33}$$

Example:

For the example (9.1.13), $\underline{F}$ consists of the second, third and fifth row of $\underline{E}$, eq. (9.3.18)

$$\underline{F}_D = \begin{bmatrix} 1.5 & -0.5 & -0.25 & 0 & 1.75 \\ 0 & -1 & 0.5 & 1 & 0.5 \\ -4 & 0 & 2.5 & 1 & -1.5 \end{bmatrix} \tag{9.3.34}$$

and for the input $\underline{u}$ by eqs. (9.3.3) and (9.1.28)

$$\underline{K}_D = \underline{MF}_D = \begin{bmatrix} 1 & -0.5 & 0 \\ 0 & 1 & 0 \\ 0 & 0 & 1 \end{bmatrix} \underline{F}_D$$

$$= \begin{bmatrix} 1.5 & 0 & -0.5 & -0.5 & 1.5 \\ 0 & -1 & 0.5 & 1 & 0.5 \\ -4 & 0 & 2.5 & 1 & -1.5 \end{bmatrix} \tag{9.3.35}$$

Check:

$$\underline{x}[0] = \begin{bmatrix} a \\ b \\ c \\ d \\ e \end{bmatrix} \qquad \underline{u}[0] = \begin{bmatrix} -1.5a + 0.5c + 0.5d - 1.5e \\ b - 0.5c - d - 0.5e \\ 4a - 2.5c - d + 1.5e \end{bmatrix}$$

$$\underline{x}[1] = \begin{bmatrix} -8.5a + 6c + 3.5d - 4e \\ -7a + 5c + 3d - 3e \\ -6.5a + 4.5c + 2.5d - 3.5e \\ -5.5a + 4c + 2.5d - 2e \\ 3.5a - 2.5c - 1.5d + 1.5e \end{bmatrix} \qquad \underline{u}[1] = \begin{bmatrix} 1.5a - c - 0.5d + e \\ 0 \\ -7a + 5c + 3d - 3e \end{bmatrix}$$

$$\underline{x}[2] = \underline{0} \qquad \underline{u}[2] = \underline{0}$$

Indeed $u_2[1] = 0$ for all initial states, i.e. the individual inputs go to zero as fast as possible.

For the design process, $\underline{K}_D$ is a finite, unique point in K space. Provided the actuator constraints are not violated this is the most desirable fast solution. Therefore it is worthwhile, in any design to determine $\underline{K}_D$; essentially it is available already in $\underline{E}$, as shown by the example. However, in many cases, the gains and actuator inputs are too high. This is particularly true for mechanical systems in which the actuators have to produce forces and torques. The deadbeat solution is more realistic in the control of slow processes.

Another special point in K space is of course the origin $\underline{K} = \underline{0}$. We assume that this open-loop behavior is unsatisfactory such that the need for feedback $\underline{K} \neq \underline{0}$ arises. But in order to obtain small or medium-sized gains we do not want to move away from the origin too much. A simple idea would be to try $\underline{K} = \alpha \underline{K}_D$, $0 < \alpha < 1$, i.e. to move directly from the origin to the deadbeat solution. In the z plane, this means moving from the open-loop eigenvalues along a root locus to a branching point of multiplicity n at the origin. In the neighborhood of such a branching point, the solution is very sensitive to parameter uncertainties. Also for small

α, the roots may first move away from $z = 0$. Thus a straight movement in K space is not a good idea. This is not surprising, if we remember the complicated nonconvex forms of stability regions studied in chapter 7. We must use the eigenvalue relation $P(z) = \underline{Q}_c\underline{Z}$ in the design.

Now there are two possible strategies

a) start at $\underline{K} = \underline{K}_D$ and find a fast control with reduced gains.

b) start at $\underline{K} = \underline{0}$ and find a soft stabilization with small gains.

2) Fast control with reduced gains

The minimum-time deadbeat requirement can be relaxed in different directions. Allowing more time for the deadbeat response can reduce the required actuator amplitudes. This effect can also be achieved by placing eigenvalues not at $z = 0$, but to the right of this point. This assumes, that the open loop eigenvalues were even further to the right, by proper choice of the sampling interval. The possibilities can be discussed by the relation

$$P(z) = \left|\begin{array}{c|c|c|c} \pi_{110}+\ldots+\pi_{11\mu_1-1}z^{\mu_1-1}+z^{\mu_1} & \pi_{210}+\ldots & & \pi_{rro}+\ldots+\pi_{rr\mu_r-1}z^{\mu_r-1} \\ \vdots & \pi_{220}+\ldots+z^{\mu_2} & \ldots & \vdots \\ \vdots & & & \vdots \\ \pi_{1ro}+\ldots+\pi_{1r\mu_1-1}z^{\mu_i-1} & \pi_{2ro}+\ldots & & \pi_{rro}+\ldots+\pi_{rr\mu_r-1}z^{\mu_r-1}+z^{\mu_r} \end{array}\right| \qquad (9.3.36)$$

For the minimum-time deadbeat solution only diag $\{z^{\mu_i}\}$ remains, i.e. $P(z) = z^n$ and the minimal polynomial is z^μ, because no coupling terms between the subsystems represented by the main diagonal occur.

$P(z) = z^n$, i.e. a deadbeat solution is preserved, if couplings between the subsystems are possible only in one direction. Thus the main diagonal is kept and nonzero π elements are admitted, either in the right upper triangle or in the left lower triangle.

Example:

For the example (9.1.13) the deadbeat gain matrix $\underline{K}_D$ of eq. (9.3.35) has its two largest gains in its third row $\underline{k}_3' = [-4\ ,\ 0\ ,\ 2.5\ ,\ 1\ ,\ -1.5]$. In order to avoid saturation of the actuator for u_3 these gains should be reduced, but all eigenvalues are to remain at $z = 0$. By eq. (9.3.17) the third row of $\underline{K}$ and $\underline{F}$ is

$$\underline{k}_3' = \underline{f}_3' = [0 \quad 0 \quad 1](\underline{H}_c - \underline{H})\underline{T}$$

$$= [\pi_{130} - \alpha_{130},\ \pi_{131} - \alpha_{131},\ \pi_{230} - \alpha_{230},\ \pi_{330} - \alpha_{330},\ \pi_{331} - \alpha_{331}]\underline{T}$$

The deadbeat character of the solution is preserved, if $\pi_{330} = \pi_{331} = 0$. The actual $\underline{T}$ of eq. (9.3.24) shows that $\gamma = \pi_{131}$ is the most efficient parameter for reducing the gains. Thus we keep $\pi_{130} = \pi_{230} = 0$. Then

$$\underline{k}_3' = [-\alpha_{130},\ \gamma - \alpha_{131},\ -\alpha_{230},\ -\alpha_{330},\ -\alpha_{331}]\underline{T}$$

$$= [\ 2\ ,\ \ \gamma - 2\ ,\ \ 1\ ,\ \ -3\ ,\ \ 3\ ]\underline{T}$$

$$= [0.5\gamma - 4\ ,\ 0\ ,\ -0.25\gamma + 2.5\ ,\ 1\ ,\ 0.75\gamma - 1.5]$$

Minimizing the largest gain in this row results in $\gamma = 4.4$ and

$$\underline{k}_3' = [-1.8\ ,\ 0\ ,\ 1.4\ ,\ 1\ ,\ 1.8]$$

The other rows of $\underline{K}$ are not changed by γ, thus

$$\underline{K} = \begin{bmatrix} 1.5 & 0 & -0.5 & -0.5 & 1.5 \\ 0 & -1 & 0.5 & 1 & 0.5 \\ -1.8 & 0 & 1.4 & 1 & 1.8 \end{bmatrix} \qquad (9.3.37)$$

The maximum gain was reduced from 4 to 1.8. The characteristic polynomial is

$$P(z) = \begin{vmatrix} z^2 & 0 & 0 \\ 0 & z & 0 \\ 4.4z & 0 & z^2 \end{vmatrix} = z^5$$

and the minimal polynomial is z^3, i.e. the solution now takes one sampling interval longer than with the minimum-time deadbeat control $\underline{K}_D$.

In the diagonal or triangular form the elements in the main diagonal may also be changed. Their product is still P(z).

Example:

For the example (9.1.13) let

$$\underline{Q}_c\underline{Z} = \begin{bmatrix} (z-0.5)^2 & 0 & 0 \\ 0 & z-0.5 & 0 \\ 2.5z & 0 & (z-0.5)^2 \end{bmatrix}$$

Then

$$\underline{K} = \begin{bmatrix} 1 & -0.125 & 0.0625 & -0.25 & 0.8125 \\ 0 & -0.5 & 0.25 & 0.5 & 0.25 \\ -1 & 0 & 0.5 & 0.25 & 0.75 \end{bmatrix} \qquad (9.3.38)$$

and $P(z) = \det \underline{Q}_c\underline{Z} = (z-0.5)^5$. Compared with the previous solution, the eigenvalues are shifted not so far to the left and the maximum gain is reduced to 1. Note, that without the off-diagonal term 2.5z in $\underline{Q}_c\underline{Z}$, a minimal polynomial $(z-0.5)^2$, is assigned, but the gains are higher.

3) Soft stabilization

In applications where control input magnitudes $|u|$ are expensive, low gain solutions are desired which just stabilize the system sufficiently. In this situation we do not start with $\underline{K}_D$ and reduce gains; instead $\underline{K} = \underline{0}$ is the starting point and nicely stabilizing solutions are sought. Strategies depend on the open-loop eigenvalues. We only assume, that they are located in the unit circle shifted to the right by $\sqrt{2}$. This is only a matter of the appropriate choice of the sampling interval, see figure 4.4. A simple rule would be to shift all eigenvalues to the left by $\sqrt{2}$. But in many cases, this is too much. It would, for example, shift a nicely located eigenvalue at $z = 0.5$ to an undesirable eigenvalue at -0.914. Anyway, the left shift by b, $0 < b < \sqrt{2}$ is a useful element of pole shifting strategies. This is easily achieved by

$$\underline{Q}_c\underline{Z}(z) = \underline{QZ}(z+b) \tag{9.3.39}$$

Another important strategy element is pole contraction. All eigenvalues inside a circle $|z| = R$ are moved inside a circle $|z| = cR$ by

$$\underline{Q}_c\underline{Z}(z) = \underline{QZ}(z/c) \tag{9.3.40}$$

and multiplying all polynomials in the ith column by c^{μ_i}, in order to scale the coefficient of z^{μ_i} to one. This is equivalent to normalizing the starting impulse of the FES to one. For $c < 1$ this is a contraction. The effect of a contraction on the FESs is, that the individual sequences are damped by a factor c, see figure 9.5.

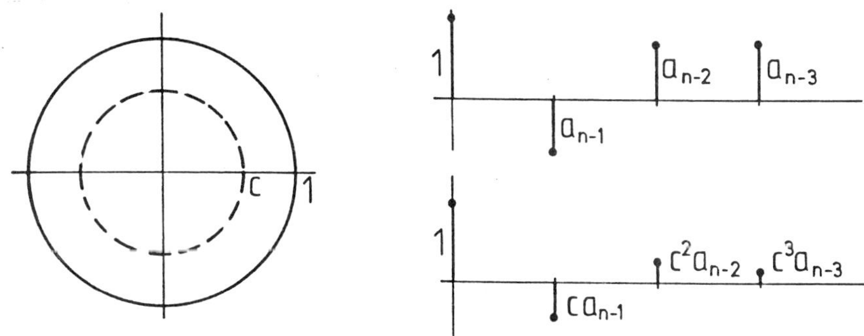

Figure 9.5 Contraction of poles ba y factor c corresponds to a damping of all FESs

Strategies can be developed using a combination of the two basic approaches above. For example a contraction of poles not towards z = 0, but towards z = 1, is achieved by shifting them to the left by one, contracting towards z = 0 and shifting back to the right by one.

Example:

The example (9.1.13) has four eigenvalues at z = 1 and one at z = 2. Contract them towards one and softly stabilize the system. In the first step all eigenvalues are shifted to the left by one, i.e.

$$\underline{Q}_c\underline{Z}^{(1)}(z) = \underline{QZ}(z+1) = \begin{bmatrix} z^2 & 0 & 0.5z \\ 0 & z & 0 \\ 2z & -1 & 1-z+z^2 \end{bmatrix}$$

In the second step the eigenvalues are contracted by a factor c = 0.2, i.e. the characteristic polynomial becomes $z^4(z-0.2)$ and

$$\underline{Q}_c\underline{Z}^{(2)}(z) = \underline{Q}_c\underline{Z}^{(1)}(z/0.2)\begin{bmatrix} c^{\mu_1} & & \\ & c^{\mu_2} & \\ & & c^{\mu_3} \end{bmatrix}$$

$$= \begin{bmatrix} 25z^2 & 0 & 2.5z \\ 0 & 5z & 0 \\ 10z & -1 & 1-5z+25z^2 \end{bmatrix}\begin{bmatrix} 0.2^2 & 0 & 0 \\ 0 & 0.2 & 0 \\ 0 & 0 & 0.2^2 \end{bmatrix}$$

$$= \begin{bmatrix} z^2 & 0 & 0.1z \\ 0 & z & 0 \\ 0.4z & -0.2 & 0.04-0.2z+z^2 \end{bmatrix}$$

The third step shifts the eigenvalues back to the right by 0.7 resulting in a characteristic polynomial $(z-0.7)^4(z-0.9)$ and

$$\underline{Q}_c\underline{Z}^{(3)}(z) = \underline{Q}_c\underline{Z}^{(2)}(z-0.7) = \begin{bmatrix} 0.49-1.4z+z^2 & 0 & -0.07+0.1z \\ 0 & -0.7+z & 0 \\ -0.28+0.4z & -0.2 & 0.67-1.6z+z^2 \end{bmatrix}$$

The corresponding feedback matrices $\underline{F}$ and $\underline{K} = \underline{MF}$ are

$$\underline{F} = (\underline{H}_c - \underline{H})\underline{T}$$

$$= \left\{ \left[\begin{array}{cc|c|cc} 0.49 & -1.4 & 0 & -0.07 & 0.1 \\ 0 & 0 & -0.7 & 0 & 0 \\ -0.28 & 0.4 & -0.2 & 0.67 & -1.6 \end{array}\right] - \begin{bmatrix} 1 & -2 & 0 & -0.5 & 0.5 \\ 0 & 0 & -1 & 0 & 0 \\ -2 & 2 & -1 & 3 & -3 \end{bmatrix} \right\} \underline{T}$$

$$= \begin{bmatrix} -0.51 & 0.6 & 0 & 0.43 & -0.4 \\ 0 & 0 & 0.3 & 0 & 0 \\ 1.72 & -1.6 & 0.8 & -2.33 & 1.4 \end{bmatrix} \begin{bmatrix} 0 & 0.5 & -0.75 & 0 & -0.25 \\ 0.5 & 0 & -0.25 & 0 & 0.75 \\ 0 & -1 & 0.5 & 1 & 0.5 \\ -1 & 0 & 0.5 & 1 & -0.5 \\ -2 & 0 & 1.5 & 1 & -0.5 \end{bmatrix}$$

$$= \begin{bmatrix} 0.67 & -0.255 & 0.3875 & 0.03 & 0.5625 \\ 0 & -0.3 & 0.15 & 0.3 & 0.15 \\ -1.27 & 0.06 & 0.445 & -0.13 & -0.765 \end{bmatrix}$$

$$\underline{K} = \underline{MF} = \begin{bmatrix} 0.67 & -0.105 & 0.3125 & -0.12 & 0.4875 \\ 0 & -0.3 & 0.15 & 0.3 & 0.15 \\ -1.27 & 0.06 & 0.445 & -0.13 & -0.765 \end{bmatrix} \tag{9.3.41}$$

9.3.3 Pole Region Assignment

For single-input systems, two methods of pole region assignment and simultaneous Γ-stability assignment for a family of plant models are derived in chapter 7 and 8. The first one is illustrated by figure 8.2. Here a nice stability region P_Γ is studied in P space. An affine mapping

$$\underline{k}_j' = [\underline{p}' \quad \underline{1}]\underline{E}_j \quad , \quad j = 1,2 \ldots J \tag{9.3.42}$$

then maps this region into K space for each plant model j. A possible generalization would now be to study Γ-stability regions for $P(z) = \det \underline{Q}_c \underline{Z}(z)$ in the space of elements π_{ijk} of $\underline{Q}_c(z)$. Also specific multi-variable properties, beyond Γ stability, could be incorporated, e.g. certain couplings between the subsystems. This leads to a generalized region P_Γ. Assuming, that all elements of the model family $(\underline{A}_j, \underline{B}_j)$ have the same controllability indices, then P_Γ can be mapped into the feedback gain space by

$$\underline{F}_j = \underline{Q}_c\underline{E}_j \tag{9.3.43}$$

or

$$\underline{K}_j = \underline{M}_j\underline{Q}_c\underline{E}_j \tag{9.3.44}$$

and intersections in the K space must be determined. This approach has not yet been pursued, because it seems very difficult to describe desired regions P_Γ in the space of π_{ijk}, i.e. of closed-loop FESs.

The second approach for single-input systems uses the inverse affine mapping

$$\underline{p}_j' = \underline{a}_j' + \underline{k}'\underline{W}_j \tag{9.3.45}$$

Here the real and complex boundaries are directly generated in K space or a cross section thereof, for graphical representation. The multivariable generalization is eq. (9.3.21)

$$\underline{H}_{cj} = \underline{H}_j + \underline{FT}_j^{-1} = \underline{H}_j + \underline{M}_j^{-1}\underline{KT}_j^{-1} \tag{9.3.46}$$

The polynomial version of this, eq. (9.3.28), produces $\underline{Q}_{cj}\underline{Z}$ and the relation with the closed-loop characteristic polynomial is

$$P_j(z) = \det \underline{Q}_{cj}\underline{Z} \tag{9.3.47}$$

In this r×r determinant, products of r elements of the r×n matrix $\underline{K}$ occur. These r elements belong to different rows and different columns of $\underline{K}$.

In graphical representations in two dimensional cross sections of Γ-stability regions K_Γ, two controller parameters must be selected in each design step. Call them k_a and k_b. These can be two selected elements of the $\underline{K}$ matrix, or two linear combinations of elements of two rows of $\underline{K}$, or the same from two columns of $\underline{K}$. In the νth design step

$$\underline{K}^{(\nu)} = \underline{K}^{(\nu-1)} + \underline{K}_\nu(k_a, k_b) \tag{9.3.48}$$

The characteristic polynomial coefficient vector is now

$$\underline{p}(\underline{K}) = \underline{a} + \underline{f}(\underline{K}) \tag{9.3.49}$$

where $f(\underline{K})$ is linear in the matrix $\underline{K}$, i.e.

$$f(c_1\underline{K}_1 + c_2\underline{K}_2) = c_1 f(\underline{K}_1) + c_2 f(\underline{K}_2) \tag{9.3.50}$$

and for $\underline{K}^{(\nu)}$ of eq. (9.3.48)

$$\begin{aligned} \underline{p}(\underline{K}^{(\nu)}) &= \underline{a} + \underline{f}(\underline{K}^{(\nu-1)}) + \underline{f}[K_\nu(k_a, k_b)] \\ &= \underline{p}(\underline{K}^{(\nu-1)}) + \underline{\alpha}\, k_a + \underline{\beta}\, k_b + \underline{\gamma}(k_a \times k_b) \end{aligned} \tag{9.3.51}$$

If k_a and k_b are linear combinations of elements from only one row or one column then $\underline{\gamma} = \underline{0}$. In general the vectors $\underline{\alpha}$, $\underline{\beta}$ and $\underline{\gamma}$ are found as follows:

1. Let $k_a = 1$, $k_b = 0$, then
$$\underline{\alpha} = \underline{p}(\underline{K}^{(\nu)}) - \underline{p}(\underline{K}^{(\nu-1)}) \tag{9.3.52}$$

2. Let $k_a = 0$, $k_b = 1$, then
$$\underline{\beta} = \underline{p}(\underline{K}^{(\nu)}) - \underline{p}(\underline{K}^{(\nu-1)}) \tag{9.3.53}$$

3. Let $k_a = 1$, $k_b = 1$, then
$$\underline{\gamma} = \underline{p}(\underline{K}^{(\nu)}) - \underline{p}(\underline{K}^{(\nu-1)}) - \underline{\alpha} - \underline{\beta} \tag{9.3.54}$$

$\underline{p}(\underline{K})$ can be determined numerically, e.g. by an eigenvalue program, from

$$P(z) = [\underline{p}(\underline{K}) \quad 1]\underline{z}_n = \det(z\underline{I}-\underline{A}+\underline{B}\underline{K}) \tag{9.3.55}$$

such that $\underline{\alpha}$, $\underline{\beta}$ and $\underline{\gamma}$ can be found numerically without evaluating $f[\underline{K}_\nu(k_a, k_b)]$.

$\underline{p}'$ of eq. (9.3.51) is now substituted into eq. (7.4.21), for the real boundaries, and eq. (7.4.30) for the complex boundary. For $\underline{\gamma} \neq \underline{0}$, the real boundary is a hyperbola in the k_a-k_b plane and for each point on the complex boundary the corresponding points in k_a-k_b plane are obtained as intersections of two hyperbolas.

Intersections for which k_a and k_b are chosen either from the same row, or the same column, are easier to calculate and clearer in the representation. The product $k_a \times k_b$ is then omitted and only straight lines, or the intersection of two straight lines, need be determined as in the single-input case.

An advantage of the second approach, eqs. (9.3.45) through (9.3.55) is, that it applies also to the output feedback case

$$\underline{u} = -\underline{K}_y\underline{y} = -\underline{K}_y\underline{Cx} \qquad (9.3.56)$$

with rank $\underline{C} = s < n$. Here $\underline{K}$ is restricted to be of the form $\underline{K} = \underline{K}_y\underline{C}$ with given $\underline{C}$.

If the system is written in sensor coordinates, then $\underline{C}$ does not depend on j and eq. (9.3.46) reads

$$\underline{H}_{cj} = \underline{H}_j + \underline{M}_j^{-1}\underline{K}_y\underline{CT}_j^{-1} \qquad (9.3.57)$$

9.4 Quadratic Optimal Control

A synthesis formalism, for both single and multi-input systems, for the calculation of a time variable state feedback

$$\underline{u}(t) = -\underline{K}(t)\underline{x}(t) \qquad (9.4.1)$$

for the plant

$$\dot{\underline{x}}(t) = \underline{Fx}(t) + \underline{Gu}(t) \qquad (9.4.2)$$

follows from the minimization of the quadratic cost functional

$$J = \underline{x}'(t_e)\underline{D}\underline{x}(t_e) + \int_0^{t_e} \{\underline{x}'(t)\underline{P}\underline{x}(t) + \underline{u}'(t)\underline{M}\underline{u}(t)\}dt \qquad (9.4.3)$$

with $\underline{D}$ and $\underline{P}$ positive semi-definite and $\underline{M}$ positive definite, $\underline{x}(t)$ should be controlled from $\underline{x}(0) = \underline{x}_o$ during the control interval $0 \leq t \leq t_e$ to $\underline{x}(t_e)$. $\underline{K}(t)$ for $0 \leq t \leq t_e$ in eq. (9.4.1) results from the solution of a matrix Riccati differential equation. See [71.8], [72.3] for example. For $t_e \to \infty$ a constant feedback gain $\underline{K}(t) = \underline{K}$ is obtained. Such a "Riccati design", performed with available general computer programs, may provide a useful starting point for design.

9.4.1 Discrete-Time Systems

The corresponding approach for discrete systems

$$\underline{x}_{k+1} = \underline{A}\underline{x}_k + \underline{B}\underline{u}_k \qquad (9.4.4)$$

is to minimize the following cost functional

$$J := \underline{x}_N'\underline{D}_N\underline{x}_N + \sum_{k=0}^{N-1} (\underline{x}_k'\underline{Q}\underline{x}_k + \underline{u}_k'\underline{R}\underline{u}_k) \qquad (9.4.5)$$

Here $\underline{D}_N$ and $\underline{Q}$ are positive semi-definite and $\underline{R}$ is positive definite. (The discrete time k is written as an index here in order to simplify the notation. Since no vector components are considered in the following, confusion should not arise.) The system (9.4.4) with initial state x_o should be transferred to the final state $\underline{x}_N$, so that the cost functional (9.4.5) becomes minimal. This problem was solved by Kalman and Koepcke [58.7]. The resulting optimal solution is the following time varying-control law

$$\underline{u}_k = -\underline{K}_k\underline{x}_k \qquad (9.4.6)$$

The calculation of $\underline{K}_k$ is performed as follows. First $\underline{K}_{N-1}$ is calculated for the last interval. The contribution of this interval to the cost functional J is by eq. (9.4.5)

$$J_{N-1} := \underline{x}_N'\underline{D}_N\underline{x}_N + \underline{x}_{N-1}'\underline{Q}\underline{x}_{N-1} + \underline{u}_{N-1}'\underline{R}\underline{u}_{N-1} \tag{9.4.7}$$

Here $\underline{x}_N$ and $\underline{u}_{N-1}$ are inserted according to eqs. (9.4.4) and (9.4.6), i.e.

$$\underline{x}_N = (\underline{A}-\underline{B}\underline{K}_{N-1})\underline{x}_{N-1}, \quad \underline{u}_{N-1} = -\underline{K}_{N-1}\underline{x}_{N-1} \tag{9.4.8}$$

The following is obtained

$$J_{N-1} = \underline{x}_{N-1}'\underline{D}_{N-1}\underline{x}_{N-1} \tag{9.4.9}$$

$$\underline{D}_{N-1} = (\underline{A}-\underline{B}\underline{K}_{N-1})'\underline{D}_N(\underline{A}-\underline{B}\underline{K}_{N-1}) + \underline{Q} + \underline{K}_{N-1}'\underline{R}\underline{K}_{N-1} \tag{9.4.10}$$

According to the Bellmann principle, the transition from $\underline{x}_{N-1}$ to $\underline{x}_N$ must be optimal by itself,if it is part of an optimal transition from $\underline{x}_o$ to $\underline{x}_N$. $\underline{K}_{N-1}$ must therefore be chosen such that J_{N-1} becomes minimal. For this $\underline{D}_{N-1}$, is first put into the form

$$\underline{D}_{N-1} = \underline{A}_o - \underline{A}_1\underline{K}_{N-1} - \underline{K}_{N-1}'\underline{A}_1' + \underline{K}_{N-1}'\underline{A}_2\underline{K}_{N-1} \tag{9.4.11}$$

$$\text{with } \underline{A}_o = \underline{A}_o' = \underline{A}'\underline{D}_N\underline{A} + \underline{Q}$$

$$\underline{A}_1 = \underline{A}'\underline{D}_N\underline{B}$$

$$\underline{A}_2 = \underline{A}_2' = \underline{B}'\underline{D}_N\underline{B} + \underline{R} \qquad \text{positive definite.}$$

The terms dependent on $\underline{K}_{N-1}$, in eq. (9.4.11), are extended to a complete quadratic form by the addition and subtraction of the terms $\underline{A}_1\underline{A}_2^{-1}\underline{A}_1'$.

$$\underline{D}_{N-1} = \underline{A}_o - \underline{A}_1\underline{A}_2^{-1}\underline{A}_1' + (\underline{K}_{N-1}'-\underline{A}_1\underline{A}_2^{-1})\underline{A}_2(\underline{K}_{N-1}'-\underline{A}_1\underline{A}_2^{-1})' \tag{9.4.12}$$

The first two terms do not depend on $\underline{K}_{N-1}$. The minimum of J_{N-1} is obtained by

$$\underline{K}_{N-1}' = \underline{A}\,\underline{A}_1{}_2^{-2} \quad \text{i.e.} \quad \underline{K}_{N-1} = \underline{A}_2^{-1}\underline{A}_1'$$

$$\underline{K}_{N-1} = (\underline{B}'\underline{D}_N\underline{B}+\underline{R})^{-1}\underline{B}'\underline{D}_N\underline{A} \tag{9.4.13}$$

From eq. (9.4.5) the portion of the cost functional for the interval before the last one is

$$J_{N-2} := \underline{x}'_{N-1}\underline{D}_{N-1}\underline{x}_{N-1} + \underline{x}'_{N-2}\underline{Qx}_{N-2} + \underline{u}'_{N-2}\underline{Ru}_{N-2} \tag{9.4.14}$$

The minimization of $J_{N-2} = \underline{x}'_{N-2}\underline{D}_{N-2}\underline{x}_{N-2}$ by $\underline{K}_{N-2}$ is the same problem as the minimization of J_{N-1} by $\underline{K}_{N-1}$ in the last interval; N need only be replaced by N-1 in eqs. (9.4.9), (9.4.10) and (9.4.13). Continuing this for k = N-2, N-3 ... 2,1,0 the minimal value $J_k = \underline{x}'_k\underline{D}_k\underline{x}_k$ of the cost function is obtained.

The sequence of gain matrices K_k, k = n-1, N-2...2,1,0 is calculated recursively, beginning with $\underline{D}_N$. Eq. (9.4.13) is then in general

$$\underline{K}_k = (\underline{B}'\underline{D}_{k+1}\underline{B}+\underline{R})^{-1}\underline{B}'\underline{D}_{k+1}\underline{A} \tag{9.4.15}$$

and eq. (9.4.10)

$$\underline{D}_k = (\underline{A}-\underline{BK}_k)'\underline{D}_{k+1}(\underline{A}-\underline{BK}_k) + \underline{Q} + \underline{K}'_k\underline{RK}_k \tag{9.4.16}$$

Eq. (9.4.16) is a Riccati difference equation. Along with eq. (9.4.15), it provides the optimal gain matrices $\underline{K}_k$. These are independent of the initial state $\underline{x}_o$ of the plant. The state feedback is therefore optimal for all initial states.

So far a finite control interval $0 \leq t \leq NT$ has been considered. N is now allowed to go to infinity and $\underline{D}_N$ is set to zero. The cost functional becomes

$$J = \sum_{k=0}^{\infty} [\underline{x}'_k\underline{Qx}_k + \underline{u}'_k\underline{Ru}_k] \tag{9.4.17}$$

In contrast to the finite control interval the unstable plant eigenvalues must be controllable so that J remains finite. Then the stationary solution $K = K_\infty$, $D = D_\infty$ of the Riccati difference eqs. (9.4.15), (9.4.16) exists and provides a constant state feedback $\underline{K}$.

The closed loop with $\underline{u} = -\underline{Kx}$ is stable, if the weighting matrix $\underline{Q} = \underline{H}'\underline{H}$ in eq. (9.4.5) was chosen such that the unstable eigenvalues of $\underline{A}$ are observable by $\underline{y}_H = \underline{Hx}$.

In the stationary case with

$\underline{D} = \underline{D}'$, $\underline{R} = \underline{R}'$ eq. (9.4.16) becomes

$$\underline{D} = (\underline{A}-\underline{BK})'\underline{D}(\underline{A}-\underline{BK}) + \underline{Q} + \underline{K}'\underline{RK}$$

$$= \underline{A}'\underline{DA} - \underline{K}'\underline{B}'\underline{DA} - \underline{A}'\underline{DBK} + \underline{Q} + \underline{K}'(\underline{B}'\underline{DB}+\underline{R})\underline{K} \tag{9.4.18}$$

Here according to eq. (9.4.15)

$$\underline{K} = (\underline{B}'\underline{DB}+\underline{R})^{-1}\underline{B}'\underline{DA} \tag{9.4.19}$$

Insertion into eq. (9.4.18) yields

$$\underline{D} = \underline{A}'\underline{DA} - 2\underline{A}'\underline{DB}(\underline{B}'\underline{DB}+\underline{R})^{-1}\underline{B}'\underline{DA} + \underline{Q} + \underline{A}'\underline{DB}(\underline{B}'\underline{DB}+\underline{R})^{-1}\underline{B}'\underline{DA}$$

$$= \underline{A}'\underline{DA} - \underline{A}'\underline{DB}(\underline{B}'\underline{DB}+\underline{R})^{-1}\underline{B}'\underline{DA} + \underline{Q}$$

$$\underline{D} = \underline{A}'[\underline{D}-\underline{DB}(\underline{B}'\underline{DB}+\underline{R})^{-1}\underline{BD}]\underline{A} + \underline{Q} \tag{9.4.20}$$

Efficient numerical procedures exist for the solution of this algebraic Riccati difference equation [79.6].

9.4.2 Sampled-Data Systems

Only the state $\underline{x}_k = \underline{x}(kT)$ at the sampling instants was weighted with the cost functional (9.4.5). If the discrete system (9.4.4) describes a sampled-data system consisting of a continuous plant with sampler and hold at the input, then also the continuous state can be weighted, i.e. the cost functional (9.4.3) is minimized [70.6]. It is reformulated as

$$J = \underline{x}'(NT)\underline{D}_N\underline{x}(NT) + \sum_{k=0}^{N-1} \{ \int_0^1 [\underline{x}'(kT+\gamma T)\underline{Px}(kT+\gamma T)]d\gamma + T\times\underline{u}'(kT)\underline{Mu}(kT) \} \tag{9.4.21}$$

According to eq. (3.7.5)

$$\underline{x}(kT+\gamma T) = \underline{A}_\gamma \underline{x}(kT) + \underline{B}_\gamma \underline{u}(kT) \tag{9.4.22}$$

$$\text{with } \underline{A}_\gamma = e^{\underline{F}\gamma T} \quad , \quad \underline{B}_\gamma = \int_0^{\gamma T} e^{\underline{F}v} dv \underline{G}$$

Inserted into eq. (9.4.21), this yields in index notation $\underline{x}_k = \underline{x}(kT)$

$$J = \underline{x}_N' \underline{D}_N \underline{x}_N + \sum_{k=0}^{N-1} (\underline{x}_k' \underline{\overline{P}} \underline{x}_k + \underline{x}_k' \underline{S} \underline{u}_k + \underline{u}_k' \underline{S}' \underline{x}_k + \underline{u}_k' \underline{\overline{M}} \underline{u}_k) \tag{9.4.23}$$

with

$$\underline{\overline{P}} := \int_0^1 \underline{A}_\gamma' \underline{P} \underline{A}_\gamma d_\gamma \qquad \text{positive semi-definite}$$

$$\underline{S} := \int_0^1 \underline{A}_\gamma' \underline{P} \underline{B}_\gamma d_\gamma$$

$$\underline{\overline{M}} := \int_0^1 \underline{B}_\gamma' \underline{P} \underline{B}_\gamma d_\gamma + T\underline{M} \qquad \text{positive definite} \tag{9.4.24}$$

The minimization of J according to eq. (9.4.23) can be easily reduced to the minimization of J according to eq. (9.4.5) by substituting

$$\underline{u}_k = -\underline{\overline{M}}^{-1} \underline{S}' \underline{x}_k + \underline{v}_k \tag{9.4.25}$$

into eq. (9.4.23)

$$J = \underline{x}_N' \underline{D}_N \underline{x}_N + \sum_{k=0}^{N-1} \{\underline{x}_k' (\underline{\overline{P}} - \underline{S}\underline{\overline{M}}^{-1} \underline{S}') \underline{x}_k + \underline{v}_k' \underline{\overline{M}} \underline{v}_k\} \tag{9.4.26}$$

The minimization of J is equivalent to the following problem: given a substitute system

$$\underline{x}_{k+1} = \underline{A}^* \underline{x}_k + \underline{B} \underline{v}_k \quad , \quad \underline{A}^* := \underline{A} - \underline{B}\underline{\overline{M}}^{-1} \underline{S}' \tag{9.4.27}$$

Minimize the cost functional

$$J = \underline{x}_N'\underline{D}_N\underline{x}_N + \sum_{k=0}^{N-1} (\underline{x}_k'\underline{Q}\underline{x}_k + \underline{u}_k'\underline{R}\underline{u}_k) \qquad (9.4.28)$$

where

$$\underline{Q} := \overline{\underline{P}} - \underline{S}\overline{\underline{M}}^{-1}\underline{S}' \quad ; \quad \underline{R} := \overline{\underline{M}} \qquad (9.4.29)$$

From the resulting control law of the substitute system

$$\underline{v}_k = -\underline{K}_k^{*}\underline{x}_k \qquad (9.4.30)$$

the corresponding control law for the original system with eq. (9.4.25) is

$$\underline{u}_k = -(\overline{\underline{M}}^{-1}\underline{S}' + \underline{K}_k^{*})\underline{x}_k = -\underline{K}_k\underline{x}_k \qquad (9.4.31)$$

As in section 9.4.1, a constant feedback law minimizing the performance functional

$$J = \int_0^{\infty} (\underline{x}'\underline{P}\underline{x} + \underline{u}'\underline{M}\underline{u})dt \qquad (9.4.32)$$

can be obtained by the limiting process to the stationary solution.

The calculation steps summarized as follows:

1. The sampling period T is chosen.

2. The weighting matrices $\underline{P}$ and $\underline{M}$ in eq. (9.4.32) are chosen.

3. $\overline{\underline{P}}$, $\underline{S}$ and $\overline{\underline{M}}$ are calculated from eq. (9.4.24) and from this $\underline{Q}$ and $\underline{R}$ in eq. (9.4.29) and $\underline{A}^{*}$ from eq. (9.4.27).

4. $\underline{A} = \underline{A}^{*}$ is substituted into the Riccati equation (9.4.20) and $\underline{D}$ is calculated from this equation with a standard computer program.

5. Eq. (9.4.19) is here

$$\underline{K}^{*} = (\underline{B}'\underline{D}\underline{B} + \underline{R})^{-1}\underline{B}'\underline{D}\underline{A}^{*} \qquad (9.4.33)$$

6. According to eq. (9.4.31), $\underline{K}$ for the original system ($\underline{A}$, $\underline{B}$) is

$$\underline{K} = \underline{K}^{*} + \underline{\bar{M}}^{-1}\underline{S}' \qquad (9.4.34)$$

7. The sampled system is tested by simulation. If it is unsatisfactory, try a new choice of $\underline{P}$ and $\underline{M}$ in step 2.

The Riccati design in continuous time provides a phase margin of $\pm 60^{o}$, an infinite gain margin and a gain reduction margin of 50 % for completely measurable states. All margins hold for the loop opened at only one of the input variables, under the assumption that no phase and gain changes arise for the other inputs. As was shown in [78.5], these margins unfortunately do not hold in the discrete-time case. Examples show, that the gain margins can be significantly smaller. This is also illustrated clearly in the example of figure 7.9. Within the dotted stability triangle, it is impossible to simultaneously halve the gains k_1 and k_2. An infinite gain margin does not exist for sampled-data systems, basically because the unit circle and its image in K space are finite.

In view of these limitations, the Riccati design for sampled-data systems is recommended only if the technical requirements are naturally formulated as a quadratic criterion, and the weighting matrices $\underline{P}$ and $\underline{M}$ in eq. (9.4.32) are given.

The advantages of continuous-time Riccati design may be preserved by fast sampling. In this case Rattan's technique [81.13] may be used. First a continuous-time design is made. Then a fast sampling discrete-time compensator with unknown coefficients is assumed. These coefficients are derived to achieve a weighted best-least-squares fit of the phase and gain characteristics (Bode plots) of the digital design to those of the base continuous-time design.

9.5 Exercises

9.1 Consider the following system

$$\underline{x}[k+1] = \begin{bmatrix} 5 & -1 & 2 \\ -2 & -2 & 6 \\ 4 & -3 & 7 \end{bmatrix} \underline{x}[k] + \begin{bmatrix} 0 & 1 \\ 1 & 5 \\ 1 & 6 \end{bmatrix} \underline{u}[k]$$

Determine

a) The controllability indices μ_1, μ_2;

b) an input sequence which transfers the system into the zero state in minimum time;

c) the basic FESs;

d) a state feedback for minimum-time deadbeat control;

e) a state feedback $\underline{K}$ which gives the minimal polynomial $(z-0.4)^2$ for $\underline{A}-\underline{BK}$.

9.2 Can arbitrary poles be assigned with a feedback matrix

$$\underline{K} = \begin{bmatrix} k_{11} & 0 & k_{13} \\ k_{21} & 0 & k_{23} \end{bmatrix}$$

for the system of exercise 9.1?

9.3 The position of the trolley x_1 and the cable angle velocity x_4 are measured on the loading bridge. Determine the observability indices and calculate an observer of reduced order with time-optimal deadbeat response of the construction error.

9.4 Assign a minimal polynomial $(z-0.5)^2$ to the system of example (9.1.13). Check the closed-loop eigenvalue locations for failure of each one of the actuators. Do the same for the solution (9.3.38) and compare the results.

9.5 Consider the system

$$\underline{x}[k+1] \begin{bmatrix} 1 & 1 \\ 0 & 1 \end{bmatrix} \underline{x}[k] + \begin{bmatrix} 1 & 1 \\ 1 & 2 \end{bmatrix} \underline{u}[k]$$

The state vector feedback can assume three configurations:

a) nominal

$$\underline{K} = \begin{bmatrix} k_{11} & k_{12} \\ k_{21} & k_{22} \end{bmatrix}$$

b) failure of actuator 1

$$\underline{K}_2 = \begin{bmatrix} 0 & 0 \\ k_{21} & k_{22} \end{bmatrix}$$

c) failure of actuator 2

$$\underline{K}_1 = \begin{bmatrix} k_{11} & k_{12} \\ 0 & 0 \end{bmatrix}$$

A $\underline{K}$ is sought such that nominally a double eigenvalue at $z = 0.4$ is assigned and the eigenvalues lie in the smallest possible circle Γ_r, from the family of circles of figure 7.5, in both failure situations.

9.6 The loading bridge has the parameter values from exercise 3.6. The state vector feedback $\underline{k}^{*\prime} = [400 \quad 1550 \quad 15450 \quad -10267]$ was found by pole assignment in the center of gravity coordinates. This feedback requires a maximal input amplitude of $|u| = 400$ in order to shift the load by 1 meter. It reduces $||\underline{x}|| = \sqrt{\underline{x}'\underline{x}}$ from 1 to 0.003 within 10 seconds. Try some weighting matrices in the quadratic optimal design, in order to find a solution which comes close to these specifications.

Appendix A Canonical Forms and Further Results from Matrix Theory

A.1 Linear Transformations

Linear transformations

$$\underline{x}^* = \underline{T}\underline{x}, \ \det \underline{T} \neq 0 \tag{A.1.1}$$

can be used to transform a system ($\underline{A}$, $\underline{B}$, $\underline{C}$) to a canonical representation

$$\underline{A}^* = \underline{TAT}^{-1}, \ \underline{B}^* = \underline{TB}, \ \underline{C}^* = \underline{CT}^{-1} \tag{A.1.2}$$

For this change of basis in state space, it is irrelevant whether ($\underline{A}$, $\underline{B}$, $\underline{C}$) describes a continuous or a discrete time system. The dimensions of the matrices and the corresponding number of elements in them is

A	:	$n \times n$
B	:	$n \times r$
C	:	$s \times n$
Total	:	$n(n+r+s)$

n^2 coefficients can be chosen in $\underline{T}$ such that n^2 coefficients in $\underline{A}^*$, $\underline{B}^*$ and $\underline{C}^*$ can be made zero or one. The remaining $n(r+s)$ coefficients suffice to describe the system in a canonical form. They are necessary in the "generic" case, i.e. if all states are controllable from each input and observable from each output. In the nongeneric case each missing connection between a state and an input or output may be expressed by a zero coefficient, thus exhibiting the controllability and observability structure of the system.

A description with the minimal number of coefficients is especially important if the mathematical model of the plant cannot be derived from physical laws, but must be determined from measured input-output-signals (identification). In the space of minimal parameters this is a search for a point. For a nonminimum description the search would be for a subspace in which arbitrarily many correct solutions lie, which are distinguished from each other by a transformation $\underline{T}$ of the basis of state space. This has a significant influence on the convergence of identification algorithms.

In addition, canonical forms are suitable for visualizing structural properties of a system, for example controllable or observable subsystems. This helps in the understanding of such properties and does not mean that the calculation of a canonical form must actually be a part of a design procedure. The Jordan and Frobenius forms for example are occasionally used in proofs. Calculations can be performed in a numerically favorable way by transformation to Hessenberg or Schur form [78.7], [79.6], [80.4]. Canonical forms are also used for controller implementation.

In this appendix, the most important canonical forms and the determination of the appropriate transformation matrix $\underline{T}$ from $(\underline{A}, \underline{B}, \underline{C})$ are presented. For controllable or observable systems the transformation matrix $\underline{T}$ can be calculated from the relation of the controllability matrices

$$[\underline{B}^{*}, \underline{A}^{*}\underline{B}^{*} \ldots] = \underline{T}[\underline{B}, \underline{A}\underline{B} \ldots] \tag{A.1.3}$$

or of the observability matrices

$$\begin{bmatrix} \underline{C}^{*} \\ \underline{C}^{*}\underline{A} \\ \vdots \end{bmatrix} \underline{T} = \begin{bmatrix} \underline{C} \\ \underline{C}\underline{A} \\ \vdots \end{bmatrix} \tag{A.1.4}$$

For the single-input case:

$$\underline{T} = [\underline{b}^{*} \ \ldots \ \underline{A}^{*n-1}\underline{b}^{*}][\underline{b} \ \ldots \ \underline{A}^{n-1}\underline{b}]^{-1} = \underline{\mathcal{C}}^{*}\underline{\mathcal{C}}^{-1} \tag{A.1.5}$$

or

$$\underline{T} = \begin{bmatrix} \underline{c}^{*\prime} \\ \vdots \\ \underline{c}^{*\prime}\underline{A}^{*n-1} \end{bmatrix}^{-1} \begin{bmatrix} \underline{c}^{\prime} \\ \vdots \\ \underline{c}^{\prime}\underline{A}^{n-1} \end{bmatrix} = \underline{\mathcal{O}}^{*-1}\underline{\mathcal{O}} \tag{A.1.6}$$

It is usually easier to invert the canonical form of the controllability matrix $\underline{\mathcal{C}}^*$ in eq. (A.1.5) and to calculate

$$\underline{S} = \underline{T}^{-1} = \underline{\mathcal{C}}\,\underline{\mathcal{C}}^{*-1} \tag{A.1.7}$$

A.2 Diagonal and Jordan Forms

The case of distinct eigenvalues $z_1, z_2 \ldots z_n$ will be treated first for which the state equations can be transformed to diagonal form, that is

$$\underline{x}_D[k+1] = \underline{\Lambda}\underline{x}_D[k] + \underline{b}_D u[k] \text{ with } \underline{\Lambda} := \begin{bmatrix} z_1 & 0 & \cdots & 0 \\ 0 & z_2 & \ddots & \vdots \\ \vdots & & \ddots & 0 \\ 0 & \cdots & 0 & z_n \end{bmatrix} \tag{A.2.1}$$

$$y[k] = \underline{c}_D'\underline{x}_D[k]$$

The most important properties of this form are:

1. The characteristic equation is

$$\det(z\underline{I}-\underline{\Lambda}) = (z-z_1)(z-z_2)\ \ldots\ (z-z_n) \tag{A.2.2}$$

2. The controllability matrix is

$$\underline{\mathcal{C}}_D = \begin{bmatrix} b_{D1} & 0 & \cdots & 0 \\ 0 & b_{D2} & & \\ \vdots & & \ddots & \\ 0 & \cdots & 0 & b_{Dn} \end{bmatrix} \begin{bmatrix} 1 & z_1 & \cdots & z_1^{n-1} \\ 1 & z_2 & \cdots & z_2^{n-1} \\ \vdots & \vdots & & \vdots \\ 1 & z_n & \cdots & z_n^{n-1} \end{bmatrix} \tag{A.2.3}$$

The second factor in eq. (A.2.3) is called the "Vandermonde matrix". It is regular, if and only if all z_i are different from each other [59.4]. The system is therefore controllable, if and only if all components of the vector $\underline{b}_D$ are nonzero. The system is decomposed into n decoupled subsystems of first order with the eigenvalues $z_1, z_2 \ldots z_n$. Such a subsystem is controllable if it can be stimulated from the input u, and the subsystem is observable if it contributes to the measured output signal y.

3. Every system ($\underline{A}$, $\underline{B}$, $\underline{C}$) with distinct eigenvalues can be transformed into the diagonal form by $\underline{x} = \underline{T}^{-1}\underline{x}_D = \underline{S}\underline{x}_D$. The columns $\underline{s}_i$ of $\underline{S}$ are determined by

$$\underline{AS} = \underline{S\Lambda} \tag{A.2.4}$$

$$\underline{A}[\underline{s}_1, \underline{s}_2 \ldots \underline{s}_n] = [\underline{s}_1, \underline{s}_2 \ldots \underline{s}_n]\begin{bmatrix} z_1 & 0 & \ldots & 0 \\ 0 & z_2 & & \\ \vdots & & \ddots & \\ 0 & & & z_n \end{bmatrix}$$

$$= [z_1\underline{s}_1, z_2\underline{s}_2 \ldots z_n\underline{s}_n] \tag{A.2.5}$$

Thus

$$\underline{A}\underline{s}_i = z_i\underline{s}_i \tag{A.2.6}$$

In other words, $\underline{s}_i$ is the eigenvector of $\underline{A}$, which corresponds to the eigenvalue z_i. $\underline{S}$ is composed of the eigenvectors. $\underline{S}$ is regular, because eigenvectors corresponding to distinct eigenvalues are always linearly independent [59.4]. Since $\det(z_i\underline{I}-\underline{A}) = 0$, only the direction of the eigenvector $\underline{s}_i$ is fixed by eq. (A.2.6). Its length can be arbitrarily chosen. It is fixed such, that only ones or zeros can arise in $\underline{b}_D$ or in $\underline{c}'_D$.

Remark A.1:

The eigenvector s_i may be viewed as a special initial state $\underline{x}[0] = \underline{s}_i$ of a system $\underline{x}[k+1] = \underline{Ax}[k]$, such that its direction is invariant under multiplication with $\underline{A}$, i.e. $\underline{x}[1] = z_i\underline{x}[0]$ is only a stretched or contracted version of $\underline{x}[0]$. $\underline{x}[0] = \underline{s}_i$ excites only the mode z_i of the system.

Example:

The system

$$\underline{x}[k+1] = \begin{bmatrix} 1 & 1 & 2 \\ -1 & 1 & -2 \\ 0 & 0 & 1 \end{bmatrix}\underline{x}[k] + \begin{bmatrix} 2 \\ 4 \\ 1 \end{bmatrix} u[k] = \underline{Ax}[k] + \underline{b}u[k]$$

$$y[k] = [\,1 \quad 2 \quad 4\,]\underline{x}[k] = \underline{c}'\underline{x}[k] \tag{A.2.7}$$

has the characteristic equation

$$\det(z\underline{I}-\underline{A}) = z^3 - 3z^2 + 4z - 2 = 0$$

with the eigenvalues

$$z_1 = 1,\ z_2 = 1 + j,\ z_3 = 1 - j$$

The eigenvector $\underline{s}_1$ which corresponds to z_1 is obtained from

$$(\underline{A}-z_1\underline{I})\underline{s}_1 = 0$$

$$\begin{bmatrix} 0 & -1 & -2 \\ 1 & 0 & 2 \\ 0 & 0 & 0 \end{bmatrix}\begin{bmatrix} s_{11} \\ s_{21} \\ s_{31} \end{bmatrix} = \begin{bmatrix} 0 \\ 0 \\ 0 \end{bmatrix}$$

From the first two lines $s_{11} = s_{21} = -2s_{31}$. The eigenvector is therefore

$$\underline{s}_1 = s_{11}\begin{bmatrix} 1 \\ 1 \\ -0.5 \end{bmatrix}$$

Correspondingly one obtains for z_2 and z_3

$$\underline{s}_2 = s_{12}\begin{bmatrix} 1 \\ j \\ 0 \end{bmatrix},\quad \underline{s}_3 = s_{13}\begin{bmatrix} 1 \\ -j \\ 0 \end{bmatrix}$$

The transformation matrix becomes

$$\underline{S} = [\underline{s}_1,\ \underline{s}_2,\ \underline{s}_3] = \begin{bmatrix} 1 & 1 & 1 \\ 1 & j & -j \\ -0.5 & 0 & 0 \end{bmatrix}\begin{bmatrix} s_{11} & 0 & 0 \\ 0 & s_{12} & 0 \\ 0 & 0 & s_{13} \end{bmatrix} \tag{A.2.8}$$

Thus

$$\underline{b}_D = \underline{S}^{-1}\underline{b} = \begin{bmatrix} 1/s_{11} & 0 & 0 \\ 0 & 1/s_{12} & 0 \\ 0 & 0 & 1/s_{13} \end{bmatrix} \begin{bmatrix} 0 & 0 & -2 \\ 0.5 & -0.5j & 1-j \\ 0.5 & 0.5j & 1+j \end{bmatrix} \begin{bmatrix} 2 \\ 4 \\ 1 \end{bmatrix} = \begin{bmatrix} -2/s_{11} \\ (2-3j)/s_{12} \\ (2+3j)/s_{13} \end{bmatrix}$$

$$\underline{c}'_D = \underline{c}'\underline{S} = [s_{11},\ (1+2j)s_{12},\ (1-2j)s_{13}]$$

Because all components of $\underline{b}_D$ and $\underline{c}'_D$ are nonzero, it can be seen immediately, that all states of the system are controllable and observable.

If, for example, only ones are wanted in $\underline{b}_D$, then $s_{11} = -2$, $s_{12} = 2-3j$ and $s_{13} = 2+3j$ are chosen. The diagonal form of the state equation is then

$$\underline{x}_D[k+1] = \begin{bmatrix} 1 & 0 & 0 \\ 0 & 1+j & 0 \\ 0 & 0 & 1-j \end{bmatrix} \underline{x}_D[k] + \begin{bmatrix} 1 \\ 1 \\ 1 \end{bmatrix} u[k]$$

$$y[k] = [-2 \quad 8+j \quad 8-j]\underline{x}_D[k] \qquad \text{(A.2.9)}$$

Remark A.2:

For controllable systems the transformation matrix $\underline{S}$ can also be determined from eqs. (A.1.7) and (A.2.3) For $b_{D1} = b_{D2} = \dots = b_{DN} = 1$, this becomes

$$\underline{S} = \underline{\mathcal{C}}\ \underline{\mathcal{C}}_D^{-1} = \underline{\mathcal{C}} \begin{bmatrix} 1 & z_1 & \dots & z_1^{n-1} \\ \vdots & \vdots & & \vdots \\ 1 & z_n & \dots & z_n^{n-1} \end{bmatrix}^{-1} \qquad \text{(A.2.10)}$$

In example (A.2.7)

$$\underline{S} = \begin{bmatrix} 2 & 8 & 10 \\ 4 & 0 & -10 \\ 1 & 1 & 1 \end{bmatrix} \begin{bmatrix} 1 & 1 & 1 \\ 1 & 1+j & 2j \\ 1 & 1-j & -2j \end{bmatrix}^{-1}$$

$$= \begin{bmatrix} 2 & 8 & 10 \\ 4 & 0 & -10 \\ 1 & 1 & 1 \end{bmatrix} \begin{bmatrix} 2 & -0.5+0.5j & -0.5-0.5j \\ -2 & 1-0.5j & 1+0.5j \\ 1 & -0.5 & -0.5 \end{bmatrix}$$

$$= \begin{bmatrix} -2 & 2-3j & 2+3j \\ -2 & 3+2j & 3-2j \\ 1 & 0 & 0 \end{bmatrix}$$

This correspondingly holds for observable systems.

There is a simple relation between the diagonal form and the z-transfer function of the system:

$$h_z(z) = \underline{c}_D'(z\underline{I}-\underline{\Lambda})^{-1}\underline{b}_D = \frac{c_{D1}b_{D1}}{z-z_1} + \frac{c_{D2}b_{D2}}{z-z_2} + \ldots + \frac{c_{Dn}b_{Dn}}{z-z_n} \qquad \text{(A.2.11)}$$

The diagonal form can therefore be easily determined for a given transfer function $h_z(z)$ with simple poles at z_1, z_2 ... z_n. If $h_z(z)$ is relatively prime (i.e. it does not have a common root in the numerator and denominator polynomials), then the system is controllable and observable. All components of the $\underline{b}_D$ vector can be normalized to one, and

$$c_{Di} = \text{res}\; h_z(z)\Big|_{z=z_i} = \lim_{z\to z_i} (z-z_i)h_z(z) \qquad \text{(A.2.12)}$$

In other words: the representation in diagonal form corresponds to a partial fractions decomposition of the transfer function.

Remark A.3:

Such a realization of a z-transfer function in the form of a state representation will be performed repeatedly in the following sections. It should be kept in mind, that these are the models only for the controllable and observable subsystem, i.e. subsystem B in figure 2.6.

An advantage of the diagonal form of the continuous system with eigenvalues $s_1, s_2 \ldots s_n$ is that the transition matrix

$$e^{\underline{\Lambda}t} = \begin{bmatrix} e^{s_1 t} & 0 & \ldots & 0 \\ 0 & e^{s_2 t} & & \\ \vdots & & \ddots & \\ 0 & \ldots & \ldots & e^{s_n t} \end{bmatrix} \qquad \text{(A.2.13)}$$

can be calculated very easily. It was already pointed out in section 3.1, that the discretization of the continuous system can thus be easily performed. The continuous system is put into diagonal form (respectively Jordan form), which simplifies the calculation of $\underline{A} = e^{\underline{F}t}$.

Now we turn to the general case in which multiple eigenvalues arise. As one sees from eqs. (A.2.5) and (A.2.6), a transformation into diagonal form is possible if and only if the $\underline{A}$ matrix has n linearly independent eigenvectors. This case can also arise for multiple eigenvalues. Then, however, the matrix $\underline{A}$ is non-cyclic, its minimal polynomial contains all eigenvalues but with lower multiplicity, and is therefore of lower degree than the characteristic polynomial. Such a system cannot be controlled from a single input or observed from a single output. An example for this case is the system from figure 2.7 c. Two parallel blocks with the same denominator z-d arise in the partial fractions decomposition. The corresponding two state variables x_1 and x_2 cannot be controlled or observed independently from each other. Thus a subsystem of type D with eigenvalue z = d arises in the canonical decomposition of figure 2.6.

Example:

The system

$$\underline{x}[k+1] = \begin{bmatrix} 1 & 0 & 0 \\ 1 & 1 & 1 \\ -1 & 0 & 0 \end{bmatrix} \underline{x}[k] + \begin{bmatrix} 1 \\ 1 \\ 1 \end{bmatrix} u[k] = \underline{A}\underline{x}[k] + \underline{b}u[k]$$

$$y[k] = [\,2 \quad 2 \quad 1\,]\underline{x}[k] = \underline{c}'\underline{x}[k] \qquad \text{(A.2.14)}$$

is transformed by

$$\underline{x} = \underline{S}\underline{x}_D = \begin{bmatrix} 0 & 1 & -2 \\ -2 & 3 & 1 \\ 2 & -1 & 2 \end{bmatrix} \underline{x}_D \tag{A.2.15}$$

into the following form:

$$\begin{aligned} \underline{x}_D[k+1] &= \begin{bmatrix} 0 & 0 & 0 \\ 0 & 1 & 0 \\ 0 & 0 & 1 \end{bmatrix} \underline{x}_D[k] + \begin{bmatrix} 1 \\ 1 \\ 0 \end{bmatrix} u[k] \\ y[k] &= [-2 \quad 7 \quad 0]\underline{x}_D[k] \end{aligned} \tag{A.2.16}$$

In this representation, the canonical decomposition of the system according to figure 2.6 can immediately be read off. The last subsystem with the eigenvalue 1 is of type D, because it is neither controllable from the input u: $b_{D3} = 0$, nor observable from the output y: $c_{D3} = 0$. For the design of a control system, all that is of interest is the second order system of type B, with the state vector

$$\begin{aligned} \underline{x}^* &= [x_{D1}, x_{D2}]' \\ \underline{x}^*[k+1] &= \begin{bmatrix} 0 & 0 \\ 0 & 1 \end{bmatrix} \underline{x}^*[k] + \begin{bmatrix} 1 \\ 1 \end{bmatrix} u[k] \\ y[k] &= [-2 \quad 7]\underline{x}^*[k] \end{aligned} \tag{A.2.17}$$

In general it follows from eq. (A.2.3), that every system in diagonal form with two equal eigenvalues $z_i = z_j$ is uncontrollable from one input variable and is also unoberservable from one output variable. In this case, the two identical subsystems cannot be influenced independently from each other at the input and their contributions to the output signal are indistinguishable.

More important is the case of a system with multiple eigenvalues which cannot be transformed into diagonal form, because the

$\underline{A}$-matrix has only m (m<n) linearly independent eigenvectors. In this case, the system can always be transformed into the Jordan form with the dynamics matrix

$$\underline{J} = \begin{bmatrix} \underline{J}_1 & \underline{0} & \cdots & \underline{0} \\ \underline{0} & \underline{J}_2 & & \\ \vdots & & \ddots & \\ \underline{0} & & & \underline{J}_m \end{bmatrix} \qquad \underline{J}_i = \begin{bmatrix} z_i & 1 & & 0 \\ 0 & z_i & \ddots & \\ & & \ddots & 1 \\ 0 & & & z_i \end{bmatrix} \tag{A.2.18}$$

The submatrices $\underline{J}_i$ in the main diagonal are referred to as "Jordan blocks". If a Jordan block has dimension one, it consists only of the eigenvalue z_i. If all z_i are different from each other, then $\underline{J}$ is called "cyclic". The basic theory is derived in detail in the book of Gantmacher [59.4]. Here it will only be shown how the Jordan form can be calculated. At least one eigenvector $\underline{s}_i$ corresponds to each eigenvalue z_i. This vector is calculated from

$$(\underline{A}-z_i\underline{I})\underline{s}_i = \underline{0} \tag{A.2.19}$$

If eq. (A.2.19) yields less than p eigenvectors for an eigenvalue of multiplicity p, then at least one of these eigenvectors is the beginning of a chain of "generalized eigenvectors" $\underline{s}_{i+1}, \underline{s}_{i+2} \cdots \underline{s}_{i+k_i-1}$, which are calculated from the relations

$$\begin{aligned} (\underline{A}-z_i\underline{I})\underline{s}_{i+1} &= \underline{s}_i \\ &\vdots \\ (\underline{A}-z_i\underline{I})\underline{s}_{i+k_i-1} &= \underline{s}_{i+k_i-2} \end{aligned} \tag{A.2.20}$$

In total, p linearly independent eigenvectors and generalized eigenvectors exist for an eigenvalue of multiplicity p. These form the columns of the transformation matrix $\underline{S}$. Every eigenvector has a Jordan block $\underline{J}_i$. The dimension of this Jordan block is equal to the number k_i of the corresponding linearly independent vectors $\underline{s}_i, \underline{s}_{i+1} \cdots \underline{s}_{i+k_i-1}$.

The transformation matrix is calculated from

$$\underline{AS} = \underline{SJ}$$

$$= \underline{A}[\underline{s}_1, \underline{s}_2 \cdots \underline{s}_{k_1} \vdots \cdots \underline{s}_n]$$

$$= [\underline{s}_1, \underline{s}_2 \cdots \underline{s}_{k_1} \vdots \cdots \underline{s}_n] \left[\begin{array}{cccc|c} \multicolumn{4}{c|}{\overbrace{\qquad\qquad}^{k_1}} & \\ z_1 & 1 & & 0 & \\ 0 & z_1 & \ddots & & \\ & & \ddots & 1 & \underline{0} \\ 0 & & & z_1 & \\ \hline & & & & J_2 \\ & & \underline{0} & & \quad \ddots \\ & & & & \qquad J_m \end{array}\right] \tag{A.2.21}$$

The columns yield - agreeing with eqs. (A.2.19) and (A.2.20) -

$$\begin{aligned} \underline{As}_1 &= z_1\underline{s}_1 \\ \underline{As}_2 &= \underline{s}_1 + z_1\underline{s}_2 \\ &\vdots \\ \underline{As}_{k_1} &= \underline{s}_{k_1-1} + z_1\underline{s}_{k_1} \end{aligned} \tag{A.2.22}$$

The structure of the Jordan form therefore depends on the length of the chains of generalized eigenvectors.

Example:

$$\underline{x}[k+1] = \begin{bmatrix} 1 & 1 & 0 & 0 & 0 \\ 0 & 1 & 1 & 0 & 1 \\ -1 & 1 & 1 & 0 & -1 \\ 0 & 0 & 1 & 1 & 1 \\ 1 & -1 & 0 & 0 & 2 \end{bmatrix} \underline{x}[k] + \begin{bmatrix} 1 \\ 2 \\ 1 \\ 1 \\ 1 \end{bmatrix} u[k]$$

$$= \underline{Ax}[k] + \underline{b}u[k]$$

$$y[k] = [1 \quad 1 \quad 1 \quad 1 \quad 3]\underline{x}[k] = \underline{c}'\underline{x}[k] \tag{A.2.23}$$

$$\det(z\underline{I}-\underline{A}) = (z-1)^4(z-2)$$

For the eigenvalue $z_1 = 1$ one obtains

$$\underline{s}_j = s_{1j}\begin{bmatrix}1\\0\\1\\0\\-1\end{bmatrix} + s_{4j}\begin{bmatrix}0\\0\\0\\1\\0\end{bmatrix}$$

i.e. there are two eigenvectors in the plane hereby fixed and consequently two Jordan blocks with the eigenvalue $z_1 = 1$. If only the dynamics matrix is to be transformed to Jordan form $\underline{J} = \underline{S}^{-1}\underline{AS}$, proceed from the two eigenvectors $[1\ \ 0\ \ 1\ \ 0\ \ -1]'$ and $[0\ \ 0\ \ 0\ \ 1\ \ 0]'$ and try to find generalized eigenvectors for them. Because we also want to put one of the vectors $\underline{b}_J$ or $\underline{c}_j$ into a simple form, however, the calculation should be done in a form as general as possible. The first generalized eigenvector $\underline{s}_2 = [v_1, v_2, v_3, v_4, v_5]'$ is determined from

$$(\underline{A}-\underline{I})\underline{s}_2 = \underline{s}_1$$

$$\begin{bmatrix}0 & 1 & 0 & 0 & 0\\0 & 0 & 1 & 0 & 1\\-1 & 1 & 0 & 0 & -1\\0 & 0 & 1 & 0 & 1\\1 & -1 & 0 & 0 & 1\end{bmatrix}\begin{bmatrix}v_1\\v_2\\v_3\\v_4\\v_5\end{bmatrix} = \begin{bmatrix}s_{1j}\\0\\s_{1j}\\s_{4j}\\-s_{1j}\end{bmatrix}$$

The equations corresponding to rows 2 and 4 can be simultaneously satisfied only if $s_{4j} = 0$, i.e. a chain of generalized eigenvectors exists for the eigenvector

$$\underline{s}_1 = s_{11}\begin{bmatrix}1\\0\\1\\0\\-1\end{bmatrix}$$

The first one is

$$\underline{s}_2 = s_{12}\begin{bmatrix}1\\0\\1\\0\\-1\end{bmatrix} + s_{11}\begin{bmatrix}0\\1\\0\\0\\0\end{bmatrix} + s_{42}\begin{bmatrix}0\\0\\0\\1\\0\end{bmatrix}$$

A solution for $(\underline{A}-\underline{I})\underline{s}_3 = \underline{s}_2$ then exists if and only if $s_{42} = s_{11}$. It is

$$\underline{s}_3 = s_{13}\begin{bmatrix}1\\0\\1\\0\\-1\end{bmatrix} + s_{11}\begin{bmatrix}0\\0\\1\\0\\0\end{bmatrix} + s_{12}\begin{bmatrix}0\\1\\0\\0\\0\end{bmatrix} + s_{43}\begin{bmatrix}0\\0\\0\\1\\0\end{bmatrix}$$

The equation $(\underline{A}-\underline{I})\underline{s}_4 = \underline{s}_3$ yields no further generalized eigenvector. A solution would only exist for $s_{11} = 0$, i.e. $\underline{s}_1 = 0$. For the eigenvalue $z_1 = 1$ with multiplicity four, there results an eigenvalue chain $\underline{s}_1$, $\underline{s}_2$, $\underline{s}_3$ and a further linearly independent eigenvector

$$\underline{s}_4 = s_{14}\begin{bmatrix}1\\0\\1\\0\\-1\end{bmatrix} + s_{44}\begin{bmatrix}0\\0\\0\\1\\0\end{bmatrix}, \quad s_{44} \neq 0$$

Two Jordan blocks with the dimensions 3 and 1 are obtained for the eigenvalue $z_1 = 1$. For the eigenvalue $z_2 = 2$, the equation $(\underline{A}-2\underline{I})\underline{s}_5 = \underline{0}$ yields the eigenvector

$$\underline{s}_5 = s_{55}\begin{bmatrix}0\\0\\-1\\0\\1\end{bmatrix}$$

The transformation matrix is thus

$$\underline{S} = \begin{bmatrix} s_{11} & s_{12} & s_{13} & s_{14} & 0 \\ 0 & s_{11} & s_{12} & 0 & 0 \\ s_{11} & s_{12} & s_{13}+s_{11} & s_{12} & -s_{55} \\ 0 & s_{11} & s_{43} & s_{44} & 0 \\ -s_{11} & -s_{12} & -s_{13} & -s_{14} & s_{55} \end{bmatrix}$$

$$= \begin{bmatrix} 1 & 0 & 0 & 0 & 0 \\ 0 & 1 & 0 & 0 & 0 \\ 1 & 0 & 1 & 0 & -1 \\ 0 & 0 & 0 & 1 & 0 \\ -1 & 0 & 0 & 0 & 1 \end{bmatrix} \begin{bmatrix} s_{11} & s_{12} & s_{13} & s_{14} & 0 \\ 0 & s_{11} & s_{12} & 0 & 0 \\ 0 & 0 & s_{11} & 0 & 0 \\ 0 & s_{11} & s_{43} & s_{44} & 0 \\ 0 & 0 & 0 & 0 & s_{55} \end{bmatrix} \tag{A.2.24}$$

Since the structure has already been established the Jordan matrix can be written down immediately

$$J = \underline{S}^{-1}\underline{AS} = \left[\begin{array}{ccc|c|c} z_1 & 1 & 0 & 0 & 0 \\ 0 & z_1 & 1 & 0 & 0 \\ 0 & 0 & z_1 & 0 & 0 \\ \hline 0 & 0 & 0 & z_1 & 0 \\ \hline 0 & 0 & 0 & 0 & z_2 \end{array}\right] \tag{A.2.25}$$

If only the Jordan form of the dynamics matrix is desired, then the coefficients of the $\underline{S}$ matrix can be arbitrarily chosen as long as $\underline{S}$ remains nonsingular. For example let $s_{12} = s_{13} = s_{14} = s_{43} = 0$, $s_{11} = s_{44} = s_{55} = 1$, then

$$\underline{x}_J[k+1] = \begin{bmatrix} 1 & 1 & 0 & 0 & 0 \\ 0 & 1 & 1 & 0 & 0 \\ 0 & 0 & 1 & 0 & 0 \\ 0 & 0 & 0 & 1 & 0 \\ 0 & 0 & 0 & 0 & 2 \end{bmatrix} \underline{x}_J[k] + \begin{bmatrix} 1 \\ 2 \\ 2 \\ -1 \\ 2 \end{bmatrix} u[k]$$

$$y[k] = [-1 \quad 2 \quad 1 \quad 1 \quad 2]\underline{x}_J[k] \tag{A.2.26}$$

It is advisable to fix the free coefficients of the $\underline{S}$ matrix such that $\underline{b}_J = \underline{S}^{-1}\underline{b}$ or $\underline{c}_J' = \underline{c}'\underline{S}$ also becomes simple. In order to recognize which forms are possible, some more general results are required.

1. As for the diagonal form, the system can be controllable from one input and observable from one output only if all Jordan blocks have different eigenvalues, i.e. if the system is cyclic. Therefore the example of eq. (A.2.26) is not completely controllable or observable. For several Jordan blocks with the same eigenvalue, one can ensure by the choice of $\underline{S}$ that all blocks, except the largest one, are not connected with the input or output. In the example, $b_{J4} = 0$ or $c_{J4} = 0$ can be specified.

   If in the characteristic polynomial of a matrix, one only considers every eigenvalue with the multiplicity corresponding to the dimension of the largest related Jordan block, one obtains the "minimal polynomial" of the matrix. In the example, the characteristic polynomial is $(z-1)^4(z-2)$ and the minimal polynomial $(z-1)^3(z-2)$. The systems, which occur in practice are almost without exception cyclic, so that characteristic and minimal polynomials are identical.

2. The subsystem characterized by a Jordan block is controllable if and only if the last element in the corresponding $\underline{b}$-vector is nonzero. This results from the controllability matrix of the system

$$\underline{x}[k+1] = \begin{bmatrix} z & 1 & & 0 \\ 0 & z & \ddots & \\ & & \ddots & 1 \\ 0 & & & z \end{bmatrix} \underline{x}[k] + \begin{bmatrix} b_1 \\ \vdots \\ b_n \end{bmatrix} u[k] = \underline{A}\underline{x}[k] + \underline{b}u[k]$$

which is

$$\underline{\mathscr{C}}_J = [b, Ab \ldots A^{n-1}\underline{b}] = \begin{bmatrix} b_1 & b_2 & & b_n \\ b_2 & b_3 & \cdot^{\cdot^{\cdot}} & 0 \\ \cdot & \cdot^{\cdot^{\cdot}} & & \cdot \\ \cdot & b_n & & \cdot \\ \cdot & & & \cdot \\ \cdot & & & \cdot \\ b_n & \cdots & & 0 \end{bmatrix} \begin{bmatrix} 1 & z & z^2 & & z^{n-1} \\ 0 & 1 & 2z & & \vdots \\ 0 & 0 & 1 & \ddots & \vdots \\ \vdots & & & \ddots & (n-1)z \\ 0 & \cdots & & 0 & 1 \end{bmatrix} \tag{A.2.27}$$

It is nonsingular if and only if $b_n \neq 0$. For a controllable Jordan block, therefore $b_1 = b_2 = \ldots = b_{n-1} = 0$, $b_n = 1$ can be specified [71.1]. Since the first and third Jordan block is controllable, in the example

$$\underline{b}_J = \begin{bmatrix} 0 \\ 0 \\ 1 \\ 0 \\ 1 \end{bmatrix} \tag{A.2.28}$$

may be chosen.

3. The corresponding situation holds for observability. A given Jordan block is observable if and only if the first element in the corresponding $\underline{c}'$-vector is nonzero. In the example

   $$\underline{c}_J' = [1 \quad 0 \quad 0 \quad 0 \quad 1]$$

   may be specified instead of (A.2.28). The example is continued now with this choice of $\underline{c}_J'$.

Example:

$\underline{c}'\underline{S} = \underline{c}_J'$ with $\underline{c}'$ from eq. (A.2.23) and $\underline{S}$ from eq. (A.2.24).

$$[-1 \quad 1 \quad 1 \quad 1 \quad 2] \begin{bmatrix} s_{11} & s_{12} & s_{13} & s_{14} & 0 \\ 0 & s_{11} & s_{12} & 0 & 0 \\ 0 & 0 & s_{11} & 0 & 0 \\ 0 & s_{11} & s_{43} & s_{44} & 0 \\ 0 & 0 & 0 & 0 & s_{55} \end{bmatrix} = [1 \quad 0 \quad 0 \quad 0 \quad 1]$$

It then follows that $s_{11} = -1$, $s_{12} = -2$, $s_{55} = 0.5$, $s_{44} = s_{14}$, $s_{13} = s_{43} - 3$. In this case of a noncyclic system, the coefficients s_{14} and s_{43} are not fixed. They produce the partition of the two subsystems with the eigenvalue $z_1 = 1$. These free parameters can be used to satisfy $b_{J4} = 0$. $\underline{b}_J$ is computed from $\underline{b} = \underline{Sb}_J$:

$$\begin{bmatrix} 1 \\ 2 \\ 1 \\ 1 \\ 1 \end{bmatrix} = \begin{bmatrix} -1 & -2 & s_{43}-3 & s_{14} & 0 \\ 0 & -1 & -2 & 0 & 0 \\ -1 & -2 & s_{43}-4 & s_{14} & -0.5 \\ 0 & -1 & s_{43} & s_{14} & 0 \\ 1 & 2 & 3-s_{43} & -s_{14} & 0.5 \end{bmatrix} \begin{bmatrix} b_{J1} \\ b_{J2} \\ b_{J3} \\ b_{J4} \\ b_{J5} \end{bmatrix}$$

For $b_{J4} = 0$, the parameter s_{14} can be arbitrarily fixed, for example $s_{14} = 1$. s_{43} is then determined such that the five equations are compatible with each other. It follows that $s_{43} = -1.5$ and $\underline{b}_J = [4 \quad 2 \quad -2 \quad 0 \quad 4]'$. The canonical form of the state equations is therefore

$$\underline{x}[k+1] = \begin{bmatrix} 1 & 1 & 0 & 0 & 0 \\ 0 & 1 & 1 & 0 & 0 \\ 0 & 0 & 1 & 0 & 0 \\ 0 & 0 & 0 & 1 & 0 \\ 0 & 0 & 0 & 0 & 2 \end{bmatrix} \underline{x}[k] + \begin{bmatrix} 4 \\ 2 \\ -2 \\ 0 \\ 4 \end{bmatrix} u[k]$$

$$y[k] = [1 \quad 0 \quad 0 \quad 0 \quad 1]\underline{x}[k] \qquad \text{(A.2.29)}$$

Figure A.1 illustrates this form. It is apparent from the figure, that $b_{J3} \neq 0$ guarantees the controllability of the state x_{J3} and therefore also of x_{J2} and x_{J1}.

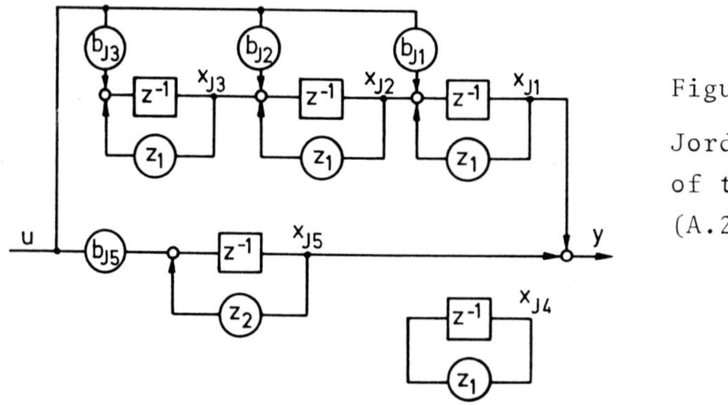

Figure A.1

Jordan form (A.2.29) of the system (A.2.23)

The canonical decomposition of figure 2.6 can be performed by transformation of the system to Jordan form. In our example a subsystem of type B with the characteristic polynomial $(z-1)^3(z-2)$ and a subsystem of type D with the characteristic polynomial $z-1$ arise. For the control, all that is of interest is the subsystem B with the state vector

$$\underline{x}^* := [x_{J1}, x_{J2}\ x_{J3}, x_{J5}]'$$

$$\underline{x}^*[k+1] = \begin{bmatrix} 1 & 1 & 0 & 0 \\ 0 & 1 & 1 & 0 \\ 0 & 0 & 1 & 0 \\ 0 & 0 & 0 & 2 \end{bmatrix} \underline{x}^*[k] + \begin{bmatrix} 4 \\ 2 \\ -2 \\ 4 \end{bmatrix} u[k]$$

$$y[k] = [1 \quad 0 \quad 0 \quad 1]\underline{x}^*[k] \tag{A.2.30}$$

As for the diagonal form, the transformation to the Jordan form also corresponds to the partial fractions decomposition of the transfer function. In the example of eq. (A.2.23)

$$h_z(z) = \underline{c}'(z\underline{I}-\underline{A})^{-1}\underline{b}$$

$$= \frac{2(z-1)(4z^3-13z^2+12z-2)}{(z-1)^4(z-2)} \tag{A.2.31}$$

As was discussed in section 2.3, the uncontrollable and unobservable part of the system with the eigenvalue $z = 1$ does not arise in the input-output description $h_z(z)$ of the system. In eq. (A.2.31), a factor $(z-1)$ is cancelled out of the numerator and denominator. The partial fractions decomposition of the remaining expression yields

$$h_z(z) = \frac{8z^3-26z^2+24z-4}{(z-1)^3(z-2)}$$

$$= \frac{4}{z-2} + \frac{4}{z-1} + \frac{2}{(z-1)^2} - \frac{2}{(z-1)^3} \tag{A.2.32}$$

This representation corresponds to the controllable and observable part of figure A.1, i.e. eq. (A.2.30).

In general the Jordan form of the controllable and observable subsystem is obtained by decomposing $h_z(z)$ into partial fractions:

$$h_z(z) = \frac{b_o+b_1 z+\ldots+b_{n-1}z^{n-1}}{(z-z_1)^{n_1}(z-z_2)^{n_2}\ldots(z-z_m)^{n_m}}\ , \quad n_1+n_2+\ldots+n_m = n$$

$$= \frac{A_{11}}{z-z_1} + \frac{A_{12}}{(z-z_1)^2} + \ldots + \frac{A_{1n_1}}{(z-z_1)^{n_1}} + \frac{A_{21}}{z-z_2} + \ldots + \frac{A_{mn_m}}{(z-z_m)^{n_m}} \qquad \text{(A.2.33)}$$

The corresponding Jordan form is for canonical $\underline{c}_J'$

$$\underline{x}[k+1] = \left[\begin{array}{ccc|c|c} \multicolumn{3}{c|}{\overbrace{\hphantom{z_1\quad 1\quad 0}}^{n_1}} & \overbrace{\hphantom{z_2}}^{n_2} & \overbrace{\hphantom{0}}^{n_m} \\ z_1 & 1 & 0 & & 0 \\ 0 & z_1 & 1 & & \\ 0 & & z_1 & & \\ \hline & & & z_2 & \\ \hline 0 & & & & z_m \end{array}\right] \underline{x}[k] + \left[\begin{array}{c} A_{11} \\ A_{12} \\ \vdots \\ A_{1n_1} \\ \hline A_{21} \\ \vdots \\ \hline \vdots \\ A_{mn_m} \end{array}\right] u[k]$$

$$y[k] = [1 \quad 0 \ldots 0 \;\vdots\; 1 \quad 0 \;\vdots\; \ldots 0]\underline{x}[k] \qquad \text{(A.2.34)}$$

or for canonical $\underline{b}_J$

$$\underline{x}[k+1] = \left[\begin{array}{ccc|c|c} z_1 & 1 \;\ddots & 0 & & 0 \\ 0 & z_1 \;\ddots & 1 & & \\ 0 & & z_1 & & \\ \hline & & & z_2 \;\ddots & \\ \hline 0 & & & & \ddots\; z_m \end{array}\right] \underline{x}[k] = \left[\begin{array}{c} 0 \\ \vdots \\ 0 \\ \vdots \\ 1 \\ \hline \vdots \\ \hline \vdots \\ 1 \end{array}\right] u[k]$$

$$y[k] = \left[A_{1n_1} \ldots\ldots A_{11} \;\middle|\; A_{2n_2} \ldots \;\middle|\; \ldots A_{m1}\right] \underline{x}[k] \qquad \text{(A.2.35)}$$

A disadvantage of the diagonal and Jordan forms is that complex coefficients arise for complex eigenvalues. This would require complex arithmetic for the computer calculations. It can be avoided, however, by combining the Jordan blocks with complex conjugate eigenvalues into a real subsystem. Let

$$\begin{bmatrix} \underline{x}_1[k+1] \\ \underline{x}_1^*[k+1] \end{bmatrix} = \begin{bmatrix} \underline{J}_1 & \underline{0} \\ \underline{0} & \underline{J}_1^* \end{bmatrix} \begin{bmatrix} \underline{x}_1[k] \\ \underline{x}_1^*[k] \end{bmatrix} = \begin{bmatrix} \underline{P}+j\underline{Q} & 0 \\ 0 & \underline{P}-j\underline{Q} \end{bmatrix} \begin{bmatrix} \underline{x}_1[k] \\ \underline{x}_1^*[k] \end{bmatrix}$$

where * denotes the complex conjugate variable.

For $\underline{x}_1 = \underline{x}+j\underline{y}$, $\underline{x}_1^* = \underline{x}-j\underline{y}$, $\underline{J}_1 = \underline{P} + j\underline{Q}$ this becomes

$$\begin{bmatrix} \underline{x}[k+1]+j\underline{y}[k+1] \\ \underline{x}[k+1]-j\underline{y}[k+1] \end{bmatrix} = \begin{bmatrix} \underline{P}+j\underline{Q} & \underline{0} \\ \underline{0} & \underline{P}-j\underline{Q} \end{bmatrix} \begin{bmatrix} \underline{x}[k]+j\underline{y}[k] \\ \underline{x}[k]-j\underline{y}[k] \end{bmatrix}$$

Forming the sum and the difference of the two rows, the real representation is obtained:

$$\begin{bmatrix} \underline{x}[k+1] \\ \underline{y}[k+1] \end{bmatrix} = \begin{bmatrix} \underline{P} & -\underline{Q} \\ \underline{Q} & \underline{P} \end{bmatrix} \begin{bmatrix} \underline{x}[k] \\ \underline{y}[k] \end{bmatrix} \qquad \text{(A.2.36)}$$

Example:

The real version of the system (A.2.9) is

$$\underline{x}_{DR}[k+1] = \begin{bmatrix} 1 & 0 & 0 \\ 0 & 1 & -1 \\ 0 & 1 & 1 \end{bmatrix} \underline{x}_{DR}[k] + \begin{bmatrix} 1 \\ 1 \\ 0 \end{bmatrix} u[k] \ , \ \underline{x}_{DR} = \begin{bmatrix} x_{D1} \\ \mathrm{Re}x_{D2} \\ \mathrm{Im}x_{D2} \end{bmatrix}$$

$$y[k] = [-2 \quad 16 \quad -2]\underline{x}_{DR}[k]$$

On the other hand, it is advantageous, for principle observations, to use only the relation

$$\underline{x}_1[k+1] = \underline{J}_1\underline{x}_1[k] \qquad \text{(A.2.37)}$$

which describes the system completely. One returns to the real and imaginary parts only after the solution of the complete control problem. An application example of this can be found in [69.2]. There the two-axes position control of a rotationally symmetric spin-stabilized satellite is treated. Two orthogonal angles are combined into one complex variable and the same is done with two orthogonal angular velocities. The transformation of this time-variable system into diagonal form admits a simple calculation of the motion of the moment of momentum and of the body-fixed nutation axis around it. The derivation of a control law is also simplified by the use of complex variables.

Consider now the case of multi-variable systems. First note, that a noncyclic system can be controllable and observable in this case [68.3]. The easiest check for this is the Hautus test [72.2]. From this, an eigenvalue is controllable if and only if

$$\text{rank}\ [\underline{A}-z_i\underline{I},\ \underline{B}] = n$$

Example:

In the example of eq. (A.2.29) for $z_1 = 1$

$$\underline{A}_J - z_1\underline{I} = \begin{bmatrix} 0 & 1 & 0 & 0 & 0 \\ 0 & 0 & 1 & 0 & 0 \\ 0 & 0 & 0 & 0 & 0 \\ 0 & 0 & 0 & 0 & 0 \\ 0 & 0 & 0 & 0 & 1 \end{bmatrix} \tag{A.2.38}$$

Two input variables are required to augment the matrix to full rank. Let

$$\underline{B}_J = \begin{bmatrix} b_{11} & b_{12} \\ b_{21} & b_{22} \\ b_{31} & b_{32} \\ b_{41} & b_{42} \\ b_{51} & b_{52} \end{bmatrix}$$

Obviously $\det\begin{bmatrix} b_{31} & b_{32} \\ b_{41} & b_{42} \end{bmatrix} = b_{31}b_{42} - b_{32}b_{41} \neq 0$ is necessary and sufficient for rank $[\underline{A}_J - z_1\underline{I},\ \underline{B}_J] = 5$.

For similar reasons a measurement matrix

$$\underline{C}_J = \begin{bmatrix} c_{11} & c_{12} & c_{13} & c_{14} & c_{15} \\ c_{21} & c_{22} & c_{23} & c_{24} & c_{25} \end{bmatrix}$$ must satisfy

$c_{11}c_{24} - c_{21}c_{12} \neq 0$ in order to make the system observable. The possible canonical forms of the C_J matrix are illustrated by the example of eq. (A.2.23) extended by the introduction of a second input and a second output variable.

$$\underline{x}[k+1] = \begin{bmatrix} 1 & 1 & 0 & 0 & 0 \\ 0 & 1 & 1 & 0 & 1 \\ -1 & 1 & 1 & 0 & -1 \\ 0 & 0 & 1 & 1 & 1 \\ 1 & -1 & 0 & 0 & 2 \end{bmatrix} \underline{x}[k] + \begin{bmatrix} 1 & 2 \\ 2 & 1 \\ 1 & 3 \\ 1 & 1 \\ 1 & 1 \end{bmatrix} \underline{u}[k]$$

$$\underline{y}[k] = \begin{bmatrix} 1 & 1 & 1 & 1 & 3 \\ 2 & 1 & 1 & 2 & 1 \end{bmatrix} \underline{x}[k] = \underline{Cx}[k] \qquad \text{(A.2.39)}$$

With the transformation matrix (A.2.24) this becomes

$$\underline{C}_J = \underline{CS} = \begin{bmatrix} -1 & 1 & 1 & 1 & 2 \\ 2 & 1 & 1 & 2 & 0 \end{bmatrix} \begin{bmatrix} s_{11} & s_{12} & s_{13} & s_{14} & 0 \\ 0 & s_{11} & s_{12} & 0 & 0 \\ 0 & 0 & s_{11} & 0 & 0 \\ 0 & s_{11} & s_{43} & s_{44} & 0 \\ 0 & 0 & 0 & 0 & s_{55} \end{bmatrix} \qquad \text{(A.2.40)}$$

As a result $c_{J25} = 0$, i.e. the subsystem with the eigenvalue $z_2 = 2$ is observable only from y_1. Normalize $c_{J15} = 1$ by fixing $s_{55} = 0.5$. The third order subsystem can be observed from y_1 as well as from y_2. Thus one can produce the coefficients 1, 0, 0 in either the first or the second row of the $\underline{C}_J$ matrix. They arise in the first row for $s_{11} = -1$, $s_{12} = -2$, $s_{13} = s_{43} - 3$. This yields

$$\underline{C}_J = \begin{bmatrix} 1 & 0 & 0 & s_{44}-s_{14} & 1 \\ -2 & -7 & 4s_{43}-9 & 2(s_{44}+s_{14}) & 0 \end{bmatrix}$$

If $s_{11} = 0.5$, $s_{12} = -0.75$, $s_{13} = 0.125 - s_{43}$ is chosen, then

$$\underline{C}_J = \begin{bmatrix} -0.5 & 1.75 & -0.625+2s_{43} & s_{44}-s_{14} & 1 \\ 1 & 0 & 0 & 2(s_{44}+s_{14}) & 0 \end{bmatrix}$$

The system is observable if the first and fourth columns of the $\underline{C}_J$ matrix are linearly independent.

$$\det \begin{bmatrix} -0.5 & s_{44}-s_{14} \\ 1 & 2(s_{44}+s_{14}) \end{bmatrix} = -2s_{44} \neq 0, \text{ because } \underline{S} \text{ is regular.}$$

The first order subsystem with eigenvalue $z_1 = 1$ is therefore observable. However, the assignment of this subsystem to y_1 or y_2 or both of them is not unique in this special case of a noncyclic system. This comes from the fact that an identical subsystem is also contained in the first Jordan block in figure A.1. Thus either $c_{J14} = 0$, $c_{J24} = 1$ with $s_{14} = s_{44} = 0.25$ or $c_{J14} = 1$, $c_{J24} = 0$ with $s_{44} = -s_{14} = 0.5$. By the same argument, s_{43} can be fixed such that $c_{J23} = 0$ or $c_{J13} = 0$.

In the generic case, n coefficients of the matrix $\underline{C}_J$ (or $\underline{B}_J$) can be fixed to zero or one. If the system has m Jordan blocks, $m \leq n$, then the number of parameters necessary for the description of the system is: m coefficients of the dynamics matrix $\underline{A}_J$, $n \times r$ coefficients of the input matrix $\underline{B}_J$ and $n(s-1)$ coefficients of the output matrix $\underline{C}_J$ - in total $n(r+s-1) + m$ parameters. The worst case regarding the number of parameters is the diagonal form with $m = n$. It follows that every linear system of order n with r input variables and s output variables can be characterized by at most

$$n(r+s) \tag{A.2.41}$$

parameters, see also section A.1. Here it is assumed, that the structure of the Jordan form, in which the parameters occur, is known. The maximal number of parameters arises if every input variable controls the entire system and every output variable observes the entire system.

The observability and controllability structures of the system can be easily seen from the Jordan form. If the last k elements of the $\underline{b}$ vector are zero in eq. (A.2.27), i.e. $b_n = b_{n-1} = \dots = b_{n-k+1} = 0$, $b_{n-k} \neq 0$, then rank $\underline{\mathcal{C}}_J = n-k$. In other words, the corresponding input variable controls a subsystem of order n-k of this Jordan block.

For computational aspects of the Jordan form see [81.15].

A.3 Frobenius Forms

The Frobenius form of the $\underline{A}$ matrix is

$$\underline{A}_F = \underline{TAT^{-1}} = \begin{bmatrix} 0 & 1 & & 0 \\ & & \ddots & \\ & & & 1 \\ -a_o & -a_1 & \dots & -a_{n-1} \end{bmatrix} \tag{A.3.1}$$

Its advantage is that it contains the coefficients of the characteristic polynomial

$$\det(z\underline{I}-\underline{A}_F) = a_o + a_1 z + \dots + a_{n-1} z^{n-1} + z^n \tag{A.3.2}$$

$\underline{A}_F$ is also called "companion matrix" of this polynomial. In the next sections the controllability and the feedback canonical form of ($\underline{A}$, $\underline{b}$) are discussed and their duals for the observability case. All of them use the Frobenius form.

A.3.1 Controllability-Canonical Form

The dynamics matrix is the transpose of $\underline{A}_F$ and the input vector has only a one in its first row.

$$\underline{A}_c = \underline{S}^{-1}\underline{AS} = \begin{bmatrix} 0 & & & -a_o \\ 1 & \ddots & & -a_1 \\ & & \ddots & \vdots \\ 0 & & 1 & -a_{n-1} \end{bmatrix}, \quad \underline{b}_c = \underline{S}^{-1}\underline{b} = \begin{bmatrix} 1 \\ 0 \\ \vdots \\ 0 \end{bmatrix} \qquad \text{(A.3.3)}$$

$$\underline{c}_c' = \underline{c}'\underline{S} = [c_{c1} \quad c_{c2} \; \dots \; c_{cn}]$$

Its controllability matrix is the unit matrix

$$\underline{\mathscr{C}}_c = [\underline{b}_c, \underline{A}_c\underline{b}_c \; \dots \; \underline{A}_c^{n-1}\underline{b}_c] = \underline{I}_n \qquad \text{(A.3.4)}$$

Therefore this form exists only for controllable systems. Every controllable pair ($\underline{A}$, $\underline{b}$) can be transformed to the controllability canonical form by $\underline{x} = \underline{T}^{-1}\underline{x}_c = \underline{Sx}_c$, where $\underline{S} = \underline{T}^{-1}$ according to eqs. (A.1.7) and (A.3.4) is the controllability matrix of the original form

$$\underline{S} = \underline{\mathscr{C}} = [\underline{b}, \underline{Ab}, \dots \underline{A}^{n-1}\underline{b}] \qquad \text{(A.3.5)}$$

The output vector is

$$\underline{c}_c' = \underline{c}'\underline{S} = [\underline{c}'\underline{b}, \underline{c}'\underline{Ab}, \dots \underline{c}'\underline{A}^{n-1}\underline{b}] \qquad \text{(A.3.6)}$$

A comparison with eq.(3.3.14) shows that the elements of $\underline{c}_c$ are elements of the weighting sequence $h[k] = \underline{c}'\underline{A}^{k-1}\underline{b}$, $k = 1,2,3\dots$, of the system $\underline{x}[k+1] = \underline{Ax}[k] + \underline{b}u[k]$, $y' = \underline{c}'\underline{x}$, i.e.

$$\underline{c}_c' = [h[1], h[2] \dots h[n]] \qquad \text{(A.3.7)}$$

The 2n parameters used for the description of the system are the n coefficients of the characteristic polynomial and n values of the weighting sequence. The controllability-canonical form is illustrated by figure A.2. This circuit with real coefficients is suitable for the simulation of the system on an analog or digital computer.

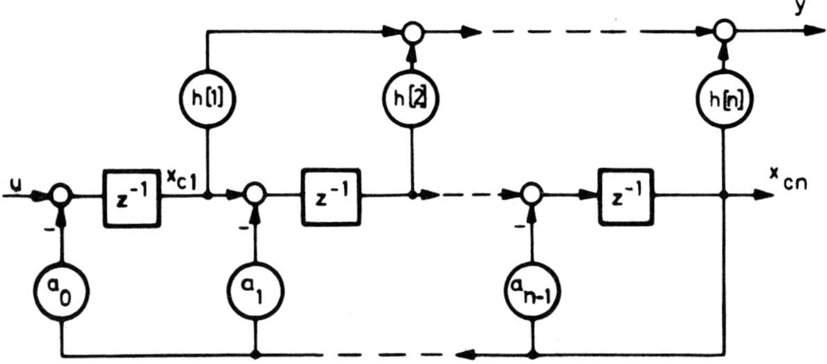

Figure A.2 Controllability-canonical form

A.3.2 Feedback-Canonical Form

$$\underline{A}_f = \underline{S}^{-1}\underline{A}\underline{S} = \begin{bmatrix} 0 & 1 & & 0 \\ & & \ddots & \\ & & & \ddots \\ & & & 1 \\ -a_0 & -a_1 & \dots & -a_{n-1} \end{bmatrix}, \quad \underline{b}_f = \underline{S}^{-1}\underline{b} = \begin{bmatrix} 0 \\ \cdot \\ \cdot \\ \cdot \\ 0 \\ 1 \end{bmatrix} \tag{A.3.8}$$

$$\underline{c}_f' = \underline{c}'\underline{S} \quad = [\, b_0 \quad b_1 \quad \dots \quad b_{n-1} \,]$$

The controllability matrix has the form

$$\mathcal{C}_f = \begin{bmatrix} 0 & \dots & 0 & 1 \\ & & \cdot & x \\ & \cdot & \cdot & \\ 0 & 1 & \cdot & \\ 1 & x & \dots & x \end{bmatrix}, \quad \mathcal{C}_f^{-1} = \begin{bmatrix} a_1 & a_2 & \dots & a_{n-1} & 1 \\ a_2 & & \cdot & \cdot & 0 \\ \cdot & \cdot & & & \cdot \\ a_{n-1} & \cdot & & & \cdot \\ 1 & 0 & \dots & \dots & 0 \end{bmatrix} \tag{A.3.9}$$

It is nonsingular, i.e. this form only exists for controllable systems. There is a simple relation between the feedback-canonical form and the z-transfer function of the system.

$$h_z(z) = \underline{c}_f'(z\underline{I}-\underline{A}_f)^{-1}\underline{b}_f = \frac{b_0+b_1z+\ldots+b_{n-1}z^{n-1}}{a_0+a_1z+\ldots+a_{n-1}z^{n-1}+z^n} \tag{A.3.10}$$

The output vector $\underline{c}_f'$ consists of the coefficients of the numerator polynomial of the z-transfer function. If all cancellations of poles and zeros have already been performed in eq. (A.3.10), then the feedback-canonical form of the state representation of the controllable and observable subsystem can be written down immediately with the parameters of $h_z(z)$. If a term b_nz^n appears in the numerator of the z-transfer function, then the output equation is

$$y[k] = [b_0-a_0b_n,\ b_1-a_1b_n,\ldots b_{n-1}-a_{n-1}b_n]\underline{x}_R[k]+b_nu[k] \tag{A.3.11}$$

The feedback-canonical form is illustrated by figure A.3. Its name results from the direct assignment of a closed-loop characteristic polynomial by state feedback, see eq. (2.6.21) for the continuous-time case.

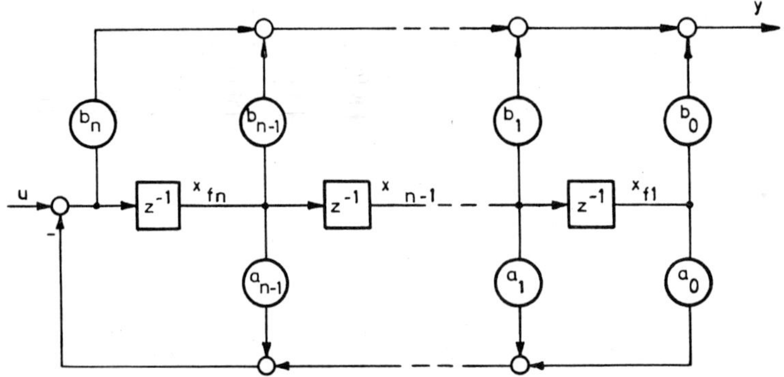

Figure A.3 Feedback-canonical form

The transformation matrix is by eqs. (A.1.7) and (A.3.9)

$$\underline{S} = \underline{T}^{-1} = \mathscr{C}\,\mathscr{C}_f^{-1} = [\underline{b},\ \underline{AB}\ldots\underline{A}^{n-1}\underline{b}]\begin{bmatrix} a_1 & a_2 & \cdots & a_{n-1} & 1 \\ a_2 & & \cdot^{\cdot^{\cdot}} & \cdot^{\cdot^{\cdot}} & 0 \\ \vdots & \cdot^{\cdot^{\cdot}} & \cdot^{\cdot^{\cdot}} & & \\ a_{n-1} & 1 & & & \\ 1 & 0 & & & 0 \end{bmatrix} \tag{A.3.12}$$

It is easier, however, to calculate the columns $\underline{s}_i$ of $\underline{S}$ recursively [66.1].

$$\underline{AS} = \underline{SA}_f$$

$$[\underline{As}_1, \underline{As}_2 \dots \underline{As}_n] = [\underline{s}_1, \underline{s}_2 \dots \underline{s}_n]\begin{bmatrix} 0 & 1 & \dots & 0 \\ \vdots & & \ddots & \\ \vdots & & & 1 \\ -a_o & -a_1 & \dots & -a_{n-1} \end{bmatrix}$$

$$= [-a_o\underline{s}_n, \underline{s}_1 - a_1\underline{s}_n, \dots \underline{s}_{n-1} - a_{n-1}\underline{s}_n] \tag{A.3.13}$$

The calculation begins with $\underline{b} = \underline{Sb}_f = \underline{s}_n$. Then $\underline{s}_{n-1}$ can be calculated from the last column of eq. (A.3.13) and so on. The first column gives a check.

The columns of the $\underline{S}$ matrix and the coefficients of the characteristic equation can also be determined simultaneously by the Leverrier algorithm, eq. (A.7.36):

$$\begin{array}{llll}
 & & \underline{D}_{n-1} = \underline{I} & \underline{s}_n = \underline{D}_{n-1}\underline{b} \\
a_{n-1} = & -\text{trace } \underline{AD}_{n-1}, & \underline{D}_{n-2} = \underline{D}_{n-1} + a_{n-1}\underline{I}, & \underline{s}_{n-1} = \underline{D}_{n-2}\underline{b} \\
a_{n-2} = & -\frac{1}{2}\text{trace } \underline{AD}_{n-2}, & \underline{D}_{n-3} = \underline{AD}_{n-2} + a_{n-2}\underline{I}, & \underline{s}_{n-2} = \underline{D}_{n-3}\underline{b} \\
\vdots & & \vdots & \vdots \\
a_1 = & -\frac{1}{n-1}\text{trace } \underline{AD}_1, & \underline{D}_o = \underline{AD}_1 + a_1\underline{I}, & \underline{s}_1 = \underline{D}_o\underline{b} \\
a_o = & -\frac{1}{n}\text{trace } \underline{AD}_o, & \underline{D}_{-1} = \underline{AD}_o + a_o\underline{I} &
\end{array} \tag{A.3.14}$$

$\underline{D}_{-1} = \underline{0}$ provides a check on the accuracy.

A.3.3 Observability-Canonical Form

This is the dual of eq. (A.3.3) with

$$\underline{A}_o = \underline{A}_c' \ , \ \underline{b}_o = \underline{c}_c \ , \ \underline{c}_o = \underline{b}_c \tag{A.3.15}$$

Its observability matrix is the unit matrix, therefore it exists only for observable systems and the transformation $\underline{T}$ for $\underline{A}_o = \underline{TAT}^{-1}$ is by eq. (A.1.4)

$$\underline{T} = \begin{bmatrix} \underline{c}' \\ \underline{c}'\underline{A} \\ \vdots \\ \underline{c}'\underline{A}^{n-1} \end{bmatrix} \tag{A.3.16}$$

Figure A.4 shows the block diagram of this form.

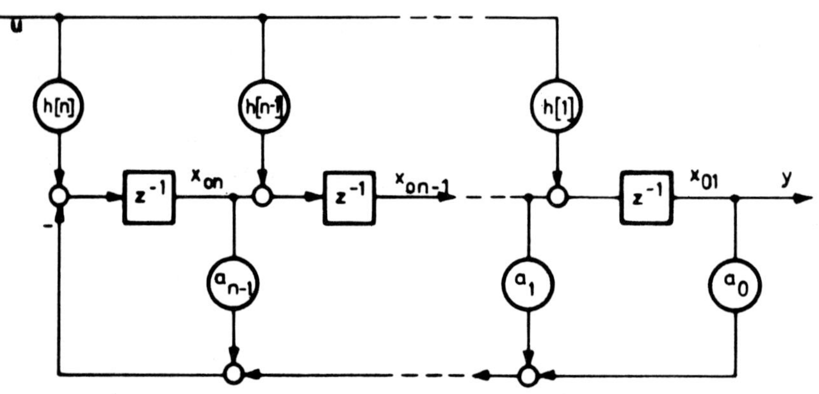

Figure A.4 Observability-canonical form

This form is particularly useful for the minimal realization problem [66.3] where the measured weighting sequence (i.e. impulse response - see eq. (3.3.14)) of a linear discrete-time system is given.

$$y[k] = h[k] = \underline{c}'\underline{A}^{k-1}\underline{b} \ , \ k = 1, 2, 3 \ldots \tag{A.3.17}$$

It may also be determined from a measured step response via eq. (3.3.21). Find a minimal realization, i.e. a controllable and observable triple $\underline{c}'$, $\underline{A}$, $\underline{b}$ such that eq. (A.3.17) holds for all k. This is easy for the observability canonical form. Immediately

$$\underline{b}_o = \begin{bmatrix} h[1] \\ h[2] \\ \vdots \\ h[n] \end{bmatrix}, \quad \underline{c}_o = [1 \quad 0 \ \dots \ 0] \tag{A.3.18}$$

$\underline{A}_o$ is defined by its characteristic polynomial. This is obtained as follows. Form the Hankel matrix

$$\underline{H}_n = \begin{bmatrix} h[1] & h[2] & \dots & h[n+1] \\ h[2] & & & \\ \vdots & & & \\ h[n] & & & h[2n] \end{bmatrix} = \begin{bmatrix} \underline{c}_o' \\ \underline{c}_o'\underline{A}_o \\ \vdots \\ \underline{c}_o'\underline{A}_o^{n-1} \end{bmatrix} [\underline{b}_o, \ \underline{A}_o\underline{b}_o \ \dots \ \underline{A}_o^{n}\underline{b}_o] \tag{A.3.19}$$

The observability matrix is the unit matrix and by Cayley-Hamilton

$$[\underline{b}_o \ \underline{A}_o\underline{b}_o \ \dots \ \underline{A}_o^{n} \ \underline{b}_o] \begin{bmatrix} \underline{a} \\ 1 \end{bmatrix} = \underline{0} \quad , \quad \underline{a} = [a_o \quad a_1 \ \dots \ a_{n-1}]' \tag{A.3.20}$$

Thus from eq. (A.3.19)

$$\underline{H}_n \begin{bmatrix} \underline{a} \\ 1 \end{bmatrix} = \underline{0} \tag{A.3.21}$$

This equation is solved for the characteristic polynomial coefficient vector $\underline{a}$.

This minimal realization in observability canonical form can be generalized to the multi-output case [71.4] and to arbitrary noisefree input-output measurements [69.5], [71.3].

A.3.4 Observer-Canonical Form

The observer or state-reconstruction canonical form is the dual of the feedback-canonical form:

$$\underline{A}_r = \underline{A}_f', \quad \underline{b}_r = \underline{c}_f, \quad \underline{c}_r = \underline{b}_f \qquad \text{(A.3.22)}$$

It can be computed by transforming the pair $(\bar{\underline{A}}, \bar{\underline{b}}) = (\underline{A}', \underline{c})$ into feedback-canonical form. Figure A.5 illustrates this form for the case with feedthrough term b_n.

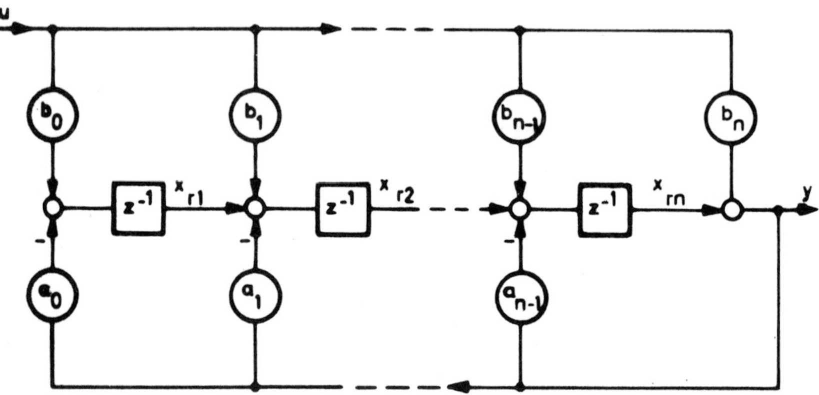

Figure A.5 Observer-canonical form

An application of this form for the explanation of a reduced-order deadbeat observer is shown in figure 5.4. This form is also used for the implementation of the controller of figure 6.7 with two inputs u and y.

A.4 Multivariable Canonical Forms

A.4.1 General Remarks

For the Jordan form the multivariable generalization is straightforward, because it can be done without affecting the $\underline{A}_J$ matrix. For the Frobenius forms, such a generalization with unchanged $\underline{A}_F$ and only one canonical column in $\underline{B}$, or only one canonical row in $\underline{C}$ would not give the desired visualization of the controllability or observability structure. This can only be achieved by decomposing the system into subsystems, which appear as Frobenius blocks in the $\underline{A}$ matrix.

Typically there are two strategies, explained here for the controllability case:

A) Pick one input, transform the system, which is controllable from this input, into controllability or feedback-canonical form. Then pick a second input and adjoin the subsystem to it, which additionally becomes controllable by the inclusion of this second input. There will be coupling terms from this second subsystem into the first subsystem. They take care of that part of the first subsystem which can also be controlled from the second input. The resulting $\underline{A}$ matrix has a block triangular structure. The main diagonal consists of Frobenius blocks. Their characteristic polynomials are factors in the characteristic polynomial of the overall system. For details see [67.4], [68.4], [70.2].

B) Decompose the system such, that a subsystem is assigned to each input. Try to make them of about the same order. This results in orders equal to the controllability indices μ_i, see eq. (9.1.19). The subsystems are represented by Frobenius blocks in the main diagonal of $\underline{A}_o$, but now the subsystems are coupled in both directions and the relation with the characteristic polynomial of the overall system is more complicated. Only one important form of this type will be discussed in this section.

A.4.2 Luenberger Feedback-Canonical Form

Given a controllable pair $(\underline{A}, \underline{B})$ with controllability indices $\mu_1, \mu_2 \dots \mu_r$. Luenberger [67.4] has derived a transformation $\underline{A}_L = \underline{TAT}^{-1}$, $\underline{B}_L = \underline{TB}$, where $\underline{A}_L$ has Frobenius blocks of dimension μ_i on its main diagonal and the submatrices $\underline{A}_{ij}$ of $\underline{A}_L$ are of dimension $\mu_i \times \mu_j$. Their elements are the α-parameters, see eq. (9.1.33). The input matrix $\underline{B}_L$ consists of submatrices of dimension $\mu_i \times r$. A factor $\underline{M}^{-1}$ of the input matrix contains the β-parameters, see eq. (9.1.23).

$$\underline{A}_L = \begin{bmatrix} \underline{A}_{11} & \dots & \underline{A}_{1r} \\ \vdots & & \vdots \\ \underline{A}_{r1} & \dots & \underline{A}_{rr} \end{bmatrix}, \quad \underline{B}_L = \begin{bmatrix} \underline{D}_1 \\ \vdots \\ \underline{D}_r \end{bmatrix} \underline{M}^{-1} = \underline{DM}^{-1} \tag{A.4.1}$$

$$\underline{A}_{ii} = \begin{bmatrix} 0 & 1 & & \\ & & \ddots & \\ & & & 1 \\ \hline & & -\underline{h}_{ii}' & \end{bmatrix}, \quad \underline{A}_{ij} = \begin{bmatrix} \underline{0} \\ \\ \hline -\underline{h}_{ij}' \end{bmatrix} \quad \text{for } i \neq j$$

$$\underline{h}_{ii}' = [\alpha_{ii0} \quad \alpha_{ii1} \quad \cdots \quad \alpha_{ii\mu_i-1}]$$

$$\underline{h}_{ij}' = \begin{cases} [\alpha_{ji0} \quad \alpha_{ji1} \quad \cdots \quad \alpha_{ji\mu_j-1}] & \mu_i \geq \mu_j \\ [\alpha_{ji0} \cdots \alpha_{ji\mu_i-1} \quad 0 \ \ldots \ 0] & \mu_i < \mu_j \end{cases}$$

$$\underline{D}_i = \begin{bmatrix} \underline{0} \\ \underline{i}_i' \end{bmatrix}, \quad \underline{M} = \begin{bmatrix} 1 & \boxed{\beta_2} & \boxed{\underline{\beta}_3} & \cdots & \boxed{\underline{\beta}_r} \\ 0 & 1 & & & \\ & & 1 & & \\ & & & \ddots & \\ 0 & & & & 1 \end{bmatrix}$$

$\underline{i}_i'$ = ith row of the r×r unit matrix

$$\underline{\beta}_i = \begin{bmatrix} \beta_{i1} \\ \vdots \\ \beta_{i\mu_i-1} \end{bmatrix} \quad \text{with } \beta_{ij} = 0 \text{ for } \mu_i \geq \mu_j$$

Essential properties of this form are:

1. It directly exhibits the controllability indices μ_i and the α- and β-parameters. Let

$$\underline{D}_L = \begin{bmatrix} \underline{D}_1 \\ \vdots \\ \underline{D}_r \end{bmatrix} = [\underline{d}_1 \ \ldots \ \underline{d}_r] \tag{A.4.2}$$

Then the α-parameters of the pair $(\underline{A}_L, \underline{D}_L)$ satisfy

$$\sum_{j=1}^{r} \sum_{k=0}^{\mu_i-1} \alpha_{ijk} \underline{A}_L^k \underline{d}_j + \underline{A}_L^{\mu_i} \underline{d}_i = 0 \tag{A.4.3}$$

This is identical with eq. (9.1.35). Now $\underline{B}_L = \underline{DM}^{-1}$, $\underline{D} = \underline{B}_L\underline{M}$ substituted into (A.4.3) generates the a's and β's in

$$\sum_{i=1}^{r} \sum_{k=0}^{\mu_i - 1} a_{ijk}\underline{A}_L^k \underline{b}_{Lj} + \sum_{j=1}^{i} \beta_{ij}\underline{A}^{\mu_i}\underline{b}_j = 0 \qquad \text{(A.4.4)}$$

Example:

Let $\mu_1 = 3$, $\mu_2 = 2$, $\mu_3 = 1$. For this monotonically decreasing sequence of controllability indices all β-parameters occur, according to eq. (9.1.24).

$$\underline{A}_L = \left[\begin{array}{ccc|cc|c} 0 & 1 & 0 & 0 & 0 & 0 \\ 0 & 0 & 1 & 0 & 0 & 0 \\ -\alpha_{110} & -\alpha_{111} & -\alpha_{112} & -\alpha_{210} & -\alpha_{211} & -\alpha_{310} \\ \hline 0 & 0 & 0 & 0 & 1 & 0 \\ -\alpha_{120} & -\alpha_{121} & 0 & -\alpha_{220} & -\alpha_{221} & -\alpha_{320} \\ \hline -\alpha_{130} & 0 & 0 & -\alpha_{230} & 0 & -\alpha_{330} \end{array}\right] \qquad \text{(A.4.5)}$$

$$\underline{B}_L = \left[\begin{array}{ccc} 0 & 0 & 0 \\ 0 & 0 & 0 \\ 1 & 0 & 0 \\ \hline 0 & 0 & 0 \\ 0 & 1 & 0 \\ \hline 0 & 0 & 1 \end{array}\right] \left[\begin{array}{ccc} 1 & \beta_{21} & \beta_{31} \\ 0 & 1 & \beta_{32} \\ 0 & 0 & 1 \end{array}\right]^{-1} = DM^{-1}$$

It is easily verified that

$$\underline{A}_L^3\underline{d}_1 = -\alpha_{110}\underline{d}_1 - \alpha_{111}\underline{A}_L\underline{d}_1 - \alpha_{112}\underline{A}_L^2\underline{d}_1 - \alpha_{120}\underline{d}_2 - \alpha_{121}\underline{A}_L\underline{d}_2 - \alpha_{130}\underline{d}_3$$

$$\underline{A}_L^2\underline{d}_2 = -\alpha_{210}\underline{d}_1 - \alpha_{211}\underline{A}_L\underline{d}_1 - \alpha_{220}\underline{d}_2 - \alpha_{221}\underline{A}_L\underline{d}_2 - \alpha_{230}\underline{d}_3$$

$$\underline{A}_L\underline{d}_1 = -\alpha_{310}\underline{d}_1 - \alpha_{320}\underline{d}_2 - \alpha_{330}\underline{d}_3$$

Now by $\underline{D} = \underline{B}_L\underline{M}$

$$\underline{D} = [\underline{d}_1 \quad \underline{d}_2 \quad \underline{d}_3] = [\underline{b}_{L1} \quad \underline{b}_{L2} \quad \underline{b}_{L3}] \begin{bmatrix} 1 & \beta_{21} & \beta_{31} \\ 0 & 1 & \beta_{32} \\ 0 & 0 & 1 \end{bmatrix}$$

$$[\underline{B}_L, \ \underline{A}_L\underline{B}_L, \ \underline{A}_L^2\underline{B}_L]\underline{a}_1 + \underline{A}_L^3\underline{b}_{L1} = 0$$

$$[\underline{B}_L, \ \underline{A}_L\underline{B}_L]\underline{a}_2 + \beta_{21}\underline{A}_L^2\underline{b}_{L1} + \underline{A}_L^2\underline{b}_{L2} = 0$$

$$\underline{B}_L\underline{a}_3 + \beta_{31}\underline{A}_L^2\underline{b}_{L1} + \beta_{32}\underline{A}_L\underline{b}_{L2} + \underline{A}_L\underline{b}_{L3} = 0$$

with

$$\underline{a}_1 = [a_{110} \quad a_{120} \quad a_{130} \quad a_{111} \quad a_{121} \quad 0 \quad a_{112} \quad 0 \quad 0]'$$

$$\underline{a}_2 = [a_{210} \quad a_{220} \quad a_{230} \quad a_{211} \quad a_{221} \quad 0]'$$

$$\underline{a}_3 = [a_{310} \quad a_{320} \quad a_{330}]'$$

The $\underline{a}_i$ are related to the $\underline{\alpha}_i$ of eq. (9.1.33) (same structure, the a_{ijk} replaced by α_{ijk}) via

$$\underline{a}_1 = \begin{bmatrix} \underline{M} & & \\ & \underline{M} & \\ & & \underline{M} \end{bmatrix} \underline{\alpha}_i \ , \quad \underline{a}_2 = \begin{bmatrix} \underline{M} & \\ & \underline{M} \end{bmatrix} \underline{\alpha}_2 \ , \quad \underline{a}_3 = \underline{M\alpha}_3$$

The $\underline{a}_i$ are the invariants defined by Popov [72.4]. These invariants are summarized in table A.1.

Under these transformations of $(\underline{A},\underline{B})$ with $\underline{T}$ and $\underline{M}$ nonsingular ...	... the following quantities constitute a complete set of independent invariants i = 1,2...m	Reference	
$\underline{T}(\underline{A}-\underline{BK})\underline{T}^{-1}$, $\underline{TBM}$	μ_i (unordered)	Brunovsky [70.1]	(A.4.6)
$\underline{T}(\underline{A}-\underline{BK})\underline{T}^{-1}$, $\underline{TB}$	μ_i , $\underline{\beta}_i$	Popov [72.4]	(A.4.7)
$\underline{TAT}^{-1}$, $\underline{TB}$	μ_i , $\underline{\beta}_i$, $\underline{a}_i$	Popov [72.4]	(A.4.8)

Table A.1 Invariants of $(\underline{A}, \underline{B})$

Since there is a unique relation between the parameters $(\mu_i, \underline{\beta}_i, \underline{a}_i)$ and $(\mu_i, \underline{\beta}_i, \underline{\alpha}_i)$, the latter also constitute a complete set of independent invariants under a transformation $(\underline{A}, \underline{B}) \rightarrow (\underline{TAT}^{-1}, \underline{TB})$. In other words: the α and β parameters can be determined for any controllable pair $(\underline{A}, \underline{B})$ from the linear relations between the columns $[\underline{B}, \underline{AB}, \ldots]$. Using these parameters, the Luenberger form, eq. (A.4.1), can be written down almost immediately. Only one inversion of the triangular matrix $\underline{M}$ is required.

2. The characteristic polynomial of $\underline{A}_L$ is obtained as follows: Let $\underline{z}_j = [1, z \ldots z^j]'$

$$\underline{Z} = \begin{bmatrix} \underline{z}_{\mu_1 - 1} & & \\ & \ddots & \\ & & \underline{z}_{\mu_m - 1} \end{bmatrix}, \quad \underline{Q} = \left[\begin{array}{cc|cc|c|cc} \underline{h}'_{11} & 1 & \underline{h}'_{12} & 0 & & \underline{h}'_{1r} & 0 \\ \underline{h}'_{21} & 0 & \underline{h}'_{22} & 1 & \cdots & & \\ \vdots & & & & & & \\ \underline{h}'_{r1} & 0 & \underline{h}'_{r2} & 0 & & \underline{h}'_{rr} & 1 \end{array}\right] \tag{A.4.9}$$

Then

$$\det(z\underline{I} - \underline{A}_L) = \det \underline{QZ} \tag{A.4.10}$$

3. The transformation matrix $\underline{T}$ for $\underline{A}_L = \underline{TAT}^{-1}$, $\underline{B}_L = \underline{TB}$ was derived by Luenberger [67.4] as follows: arrange the columns of $\mathrm{reg}[\underline{B}, \underline{AB} \ldots]$ into the $n \times n$ nonsingular matrix

$$\underline{R} := [\underline{b}_1, \underline{AB}_1 \ldots \underline{A}^{\mu_1 - 1}\underline{b}_1, \underline{b}_2, \ldots, \ldots, \ldots \underline{A}^{\mu_r - 1}\underline{b}_r] \tag{A.4.11}$$

Its inverse has the structure

$$\underline{R}^{-1} = \begin{bmatrix} \underline{S}_1 \\ \vdots \\ \underline{S}_r \end{bmatrix} \tag{A.4.12}$$

Let $\underline{e}_i'$ be the last row of the $\mu_i \times n$ matrix $\underline{S}_r$. Then

$$\underline{T} = \begin{bmatrix} \underline{T}_1 \\ \vdots \\ \underline{T}_r \end{bmatrix}, \quad \underline{T}_i = \begin{bmatrix} \underline{e}_i' \\ \underline{e}_i'\underline{A} \\ \vdots \\ \underline{e}_i'\underline{A}^{\mu_i - 1} \end{bmatrix} \tag{A.4.13}$$

4. For the single-input case, eq. (A.3.8), it was convenient to construct the columns of the inverse matrix $\underline{S} = \underline{T}^{-1}$. A generalized procedure can be developed for the Luenberger feedback canonical form, see [73.4]. It is similar to eq. (A.3.13), but it assumes, that the α and β parameters are known. Thus it does not generalize eq. (A.3.14).

A.4.3 Brunovsky Canonical Form

Previously transformations of the form $(\underline{A}, \underline{B}) \to (\underline{TAT}^{-1}, \underline{TB})$ were considered. Now also an input transformation $\underline{B} \to \underline{BM}$ and state feedback $\underline{A} \to \underline{A}-\underline{BK}$ will be allowed. By eq. (A.4.6) then only the controllability indices μ_i are invariant and a canonical form can be specified only by this set of integers. Strictly speaking, only the unordered set of the μ_i is invariant, because the input transformation $\underline{M}$ also allows the inputs to be permuted. The possibility of permutation is obvious and will not be discussed here.

Starting with the Luenberger feedback canonical form (A.4.1), proceed in two steps:

1. Perform an input transformation with the matrix $\underline{M}$ of eq. (A.4.1). For the normalized input $\underline{v} := \underline{M}^{-1}\underline{u}$, $\underline{u} = \underline{Mv}$, the input matrix is $\underline{B}_L\underline{M} = \underline{D}$. This changes all β parameters to zero.

2. Let $\underline{u} = -\underline{Kx}$, i.e. $\underline{v} = \underline{M}^{-1}\underline{Kx} = \underline{Fx}$ with

$$\underline{F} = \underline{H} := \begin{bmatrix} \underline{h}'_{11} & \cdots & \underline{h}'_{1r} \\ \vdots & & \vdots \\ \underline{h}'_{r1} & \cdots & \underline{h}'_{rr} \end{bmatrix} \tag{A.4.14}$$

This changes all α parameters to zero. The resulting form is

$$\underline{A}_B = \begin{bmatrix} A_{11} & \cdots & A_{1r} \\ \vdots & & \vdots \\ A_{r1} & \cdots & A_{rr} \end{bmatrix}, \quad \underline{B}_B = \begin{bmatrix} \underline{D}_1 \\ \vdots \\ \underline{D}_r \end{bmatrix} \tag{A.4.15}$$

$$\underline{A}_{ii} = \begin{bmatrix} 0 & 1 & & 0 \\ & & \ddots & \\ & & & 1 \\ 0 & \cdots & & 0 \end{bmatrix}, \quad \underline{A}_{ij} = \underline{0} \text{ for } i \neq j, \quad \underline{D}_i = \begin{bmatrix} \underline{0} \\ \underline{i}_i' \end{bmatrix}$$

$\underline{i}_i'$ = ith row of the r×r unit matrix.

This is called the "Brunovsky canonical form". It is fully specified by the dimensions μ_i of the submatrices. These may be ordered, according to size, by an input permutation matrix. The above procedure for generation of the Brunovsky form was introduced in [68.2], in the context of minimum-time deadbeat control.

The Brunovsky form was here derived as a special case of Luenberger's block Frobenius form. It may also be considered as a special case of the Jordan form, eq. (A.2.18), with all eigenvalues at zero and r Jordan blocks of dimensions $\mu_1, \mu_2 \ldots \mu_r$. This directly shows that the minimal polynomial of $\underline{A}_B$ is z^μ, where μ is the maximal controllability index.

A.5 Computational Aspects

So far we have used the canonical forms mainly for the illustration of structural properties and parameters. For design purposes, we do not need the canonical form itself, but the transformation matrix leading to this form. Efficient procedures determine both the form and the transformation. The numerical aspects become important if the transformation matrices must be computed for nontrivial examples.

A.5.1 Elementary Transformations to Hessenberg Form

The basic idea is to generate the transformation matrix $\underline{T}$ and its inverse $\underline{T}^{-1}$ by numerically stable elementary transformations $\underline{T}_i$ as $\underline{T} = \underline{T}_m \dots \underline{T}_2\underline{T}_1$, $\underline{T}^{-1} = \underline{T}_1^{-1}\underline{T}_2^{-1} \dots \underline{T}_m^{-1}$.

Consider the problem of reducing $[\underline{A}, \underline{b}]$ to a Hessenberg form $\underline{A}_H$ of the dynamic matrix

$$\left[\begin{array}{c|c} \underline{A}_H & \underline{b}_H \end{array}\right] = \left[\begin{array}{ccccc|c} x & \textcircled{n} & 0 & & \dots & 0 \\ x & x & \textcircled{n-1} & & & \vdots \\ \vdots & & & \ddots & \textcircled{2} & 0 \\ x & \cdot & \cdot & \cdot & x & \textcircled{1} \end{array}\right] \tag{A.5.1}$$

Here x denotes an arbitrary element and the encircled elements are nonzero. The controllability matrix of (A.5.1) is

$$\left[\underline{b}_H, \underline{A}_H\underline{b}_H \dots \underline{A}_H^{n-1}\underline{b}_H\right] = \begin{bmatrix} 0 & & & \cdot^{\cdot^{\cdot}} & \textcircled{n!} \\ \vdots & 0 & \textcircled{123} & & x \\ 0 & \textcircled{12} & x & & \\ \textcircled{1} & x & x & \dots & x \end{bmatrix} \tag{A.5.2}$$

Notation: $\textcircled{123} = \textcircled{1} \times \textcircled{2} \times \textcircled{3}$, $\textcircled{n!} = \textcircled{1} \times \textcircled{2} \dots \textcircled{n}$.

The system is controllable, if and only if all elements $\textcircled{1}$, $\textcircled{2} \dots \textcircled{n}$ are nonzero. Assume that the elements $\textcircled{1}$, $\textcircled{2} \dots \textcircled{i}$ are nonzero and $\textcircled{i+1} = 0$, then an ith order subsystem is controllable and the reduction in (A.5.1) can be terminated with

$$\left[\begin{array}{ccc|cccc|c} x & \dots & x & 0 & \dots & & & 0 \\ \vdots & & \vdots & & & & & \vdots \\ x & & x & 0 & & 0 & \ddots & \vdots \\ \hline x & \dots & x & x & & \textcircled{i} & \ddots & \vdots \\ \vdots & & \vdots & & & & \textcircled{2} & 0 \\ x & \dots & x & x & \dots & & x & \textcircled{1} \end{array}\right] \tag{A.5.3}$$

This directly separates the controllable subsystem with states $x_{n-i+1} \ldots x_n$ from the uncontrollable subsystem with states $x_1 \ldots x_{n-i}$.

The form (A.5.1) is obtained from $[\underline{A}, \underline{b}]$ in the same way as for the Gaussian elimination procedure [74.4], [77.6], [80.4]:

1. Select the largest element b_i of b as pivot element. (It is assumed that $[\underline{A},\underline{b}]$ is "balanced", i.e. the orders of magnitudes of the elements are roughly equal.)

2. Bring it to the ① position in eq. (A.5.3) by a permutation step with

$$\underline{P}_1 = \begin{bmatrix} 1 & & & & & & \\ & \ddots & & & & & \\ & & 1 & & & & \\ & & & 0 & & & 1 \\ & & & & 1 & & \\ & & & & & \ddots & \\ & & & & & & 1 \\ & & & 1 & & & & 0 \end{bmatrix} \text{ ith row, } \underline{P}_1^{-1} = \underline{P}_1 \tag{A.5.4}$$

3. In $[\underline{P}_1\underline{A}\underline{P}_1, \underline{P}_1\underline{b}]$ subtract an appropriate multiple of the last row from all other rows, such that their elements in the last column vanish:

$$\left[\begin{array}{c|c} & 0 \\ & \vdots \\ \underline{A}^{(1)} & 0 \\ & b_i \end{array}\right] \tag{A.5.5}$$

This is accomplished by an elimination matrix

$$\underline{G}_1 = \begin{bmatrix} 1 & & & & -l_1 \\ & \ddots & & & \vdots \\ & & \ddots & & -l_{n-1} \\ 0 & & & & 1 \end{bmatrix}, \quad \underline{G}_1^{-1} = \begin{bmatrix} 1 & & & & l_1 \\ & \ddots & & & \vdots \\ & & \ddots & & l_{n-1} \\ 0 & & & & 1 \end{bmatrix} \tag{A.5.6}$$

with all $|l_i| \leq 1$. Now $\underline{A}^{(1)} = \underline{G}_1\underline{P}_1\underline{A}\underline{P}_1\underline{G}_1^{-1}$, $\underline{b}^{(1)} = \underline{G}_1\underline{P}_1\underline{b}$

The last row of $[\underline{A}^{(1)}, \underline{b}^{(1)}]$ is fixed from now on.

4. Select the largest element of the last column of $\underline{A}^{(1)}$ (except the fixed element in the last row) as pivot element and proceed as before. This fixes the row before the last one.

5. If the largest element is below a small numerical threshold in one of the steps, then the process terminates with the form of eq. (A.5.3), otherwise it continues until the form (A.5.1) is reached, where all ⓧ elements are nonzero and the system is controllable.

The transformations proceed as

$$\begin{aligned} \underline{A}^{(o)} &= \underline{A}\ , \quad \underline{b}^{(o)} = \underline{b} \\ \underline{A}^{(i)} &= \underline{T}_i\underline{A}^{(i-1)}\underline{T}_i^{-1}\ , \quad \underline{b}^{(i)} = \underline{T}_i\underline{b}^{(i-1)} \\ \underline{T}_i &= \underline{G}_i\underline{P}_i \end{aligned} \tag{A.5.7}$$

where $\underline{P}_i^{-1} = \underline{P}_i$ and the inversion of $\underline{G}_i$ is trivial, see eq. (A.5.6).

If also $\underline{C}$ is to be transformed, then

$$\underline{C}^{(o)} = \underline{C}\ , \quad \underline{C}^{(i)} = \underline{C}^{(i-1)}\underline{T}_i^{-1} \tag{A.5.8}$$

The Hessenberg form $\underline{A}_H$ may be further reduced to Frobenius form $\underline{A}_F$, see eq. (A.3.1). In this reduction, we have no alternative but to pivot with elements ①, ② ... ⓝ in eq. (A.5.1) so that the transformations cannot be numerically stabilized. In the literature on numerical eigenvalue problems [65.1], [80.4] it is emphasized that transformation to Frobenius form is not a helpful step in determining eigenvalues. The zeros of a polynomial may be very sensitive to small changes in the polynomial coefficients in the Frobenius form. Thus the eigenvalue problem for $\underline{A}_F$ is usually much more badly conditioned than that for the original $\underline{A}$ matrix or its Hessenberg form $\underline{A}_H$.

The control engineer is not only interested in the exact computation of eigenvalues, but also in the inverse problem of assigning the closed-loop eigenvalues z_i by appropriate choice of the feedback gains. In this case, it is not advisable to multiply

$$P(z) = (z-z_1)(z-z_2)\dots(z-z_n) = p_o + p_1 z + \dots + p_{n-1} z^{n-1} + z^n \tag{A.5.9}$$

because the p_i frequently have extremely different orders of magnitude for higher n. (Example: Let $P(z) = (z-0.1)^{10}$). In this case the pole placement formula, eq. (4.4.5), may be used in its factorized form, see eq. (2.6.29),

$$\begin{aligned}\underline{k}' = \underline{e}'P(\underline{A}) &= \underline{e}'(\underline{A}-z_1\underline{I})(\underline{A}-z_2\underline{I}) \dots (\underline{A}-z_n\underline{I}) \\ &= \underline{e}'(a_o\underline{I}+b_o\underline{A}+\underline{A}^2)(a_1\underline{I}+b_1\underline{A}+\underline{A}^2)\dots(c\underline{I}+\underline{A})\end{aligned} \tag{A.5.10}$$

where the last factor $(c\underline{I}+\underline{A})$ occurs only for n odd. Accuracy is then required in the evaluation of $\underline{e}'$, the last row of the inverted controllability matrix. Using the Hessenberg form and eq. (A.5.2), this is

$$\underline{k}_H' = \underline{e}_H'P(\underline{A}_H) \quad , \quad \underline{e}_H' = [e_1 \quad e_2 \quad \dots \quad e_n] \tag{A.5.11}$$

$$[e_1 \quad e_2 \quad \dots \quad e_n]\begin{bmatrix} 0 & & & & \textcircled{n!} \\ \cdot & & & \cdot^{\cdot^{\cdot}} & x \\ \cdot & & & & \\ \cdot & & \textcircled{123} & & \\ 0 & \textcircled{12} & & & \\ \textcircled{1} & x & & & x \end{bmatrix} = [0 \quad \dots \quad 0 \quad 1] \tag{A.5.12}$$

It follows that

$$\begin{aligned}\underline{e}_H' &= [e_1 \quad 0 \quad \dots \quad 0] \\ e_1 &= 1/\textcircled{n!} = 1/\textcircled{1}\,\textcircled{2}\dots\textcircled{n}\end{aligned} \tag{A.5.13}$$

Thus the transformation to Hessenberg form $\underline{A}_H = \underline{T}_H\underline{A}\underline{T}_H^{-1}$ immediately yields $\underline{e}_H'$ and thereby $\underline{k}_H'$, for a desired set of eigenvalues $z_1, z_2 \dots z_n$, and in the original coordinates

$$\underline{k}' = \underline{k}_H'\underline{T}_H = \underline{e}_H'P(\underline{A}_H)\underline{T}_H = \underline{e}_H'\underline{T}_H P(\underline{A}) \tag{A.5.14}$$

where $\underline{T}_H$ consists of numerically stable elementary transformations, see eq. (A.5.7). This procedure avoids the usual transformation to feedback-canonical form (= Frobenius form).

If only $\underline{e}'$ is needed, this can be calculated as $\underline{e}' = \underline{e}_H'\underline{T}_H$; for the calculation of $\underline{k}'$ the form $\underline{k}' = \underline{e}_H'P(\underline{A}_H)\underline{T}_H$ is better suited, see eq. (A.5.24).

A.5.2 HN Form

A multi-variable generalization of eq. (A.5.1) was derived by Nour Eldin [81.12]. He calls it the first HN form (Hessenberg-Nour Eldin form). The second HN form is a generalization of the Frobenius forms and the third HN form introduces the structural zero elements into it, which are the same zero elements, that occur in the Luenberger form (A.4.1). For the same reason as in the single-input case, only the first HN form will be discussed, and it is briefly called HN form. Here the matrix $[\underline{A}, \underline{b}_r \ldots \underline{b}_1]$ is reduced to Hessenberg form:

$$[\underline{A}_H, \underline{b}_{Hr} \ldots \underline{b}_{H1}] = \left[\begin{array}{ccccccc|cccc}
x & \textcircled{n} & 0 & & & & 0 & 0 & & & 0 \\
\vdots & & \ddots & & & & & & & & \\
\vdots & & & \textcircled{x} & 0 & & & & & & \\
\vdots & & & & \textcircled{x} & \ddots & & & & & \\
\vdots & & & & & \ddots & 0 & & & & \\
\vdots & & & & & & \textcircled{r+1} & & & & \\
 & & & & & & x & \textcircled{r} & & & \\
 & & & & & & \vdots & x & \ddots & & 0 \\
x & & & & & & x & x & \cdots & x & \textcircled{1}
\end{array}\right] \tag{A.5.15}$$

Assuming rank $\underline{B} = r$, the elements $\textcircled{1}\ \textcircled{2} \ldots \textcircled{r}$ are in a diagonal. The last column of $\underline{A}$ is reduced as in the single-input case, and so on. In the generic case where the system is controllable from each individual input, all $\textcircled{x}$ elements are in the same diagonal. For a general controllability structure, however, it turns out, that the upper part of a column is already zero and we must proceed to the column left of it to search for a pivot element. The interpretation of this effect follows directly from the controllability matrix

$$[\underline{B}_H,\ \underline{A}_H\underline{B}_H \ldots] = \left[\begin{array}{ccc|cc|c} 0 & & 0 & 0 & 0 & \\ & & & & & \\ & & & & \textcircled{x} & \\ & & & & x & \ldots \\ & & 0 & \textcircled{1(r+1)} & & \\ & & \textcircled{r} & x & & \\ 0 & & & & & \\ \textcircled{1} & x & x & x & x & \end{array}\right] \quad (A.5.16)$$

If the element in the location $\textcircled{r+1}$ in eq. (A.5.15) is zero, then rank $[\underline{B}_H, \underline{A}_H\underline{b}_{H1}]$ = rank $\underline{B}_H$ and only a first order subsystem is assigned to the input 1, i.e. $\mu_1 = 1$. In this case this part of the $\underline{A}_H$ matrix is

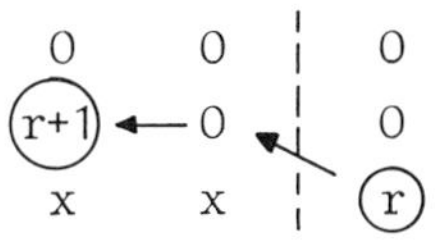

$$\begin{array}{cc|c} 0 & 0 & 0 \\ \textcircled{r+1} \leftarrow & 0 \nwarrow & 0 \\ x & x & \textcircled{r} \end{array} \quad (A.5.17)$$

Then $\underline{A}_H\underline{b}_{H1}$ does not belong to reg$[\underline{B}_H, \underline{A}_H\underline{B}_H \ldots]$ and can be omitted. ("reg" is the operation of selecting linear independent columns from the left.) The same is true for $\underline{A}_H{}^k\underline{b}_{H1}$, $k = 2, 3 \ldots$. The structure of eq. (A.5.17) with $\textcircled{r+1} \neq 0$ applies to the case that $\underline{A}_H\underline{b}_{H2}$ is linearly independent of its predecessors and

$$\text{reg}[\underline{B}_H,\ \underline{A}_H\underline{B}_H \ldots] = [\underline{B}_H,\ \underline{A}_H\underline{b}_{H2} \ldots] = \left[\begin{array}{ccc|c|c} 0 & & & 0 & \\ & & & 0 & \\ & & 0 & \textcircled{2(r+1)} & \ldots \\ & & \textcircled{r} & x & \\ 0 & \cdot^{\cdot^{\cdot}} & & \vdots & \\ \textcircled{1} & x & & x & \end{array}\right]$$

(A.5.18)

The elements indicated by arrows in eq. (A.5.17) are called the "Hessenberg chain". In the above example it is

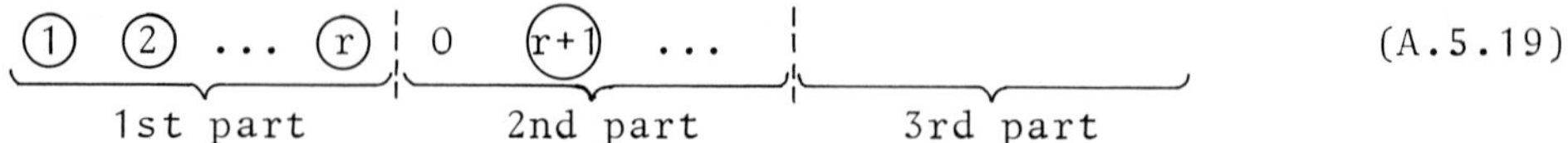

(A.5.19)

indicating $\mu_1 = 1$. In the third part of the chain the first element is not repeated. If the system is controllable, then there can be at most r-1 zeros in the chain from ① to ⓝ. If the rth zero is reached before ⓝ, then the process terminates as in eq. (A.5.3) and the columns to the left describe the uncontrollable subsystem. The zero locations in the chain determine the controllability indices of the controllable subsystem.

Example:

Let n = 8, r = 3

Input	1	2	3	1	2	3	1	3	1	1
Column of contr. matr.	b_1	b_2	b_3	Ab_1	Ab_2	Ab_3	A^2b_1	A^2b_3	A^3b_1	A^4b_1
Hessenberg chain	①	②	③	④	0	⑤	⑥	0	⑦	⑧
Controllability indices					$\mu_2=1$			$\mu_3=2$		$\mu_1=5$

(A.5.20)

In this table, the number of nonzero elements below an input number is the corresponding controllability index. The form of (A_H, B_H) for this case is completely specified by the chain.

$$\underline{A}_H = \begin{bmatrix} x & \textcircled{8} & 0 & 0 & 0 & 0 & 0 & 0 \\ x & x & \textcircled{7} & 0 & 0 & 0 & 0 & 0 \\ x & x & x & x & \textcircled{6} & 0 & 0 & 0 \\ x & x & x & x & x & \textcircled{5} & 0 & 0 \\ x & x & x & x & x & x & x & \textcircled{4} \\ x & x & x & x & x & x & x & x \\ x & x & x & x & x & x & x & x \\ x & x & x & x & x & x & x & x \end{bmatrix} \qquad \underline{B}_H = \begin{bmatrix} 0 & 0 & 0 \\ 0 & 0 & 0 \\ 0 & 0 & 0 \\ 0 & 0 & 0 \\ 0 & 0 & 0 \\ 0 & 0 & \textcircled{3} \\ 0 & \textcircled{2} & x \\ \textcircled{1} & x & x \end{bmatrix}$$

(A.5.21)

Eq. (A.5.16) also allows a numerically efficient determination of the α- and β-parameters of the system, which constitute the finite effect sequences (FESs), see section 9.2. For FES assignment the matrix $\underline{E}$ of eq. (9.3.9) is needed. Its accuracy depends primarily on the accuracy of the vectors $\underline{e}_i'$, i = 1, 2 ... r, according to eq. (9.3.11). This can be obtained again from the HN form.

Example:

Consider the example of eq. (A.5.21)

$$\text{reg}[\underline{B}_H,\ \underline{A}_H\underline{B}_H \ldots] =$$

$$= [\underline{B}_H \mid \underline{Ab}_{H1},\ \underline{A}_H\underline{b}_{H3} \mid \underline{A}_H^2\underline{b}_{H1} \mid \underline{A}_H^3\underline{b}_{H1} \mid \underline{A}_H^4\underline{b}_{H1}]$$

$$= \left[\begin{array}{ccc|cc|c|c|c}
0 & 0 & 0 & 0 & 0 & 0 & 0 & \textcircled{14678} \\
0 & 0 & 0 & 0 & 0 & 0 & \textcircled{1467} & x \\
0 & 0 & 0 & 0 & 0 & \textcircled{146} & x & x \\
0 & 0 & 0 & 0 & \textcircled{35} & x & x & x \\
0 & 0 & 0 & \textcircled{14} & x & x & x & x \\
0 & 0 & \textcircled{3} & x & x & x & x & x \\
0 & \textcircled{2} & x & x & x & x & x & x \\
\textcircled{1} & x & x & x & x & x & x & x
\end{array}\right] \qquad (A.5.22)$$

The vectors $\underline{e}_{H1}'$, $\underline{e}_{H2}'$, $\underline{e}_{H3}'$ are the rows # 5, 6 and 8 of $\underline{R}_H^{-1}$, where

$$\underline{R}_H = [\underline{b}_{H1},\ \underline{Ab}_{H1} \ \ldots \ \underline{A}^4\underline{b}_{H1} \mid \underline{b}_{H2} \mid \underline{b}_{H3} \ \ \underline{Ab}_{H3}]$$

Therefore

$$\begin{bmatrix} \underline{e}_{H1}' \\ \underline{e}_{H2}' \\ \underline{e}_{H3}' \end{bmatrix} \times \left[\begin{array}{ccccc|c|cc}
0 & 0 & 0 & 0 & \textcircled{14678} & 0 & 0 & 0 \\
0 & 0 & 0 & \textcircled{1467} & x & 0 & 0 & 0 \\
0 & 0 & \textcircled{146} & x & x & 0 & 0 & 0 \\
0 & 0 & x & x & x & 0 & 0 & \textcircled{35} \\
0 & \textcircled{14} & x & x & x & 0 & 0 & x \\
0 & x & x & x & x & 0 & \textcircled{3} & x \\
0 & x & x & x & x & \textcircled{2} & x & x \\
\textcircled{1} & x & x & x & x & x & x & x
\end{array}\right] = \begin{bmatrix}
0 & 0 & 0 & 0 & 1 & 0 & 0 & 0 \\
0 & 0 & 0 & 0 & 0 & 1 & 0 & 0 \\
0 & 0 & 0 & 0 & 0 & 0 & 0 & 1
\end{bmatrix}$$

$$\begin{bmatrix} \underline{e}_{H1}' \\ \underline{e}_{H2}' \\ \underline{e}_{H3}' \end{bmatrix} = \begin{bmatrix} 1/\textcircled{14678} & 0 & 0 & 0 & 0 & 0 & 0 & 0 & 0 \\ 0 & 0 & 0 & 0 & 0 & 0 & 0 & 1/\textcircled{2} & 0 \\ 0 & 0 & 0 & 1/\textcircled{35} & 0 & 0 & 0 & 0 & 0 \end{bmatrix} \tag{A.5.23}$$

Then the FES assignment matrix is

$$\underline{E}_H = \begin{bmatrix} \underline{e}_{H1}' \\ \underline{e}_{H1}'\underline{A}_H \\ \underline{e}_{H1}'\underline{A}_H^2 \\ \underline{e}_{H1}'\underline{A}_H^3 \\ \underline{e}_{H1}'\underline{A}_H^4 \\ \underline{e}_{H1}'\underline{A}_H^5 \\ \hline \underline{e}_{H2}' \\ \underline{e}_{H2}'\underline{A}_H \\ \hline \underline{e}_{H3}' \\ \underline{e}_{H3}'\underline{A}_H \\ \underline{e}_{H3}'\underline{A}_H^2 \end{bmatrix} = \left[\begin{array}{cccccccc} 1/\textcircled{14678} & 0 & 0 & 0 & 0 & 0 & 0 & 0 \\ x & 1/\textcircled{1467} & 0 & 0 & 0 & 0 & 0 & 0 \\ x & x & 1/\textcircled{146} & 0 & 0 & 0 & 0 & 0 \\ x & x & x & x & 1/\textcircled{14} & 0 & 0 & 0 \\ x & x & x & x & x & x & x & 1/\textcircled{1} \\ x & x & x & x & x & x & x & x \\ \hline 0 & 0 & 0 & 0 & 0 & 0 & 1/2 & 0 \\ x & x & x & x & x & x & x & x \\ \hline 0 & 0 & 0 & 1/\textcircled{35} & 0 & 0 & 0 & 0 \\ x & x & x & x & x & 1/\textcircled{3} & 0 & 0 \\ x & x & x & x & x & x & x & x \end{array}\right]$$

(A.5.24)

The generalization of eq. (A.5.10) can be achieved if the polynomial version of the closed-loop FES matrix

$$\underline{Q}_c\underline{Z} = \begin{bmatrix} q_{11}(z) & \dots & q_{1r}(z) \\ \vdots & & \vdots \\ q_{r1}(z) & \dots & q_{rr}(z) \end{bmatrix} \tag{A.5.25}$$

is assigned in factorized form

$$\underline{Q}_c\underline{Z} = (\underline{Q}_c\underline{Z})_1 \times (\underline{Q}_c\underline{Z})_2 \times \dots (\underline{Q}_c\underline{Z})_\mu \tag{A.5.26}$$

and $\underline{Q}_c\underline{E}_H$ of eq. (9.3.8) is rewritten as [77.4]

$$\underline{Q}_c\underline{E}_H = \begin{bmatrix} \underline{e}_{H1}'q_{11}(\underline{A}_H) + \dots + \underline{e}_{Hr}'q_{1r}(\underline{A}_H) \\ \vdots \\ \underline{e}_{H1}'q_{r1}(\underline{A}_H) + \dots + \underline{e}_{Hr}'q_{rr}(\underline{A}_H) \end{bmatrix} \tag{A.5.27}$$

$$\underline{K} = \underline{M}\underline{Q}_c\underline{E} = \underline{M}\underline{Q}_c\underline{E}_H\underline{T}_H \tag{A.5.28}$$

A.6 Sensor Coordinates

For the design of reduced order observers and for design for robustness with respect to sensor failures, it is convenient to choose the state vector such that it contains the measured variables directly, i.e. such that the $\underline{C}$ matrix contains only one "one" element in each row and zeros otherwise. If the plant is not modelled in this form, then it can be transformed as follows. Let rank $\underline{C} = s$, then $\begin{bmatrix}\underline{A}\\ \underline{B}\end{bmatrix}$ can be reduced by elementary transformations to

$$\begin{bmatrix}\underline{A}_E\\ \underline{C}_E\end{bmatrix} = \left[\begin{array}{c|c} \underline{A}_{11} & \underline{A}_{12} \\ \underline{A}_{21} & \underline{A}_{22} \\ \hline \underline{0} & \begin{matrix}\otimes & x\dots x\\ & \ddots \quad \vdots\\ & \quad x \\ 0 & \quad \otimes\end{matrix}\end{array}\right] \begin{matrix}\}\,n-s\\ \}\,s \\ \\ \leftarrow C_2 \\ \end{matrix} \qquad \begin{matrix}\underbrace{\qquad\qquad}_{n-s} & \underbrace{\qquad\qquad}_{s}\end{matrix} \tag{A.6.1}$$

$$\text{Then } \underline{T} = \begin{bmatrix}\underline{I}_{n-s} & \underline{0}\\ \underline{0} & \underline{C}_2\end{bmatrix}, \quad \underline{T}^{-1} = \begin{bmatrix}\underline{I}_{n-s} & \underline{0}\\ \underline{0} & \underline{C}_2^{-1}\end{bmatrix} \tag{A.6.2}$$

transforms this system to

$$\underline{A}_S = \underline{T}\underline{A}_E\underline{T}^{-1} = \begin{bmatrix}\underline{A}_{11} & \underline{A}_{12}\underline{C}_2^{-1}\\ \underline{C}_2\underline{A}_{21} & \underline{C}_2\underline{A}_{22}\underline{C}_2^{-1}\end{bmatrix} = \begin{bmatrix}\underline{P} & \underline{Q}\\ \underline{R} & \underline{S}\end{bmatrix} \tag{A.6.3}$$

$$\underline{B}_S = \underline{T}\begin{bmatrix}\underline{B}_1\\ \underline{B}_2\end{bmatrix} = \begin{bmatrix}\underline{B}_1\\ \underline{C}_2\underline{B}_2\end{bmatrix} = \begin{bmatrix}D\\ E\end{bmatrix}$$

$$\underline{C}_S = [0 \quad \underline{C}_2]\underline{T}^{-1} = [0 \quad \underline{I}_S]$$

A.7 Further results from Matrix Theory

This section is a collection of some frequently used results from matrix theory. For the mathematical foundation the reader is referred to the literature on applied linear algebra and matrix analysis (e.g. [59.4], see also the appendix in [80.2]).

A.7.1 Notations

Consider a matrix

$$\underline{A} = \begin{bmatrix} a_{11} & \dots & a_{1m} \\ \vdots & & \\ a_{n1} & \dots & a_{nm} \end{bmatrix} \tag{A.7.1}$$

$\underline{A}'$ denotes the transpose of $\underline{A}$, the elements of $\underline{A}'$ are $\tilde{a}_{ij} = a_{ji}$. (If the a_{ij} are complex, we must use the Hermitian transpose, i.e. we must take complex conjugates in the transposed matrix. Usually we deal with real valued vectors in this book.)

We are particularly interested in "square matrices" (m = n) and "column matrices" or "column vectors" (m = 1). Strictly speaking a column vector is a representation of a vector with respect to a particular basis in the vector space. Introducing a different set of basis vectors is equivalent to a linear transformation (e.g. eq. A.1.1).

Column vectors $\underline{a}$ and row vectors $\underline{a}'$ are written as

$$\underline{a} = \begin{bmatrix} a_1 \\ \vdots \\ a_n \end{bmatrix}, \quad \underline{a}' = [a_1 \ \dots \ a_n] \tag{A.7.2}$$

We will call this an n-vector.

We are also interested in block matrices

$$\tilde{A} = \begin{bmatrix} \underline{A} & \underline{D} \\ \underline{C} & \underline{B} \end{bmatrix} \qquad (A.7.3)$$

where $\underline{A}$ and $\underline{B}$ are square matrices.

A.7.2 Vector Operations

The "scalar product" of two n-vectors $\underline{a}$ and $\underline{b}$ is defined as the scalar

$$\underline{a}'\underline{b} = [a_1 \; \dots \; a_n] \begin{bmatrix} b_1 \\ \vdots \\ b_n \end{bmatrix} = a_1 b_1 + a_2 b_2 + \dots + a_n b_n \qquad (A.7.4)$$

The scalar product is commutative, i.e. $\underline{a}'\underline{b} = \underline{b}'\underline{a}$. The "Euclidean Norm" of a vector is

$$||\underline{a}|| = \sqrt{\underline{a}'\underline{a}} = \sqrt{a_1^{\,2} + a_2^{\,2} + \dots + a_n^{\,2}} \qquad (A.7.5)$$

$\underline{a}$ and $\underline{b}$ are "orthogonal" if $\underline{a}'\underline{b} = 0$.

The "dyadic product" of an n-vector $\underline{a}$ and an m-vector $\underline{b}$ is the n×m matrix

$$\underline{ab}' = \begin{bmatrix} a_1 \\ \vdots \\ a_n \end{bmatrix} [b_1 \; \dots \; b_m] = \begin{bmatrix} a_1 b_1 & \dots & a_1 b_m \\ \vdots & & \\ a_n b_1 & \dots & a_n b_m \end{bmatrix} \qquad (A.7.6)$$

A.7.3 Determinant of a Matrix

The determinant of a 2×2 matrix is

$$\det \begin{bmatrix} a_{11} & a_{12} \\ a_{21} & a_{22} \end{bmatrix} = \begin{vmatrix} a_{11} & a_{12} \\ a_{21} & a_{22} \end{vmatrix} = a_{11} a_{22} - a_{12} a_{21} \qquad (A.7.7)$$

Determinants of higher dimensional square matrices can be reduced by Laplace's expansion

$$\det \underline{A} = \sum_{j=1}^{n} a_{ij}\gamma_{ij} \quad \text{for any } i = 1,2 \ldots n \tag{A.7.8}$$

The "cofactor" γ_{ij} corresponding to a_{ij} is defined as

$$\gamma_{ij} = (-1)^{i+j}\det \underline{M}_{ij} \tag{A.7.9}$$

where M_{ij} is the $(n-1)\times(n-1)$ matrix obtained by deleting the ith row and jth column of det $\underline{A}$.

The computation of det $\underline{A}$ is simplified by the following elementary operations:

1. If $\tilde{\underline{A}}$ is obtained from $\underline{A}$ by adding a multiple of any one row to another, then det $\tilde{\underline{A}}$ = det $\underline{A}$.

2. If $\tilde{\underline{A}}$ is obtained from $\underline{A}$ by interchanging two rows of $\underline{A}$, then det $\tilde{\underline{A}}$ = -det $\underline{A}$.

3. If $\tilde{\underline{A}}$ is obtained from $\underline{A}$ by multiplying a row by a scalar factor c, then det $\tilde{\underline{A}}$ = c det $\underline{A}$. Thus det $[c\underline{A}] = c^n$det $\underline{A}$.

4. det $\underline{A}'$ = det $\underline{A}$, thus the above operations may be applied to columns instead of rows.

The determinant of a lower triangular matrix ($a_{ij} = 0$ for $i < j$) or of an upper triangular matrix ($a_{ij} > 0$ for $i > j$) is the product of the diagonal elements, i.e. $a_{11}a_{22}\ldots a_{nn}$. If $\underline{A}$ is block triangular then this result applies to the determinants of the diagonal blocks, e.g.

$$\det \begin{bmatrix} \underline{A} & \underline{D} \\ \underline{0} & \underline{B} \end{bmatrix} = \det \underline{A} \times \det \underline{B} \tag{A.7.10}$$

The determinant of the product of two square matrices is the product of the determinants of the two matrices, i.e.

$\underline{\det AB} = \det \underline{A} \times \det \underline{B}$ (A.7.11)

If $\underline{A}$ is an n×m matrix, and $\underline{B}$ is an m×n matrix, then

$$\det[\underline{I}_n+\underline{AB}] = \det[\underline{I}_m+\underline{BA}] \qquad (A.7.12)$$

where $\underline{I}_n$ ($\underline{I}_m$) is the n×n (m×m) identity matrix ($a_{ij} = 1$ for i = j, $a_{ij} = 0$ for i ≠ 0). In the special case m = 1

$$\det(\underline{I}_n+\underline{ab}') = 1+\underline{b}'\underline{a} = 1+\underline{a}'\underline{b} \qquad (A.7.13)$$

A.7.4 Trace of a Matrix

The "trace" of a matrix is the sum of the main diagonal elements, i.e.

$$\text{trace } \underline{A} = a_{11}+a_{22}+\ldots+a_{nn} \qquad (A.7.14)$$

For a dyadic product holds

$$\text{trace } \underline{ab}' = \underline{a}'\underline{b} \qquad (A.7.15)$$

Substituting $\underline{a}'\underline{b}$ into eq. (A.7.13)

$$\det(\underline{I}_n+\underline{ab}') = 1 + \text{trace } \underline{ab}' \qquad (A.7.16)$$

If $\underline{A}$ is an n×m matrix, and $\underline{B}$ is an m×n matrix, then

$$\text{trace } \underline{AB} = \text{trace } \underline{BA} \qquad (A.7.17)$$

A.7.5 Rank of a Matrix

The "null-space" or "kernel" of an n×m matrix $\underline{A}$ is the space of all solutions $\underline{x}$ of the equation

$$\underline{Ax} = \underline{0} \qquad (A.7.18)$$

The m columns of $\underline{A}$ are "linearly independent" if the null space is empty. Then $\underline{A}$ has "full column rank". The "column rank" of $\underline{A}$ is equal to the number of linearly independent columns. Corresponding definitions for the "row rank" apply to the equation

$$\underline{x}'\underline{A} = \underline{0} \tag{A.7.19}$$

For any matrix $\underline{A}$ the column rank is equal to the row rank. Therefore it suffices to talk only about the rank of a matrix.

A square matrix $\underline{A}$ of dimension n×n has rank n (i.e. "full rank"), if and only if det $\underline{A} \neq 0$. The rank of a product of two matrices satisfies

$$\text{rank } \underline{AB} \leq \min \{ \text{ rank } \underline{A} \text{ rank } \underline{B} \ \} \tag{A.7.20}$$

$$\text{rank } \underline{AB} = \text{rank } \underline{A} \text{ if } \det \underline{B} \neq 0 \tag{A.7.21}$$

A dyadic product $\underline{ab}'$ has rank one.

A.7.6 Inverse Matrix

The inverse of a nonsingular matrix $\underline{A}$ is denoted $\underline{A}^{-1}$, it satisfies

$$\underline{AA}^{-1} = \underline{A}^{-1}\underline{A} = \underline{I} \tag{A.7.22}$$

The inverse of a product of two nonsingular matrices $\underline{A}$ and $\underline{B}$ is

$$(\underline{AB})^{-1} = \underline{B}^{-1}\underline{A}^{-1} \tag{A.7.23}$$

The inverse of the block matrix $\underline{\tilde{A}}$ of eq. (A.7.3) is (if $\underline{A}^{-1}$ exists)

$$\begin{bmatrix} \underline{A} & \underline{D} \\ \underline{C} & \underline{B} \end{bmatrix}^{-1} = \begin{bmatrix} \underline{A}^{-1}+\underline{E\Delta}^{-1}\underline{F} & -\underline{E\Delta}^{-1} \\ -\underline{\Delta}^{-1}\underline{F} & \underline{\Delta}^{-1} \end{bmatrix} \tag{A.7.24}$$

where $\underline{\Delta} = \underline{B}-\underline{CA}^{-1}\underline{D}$, $\underline{E} = \underline{A}^{-1}\underline{D}$, and $\underline{F} = \underline{CA}^{-1}$.

The determinant of the above block matrix is

$$\det \begin{bmatrix} \underline{A} & \underline{D} \\ \underline{C} & \underline{B} \end{bmatrix} = \det \underline{A} \det [\underline{B}-\underline{CA}^{-1}\underline{D}]$$

$$= \det \underline{B} \times \det[\underline{A}-\underline{DB}^{-1}\underline{C}] \qquad (A.7.25)$$

For nonsingular matrices $\underline{A}$ and $\underline{C}$ the following inversion formula holds

$$(\underline{A}+\underline{BCD})^{-1} = \underline{A}^{-1}-\underline{A}^{-1}\underline{B}(\underline{DA}^{-1}\underline{B}+\underline{C}^{-1})^{-1}\underline{DA}^{-1} \qquad (A.7.26)$$

The inverse of a matrix $\underline{A}$ may be calculated by Cramer's rule

$$\underline{A}^{-1} = \frac{\text{Adj } \underline{A}}{\det \underline{A}} \qquad (A.7.27)$$

where Adj $\underline{A}$ is the adjugate matrix of $\underline{A}$ with elements γ_{ji}, i.e. Adj $\underline{A} = [\gamma_{ij}]'$ (note the transpose!), and γ_{ij} is the cofactor of a_{ij} as defined in eq. (A.7.9).

A.7.7 Eigenvalues of a Matrix

The "characteristic polynomial" of a square matrix $\underline{A}$ is defined as

$$P_A(\lambda) = \det(\lambda\underline{I}-\underline{A}) = a_o+a_1\lambda+\ldots+a_{n-1}\lambda^{n-1}+\lambda^n \qquad (A.7.28)$$

The complex, scalar variable λ is identified with s for continuous-time systems and with z for discrete-time systems.

The "eigenvalues" λ_i (i = 1,2...n) of $\underline{A}$ are the roots of the characteristic polynomial, i.e. they satisfy

$$P_A(\lambda) = (\lambda-\lambda_1)(\lambda-\lambda_2)\ldots(\lambda-\lambda_n) \qquad (A.7.29)$$

$P_A(\lambda) = 0$ is called the "characteristic equation".

The "Cayley-Hamilton theorem" states that every square matrix satisfies its characteristic equation:

$$
\begin{aligned}
P_A(\underline{A}) &= a_o\underline{I}+a_1\underline{A}+\ldots+a_{n-1}\underline{A}^{n-1}+\underline{A}^n \\
&= (\underline{A}-\lambda_1\underline{I})(\underline{A}-\lambda_2\underline{I})\ldots(\underline{A}-\lambda_n\underline{I}) \\
&= \underline{0}
\end{aligned}
\tag{A.7.30}
$$

(Note that $\lambda^o = 1$ is replaced by $\underline{A}^o = \underline{I}$).

There may be polynomials $Q_A(s)$ of lower degree than n such that $Q_A(s) = 0$. The polynomial of lowest degree with this property is called "minimal polynomial".

Characteristic polynomial and minimal polynomial are invariant under a nonsingular transformation $\underline{J} = \underline{TAT}^{-1}$. For J in Jordan (see section A.2) the minimal polynomial can be easily determined. If there is only one Jordan block associated with each eigenvalue, then the minimal polynomial is equal to the characteristic polynomial. A matrix with this property is called "cyclic". (See also the discussion after eq. (2.3.24).) If there are several Jordan blocks associated with an eigenvalue, then this eigenvalue appears in the minimal polynomial with the multiplicity equal to the dimension of the largest associated Jordan block.

The matrix $\underline{A}$ is "nilpotent" of index α if

$$\underline{A}^\alpha = \underline{0} \tag{A.7.31}$$

By the Cayley-Hamilton theorem, eq. (A.7.30), $\underline{A}^n = \underline{0}$ if all eigenvalues of $\underline{A}$ are zero. For noncyclic matrices with all eigenvalues zero, eq. (A.7.31) can be satisfied with $\alpha < n$ where α is the degree of the minimal polynomial.

The "singular values" σ_i (i = 1,2 ... n) of a square matrix are the nonnegative square roots of the eigenvalues of $\underline{A}'\underline{A}$. Let σ_{min} and σ_{max} be the smallest and largest singular values of a matrix $\underline{A}$. Then all eigenvalues λ_i of $\underline{A}$ satisfy

$$\sigma_{min} \leq |\lambda_i| \leq \sigma_{max} \tag{A.7.32}$$

Eigenvectors and generalized eigenvectors are discussed in the context of the Jordan form in section A.2.

The characteristic polynomial of the Jordan form shows that

$$\det \underline{A} = \lambda_1 \lambda_2 \ldots \lambda_n = (-1)^n a_o \tag{A.7.33}$$

$$\text{trace } \underline{A} = \lambda_1 + \lambda_2 + \ldots + \lambda_n = -a_{n-1} \tag{A.7.34}$$

A.7.8 Resolvent of a Matrix

The resolvent of a square matrix $\underline{A}$ is

$$(\lambda \underline{I} - \underline{A})^{-1} = \frac{\text{Adj}(\lambda \underline{I} - \underline{A})}{\det(\lambda \underline{I} - \underline{A})} = \frac{\underline{D}(\lambda)}{P_A(\lambda)} \tag{A.7.35}$$

The coefficients of the characteristic polynomial $P_A(\lambda) = a_o + a_1\lambda + \ldots + a_{n-1}\lambda^{n-1} + \lambda^n$ and the polynomial matrix $\underline{D}(\lambda) = \underline{D}_o + \underline{D}_1\lambda + \ldots + \underline{D}_{n-1}\lambda^{n-1}$ can be computed by the "Leverrier algorithm" (also known as algorithm of Souriau or Faddeeva or Frame).

$$\begin{array}{lll} & & \underline{D}_{n-1} = \underline{I} \\ a_{n-1} = -\text{ trace } \underline{AD}_{n-1} & & \underline{D}_{n-2} = \underline{AD}_{n-1} + a_{n-1}\underline{I} \\ a_{n-2} = -\frac{1}{2}\text{ trace } \underline{AD}_{n-2} & & \underline{D}_{n-3} = \underline{AD}_{n-2} + a_{n-2}\underline{I} \\ \vdots & & \vdots \\ a_1 = -\frac{1}{n-1}\text{ trace } \underline{AD}_1 & & \underline{D}_o = \underline{AD}_1 + a_1\underline{I} \\ a_o = -\frac{1}{n}\text{ trace } \underline{AD}_o & & \text{"}\underline{D}_{-1}\text{"} = \underline{AD}_o + a_o\underline{I} = \underline{0} \text{ (check)} \end{array} \tag{A.7.36}$$

The last matrix (formally written "D_{-1}") is the zero matrix. This provides a check on the calculations. The equations for the matrices $\underline{D}_i$ may also be written as follows:

$$
\begin{aligned}
\underline{D}_{n-1} &= \underline{I} \\
\underline{D}_{n-2} &= \underline{A}+a_{n-1}\underline{I} \\
\underline{D}_{n-3} &= \underline{A}^2+a_{n-1}\underline{A}+a_{n-2}\underline{I} \\
&\vdots \\
\underline{D}_{o} &= \underline{A}^{n-1}+a_{n-1}\underline{A}^{n-2}+\ldots+a_1\underline{I} \\
"\underline{D}_{-1}" &= \underline{A}^{n}+a_{n-1}\underline{A}^{n-1}+\ldots+a_o\underline{I} = \underline{0}
\end{aligned}
\tag{A.7.37}
$$

The last row is identical with the Cayley-Hamilton theorem (see eq. A.7.30)).

A.7.9 Orbit and Controllability of (A, b)

The "orbit" of an n-vector $\underline{b}$ with respect to the n×n matrix $\underline{A}$ is the sequence of vectors

$$\underline{O}(\underline{A}, \underline{b}) = [\underline{b}, \underline{Ab}, \underline{A}^2\underline{b}\ldots] \tag{A.7.38}$$

If $\underline{A}^i\underline{b}$ is linearly dependent on its predecessors $\underline{b}, \underline{Ab}\ldots\underline{A}^{i-1}\underline{b}$ then also all further vectors $\underline{A}^j\underline{b}$, $j > i$ are linearly dependent on the predecessors. Thus the rank of $\underline{O}(\underline{A}, \underline{b})$ is equal to the number of linearly independent vectors in the sequence $\underline{b}, \underline{Ab}, \underline{A}^2\underline{b}\ldots$ before the first linearly dependent one. (See also eq. (4.1.6).)

The pair $(\underline{A}, \underline{b})$ is "controllable" if $\underline{O}(\underline{A}, \underline{b})$ has full rank. Then

$$\text{rank}\,[\underline{b}, \underline{Ab} \ldots \underline{A}^{n-1}\underline{b}] = n \tag{A.7.39}$$

or

equivalently

$$\det[\underline{b}, \underline{Ab} \ldots \underline{A}^{n-1}\underline{b}] \neq 0 \tag{A.7.40}$$

A.7.10 Eigenvalue Assignment

For a given controllable pair ($\underline{A}$, $\underline{b}$) consider the eigenvalues of $\underline{A}-\underline{b}\underline{k}'$ where $\underline{k}$ is an arbitrary n-vector. The characteristic polynomial of $\underline{A}-\underline{b}\underline{k}'$ is

$$\begin{aligned} P(\lambda) &= \det(\lambda\underline{I}-\underline{A}+\underline{b}\underline{k}') \\ &= p_o+p_1\lambda+\ldots+p_{n-1}\lambda^{n-1}+\lambda^n \qquad (A.7.41) \\ &= [\underline{p}' \quad 1]\begin{bmatrix} 1 \\ \lambda \\ \vdots \\ \lambda \end{bmatrix} \end{aligned}$$

The relation $\underline{p} = \underline{p}(\underline{k})$ is given by

$$\underline{p}' = \underline{a}'+\underline{k}'\underline{W} \qquad (A.7.42)$$

where $\underline{a}'$ is the coefficient vector of $P_A(\lambda) = \det(\lambda\underline{I}-\underline{A})$ and

$$\underline{W} = [\underline{D}_o\underline{b},\ \underline{D}_1\underline{b}\ldots\underline{D}_{n-1}\underline{b}] \qquad (A.7.43)$$

Substituting eq. (A.7.37) yields

$$\underline{W} = [\underline{b},\ \underline{A}\underline{b}\ldots\underline{A}^{n-1}\underline{b}]\begin{bmatrix} a_1 & a_2 & & a_{n-1} & 1 \\ a_2 & & & \cdot^{\cdot^{\cdot}} & 0 \\ & & \cdot^{\cdot^{\cdot}} & & \\ & \cdot^{\cdot^{\cdot}} & & & \\ a_{n-1} & 1 & & & \\ 1 & 0 & & & 0 \end{bmatrix} \qquad (A.7.44)$$

Thus $\underline{W}^{-1}$ exists for controllable pairs ($\underline{A}$, $\underline{b}$) and the affine mapping of eq. (A.7.42) is unvertible

$$\underline{k}' = (\underline{p}' - \underline{a}')\underline{W}^{-1} \qquad (A.7.45)$$

$\underline{k} = \underline{k}(\underline{p})$ may also be calculated by "Ackermann's formula" [72.1], [72.5]

$$\underline{k}' = \underline{e}'P(\underline{A}) \tag{A.7.46}$$

$$\underline{e}' = [0 \ \dots \ 0 \quad 1]\ [\underline{b},\ \underline{A}\underline{b} \dots \underline{A}^{n-1}\underline{b}]^{-1} \tag{A.7.47}$$

Proof:

Expand $\underline{H}^k = (\underline{A}-\underline{b}\underline{k}')^k$ into expressions of the form $\underline{A}^k$ and $\underline{A}^i\underline{g}\underline{k}'\underline{H}^j$ and determine $P(\underline{H}) = p_o\underline{H}^o + p_1\underline{H}^1 + \dots + p_{n-1}\underline{H}^{n-1} + \underline{H}^n$

$$\begin{array}{ll|l} \underline{H}^o & = \underline{A}^o & \times\ p_o \\ \underline{H}^1 & = \underline{A}^1 - \underline{b}\underline{k}' & \times\ p_1 \\ \underline{H}^2 & = \underline{A}^2 - \underline{A}\underline{b}\underline{k}' - \underline{b}\underline{k}'\underline{H} & \times\ p_2 \\ & \vdots & \\ \underline{H}^n & = \underline{A}^n - \underline{A}^{n-1}\underline{b}\underline{k}' - \underline{A}^{n-2}\underline{b}\underline{k}'\underline{H} - \dots - \underline{b}\underline{k}'\underline{H}^{n-1} & \times\ 1 \\ \hline \end{array}$$

$$P(\underline{H}) = P(\underline{A}) - [\underline{b},\ \underline{A}\underline{b} \dots \underline{A}^{n-1}\underline{b}] \begin{bmatrix} \vdots \\ \underline{k}' \end{bmatrix} \tag{A.7.48}$$

If $P(\lambda)$ is the characteristic polynomial to be assigned to $\underline{A}-\underline{b}\underline{k}'$, then $P(\underline{H}) = \underline{0}$ by the Cayley-Hamilton theorem, eq. (A.7.30). Then the last row of eq. (A.7.48) yields

$$\underline{k}' = [0 \dots 0 \quad 1]\ [\underline{b},\ \underline{A}\underline{b} \dots \underline{A}^{n-1}\underline{b}]^{-1} P(\underline{A}) \tag{A.7.49}$$

(See also the discussion after eq. (2.6.24).)

Let $\lambda_1, \lambda_2 \dots \lambda_n$ be the eigenvalues to be assigned to $\underline{A}-\underline{b}\underline{k}'$, then eq. (A.7.46) becomes

$$\underline{k}' = \underline{e}'(\underline{A}-\lambda_1\underline{I})(\underline{A}-\lambda_2\underline{I}) \dots (\underline{A}-\lambda_n\underline{I}) \tag{A.7.50}$$

Equivalently the polynomial coefficient vector $\underline{p}'$ of eq. (A.7.41) may be specified and eq. (A.7.46) yields

$$\underline{k}' = [\underline{p}' \quad 1]\underline{E} \tag{A.7.51}$$

where

$$\underline{E} = \begin{bmatrix} \underline{e}' \\ \underline{e}'\underline{A} \\ \vdots \\ \underline{e}'\underline{A}^n \end{bmatrix} \tag{A.7.52}$$

in the "pole assignment matrix".

A.7.11 Functions of a Matrix

Consider a function f(a) which has a power series expansion

$$f(a) = \sum_{i=0}^{\infty} \alpha_i a^i \tag{A.7.53}$$

The function of a square matrix $\underline{A}$ is defined by

$$f(\underline{A}) := \sum_{i=0}^{\infty} \alpha_i \underline{A}^i \tag{A.7.54}$$

For an n×n matrix $\underline{A}$ there exist coefficients c_i such that

$$f(\underline{A}) = c_0\underline{I}+c_1\underline{A}+\ldots+c_{n-1}\underline{A}^{n-1} \tag{A.7.55}$$

($\underline{A}^n$ may be expressed in these terms by the Cayley-Hamilton theorem, similarly $\underline{A}^{n+1}$ etc. Thus this finite series is equal to the infinite sequence of eq. (A.7.54).)

The coefficients c_i, i = 0, 1...n are determined from the condition that eq. (A.7.55) is satisfied not only by the matrix $\underline{A}$ but also by its eigenvalues [59.4], [63.1]. For each distinct eigenvalue λ_j an equation of the following form is obtained

$$f(\lambda_j) = c_0+c_1\lambda_j+\ldots+c_{n-1}\lambda_j^{n-1} \tag{A.7.56}$$

If an eigenvalue λ_j of multiplicity p arises, eq. (A.7.56) is differentiated with respect to λ_j:

$$\frac{df}{d\lambda_j}(\lambda_j) = c_1 + 2c_2\lambda_j + 3c_3\lambda_j^2 + \ldots + (n-1)c_{n-1}\lambda_j^{n-2}$$

$$\frac{d^2f}{d\lambda_j^2}(\lambda_j) = 2c_2 + 3\times 2c_3\lambda_j + \ldots + (n-1)(n-2)c_{n-1}\lambda_j^{n-3} \qquad (A.7.57)$$

etc.

Appendix B The z-Transform

B.1 Notation and Assumptions

The z-transform is defined by

$$f_z(z) = \mathfrak{Z}\{f_k\} := \sum_{k=0}^{\infty} f_k z^{-k} \tag{B.1.1}$$

(It is also called "one-sided z-transform" if a distinction from the "two-sided z-transform" with k starting at $-\infty$ is necessary.) We will now state and verify from the definition the most important theorems on z-transforms. More comprehensive treatments can be found in the literature, for example [64.1]. For purposes of comparison, the rules of the Laplace transform will be given in addition to the rules of the z-transform.

By eq. (B.1.1) the z-tranform is a power series in z^{-1}, also known in the literature as a "discrete Laplace transform", "generating function", or "Dirichlet transform". The elements of the transformed sequence f_k can be complex. It will be assumed throughout that $f_k = 0$ for $k < 0$. In the application of the z-transform to sampled-data control systems the sequences are generated by equidistant sampling of continuous time functions f(t) with a sampling period T. The sequences may be $f_k = f(kT)$ or $f_k = f(kT+\gamma T)$, $0 \le \gamma < 1$. The sampling period T is fixed, therefore the notation

$$f(kT) = f[k], \quad f(kT+\gamma T) = f[k+\gamma] \tag{B.1.2}$$

will be used. Thus

$$f_z(z) = \sum_{k=0}^{\infty} f_k z^{-k} = \sum_{k=0}^{\infty} f[k] z^{-k} \tag{B.1.3}$$

$$f_{z\gamma}(z) = \sum_{k=0}^{\infty} f_k(\gamma) z^{-k} = \sum_{k=0}^{\infty} f[k+\gamma] z^{-k} \tag{B.1.4}$$

Example 1:

$$f_k = e^{akT}$$

$$f_z(z) = \sum_{k=0}^{\infty} e^{akT} z^{-k} = \sum_{k=0}^{\infty} c^k \text{ with } c = e^{aT} z^{-1} \tag{B.1.5}$$

The sum of the geometric series is obtained from

$$x(N) = \sum_{k=0}^{N} c^k = 1 + c + c^2 + \ldots + c^N \qquad \text{(a)}$$

$$cx(N) = c + c^2 + \ldots + c^N + c^{N+1} \qquad \text{(b)}$$

$$(1-c)x(N) = 1 - c^{N+1} \qquad \text{(a) - (b)}$$

and for $c \neq 1$

$$x(N) = \frac{1-c^{N+1}}{1-c}$$

For $|c| < 1$

$$\sum_{k=0}^{\infty} c^k = \lim_{N\to\infty} x(N) = \frac{1}{1-c}$$

Therefore

$$f_z(z) = \mathfrak{Z}\{e^{akT}\} = \frac{z}{z-e^{aT}} \text{ for } |z| > |e^{aT}| \tag{B.1.6}$$

If a is real, then a pole appears on the positive real axis of the z-plane. For $a > 0$ the sequence f_k increases and the pole is at $z > 1$, for $a = 0$ the sequence is constant

$$f_z(z) = \mathfrak{Z}\{1(kT)\} = \frac{z}{z-1} \tag{B.1.7}$$

i.e. the pole is at $z = 1$, and for decreasing sequences with $a < 0$, the pole lies in the interval $0 < z < 1$. Examples are given in figures 3.2 a, b, c.

Example 2:

$$f_k = e^{akT}\cos(\omega kT+\alpha) = \left(\frac{1}{2}\ e^{akT+j(\omega kT+\alpha)} + e^{akT-j(\omega kT+\alpha)}\right)$$

From eq. (B.1.6)

$$f_z(z) = \frac{1}{2}\left(\frac{ze^{j\alpha}}{z-e^{aT+j\omega T}} + \frac{ze^{-j\alpha}}{z-e^{aT-j\omega T}}\right)$$

$$f_z(z) = \frac{z^2\cos\alpha - ze^{aT}\cos(\omega T-\alpha)}{(z-e^{aT+j\omega T})(z-e^{aT-j\omega T})} \tag{B.1.8}$$

Two conjugate complex poles arise with absolute value $|z| = e^{aT}$ and lying at an angle $\pm\omega T$ with respect to the positive real axis. For $a > 0$ the sequence f_k represents an oscillation of increasing magnitude and the pole lies outside the unit circle of the z-plane. An oscillation of constant amplitude occurs for $a = 0$ and a pair of poles on the unit circle are obtained. A pair of poles inside the unit circle are associated with a decreasing oscillation with $a < 0$. Figures 3.2 h and i illustrate two cases.

For $\omega T = \pi$ a pole and a zero are cancelled and there remains

$$f_z(z) = \mathfrak{Z}\{e^{akT}\cos(k\pi+\alpha)\} = \cos\alpha\,\mathfrak{Z}\{(-e^{aT})^k\} = \frac{z\cos\alpha}{z+e^{aT}} \tag{B.1.9}$$

as represented in figure 3.2 d.

For $|\omega T| > \pi$ a sequence arises, which is identical with a sequence for $|\omega T| < \pi$, figure 3.2 j. Identical series have identical transforms, even if they originate from the sampling of different continuous-time functions.

Example 3:

Only the first terms 0 to N of the sequence arise, i.e., $f_k = 0$ for $k > N$, the z-transform is

$$f_z(z) = f_o+f_1 z^{-1}+\ldots+f_N z^{-N} = \frac{f_o z^N+f_1 z^{N-1}+\ldots+f_N}{z^N} \qquad \text{(B.1.10)}$$

It has a pole of multiplicity N at z = 0, figure 3.2 k.

B.2 Linearity

Theorem:

$$\mathcal{Z}\{af_k+bg_k\} = a\mathcal{Z}\{f_k\} + b\mathcal{Z}\{g_k\}, \quad a,\ b \text{ constant} \qquad \text{(B.2.1)}$$

Proof:

Follows immediately by substituting into the defining eq. (B.1.1)

Laplace-transform:

$$\mathcal{L}\{af(t) + bg(t)\} = a\mathcal{L}\{f(t)\} + b\mathcal{L}\{g(t)\} \qquad \text{(B.2.2)}$$

Example:

From eq. (B.2.1) it follows, that the two block diagrams B.1a and b are equivalent.

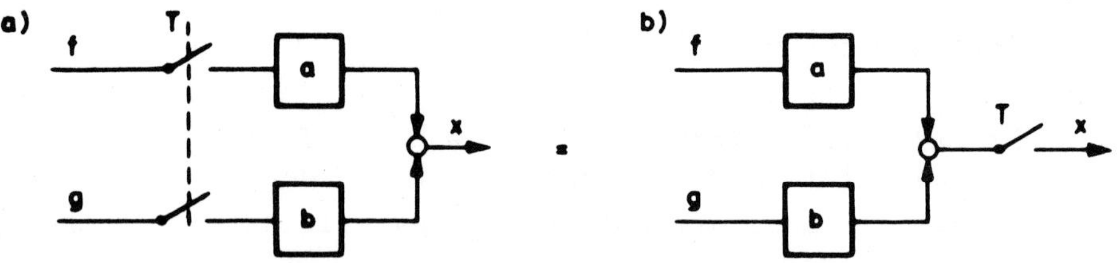

Figure B.1 Linearity of sampling

B.3 Right Shifting Theorem

If a sequence is shifted by n sampling intervals to the right on the time axis, this corresponds to a multiplication of its z-transform by z^{-n}

$$\mathcal{Z}\{f_{k-n}\} = z^{-n}\mathcal{Z}\{f_k\}, \quad n \geq 0 \tag{B.3.1}$$

Proof:

$$\mathcal{Z}\{f_{k-n}\} = \sum_{k=0}^{\infty} f_{k-n}z^{-k} = z^{-n} \sum_{k=0}^{\infty} f_{k-n}z^{-(k-n)}$$

substitute m = k-n

$$\mathcal{Z}\{f_{k-n}\} = z^{-n} \sum_{m=-n}^{\infty} f_m z^{-m} \tag{B.3.2}$$

Since $f_m = 0$ for $m < 0$,

$$\mathcal{Z}\{f_{k-n}\} = z^{-n} \sum_{m=0}^{\infty} f_m z^{-m} = z^{-n}\mathcal{Z}\{f_k\} \tag{B.3.3}$$

Laplace-transform:

$$\mathcal{L}\{f(t-a)\} = f_s(s)e^{-as}, \quad a \geq 0 \tag{B.3.4}$$

Examples: eqs. (3.3.28) and (3.4.28).

B.4 Left Shifting Theorem

If a sequence is shifted by n sampling intervals to the left on the time axis, this corresponds to a multiplication by z^n and truncation of those terms, which are shifted to $k < 0$.

$$\mathcal{Z}\{f_{k+n}\} = z^n \left[\mathcal{Z}\{f_k\} - \sum_{m=0}^{n-1} f_m z^{-m}\right], \quad n \geq 0 \tag{B.4.1}$$

Proof:

$$\mathcal{Z}\{f_{k+n}\} = z^n \sum_{k=0}^{\infty} f_{k+n} z^{-(k+n)} \quad , \; m := k+n$$

$$= z^n \sum_{m=n}^{\infty} f_m z^{-m}$$

$$= z^n \left[\sum_{m=0}^{\infty} f_m z^{-m} - \sum_{m=0}^{n-1} f_m z^{-m} \right]$$

$$= z^n \left[\mathcal{Z}\{f_k\} - \sum_{m=0}^{n-1} f_m z^{-m} \right] \tag{B.4.2}$$

Laplace-transform:

$$\mathcal{L}\{f(t+a)\} = e^{as} \left[\mathcal{L}\{f(t)\} - \int_0^a f(t) e^{-st} dt \right] , \; a \geq 0 \tag{B.4.3}$$

Example: eq. (3.3.3).

B.5 Damping Theorem

$$\mathcal{Z}\{f[k+\gamma] e^{-a(k+\gamma)T}\} = f_{z\gamma}(ze^{aT}) e^{-a\gamma T}, \text{ a constant} \tag{B.5.1}$$

in particular for $\gamma = 0$

$$\mathcal{Z}\{f[k] e^{-akT}\} = f_z(ze^{aT}) \tag{B.5.2}$$

Proof:

$$\mathcal{Z}\{f[k+\gamma] e^{-a(k+\gamma)T}\} = \sum_{k=0}^{\infty} f[k+\gamma] (e^{aT} z)^{-k} e^{-a\gamma T}$$

substituting $z_1 := e^{aT} z$

$$\mathcal{Z}\{f[k+\gamma] e^{-a(k+\gamma)T}\} = e^{-a\gamma T} \sum_{k=0}^{\infty} f[k+\gamma] z_1^{-k} = e^{-a\gamma T} f_{z\gamma}(z_1)$$

$$= e^{-a\gamma T} f_{z\gamma}(ze^{aT}) \tag{B.5.3}$$

Laplace-transform:

$$\mathcal{L}\{f(t)e^{-at}\} = f_s(s+a) \quad , \quad a \text{ constant} \tag{B.5.4}$$

Example: eq. (9.3.40).

B.6 Differentation Theorem

Differentiation of a sequence with respect to a parameter is equivalent to differentiation of its z-transform

$$\mathfrak{z}\{\frac{\partial}{\partial a} f_k(a)\} = \frac{\partial}{\partial a} \mathfrak{z}\{f_k(a)\} \tag{B.6.1}$$

Proof:

A power series can be differentiated term by term in its domain of convergence.

Laplace-transform:

$$\mathcal{L}\{\frac{\partial}{\partial a} f(t,a)\} = \frac{\partial}{\partial a} \mathcal{L}\{f(t,a)\} \tag{B.6.2}$$

Example:

$$\mathfrak{z}\{kTe^{akT}\} = \mathfrak{z}\{\frac{\partial}{\partial a} e^{akT}\} = \frac{\partial}{\partial a} \mathfrak{z}\{e^{akT}\}$$

$$= \frac{\partial}{\partial a}\left(\frac{z}{z-e^{aT}}\right) = \frac{Tze^{aT}}{(z-e^{aT})^2}$$

and in the notation of eq. (3.3.29)

$$\mathcal{Z}\left\{\frac{1}{(s+a)^2}\right\} = \frac{Tze^{aT}}{(z-e^{aT})^2} \tag{B.6.3}$$

This also applies to multiple differentiation

$$\mathfrak{z}\{(kT)^r e^{akT}\} = \frac{\partial^r}{\partial a^r}\left(\frac{z}{z-e^{aT}}\right)$$

$$\mathcal{Z}\left\{\frac{n!}{(s+a)^{n+1}}\right\} = \frac{\partial^r}{\partial a^r}\left(\frac{z}{z-e^{aT}}\right) \tag{B.6.4}$$

B.7 Initial Value Theorem

$$f_o = \lim_{z\to\infty} f_z(z)$$

Proof:

$$\lim_{z\to\infty} f_z(z) = \lim_{z\to\infty}[f_o+f_1 z^{-1}+f_2 z^{-2}+\ldots] = f_o \qquad \text{(B.7.1)}$$

Laplace-transform:

If $\lim_{t\to 0} f(t)$ exists, then

$$\lim_{t\to 0} f(t) = \lim_{s\to\infty} sf_s(s) \qquad \text{(B.7.2)}$$

If $f_k = 0$ for $0 \le k < n$, then the sequence can be shifted to the left by n sampling intervals by multiplication with z^n. Then

$$f_n = \lim_{z\to\infty} z^n f_z(z) \qquad \text{(B.7.3)}$$

Example:

Consider the z-transform

$$f_z(z) = \frac{b_n z^n + b_{n-1} z^{n-1} + \ldots + b_o}{z^n + a_{n-1} z^{n-1} + \ldots + a_o} \qquad \text{(B.7.4)}$$

The initial value is $f_o = b_n$. If it is zero, then $f_1 = b_{n-1}$. If also this term is zero, then $f_2 = b_{n-2}$ etc. This means, that the difference defined by denominator degree minus numerator degree ("pole excess") is equal to the number of initial zero elements of the sequence.

B.8 Final Value Theorem

If $\lim_{k\to\infty} f_k$ exists, then

$$\lim_{k\to\infty} f_k = \lim_{z\to 1}(z-1) f_z(z) \qquad \text{(B.8.1)}$$

Proof:

The first difference of a sequence f_k is defined as

$$\Delta f_n := f_n - f_{n-1} \qquad (B.8.2)$$

the inversion of (B.8.2) is the sum

$$f_k = \sum_{n=0}^{k} \Delta f_n \qquad (B.8.3)$$

The relation of the z-transforms is

$$\mathfrak{Z}\{\Delta f_n\} = \sum_{n=0}^{\infty} \Delta f_n z^{-n} = (1-z^{-1}) f_z(z) = \frac{z-1}{z} f_z(z) \qquad (B.8.4)$$

If f_k approaches a finite value for $k \to \infty$, then from eq. (B.8.3)

$$\lim_{k\to\infty} f_k = \sum_{n=0}^{\infty} \Delta f_n = \lim_{z\to 1} \sum_{n=0}^{\infty} \Delta f_n z^{-n} = \lim_{z\to 1} (z-1) f_z(z) \qquad (B.8.5)$$

Laplace-transform:

If $\lim_{t\to\infty} f(t)$ exists, then

$$\lim_{t\to\infty} f(t) = \lim_{s\to 0} s f_s(s) \qquad (B.8.6)$$

Remark B.1:

It is important that theorem (B.8.1) be only applied, if a final value exists. This limit does not exist, for example, for $f(kT) = \sin \omega_0 kT$. The formal application of eq. (B.8.1) here leads to the false result $\lim_{k\to\infty} f(kT) = 0$.

For $f_z(z)$ rational, $\lim_{k\to\infty} f_k$ exists, if all poles z_i of $f_z(z)$ satisfy the condition $|z_i| < 1$, with the exception of a single pole at $z = 1$.

Example: eq. (3.4.24).

B.9 The Inverse z-Transform

The coefficients of a Laurent series

$$f_z(z) = \sum_{k=0}^{\infty} f_k z^{-k} \qquad \text{(B.9.1)}$$

can be calculated from a theorem of complex variable theory as follows:

$$f_k = \mathcal{Z}^{-1}\{f_z(z)\} = \frac{1}{2\pi j} \oint f_z(z) z^{k-1} dz \qquad \text{(B.9.2)}$$

The contour of the integral encloses all singularities of $f_z(z)z^{k-1}$ and runs counterclockwise. In the applications considered in this book, $f_z(z)$ is always rational. The integral can then be calculated with Cauchy's residue theorem:

$$f_k = \sum_i \operatorname{res}[f_z(z) z^{k-1}]_{z=z_i} \qquad \text{(B.9.3)}$$

Here the terms z_i are the poles of $f_z(z)z^{k-1}$. The residues are determined as follows:

1. For a single pole at $z = a$

$$\operatorname{res}[f_z(z) z^{k-1}]_{z=a} = \lim_{z \to a} [(z-a) f_z(z) z^{k-1}] \qquad \text{(B.9.4)}$$

2. For a pole of multiplicity p at $z = a$

$$\operatorname{res}[f_z(z) z^{k-1}]_{z=a} = \frac{1}{(p-1)!} \lim_{z \to a} \frac{d^{p-1}}{dz^{p-1}} \{(z-a)^p f_z(z) z^{k-1}\} \qquad \text{(B.9.5)}$$

The inversion of irrational functions $f_z(z)$ is treated in [64.1].

Laplace-transform:

$$f(t) = \mathcal{L}^{-1}\{f_s(s)\} = \frac{1}{2\pi j} \int_{\sigma - j\infty}^{\sigma + j\infty} f_s(s) e^{st} ds \qquad \text{(B.9.6)}$$

σ is the abscissa of convergence of f_s.

Example 1:

$$f_k = \mathcal{Z}^{-1}\left\{\frac{-0.4z^2+1.08z}{(z+0.5)(z-0.3)^2}\right\} \qquad (B.9.7)$$

$$= \text{res}[f_z(z)z^{k-1}]_{z=-0.5} + \text{res}[f_z(z)z^{k-1}]_{z=0.3}$$

$$\text{res}[f_z(z)z^{k-1}]_{z=-0.5} = \frac{-0.4\times0.25-1.08\times0.5}{(-0.5-0.3)^2}(-0.5)z^{k-1} = 2(-0.5)^k$$

$$\text{res}[f_z(z)z^{k-1}]_{z=0.3} = \lim_{z\to0.3}\frac{d}{dz}[\frac{-0.4z^2+1.08z}{z+0.5}z^{k-1}] = (4k-2)0.3^k$$

$$f_k = 2(-0.5)^k + (4k-2)\times0.3^k \qquad (B.9.8)$$

Example 2:

For the control loop of figure 3.12, $k_y = 0.326$ was determined as a suitable value from the root locus in figure 3.13. For a step input the z-transform of the error is by eq. (3.6.8)

$$e_z(z) = \frac{z^2-0.368z}{z^2-1.248z+0.454} \qquad (B.9.9)$$

$$e_z(z) = \frac{z(z-0.368)}{(z-0.624-0.255j)(z-0.624+0.255j)} \qquad (B.9.10)$$

$$\text{res}\left[e_z(z)z^{k-1}\right]_{z=0.624+0.255j} =$$

$$= \frac{(0.624+0.255j-0.368)(0.624+0.255j)^k}{2\times0.255j}$$

$$= -0.714e^{j2.353}(0.674e^{j0.393})^k \qquad (B.9.11)$$

For $z = 0.624 - 0.255j$, the complex conjugate values are obtained. The step response sequence is therefore

$$y(kT) = 1-e(kT) = 1+0.714e^{j2.353}(0.674e^{j0.393})^k +$$

$$+ 0.714e^{-j2.353}(0.674e^{-j0.393})^k$$

$$y(kT) = 1+1.42\times 0.674^k\cos(134.7^o+k\times 22.5^o) \qquad (B.9.12)$$

The response y(t) is shown in figure 3.15. The numerical inversion of eq. (3.6.8) is useful for practical calculations, but the general expression (B.9.12) for the inverse is not obtained.

The general expression for the inverse can also be determined by the table at the end of this section. If the given expression is not contained in the table, then $f_s(s)$ or $f_z(z)$ must be decomposed into expressions, which are entries of the table. For example, $f_z(z)/z$ is broken down into partial fractions. Then both sides are multiplied by z, such that expressions with a factor z in the numerator arise, as they appear in the table. Also it is recommended to leave complex conjugate terms together, so that only real coefficients need be calculated.

Example 1: Same as eq. (B.9.7)

$$f_z(z) = \frac{-0.4z^2+1.08z}{(z+0.5)(z-0.3)^2}$$

$$\frac{f_z(z)}{z} = \frac{-0.4z+1.08}{(z+0.5)(z-0.3)^2} = \frac{2}{z+0.5} - \frac{2}{z-0.3} + \frac{1.2}{(z-0.3)^2}$$

$$f_z(z) = \frac{2z}{z+0.5} - \frac{2z}{z-0.3} + \frac{4\times 0.3z}{(z-0.3)^2}$$

The first expression with a pole on the negative real axis can be found as a special case of $e^{-at}\cos\omega_o t$, the other two expressions appear under e^{-at} and $t\times e^{-at}$ in the table. In agreement with eq. (B.9.8) the sequence is given by

$$f_k = 2\times(-0.5)^k - 2\times 0.3^k + 4k\times 0.3^k$$

Example 2: Same as eq. (B.9.9)

$e_z(z)$ must be put into the form contained in the table.

$$e_z(z) = \frac{z^2-0.368z}{z^2-1.248z+0.454} =$$

$$= \frac{\alpha z e^{-aT}\sin\omega_o T+\beta(z^2-ze^{-aT}\cos\omega_o T)}{z^2-2ze^{-aT}\cos\omega_o T+e^{-2aT}}$$

The comparison yields:

$$e^{-aT} = \sqrt{0.454} = 0.674$$

$$\cos\omega_o T = \frac{1.248}{2\times 0.674} = 0.926$$

$$\sin\omega_o T = \sqrt{1-\cos^2\omega_o T} = 0.378$$

$$\omega_o T = 22.5^o$$

$$\beta = 1$$

$$\alpha = \frac{-0.368+0.674\times 0.926}{0.674\times 0.378} = 1.01$$

From the table

$$y(kT) = 1 - e(kT) = 1 - 1.01 \times 0.674^k \sin(k \times 22.5^o) - 0.674^k \cos(k \times 22.5^o)$$

$$y(kT) = 1 + 1.42 \times 0.674^k \cos(134.7^o + k \times 22.5^o)$$

This agrees with eq. (B.9.12).

B.10 Real Convolution Theorem

The convolution of sequences u and h in

$$y[n+\gamma] = \sum_{k=0}^{\infty} u[k]h[n-k+\gamma] \qquad \text{(B.10.1)}$$

corresponds to a multiplication of the z-transforms:

$$\mathcal{Z}\{y[k+\gamma]\} = \mathcal{Z}\{u[k]\}\,\mathcal{Z}\{h[k+\gamma]\} \qquad \text{(B.10.2)}$$

Proof:

$$\mathcal{Z}\{y[k+\gamma]\} = \sum_{n=0}^{\infty} y[n+\gamma]z^{-n}$$

$$= \sum_{n=0}^{\infty} \sum_{k=0}^{\infty} u[k]h[n-k+\gamma]z^{-n}$$

$$= \sum_{k=0}^{\infty} \sum_{n=0}^{\infty} u[k]h[n-k+\gamma]z^{-n}$$

$$= \sum_{k=0}^{\infty} \sum_{m=-k}^{\infty} u[k]h[m+\gamma]z^{-k}z^{-m}, \quad m = n-k$$

$$= \sum_{k=0}^{\infty} \sum_{m=0}^{\infty} u[k]h[m+\gamma]z^{-k}z^{-m} \quad \text{(causality)}$$

$$= \sum_{k=0}^{\infty} u[k]z^{-k} \times \sum_{m=0}^{\infty} h[m+\gamma]z^{-m}$$

$$= \mathcal{Z}\{u[k]\} \times \mathcal{Z}\{h[k+\gamma]\}$$

$$y_{z\gamma}(z) = u_z(z) \times h_{z\gamma}(z) \tag{B.10.3}$$

see also eq. (3.7.18).

Laplace-transform:

The convolution of signals

$$y(t) = \int_0^{\infty} u(\tau)g(t-\tau)d\tau \tag{B.10.4}$$

corresponds to a multiplication of the Laplace-transforms

$$\mathcal{L}\{y(t)\} = \mathcal{L}\{u(t)\} \times \mathcal{L}\{g(t)\}$$

$$y_s(s) = u_s(s) \times g_s(s) \tag{B.10.5}$$

B.11 Complex Convolution Theorem, Parseval Equation

$$\mathcal{Z}\{f_k\, g_k\} = \frac{1}{2\pi j} \oint_c f_z(w) g_z(z/w) w^{-1} dw \tag{B.11.1}$$

where the integration path c is a circle separating the poles $f_z(w)$ from those of $g_z(z/w)$. We limit ourselves to rational functions f_z and g_z.

Proof:

From eq. (B.9.2)

$$\mathcal{Z}\{f_k\, g_k\} = \sum_{k=0}^{\infty} f_k\, g_k z^{-k} = \sum_{k=0}^{\infty} \frac{1}{2\pi j} \oint f_z(w) w^{k-1} dw\, g_k z^{-k}$$

$$= \frac{1}{2\pi j} \oint f_z(w) w^{-1} \sum_{k=0}^{\infty} g_k (z/w)^{-k} dw$$

$$= \frac{1}{2\pi j} \oint f_z(w) g_z(z/w) w^{-1} dw \tag{B.11.2}$$

Laplace-transform:

$$\mathcal{L}\{f(t) g(t)\} = \frac{1}{2\pi j} \int_{\sigma - j\infty}^{\sigma + j\infty} f_s(p) g_s(s-p)\, dp \tag{B.11.3}$$

The abscissa of integration σ is chosen such that the path of integration separates the poles of $f_s(p)$ from those of $g_s(s-p)$.

Special case: Parseval equation for $f_k = g_k$

If the poles of $f_z(w)$ lie in the unit circle, by changing the variables of integration from w to z, one obtains for $z = 1$, $f_k = g_k$ from eq. (B.11.2)

$$\sum_{k=0}^{\infty} f_k^2 = \frac{1}{2\pi j} \oint f_z(z) f_z(z^{-1}) z^{-1} dz \tag{B.11.4}$$

The path of integration is the unit circle and runs in the counterclockwise direction. A numerical method for evaluation of eq. (B.11.4) is given in [70.8].

B.12 Other Representations of Sampled Signals in Time and Frequency Domain

In section 1.2 the sampled signal $f^*(t)$ was represented by

$$f^*(t) = \sum_{k=-\infty}^{\infty} f(kT)\,\delta(t-kT) \qquad \text{(B.12.1)}$$

With $f(kT) = 0$ for $k < 0$ the Laplace-transform yields

$$f_s^*(s) = \sum_{k=0}^{\infty} f(kT)e^{-kTs} = \sum_{k=0}^{\infty} f(kT)z^{-k} \qquad \text{(B.12.2)}$$

A second possible representation in the time domain was given in eq. (1.2.14)

$$f^*(t) = f(t) \times \sum_{k=-\infty}^{\infty} \delta(t-kT) \qquad \text{(B.12.3)}$$

From eq. (B.11.3), $f_s(s)$ can be calculated by complex convolution of the two Laplace-transforms $f_s(s)$ and

$$\mathcal{L}\left\{\sum_{k=0}^{\infty} \delta(t-kT)\right\} = \sum_{k=0}^{\infty} e^{-sT} = \frac{1}{1-e^{-sT}} \qquad \text{(B.12.4)}$$

Thus

$$f_s^*(s) = f_s(s) * \frac{1}{1-e^{-sT}}$$

$$= \frac{1}{2\pi j}\int_{\sigma-j\infty}^{\sigma+j\infty} f_s(p)\frac{1}{1-e^{-T(s-p)}}\,dp$$

For $e^{Ts} = z$ and replacement of p by s, it follows that

$$f_z(z) = \mathcal{Z}\{f_s(s)\} = \sum_i \text{res}\left[\frac{f_s(s)\,z}{z-e^{Ts}}\right]_{s=s_i}, \quad s_i = \text{poles of } f_s(s) \qquad \text{(B.12.5)}$$

This is a second method of calculating z-transforms.

Finally, yet a third possible representation was given in eq. (1.2.15), where the periodic pulse of δ-functions in eq. (B.12.3) was expressed by its Fourier-series [60.1]

$$\delta_T(t) = \sum_{k=-\infty}^{\infty} \delta(t-kT) = \frac{1}{T} \sum_{m=-\infty}^{\infty} e^{-jm\omega_A t}, \quad \omega_A = 2\pi/T \qquad \text{(B.12.6)}$$

Therefore

$$f^*(t) = \frac{1}{T} f(t) \times \sum_{m=-\infty}^{\infty} e^{-jm\omega_A t} \qquad \text{(B.12.7)}$$

and from the damping theorem of the Laplace-transform, eq. (B.5.4)

$$f_s^*(s) = \frac{1}{T} \sum_{m=-\infty}^{\infty} f_s(s+jm\omega_A) + \frac{f(+0)}{2} \qquad \text{(B.12.8)}$$

The term $f(+0)/2$ is added because the Laplace-transform of $f(t)$ in eq. (B.12.7) represents the value $f(+0)/2$ at the point $t = 0$. The sampling process, however, was defined such that the value $f(+0)$ arises at time $t = 0$ in $f^*(t)$ and $f_s^*(s)$, see remark 1.2.

B.13 Table of Laplace and z-Transforms

$f(t)$	$f_s(s)=\mathcal{L}\{f(t)\}$	$f_z(z)=\mathcal{Z}\{f(kT)\}$	$f_{z\gamma}(z)=\mathcal{Z}\{f(kT+\gamma T)\},\quad 0 \le \gamma < 1$
1	$\frac{1}{s}$	$\frac{z}{z-1}$	$\frac{z}{z-1}$
t	$\frac{1}{s^2}$	$\frac{Tz}{(z-1)^2}$	$\frac{Tz[\gamma z+(1-\gamma)]}{(z-1)^2}$
t^2	$\frac{2}{s^3}$	$\frac{T^2 z(z+1)}{(z-1)^3}$	$\frac{T^2 z[\gamma^2 z^2+(1+2\gamma-2\gamma^2)z+(1-\gamma)^2]}{(z-1)^3}$
t^3	$\frac{6}{s^4}$	$\frac{T^3 z(z^2+4z+1)}{(z-1)^4}$	$\frac{T^3 z[\gamma^3 z^3+(1+3\gamma+3\gamma^2-3\gamma^3)z^2+(4-6\gamma^2+3\gamma^3)z+(1-\gamma)^3]}{(z-1)^4}$
t^n	$\frac{n!}{s^{n+1}}$	$\lim_{a\to o} \frac{\partial^n}{\partial a^n}\left\{\frac{z}{z-e^{aT}}\right\}$	$\lim_{a\to o} \frac{\partial^n}{\partial a^n}\left\{\frac{z\cdot e^{a\gamma T}}{z-e^{aT}}\right\}$
e^{-at}	$\frac{1}{s+a}$	$\frac{z}{z-e^{-aT}}$	$\frac{z\cdot e^{-a\gamma T}}{z-e^{-aT}}$
$t\cdot e^{-at}$	$\frac{1}{(s+a)^2}$	$\frac{Tze^{-aT}}{(z-e^{-aT})^2}$	$\frac{Tze^{-a\gamma T}[\gamma z+(1-\gamma)e^{-aT}]}{(z-e^{-aT})^2}$

$f(t)$	$f_s(s)=\mathcal{L}\{f(t)\}$	$f_z(z)=\mathcal{Z}\{f(kT)\}$	$f_{z\gamma}(z)=\mathcal{Z}\{f(kT+\gamma T)\}$, $0 \le \gamma < 1$
$t^2 \cdot e^{-at}$	$\frac{2}{(s+a)^3}$	$\frac{T^2 z e^{-aT}(z+e^{-aT})}{(z-e^{-aT})^3}$	$\frac{T^2 z \cdot e^{-a\gamma T}}{(z-e^{-aT})^3} \cdot [\gamma^2 z^2+(1+2\gamma-2\gamma^2)e^{-aT}z+(1-\gamma)^2 \cdot e^{-2aT}]$
$t^n \cdot e^{at}$	$\frac{n!}{(s-a)^{n+1}}$	$\frac{\partial^n}{\partial a^n}\left\{\frac{z}{z-e^{aT}}\right\}$	$\frac{\partial^n}{\partial a^n}\left\{\frac{z e^{a\gamma T}}{z-e^{aT}}\right\}$
$1-e^{-at}$	$\frac{a}{s(s+a)}$	$\frac{(1-e^{-aT})z}{(z-1)(z-e^{-aT})}$	$\frac{(1-e^{-a\gamma T})z^2+(e^{-a\gamma T}-e^{-aT})z}{(z-1)(z-e^{-aT})}$
$at-1+e^{-at}$	$\frac{a^2}{s^2(s+a)}$	$\frac{(aT-1+e^{-aT})z^2+(1-aTe^{-aT}-e^{-aT})z}{(z-1)^2(z-e^{-aT})}$	$\frac{z}{(z-1)^2(z-e^{-aT})}\Big\{(a\gamma T-1+e^{-a\gamma T})z^2 +$ $+ [aT(1-\gamma-\gamma e^{-aT}) + 1-2e^{-a\gamma T} + e^{-aT}]z +$ $+ [e^{-a\gamma T}-aTe^{-aT}(1-\gamma)-e^{-aT}]\Big\}$
$e^{-at}-e^{-bt}$	$\frac{b-a}{(s+a)(s+b)}$	$\frac{z(e^{-aT}-e^{-bT})}{(z-e^{-aT})(z-e^{-bT})}$	$\frac{(e^{-a\gamma T}-e^{-b\gamma T})z^2+(e^{-T(a+b\gamma)}-e^{-T(b+a\gamma)})z}{(z-e^{-aT})(z-e^{-bT})}$
$(a-b)+b\cdot e^{-at}$ $-a\cdot e^{-bt}$	$\frac{ab(a-b)}{s(s+a)(s+b)}$	$\frac{z}{(z-1)(z-e^{-aT})(z-e^{-bT})}$ $\Big\{(a-b-a\cdot e^{-bT}+b\cdot e^{-aT})z +$ $+ [(a-b)\cdot e^{-(a+b)T}-ae^{-aT}+b\cdot e^{-bT}]\Big\}$	$\frac{(a-b)z}{z-1} + \frac{bze^{-a\gamma T}}{z-e^{-aT}} - \frac{aze^{-b\gamma T}}{z-e^{-bT}}$

$f(t)$	$f_z(z)=\mathcal{L}\{f(t)\}$	$f_s(s)=\mathcal{L}\{f(t)\}$	$f_{z\gamma}(z)=\mathcal{Z}\{f(kT+\gamma T)\}\ ,\ 0 \le \gamma < 1$
$ab(a-b)\cdot t + (b^2-a^2)-b^2e^{-at} + a^2\cdot e^{-bt}$	$\frac{a^2b^2(a-b)}{s^2(s+a)(s+b)}$	$\frac{ab(a-b)Tz}{(z-1)^2} + \frac{(b^2-a^2)z}{z-1} - \frac{b^2z}{z-e^{-aT}} + \frac{a^2z}{z-e^{-bT}}$	$\frac{ab(a-b)Tz}{(z-1)^2} + \frac{[ab(a-b)\gamma T + b^2-a^2]z}{z-1} - \frac{b^2ze^{-a\gamma T}}{z-e^{-aT}} + \frac{a^2ze^{-b\gamma T}}{z-e^{-bT}}$
$\sin\omega_o t$	$\frac{\omega_o}{s^2+\omega_o^2}$	$\frac{z\cdot\sin\omega_o T}{z^2-2z\cos\omega_o T+1}$	$\frac{z^2\sin\cdot\gamma\omega_o T + z\cdot\sin(1-\gamma)\omega_o T}{z^2-2z\cos\omega_o T+1}$
$\cos\omega_o t$	$\frac{s}{s^2+\omega_o^2}$	$\frac{z(z-\cos\omega_o T)}{z^2-2z\cos\omega_o T+1}$ special case $\omega_o T=\pi$: $\mathcal{Z}\{(-1)^k\} = \frac{z}{z+1}$	$\frac{z^2\cdot\cos\gamma\omega_o T - z\cdot\cos(1-\gamma)\omega_o T}{z^2-2z\cos\omega_o T+1}$
$e^{-at}\sin\omega_o t$	$\frac{\omega_o}{(s+a)^2+\omega_o^2}$	$\frac{z\cdot e^{-aT}\sin\omega_o T}{z^2-2z\cdot e^{-aT}\cos\omega_o T+e^{-2aT}}$	$\frac{[z\cdot\sin\gamma\omega_o T + e^{-aT}\sin(1-\gamma)\omega_o T]\ ze^{-a\gamma T}}{z^2 - 2ze^{-aT}\cos\omega_o T + e^{-2aT}}$
$e^{-at}\cdot\cos\omega_o t$	$\frac{s+a}{(s+a)^2+\omega_o^2}$	$\frac{z^2-z\cdot e^{-aT}\cos\omega_o T}{z^2-2z\cdot e^{-aT}\cos\omega_o T + e^{-2aT}}$ special case $\omega_o T=\pi$: $\mathcal{Z}\{(-e^{-aT})^k\} = \frac{z}{z+e^{-aT}}$	$\frac{[z\cdot\cos\gamma\omega_o T - e^{-aT}\cos(1-\gamma)\omega_o T]z\cdot e^{-a\gamma T}}{z^2-2ze^{-aT}\cos\omega_o T + e^{-2aT}}$

Appendix C Stability Criteria

A necessary and sufficient condition for the asymptotic stability of the discrete system $\underline{x}[k+1] = \underline{Px}[k]$ is, that all eigenvalues of $\underline{P}$ have an absolute value smaller than one, i.e. the roots z_i of the characteristic polynomial

$$\begin{aligned} P(z) &= \det(z\underline{I}-\underline{P}) \\ &= p_0 + p_1 z + \ldots + p_{n-1} z^{n-1} + z^n \\ &= (z-z_1)(z-z_2) \ldots (z-z_n) \end{aligned} \tag{C.1}$$

must be located inside the unit circle. Polynomials with this property are called Schur-Cohn polynomials [1917], [1922].

In this appendix several stability criteria are compiled from the literature without proofs.

C.1 Bilinear Transformation to a Hurwitz Problem

Hermite [1854] noted that his algorithm for testing the number of zeros in a half plane can also be applied to the problem of zeros in a circle if the polynomial is first transformed. In the control engineering literature this idea appears in Oldenbourg and Sartorius [1944] for the stability test of sampled-data control systems.

The bilinear transformation

$$w := \frac{z-1}{z+1}, \quad z = \frac{1+w}{1-w} \tag{C.1.1}$$

maps the unit disk of the z plane onto the left half w plane.

Substituting z into P(z), a new polynomial

$$
\begin{aligned}
Q(w) &= (1-w)^n \times P\left(\frac{1+w}{1-w}\right) \\
&= p_o(1-w)^n + p_1(1+w)(1-w)^{n-1} + \ldots + p_{n-1}(1+w)^{n-1}(1-w) + (1+w)^n \\
&= q_o + q_1 w + \ldots + q_{n-1}w^{n-1} + q_n w^n
\end{aligned}
\tag{C.1.2}
$$

is obtained. P(z) is a Schur-Cohn polynomial if and only if Q(w) is a Hurwitz polynomial. This can be tested by the criteria of Routh [1877] and Hurwitz [1895], [59.4]. First $q_n > 0$ is assumed, if necessary Q(w) is multiplied by minus one to achieve this. Then the Hurwitz determinants

$$
\Delta_i = \begin{bmatrix}
q_{n-1} & q_{n-3} & q_{n-5} & \cdots & \\
q_n & q_{n-2} & q_{n-4} & & \\
0 & q_{n-1} & q_{n-3} & & \\
0 & q_n & q_{n-2} & q_{n-4} & \\
 & & & \ddots & \\
0 & & & & q_{n-i}
\end{bmatrix}, \quad q_k = 0 \text{ for } k < 0
\tag{C.1.3}
$$

are formed. All roots of Q(w) have negative real parts if and only if

$$
\Delta_1 > 0,\ \Delta_2 > 0 \ \ldots \ \Delta_n > 0
\tag{C.1.4}
$$

For first and second degree polynomials the necessary and sufficient conditions are

$Q_1(w) = (w+\alpha)$ is Hurwitz for $\alpha > 0$

$Q_2(v) = (v^2+\beta v+\gamma)$ is Hurwitz for $\beta > 0$, $\gamma > 0$.

All Hurwitz polynomials of higher degree can be formed by multiplying factors of type Q_1 and Q_2, i.e. they have only positive coefficients. A necessary stability condition for the polynomial (C.1.2) with $q_n = 1$ is consequently

$$
q_o > 0,\ q_1 > 0 \ \ldots \ q_{n-1} > 0
\tag{C.1.5}
$$

In fact there are no "better" linear inequalities in the polynomial coefficients p_i. It is shown in section 7.3, that the inequalities (C.1.5) describe the convex hull of the stability region of the discrete-time system and therefore cannot be tightened or augmented by additional linear inequalities.

If these simple linear conditions (C.1.5) are satisfied then the nonlinear inequalities (C.1.4) can be reduced. Lienard and Chipart [1914], [59.4] have given four alternative sets of necessary and sufficient conditions for Q(w) to be a Hurwitz polynomial:

1) $q_o > 0,\ q_2 > 0 \ldots \qquad ,\ \Delta_1 > 0,\ \Delta_3 > 0 \ldots$ (C.1.6)

2) $q_o > 0,\ q_2 > 0 \ldots \qquad ,\ \Delta_2 > 0,\ \Delta_4 > 0 \ldots$ (C.1.7)

3) $q_o > 0,\ q_1 > 0,\ q_3 > 0 \ldots,\ \Delta_1 > 0,\ \Delta_3 > 0 \ldots$ (C.1.8)

4) $q_o > 0,\ q_1 > 0,\ q_3 > 0 \ldots,\ \Delta_2 > 0,\ \Delta_4 > 0 \ldots$ (C.1.9)

Approximately only half as many determinant inequalities as for the Hurwitz conditions are needed. $\Delta_{n-1} > 0$ can always be chosen as the inequality of the highest degree by using the first or third form for n even and the second or fourth form for n odd. The "critical" Hurwitz determinant Δ_{n-1} is related with the zeros of $Q(w) = (w-w_1)(w-w_2)\ldots(w-w_n)$ by Orlando's formula [1911], [59.4]

$$\Delta_{n-1} = (-1)^{\frac{n(n-1)}{2}} \times q_n^{n-1} \times \prod_{i<k}^{1\ldots n} (w_i+w_k) \qquad \text{(C.1.10)}$$

If a complex conjugate root pair crosses the imaginary axis, then

$$\Delta_{n-1} = 0 \qquad \text{(C.1.11)}$$

If a real root crosses the imaginary axis, then

$$q_o = 0 \qquad \text{(C.1.12)}$$

Thus the stability region in the space of polynomial coefficients q_i is bounded by two surfaces satisfying (C.1.11) and (C.1.12). All other inequalities serve to classify the regions separated by these boundary surfaces.

C.2 Schur-Cohn Criterium and its Reduced Forms

Schur [1917], Cohn [1922], and Fujiwara [1926] have studied the conditions under which the zeros of a polynomial lie in a circle. Only the simplified versions for polynomials with real coefficients are discussed here.

In the formulation of Wilf [59.3] the Schur-Cohn criterion is: all zeros of the polynomial

$$P(z) = p_o + p_1 z + \dots + p_{n-1} z^{n-1} + p_n z^n, \; p_n > 0 \qquad (C.2.1)$$

lie in the unit circle if and only if the symmetric n×n matrix $\underline{R}$ with the elements

$$r_{ij} = \sum_{k=1}^{\min(i,j)} p_{n+k-i}\, p_{n+k-j} - p_{i-k}\, p_{j-k} \;, \quad i,j = 1,2 \dots n \qquad (C.2.2)$$

is positive definite.

Tsypkin [58.5], Jury and Blanchard [61.3] and Thoma [62.3] have given a stability criterion in the form of the following reduction procedure.

The zeros of the polynomial (C.2.1) lie in the unit circle if and only if

$$|p_o| < p_n$$

and the polynomial of degree n-1

$$P_1(z) = \frac{1}{z}\,[p_n P(z) - p_o P(z^{-1}) z^n] = b_o + b_1 z + \dots + b_{n-1} z^{n-1} \qquad (C.2.3)$$

also has roots only in the unit circle.

The coefficients of P_1 are

$$b_{n-1-k} = p_n p_{n-k} - p_o p_k \;, \quad k = 0, 1 \dots n-1 \qquad (C.2.4)$$

The polynomial $P_1(z)$ is correspondingly reduced to a polynomial of degree n-2, etc.

The reduction process is especially advantageous for the testing of numerically given polynomials and is suitable for programming on a digital computer.

In the design of sampled-data control systems, the coefficients of the characteristic equation depend upon free parameters, for example the feedback gains in the control loop. In this case the conditions for the coefficients $p_o, p_1 \dots p_n$ should be simplified as far as possible without inserting numerical values.

A simple form is obtained with the determinant formulation derived by Jury [62.2], [62.4], [64.1], [71.5] from the Schur-Cohn criterion. For this, one forms the (n-1)×(n-1) matrices

$$\underline{X} = \begin{bmatrix} p_n & p_{n-1} & \cdots & p_2 \\ 0 & p_n & \ddots & \vdots \\ \vdots & & \ddots & \vdots \\ \vdots & & p_n & p_{n-1} \\ 0 & \cdots & 0 & p_n \end{bmatrix}, \quad \underline{Y} = \begin{bmatrix} 0 & \cdots & 0 & p_o \\ \vdots & & p_o & p_1 \\ \vdots & & \iddots & \vdots \\ 0 & p_o & & \vdots \\ p_o & p_1 & \cdots & p_{n-2} \end{bmatrix} \quad (C.2.5)$$

The zeros of the polynomial (C.2.1) lie in the unit circle if and only if

$$\begin{aligned} P(1) &> 0 \\ (-1)^n P(-1) &> 0 \end{aligned} \quad (C.2.6)$$

and the determinant and the inner subdeterminants of the matrices

$\underline{X} + \underline{Y}$ and $\underline{X} - \underline{Y}$ are positive (C.2.7)

The first inner submatrix (= "inner") of a k×k matrix is obtained by omitting the first and last rows, as well as the first and last columns. The second inner results from repeating this procedure with the first inner, and so on. The last remaining inner consists of the central element for k odd, or of the central

2×2 matrix for n even. If the determinants of all inners of a matrix are positive, then the matrix is called "positive innerwise".

In analogy to Orlando's formula (C.1.10), Jury [63.4] has shown that

$$\det(\underline{X}-\underline{Y}) = p_n^{\,n-1} \prod_{\substack{k=1\\ i<k}}^{n} (1-z_i z_k) \tag{C.2.8}$$

Thus, if a complex conjugate pair of poles crosses the unit circle, then

$$\det(\underline{X}-\underline{Y}) = 0 \tag{C.2.9}$$

This condition is equivalent to $\Delta_{n-1} = 0$ of eq. (C.1.11). However, its general evaluation for non-numerical values is easier.

Example: $P(z) = p_o + p_1 z + p_2 z^2 + z^3$

On the complex root boundary

$$|\underline{X}-\underline{Y}| = \begin{vmatrix} 1 & p_2-p_o \\ -p_o & 1-p_1 \end{vmatrix} = 1 - p_1 + p_o p_2 - p_o^2 = 0 \tag{C.2.10}$$

with $|\underline{X}-\underline{Y}| > 0$ on the stable side.

For comparison the polynomial Q(w) of eq. (C.1.2) is

$$\begin{aligned} Q(w) &= p_o(1-w)^3+p_1(1+w)(1-w)^2+p_2(1+w)^2(1-w)+(1+w)^3 \\ &= q_o+q_1 w+q_2 w^2+q_3 w^3 \\ q_o &= P(1) = p_o+p_1+p_2+1 \\ q_1 &= -3p_o-p_1+p_2+3 \\ q_2 &= 3p_o-p_1-p_2+3 \\ q_3 &= -P(-1) = -p_o+p_1-p_2+1 \end{aligned}$$

It is more tedious to calculate

$$\Delta_2 = \begin{vmatrix} q_2 & q_0 \\ q_3 & q_1 \end{vmatrix} = 8(1-p_1+p_0p_2-p_0{}^2)$$

Thus $\Delta_2 = 8|\underline{X}-\underline{Y}|$

The simplest stability criteria combine the simplicity of the nonlinear inequalities, derived from $\underline{X} - \underline{Y}$ with the tight linear inequalities $q_i > 0$. Anderson and Jury [73.1], [81.8] have derived the following criterion:

A necessary and sufficient stability condition is that

1) $\underline{X} - \underline{Y}$ is positive innerwise (C.2.11)

and

2) a) $q_0 > 0$, $q_1 > 0$, $q_3 > 0 \ldots q_n > 0$ (C.2.12)

or

b) $q_0 > 0$, $q_2 > 0$, $q_4 > 0 \ldots q_n > 0$ (C.2.13)

Note that the conditions 2) always include the real root boundaries $q_0 = P(1) > 0$ and $q_n = (-1)^n P(-1) > 0$.

Examples:

<u>n = 2</u> $P(z) = p_0+p_1z+z^2$ is stable, if and only if

1) $X-Y = 1-p_0 > 0$

2) $q_0 = P(1) = p_0+p_1+1 > 0$ (C.2.14)

3) $q_2 = P(-1) = p_0-p_1+1 > 0$

<u>n = 3</u> $P(z) = p_0+p_1z+p_2z^2+z^3$ is stable, if and only if

1) $|\underline{X}-\underline{Y}| = 1-p_1+p_0p_2-p_0{}^2 > 0$

2) $q_0 = P(1) = p_0+p_1+p_2+1 > 0$ (C.2.15)

3) $q_3 = -P(-1) = -p_0+p_1-p_2+1 > 0$

4) One of the following conditions is satisfied

a) $q_1 = -3p_0-p_1+p_2+3 > 0$

or b) $q_2 = 3p_0-p_1-p_2+3 > 0$

Remark C.1:

Note that figure 7.2 may mislead one to assume that only the conditions 1), 2) and 3) should be checked. But the condition $|\underline{X}-\underline{Y}| = 0$ includes also cases, where $z_i = 1/z_k$ and $|\underline{X}-\underline{Y}| > 0$ may be satisfied in the region $P(1) > 0$, $-P(-1) > 0$, but outside the stability region of figure 7.2. As an example consider the unstable polynomial

$$P(z) = (z-0.9)(z-1.2)(z-1.1)$$
$$= -1.188 + 3.39z - 3.2z^2 + z^3$$

It is located in figure 7.2 in the neigborhood of vertex 0 outside all faces of the dotted box, because $p_0 < -1$, $p_1 > 3$, $p_2 < -3$. The conditions (C.2.15) yield

1) $|\underline{X}-\underline{Y}| = 0.000256 > 0$
2) $P(1) = 0.002 > 0$
3) $-P(-1) = 8.778 > 0$

Only the fourth condition indicates the instability:

4) a) $q_1 = -0.026 < 0$
or
b) $q_2 = -0.754 < 0$

n = 4 $\quad P(z) = p_0+p_1z+p_2z^2+p_3z^3+z^4$ is stable, if and only if

1) $$|\underline{X}-\underline{Y}| = \begin{vmatrix} 1 & p_3 & p_2-p_0 \\ 0 & 1-p_0 & p_3-p_1 \\ -p_0 & -p_1 & 1-p_2 \end{vmatrix} > 0$$

$$= (1-p_0)^2(1-p_2+p_0)+(p_3-p_1)(p_1-p_0p_3) > 0$$

2) $1 - p_0 > 0$

3) $q_0 = P(1) = p_0+p_1+p_2+p_3+1 > 0$ (C.2.16)

4) $q_4 = P(-1) = p_0-p_1+p_2-p_3+1 > 0$

5) a) $q_1 = -4p_0-2p_1+2p_3+4 > 0$ and

$q_3 = -4p_0+2p_1-2p_3+4 > 0$

or

b) $q_2 = 6p_0-2p_2+6 > 0$

conditions 2) and 5a) can be combined into
$|p_1-p_3| < 2(1-p_0)$

<u>n = 5</u> $P(z) = p_0+p_1z+p_2z^2+p_3z^3+p_4z^4+z^5$ is stable, if and only if

$$1) \quad |\underline{X}-\underline{Y}| = \begin{vmatrix} 1 & p_4 & p_3 & p_2-p_0 \\ 0 & 1 & p_4-p_0 & p_3-p_1 \\ 0 & -p_0 & 1-p_1 & p_4-p_2 \\ -p_0 & -p_1 & -p_2 & 1-p_3 \end{vmatrix} > 0$$

2) The determinant of the first inner is

$$\begin{vmatrix} 1 & p_4-p_0 \\ -p_0 & 1-p_1 \end{vmatrix} = 1-p_1+p_0p_4-p_0^2 > 0 \qquad (C.2.17)$$

3) $P(1) = p_0+p_1+p_2+p_3+p_4+1 > 0$

4) $-P(-1) = -p_0+p_1-p_2+p_3-p_4+1 > 0$

5) a) $q_1 = -5p_0-3p_1-p_2+p_3+3p_4 > 0$ and
$q_3 = -10p_0+2p_1-2p_2-2p_3+2p_4 > 0$
or
b) $q_2 = 10p_0+2p_1-2p_2-2p_3+2p_4 > 0$ and
$q_4 = 5p_0 -3p_1+p_2+p_3-3p_4 > 0$

Remark C.2:

Note that the second nonlinear condition 2) cannot be replaced by linear inequalities. This is illustrated by the following example:

$$P_\varepsilon(z) = z(z^2+1+\varepsilon)(z^2+z+1+\varepsilon) \qquad (C.2.18)$$

For $\varepsilon = 0$, the polynomial has two pairs of roots on the unit circle at $z_{1,2} = \pm j$ and $z_{3,4} = -0.5 \pm j\sqrt{0.75}$. It is stable for $\varepsilon < 0$ and unstable for $\varepsilon > 0$. By eq. (C.1.2)

$$Q_\varepsilon(w) = (1-w)^5 P\left(\frac{1+w}{1-w}\right)$$

$$= (1+w)[(1+w)^2+(1+\varepsilon)(1-w)^2][(1+w)^2+(1+w)(1-w)+(1+\varepsilon)(1-w)^2]$$

$$= (6+5\varepsilon+\varepsilon^2)+(6-5\varepsilon-3\varepsilon^2)w+(8-2\varepsilon+2\varepsilon^2)w^2+(8+2\varepsilon+2\varepsilon^2)w^3+$$

$$+(2-3\varepsilon-3\varepsilon^2)w^4+(2+3\varepsilon+\varepsilon^2)w^5$$

For $\varepsilon = 0$

$$Q_o(w) = 6+6w+8w^2+8w^3+2w^4+2w^5$$

All $q_i > 0$, i.e. this point and its neighborhood - also for positive ε - is inside the convex hull of the stability region, see eqs. (7.2.17) and (7.3.12). Thus no linear inequality can show instability. The complex root condition is

$$|\underline{X}-\underline{Y}| = \varepsilon^2(1+\varepsilon+4\varepsilon^2+4\varepsilon^3+\varepsilon^4)$$

It becomes zero for $\varepsilon = 0$, but it is positive for $\varepsilon \neq 0$ and does not indicate instability. Only the first inner condition

$$\begin{vmatrix} 1 & p_4-p_o \\ -p_o & 1-p_1 \end{vmatrix} = -\varepsilon(2+\varepsilon)$$

allows the distinction between stability for $\varepsilon < 0$ and instability for $\varepsilon > 0$.

For larger n, an increasing number of nonlinear inequalities must be satisfied for stability. In order to avoid them, one can try first to check stability only with the critical constraints

$$\begin{aligned} |\underline{X}-\underline{Y}| &= 0 \quad \text{complex root boundary} \\ \left.\begin{aligned} P(1) &= 0 \\ P(-1) &= 0 \end{aligned}\right\} &\quad \text{real root boundaries} \end{aligned} \qquad \text{(C.2.19)}$$

For this the polynomial

$$P(z) = [\underline{p}' \quad 1]\underline{z}_n \, , \quad \underline{z}_n = [1, z \dots z^n]' \qquad \text{(C.2.20)}$$

is imbedded into a family of polynomials

$$P_\alpha(z) = [\alpha \underline{p}' \; 1]\underline{z}_n \tag{C.2.21}$$

$P_\alpha(z)$ is stable for $\alpha = 0$, and there are only the three possibilities in eq. (C.2.19) of how a root can cross the unit circle with increasing α. If no boundary has been crossed up to $\alpha = 1$, then the original polynomial is stable. The critical constraints are particularly useful, if $P(z,\underline{\Theta})$ is the closed loop characteristic polynomial which is stable for nominal values $\underline{\Theta}_o$ of the plant parameters. The three stability boundaries in $\underline{\Theta}$ space are then determined by eq. (C.2.19). In root locus representation the parameter values for which the root locus crosses the unit circle are given by eq. (C.2.19).

Example:

The control loop with time delay of figure 3.21 has the characteristic polynomial

$$P(z) = 0.0125k_y + 0.3972k_y + (0.368+0.2224k_y)z^2 - 1.368z^3 + z^4 \tag{C.2.22}$$

The critical stability conditions yield the following values of k_y:

$P(1) = 0$ for $k_y = 0$
$P(-1) = 0$ for $k_y = 16.75$
$|\underline{X}-\underline{Y}| = 0$ for $k_y = -247$ and
$k_y = -5.5$ and
$k_y = 0.72$

By inserting a numerical value for k_y between 0 and 0.72, it can be verified that the control loop is stable for this interval. For $k_y = 0$, a root crosses the unit circle at $z = 1$ and for $k_y = 0.72$ a complex conjugate pole pair crosses the unit circle. This result agrees with the root locus of figure 3.22.

Because the necessary and sufficient stability conditions are nonlinear, and therefore difficult to handle in general, one is interested in simpler linear inequalities which are either only necessary, or only sufficient. Such conditions will be compiled in the next two sections.

C.3 Necessary Stability Conditions

Linear conditions in all polynomial coefficients are handled in chapter 7, see eqs. (7.3.5) and (7.3.9) and are summarized here.

Form the polynomials

$$P_i(z) = [\underline{p}_i' \quad 1]\underline{z}_n \qquad , \quad \underline{z}_n = [1 \quad z \ \dots \ z^n]'$$

$$= (z+1)^i (z-1)^{n-i} \quad , \quad i = 0, 1 \ \dots \ n$$

and from their coefficients the matrix

$$\underline{P}_n = \begin{bmatrix} \underline{p}_o' & 1 \\ \underline{p}_1' & 1 \\ \vdots & \vdots \\ \underline{p}_n' & 1 \end{bmatrix}$$

The tightest linear necessary stability conditions for a polynomial

$$P(z) = [\underline{p}' \quad 1]\underline{z}_n$$

require that all elements of the vector

$$\underline{m}' = [\underline{p}' \quad 1]\underline{P}_n \tag{C.3.1}$$

are positive. These conditions represent the convex hull of the stability region. They are identical to the conditions (C.2.12) and (C.2.13). Less tight conditions are obtained if inequalities in the individual polynomial coefficients are formulated. Figuratively speaking, the stability region is enclosed by the smallest box with edges parallel to the axes. The example of figure 7.2 shows, that this box is relatively large, due to the pronounced peaks of the stability region.

By multiplication of $P(z) = (z-z_1)\times(z-z_2)\times\ldots\times(z-z_n)$, $p_n = 1$, with $|z_i| < 1$, the following necessary conditions are obtained:

$$|p_i| < \binom{n}{i} \quad i = 0,\ 1\ \ldots\ n-1 \tag{C.3.2}$$

Some of these conditions can be tightened as follows [64.5], [64.6].

$$\begin{aligned} n = 2&: |p_o| < 1\ ,\ |p_1| < 2 \\ n = 3&: |p_o| < 1\ ,\ -1 < p_1 < 3\ ,\ |p_2| < 3 \\ n = 4&: |p_o| < 1\ ,\ |p_1| < 4\ ,\ -2 < p_2 < 6\ ,\ |p_3| < 4 \\ n = 5&: |p_o| < 1\ ,\ -3 < p_1 < 5\ ,\ |p_2| < 10\ ,\ -2 < p_3 < 10\ ,\ |p_4| < 5 \end{aligned} \tag{C.3.3}$$

For a matrix $\underline{A}$ conditions can be useful which do not require the evaluation of the characteristic polynomial $P(z) = \det(z\underline{I}-\underline{A})$. The product of the eigenvalues must be smaller (in absolute value) than one for a stable matrix, consequently

$$|\det \underline{A}| = |p_o| < 1 \tag{C.3.4}$$

The sum of the eigenvalues must be smaller (in absolute value) than n, therefore

$$|\mathrm{trace}\ \underline{A}| = |p_{n-1}| < n \tag{C.3.5}$$

From $\det(\underline{I}\pm\underline{A}^k) = (1\pm z_1^{\,k})(1\pm z_2^{\,k})\ \ldots\ (1\pm z_n^{\,k})$ with $|z_i| < 1$ the following necessary conditions are obtained

$$0 < \det(\underline{I}\pm\underline{A}^k) < 2^n\ ,\quad k = 1,\ 2,\ 3\ \ldots \tag{C.3.6}$$

For k = 1, this can be written in the form

$$0 < P(1) < 2^n\ ,\quad 0 < (-1)^n P(-1) < 2^n \tag{C.3.7}$$

C.4 Sufficient Stability Conditions

Two simple, sufficient, stability conditions are given as follows:

1. $$|p_n| > \sum_{k=0}^{n-1} |p_k| \qquad (C.4.1)$$

   Proof in [1922], [62.3].

2. If all coefficients of the polynomial

   $$P(z) = p_o + p_1 z + \dots + p_n z^n$$

   are positive, then their zeros lie in the annulus $m \le |z| \le M$, where m and M are the smallest and largest of the following numbers [51.2]:

   $$\frac{p_{n-1}}{p_n}, \frac{p_{n-2}}{p_{n-1}}, \dots \frac{p_o}{p_1} \qquad (C.4.2)$$

   For the case $m = 0$, $M = 1$, the sufficient stability condition is

   $$0 < p_o < p_1 < \dots < p_n \qquad (C.4.3)$$

For third order systems, the sufficient stability condition

$$|p_o+p_2| - 1 < p_1 < 0.5|3p_o-p_2| + 1 \qquad (C.4.4)$$

is verified in eq. (7.3.6).

Appendix D Application Examples

D.1 Aircraft Stabilization

The linearized model for the longitudinal short period mode of an F4-E aircraft is given in this appendix. A controller design for this example is described in chapter 8.

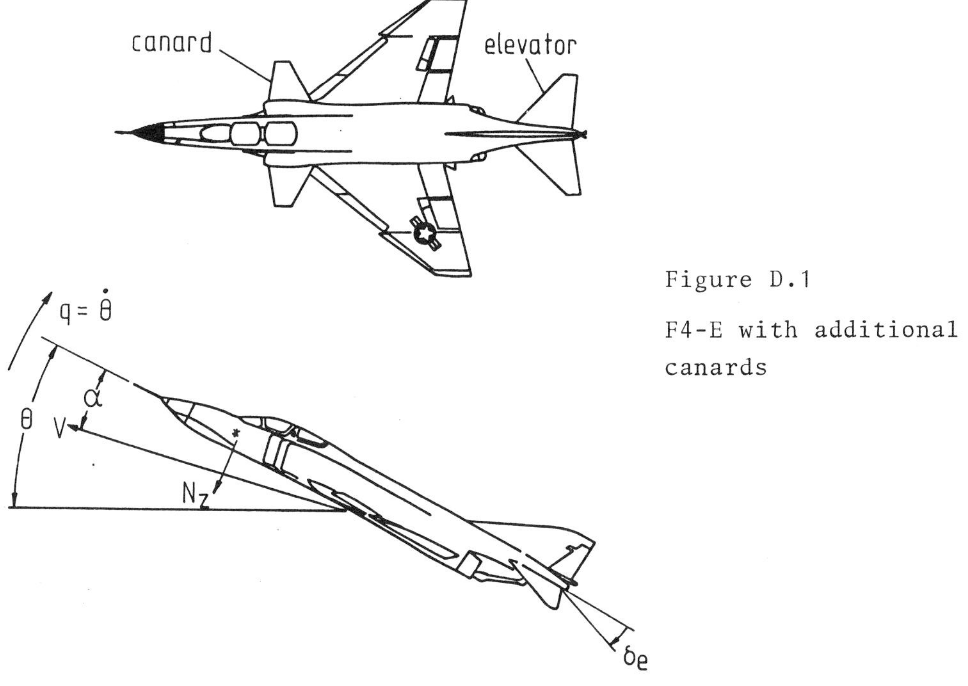

Figure D.1

F4-E with additional canards

An F4-E jet is modified as an experimental aircraft. In particular, the maneuverability is increased by additional horizontal canards. This results in a loss of longitudinal stability, however. The short period mode is unstable in subsonic flight and only weakly damped in supersonic flight.

The equations of motion can be linearized for small deviations from a stationary flight (i.e. constant altitude and velocity, small angle of attack α). In flight mechanics, it is usual to take α and the pitch rate q as state variables. Here we transform the equations to sensor coordinates, i.e. q and the normal acceleration N_z are introduced as states. This simplifies the design for robustness with respect to accelerometer failure [80.10], [81.6]. The actuator for the elevator is modelled as a low pass filter with the transfer function 14/(s+14). Its state variable is δ_e, the deviation of the elevator deflection from its trim position. δ_e is not fed back, because this would require an estimation of the trim position. For the state vector

$$\underline{x}' = [N_z \quad q \quad \delta_e] \tag{D.1.1}$$

the linearized state equation is

$$\dot{\underline{x}} = \underline{Ax} + \underline{b}u$$

$$\underline{A} = \begin{bmatrix} a_{11} & a_{12} & a_{13} \\ a_{21} & a_{22} & a_{23} \\ 0 & 0 & -14 \end{bmatrix} \qquad \underline{b} = \begin{bmatrix} b_1 \\ 0 \\ 14 \end{bmatrix} \tag{D.1.2}$$

Elevator (δ_e) and the canard rudder (δ_c) are not used independently of each other in stationary flight. The two commanded input variables are coupled by

$$\begin{aligned} \delta_{e\ com} &= u \\ \delta_{c\ com} &= -0.7\ u \end{aligned} \tag{D.1.3}$$

The factor -0.7 was chosen for minimum drag. Structural vibrations are not modelled. The bandwidth for the rigid body control is limited below the first structural mode frequency of 85 rad/sec, however, in order to avoid excitation of structural vibrations.

The possible flight conditions of this aircraft are represented in the altitude-Mach number diagram D.2 by a flight envelope.

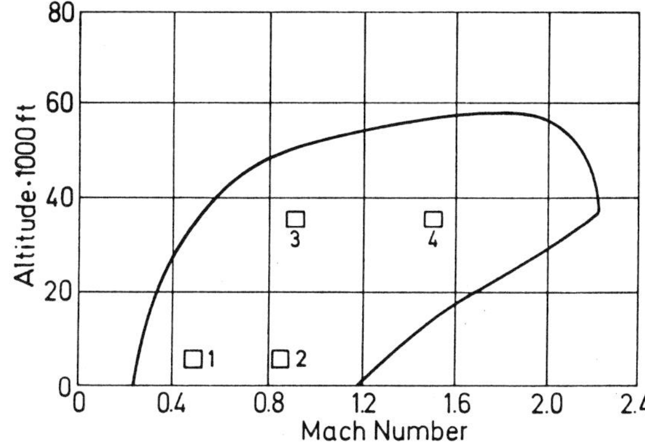

Figure D.2

Envelope of the possible flight conditions and four representative cases

Numerical values for four typical flight conditions (FC) have been taken from [73.2] and were transformed to the state equation (D.1.2). They are listed together with the respective eigenvalues s_1, s_2 in the following table.

	FC 1	FC 2	FC 3	FC 4
Mach	0.5	0.85	0.9	1.5
Altitude (ft)	5000	5000	35000	35000
a_{11}	-0.9896	-1.702	-0.667	-0.5162
a_{12}	17.41	50.72	18.11	26.96
a_{13}	96.15	263.5	84.34	178.9
a_{21}	0.2648	0.2201	0.08201	-0.6896
a_{22}	-0.8512	-1.418	-0.6587	-1.225
a_{23}	-11.39	-31.99	-10.81	-30.38
b_1	-97.78	-272.2	-85.09	-175.6
s_1	-3.07	-4.90	-1.87	-0.87±j4.3
s_2	1.23	1.78	0.56	

A common variable in flight mechanics for the evaluation of the step response is

$$C^* = (N_z + 12.43q)/C_\infty$$

The stationary value C_∞ is used for normalization. The C^*-step response should lie in the envelope shown in figure D.3. This can be obtained by a prefilter for the pilot command if the eigenvalues are suitably located according to eq. (8.5.2)

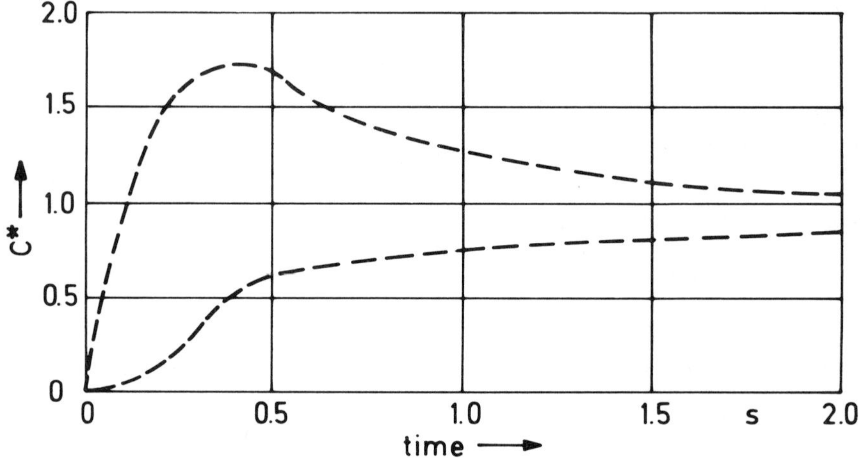

Figure D.3 Admissible envelope for the C*-response

D.2 Track-Guided Bus

A bus is guided by the electric field generated by a wire in the street. For a straight guideline and small deviations of the bus from this guideline the dynamic equations can be linearized.

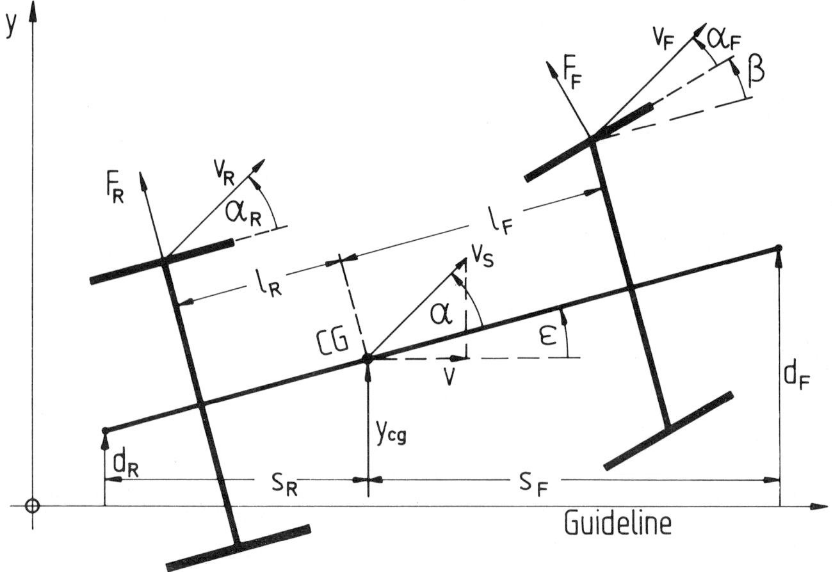

Figure D.4 Definition of variables for the track-guided bus

Input is the commanded steering angle u. It is transmitted to the true steering angle β by an actuator

$$\dot{\beta} = -4.7\beta + 4.7u \tag{D.2.1}$$

The bus has mass m, moment of inertia Θ around the center of gravity, velocity v in the direction of the guideline.

The side force of the rear wheels is

$$F_R = -\mu\delta_R\alpha_R \tag{D.2.2}$$

where the constant δ_R depends on the tire and its load, and the adhesion coefficient μ varies between $\mu = 1$ (dry road) and $\mu = 0.5$ (slippery road). For small α and $\dot{\varepsilon}$ the angle α_R is

$$\alpha_R = \alpha - \frac{\ell_R}{v} \times \dot{\varepsilon} \tag{D.2.3}$$

For the front wheels the side force is

$$F_F = -\mu\delta_F(\alpha_F+\beta) \tag{D.2.4}$$

with

$$\alpha_F = \alpha + \frac{\ell_F}{v} \times \dot{\varepsilon} \tag{D.2.5}$$

Then for small β

$$\begin{aligned} \Theta\ddot{\varepsilon} &= -F_R\ell_R + F_F\ell_F \\ &= \mu[\alpha(\delta_R\ell_R-\delta_F\ell_F)-\beta\delta_F\ell_F-\dot{\varepsilon}(\delta_R\ell_R^2+\delta_F\ell_F^2)/v] \end{aligned} \tag{D.2.6}$$

and with $\dot{y}_{cg} \approx v(\alpha+\varepsilon)$

$$m\ddot{y}_{cg} = mv(\dot{\alpha}+\dot{\varepsilon}) = F_R + F_F$$

$$mv\dot{\alpha} = [-mv+(\delta_R\ell_R-\delta_F\ell_F)\mu/v]\dot{\varepsilon}-\mu(\delta_F+\delta_R)\alpha-\mu\delta_F\beta \tag{D.2.7}$$

Rewriting eqs. (D.2.6) and (D.2.7) in state form and combining them with eq. (D.2.1) the dynamic equations in the state vector

$\underline{x} = [\alpha \quad \dot{\varepsilon} \quad \varepsilon \quad y_{cg} \quad \beta]'$ are

$$\dot{\underline{x}} = \left[\begin{array}{cc|cc|c} f_{11} & f_{12} & 0 & 0 & f_{15} \\ f_{21} & f_{22} & 0 & 0 & f_{25} \\ \hline 0 & 1 & 0 & 0 & 0 \\ v & 0 & v & 0 & 0 \\ \hline 0 & 0 & 0 & 0 & -4.7 \end{array}\right] \underline{x} + \begin{bmatrix} 0 \\ 0 \\ 0 \\ 0 \\ 4.7 \end{bmatrix} u \tag{D.2.8}$$

with $f_{11} = -\dfrac{\mu(\delta_R+\delta_F)}{mv}$, $f_{12} = -1 + \dfrac{\mu(\delta_R\ell_R-\delta_F\ell_F)}{mv^2}$

$f_{21} = -\dfrac{\mu(\delta_R\ell_R-\delta_F\ell_F)}{\Theta}$, $f_{22} = -\dfrac{\mu(\delta_R\ell_R{}^2+\delta_F\ell_F{}^2)}{\Theta v}$

$f_{15} = -\dfrac{\mu\delta_F}{mv}$, $f_{25} = -\dfrac{\mu\delta_F\ell_F}{\Theta}$

The partition of the dynamic matrix, indicated by dashed lines, factorizes the open loop characteristic polynomial into

$$P_F(s) = s^2(s+4.7)(s^2+a_1 s+a_o) \tag{D.2.9}$$

with

$$a_1 = -f_{11} - f_{22} = \frac{\mu}{vm\Theta}[\Theta(\delta_R+\delta_F)+m(\delta_R\ell_R{}^2+\delta_F\ell_F{}^2)]$$

$$a_o = f_{11}f_{22} - f_{12}f_{21}$$

$$= \frac{\mu}{\Theta}\left[\delta_R\ell_R-\delta_F\ell_F + \frac{\mu\delta_R\delta_F(\ell_F+\ell_R)^2}{mv^2}\right]$$

For the Daimler-Benz 0305 bus the data are

$\delta_F = 198000\ N/rad$, $\delta_R = 470400\ N/rad$

(for left and right wheel together)

$\ell_F = 3.67\ m$, $\ell_R = 1.93\ m$

i.e.

$$a_1 = \frac{\mu}{v} \; \frac{668400}{m} + \frac{4419035}{\Theta}$$

$$a_o = \frac{\mu}{\Theta} \; 181212 + \frac{\mu}{mv^2} \times 29208 \times 1012$$

Varying parameters are

$1\ ms^{-1} \le v \le 20\ ms^{-1}$
$0.5 \le \mu \le 1$
$m_o = 9950\ kg$ (empty bus) $\le m \le 16000\ kg$ (full bus).

The moment of inertia of the empty bus is $\Theta_o = 105700\ kgm^2$. For the nonempty bus the following assumptions are made: a) The passengers do not shift the center of gravity, b) the passengers distribute uniformly over the bus. Then for this bus

$$\Theta = \Theta_o + (m-m_o)(10.85m^2) \tag{D.2.10}$$

and $\Theta_o = 105700\ kgm^2 \le \Theta \le 171342\ kgm^2$. Thus only three uncertain parameters v, μ and m remain.

Measured variables are:

β: the steering angle,
d_F: displacement of a front antenna from the guideline, see figure D.4. Distance between sensor and center of gravity is $s_F = 6.12\ m$,
d_R: displacement of a rear antenna at $s_R = 4.99\ m$.

They are related with ε and y_{cg} by

$$\underline{y} = \begin{bmatrix} d_F \\ d_R \\ \beta \end{bmatrix} = \underline{Cx} = \begin{bmatrix} 0 & 0 & s_F & 1 & 0 \\ 0 & 0 & -s_R & 1 & 0 \\ 0 & 0 & 0 & 0 & 1 \end{bmatrix} \underline{x} \tag{D.2.11}$$

For the design of output feedback, it is helpful to write the state equations in sensor coordinates, i.e. to introduce a state vector

$$\underline{x}_s = \begin{bmatrix} \dot{d}_F \\ \dot{d}_R \\ d_F \\ d_R \\ \beta \end{bmatrix} = \begin{bmatrix} v & s_F & v & 0 & 0 \\ v & -s_R & v & 0 & 0 \\ 0 & 0 & s_F & 1 & 0 \\ 0 & 0 & -s_R & 1 & 0 \\ 0 & 0 & 0 & 0 & 1 \end{bmatrix} \begin{bmatrix} \alpha \\ \dot{\varepsilon} \\ \varepsilon \\ y_{cg} \\ \beta \end{bmatrix} = \underline{Tx} \qquad (D.2.12)$$

This transforms eq. (D.2.8) into

$$\dot{\underline{x}}_s = \underline{F}_s \underline{x}_s + \underline{g}_s u$$

where

$$\underline{F}_s = \underline{TFT}^{-1} = \begin{bmatrix} h_{11} & h_{12} & h_{13} & h_{14} & h_{15} \\ h_{21} & h_{22} & h_{23} & h_{24} & h_{25} \\ 1 & 0 & 0 & 0 & 0 \\ 0 & 1 & 0 & 0 & 0 \\ 0 & 0 & 0 & 0 & -4.7 \end{bmatrix}, \quad \underline{g}_s = \underline{Tg} = \begin{bmatrix} 0 \\ 0 \\ 0 \\ 0 \\ 4.7 \end{bmatrix} \qquad (D.2.13)$$

$$h_{11} = (s_R f_{11} + s_R s_F f_{21}/v + f_{12} v + s_F f_{22} + v)/(s_R + s_F)$$
$$h_{21} = (s_R f_{11} - s_R^2 f_{21}/v + f_{12} v - s_R f_{22} + v)/(s_R + s_F)$$
$$h_{12} = (s_F f_{11} + s_F^2 f_{21}/v - f_{12} v - s_F f_{22} - v)/(s_R + s_F)$$
$$h_{22} = (s_F f_{11} - s_R s_F f_{21}/v - f_{12} v + s_R f_{22} - v)/(s_R + s_F)$$
$$h_{14} = -h_{13} = (v f_{11} + s_F f_{21})/(s_R + s_F)$$
$$h_{24} = -h_{23} = (v f_{11} - s_R f_{21})/(s_R + s_F)$$
$$h_{15} = f_{15} v + s_F f_{25}$$
$$h_{25} = f_{15} v - s_R f_{25}$$

For this form the measurement equation is

$$\underline{y} = \begin{bmatrix} d_F \\ d_R \\ \beta \end{bmatrix} = \begin{bmatrix} 0 & 0 & 1 & 0 & 0 \\ 0 & 0 & 0 & 1 & 0 \\ 0 & 0 & 0 & 0 & 1 \end{bmatrix} \underline{x}_s = \underline{Cx}_s \qquad (D.2.14)$$

Literature

1854

Hermite, C.: Sur le nombre de racines d'une équation algébrique comprise entre des limites données. Journal Math. (1854), vol. 52, 39-51 and 397-414.

1876

Vishnegradsky, I.A.: Sur la theorie generale des regulateurs. Compt. Rend. Acad. Sci., vol. 83, Paris 1876, 318-321.

1877

Routh, E.J.: Stability of a given state of motion. London: 1877, reprinted 1975, Taylor and Francis.

1893

Ljapunov, A.M.: Problème général de la stabilité du mouvement. 1893. New York: Nachdruck Academic Press (1966) Vol. 30.

1895

Hurwitz, A.: Über die Bedingungen, unter welchen eine Gleichung nur Wurzeln mit negativen reellen Teilen besitzt. Math. Ann. (1895) vol. 46, 273-284.

1897

Gouy, G.: On an oven with constant temperature. J. Physique (1897) vol. 6, Series 3, 479-483.

1911

Orlando, L.: Sul probleme di Hurwitz relative alle parti realli delle radici di un'equazione algebraica. Math. Ann. (1911), vol. 71, 233.

1914

Liénard, A.M., Chipart, A.H.: On the sign of the real part of the roots of an algebraic equation. J. Math. Pures et Appl. (1914) vol. 10, 291-346.

1917

Schur, I.: Über Potenzreihen, die im Inneren des Einheitskreises beschränkt sind. Journal für Math. vol. 147, 205-232. (1918) vol. 148, 122-145.

1922

Cohn, A.: Über die Anzahl der Wurzeln einer algebraischen Gleichung in einem Kreise. Math. Zeitschrift (1922) vol. 14, 110-148.

1926

Fujiwara, M.: Über die algebraischen Gleichungen, deren Wurzeln in einem Kreise oder in einer Halbebene liegen. Math. Zeitschrift (1926) vol. 24, 161-169.

1944

Oldenbourg, R.C., Sartorius, H.: Dynamik selbsttätiger Regelungen. München: Oldenbourg 1944 und 1951.

1945

Bode, H.W.: Network analysis and feedback amplifier design. New York: Van Nostrand 1945.

1947

1 James, H.M., Nichols, N.B., Phillips, R.S.: Theory of servomechanisms. New York: McGraw Hill 1947.

2 Neimark, Y.I.: On the root distribution of polynomials. Dokl. Akad. Nauk. USSR (1947) vol. 58, 357.

1948

1 Evans, W.R.: Graphical analysis of control systems. Trans. AIEE (1948) vol. 67, 547-551.

2 Neimark, Y.I.: Structure of D-decomposition of the polynom space and the diagrams of Vishnegradsky and Nyquist, Dokl. Akad. Nauk. USSR (1948) vol. 59, 853.

1949

1 Tsypkin, Ya.S.: Theory of intermittent control. Avtomatika i Telemekhanika (1949) vol. 10, 189-224 and (1950) vol. 11, 300.

2 Shannon, C.E.: Communication in the presence of noise. Proc. IRE (1949) vol. 37, 10-21.

3 Marden, M.: The geometry of the zeros of a polynomial in a complex variable. New York: Americ. Math. Soc. 1949, 152-157.

1950

1 Evans, W.R.: Control system synthesis by the root locus method. Trans. AIEE (1950), vol. 69, 66-69.

1951

1 Lawden, D.F.: A general theory of sampling servo systems. Proc. IEE (1951) vol. 98, pt. IV, 31-36.

2 Perron, O.: Algebra II, Theorie der algebraischen Gleichungen. Berlin: Walter de Gruyter 1951.

1952

1 Ragazzini, J.R., Zadeh, L.H.: The analysis of sampled-data systems. Trans. AIEE (1952) vol. 71, pt. II, 225-234.

2 Barker, R.H.: The pulse transfer function and its applications to sampling servo systems. Proc. IEE (1952) vol. 99, pt. IV, 302-317.

1953

1 Linvill, W.K., Salzer, J.M.: Analysis of control systems involving digital computers. Proc. IRE (1953) vol. 41, 901-906.

2 Linvill, W.K., Sittler, R.W.: Extension of conventional techniques to the design of sampled-data systems. IRE Conv. Rec. (1953), pt. I, 99-104.

3 Birkhoff, G., MacLane, S.: A survey of modern algebra. New York: McMillan 1953.

1954

1 Salzer, J.M.: The frequency analysis of digital computers operating in real time. Proc. IRE (1954) vol. 42, 457-466.

2 Jury, E.I.: Analysis and synthesis of sampled-data control systems. Trans. AIEE (1954) vol. 73, pt. I, 332-346.

1955

1 Coddington, E.A., Levinson, N.: Theory of ordinary differential equations. New York: McGraw Hill 1955.

2 Truxal, J.G.: Automatic feedback control systems synthesis. New York: McGraw Hill 1955.

1956

1 Tsypkin, Ya.S.: Differenzengleichungen der Impuls- und Regeltechnik. Berlin: VEB-Verlag Technik 1956.

2 Simon, H.A.: Dynamic programming under uncertainty with a quadratic criterion function. Econometrica (1956), vol. 24, 74-81.

1957

1 Kalman, R.E.: Optimal nonlinear control of saturating systems by intermittent action. Wescon IRE Convention Record 1957.

2 Gould, L.A., Kaiser, J.F., Newton, G.C.: Analytical design of linear feedback controls. New York: John Wiley 1957.

3 Jury, E.I.: Hidden oscillations in sampled-data control systems. Trans AIEE (1957), 391-395.

1958

1 Jury, E.I.: Sampled-data control systems. New York: John Wiley 1958.

2 Franklin, G.F., Ragazzini, J.R.: Sampled-Data control systems. New York: McGraw Hill 1958.

3 Hahn, W.: Über die Anwendung der Methode von Ljapunov auf Differenzengleichungen. Math. Ann. (1958) vol. 136, 430-441.

4 Bertram, J.E., Kalman, R.E.: General synthesis procedure for computer control of single-loop and multi-loop linear systems. Trans. AIEE (1958) vol. 77, pt. II, 602-609.

5 Tsypkin, Ya.S.: Theory of pulse systems. Moskow: State press for Physics and Mathematical Literature 1958 (in Russian).

6 Mitrovic, D.: Graphical analysis and synthesis of feedback control systems, I. Theory and Analysis, II. Synthesis, III. Sampled-data feedback control systems. AIEE Trans. pt.II (Applications and Industry) 77 (1958), 476-503.

7 Kalman, R.E., Koepke, R.W.: Optimal synthesis of linear sampling control systems using generalized performance indexes. Trans. ASME (1958) vol. 80, 1820-1826.

1959

1 Tou, J.T.: Digital and sampled-data control systems. New York: McGraw Hill 1959.

2 Bertram, J.E., Kalman, R.E.: A unified approach to the theory of sampling systems. J. of the Franklin Inst. (1959), 405-436.

3 Wilf, H.S.: A stability criterion for numerical integration. J. Assoc. Comp. Mach. (1959) vol. 6, 363-365.

4 Gantmacher, F.R.: The theory of matrices, New York: Chelsea 1959.

1960

1 Gelfand, I.M., Schilow, G.E.: Verallgemeinerte Funktionen (Distributionen). Berlin: VEB Deutscher Verlag der Wissenschaften 1960.

2 Kalman, R.E.: A new approach to linear filtering and prediction problems. Trans. ASME, J. Basic Eng. (1960), 35-45.

3 Bertram, J.E., Kalman, R.E.: Control systems analysis and design via the second method of Ljapunov. Trans. ASME, J. Basic Engineering (1960) vol. 82, 371-400.

4 Kalman, R.E.: On the general theory of control systems. Proc. First International Congress on Automatic Control, Moskow 1960. London: Butterworth 1961, vol. 1, 481-492.

5 Rissanen, J.: Control system synthesis by analogue computer based on the generalized linear feedback concept. Proc. int. seminar on analog computation applied to the study of chemical processes. Brüssel: Presses Acadêmiques Europeènnes, Nov. 1960.

6 Kalman, R.E.: Contributions to the theory of optimal control. Bol. Soc. Mat. Mexicana (1960) vol. 5, 102-119.

1961

1 Kalman, R.E., Bucy, R.S.: New results in linear filtering and prediction theory. Trans. ASME, J. Basic Eng. (1961), vol. 83 D, 95-108.

2 Joseph, P.D., Tou, J.T.: On linear control theory. AIEE Trans. Appl. and Industry (1961) vol. 81, 193-196.

3 Blanchard, J., Jury, E.I.: A stability test for linear discrete systems in table form. Proc. IRE (1961) vol. 49, 1947-1949.

4 Ackermann, J.: Über die Prüfung der Stabilität von Abtast-Regelungen mittels der Beschreibungsfunktion. Regelungstechnik (1961) vol. 9, 467-471.

1962

1 Bass, R.W., Mendelsohn, P.: Aspects of general control theory. Final Report AFOSR 2754, 1962.

2 Jury, E.I.: On the evaluation of the stability determinants in linear discrete systems. IRE Trans. AC (1962) vol. 7, 51-55.

3 Thoma, M.: Ein einfaches Verfahren zur Stabilitätsprüfung von linearen Abtastsystemen. Regelungstechnik (1962) vol. 10, 302-306.

4 Jury, E.I.: A simplified stability criterion for linear discrete systems. Proc. IRE (1962) vol. 50, 1493-1500 and 1973.

1963

1 Zadeh, L.A., Desoer, C.A.: Linear system theory - the state space approach. New York: McGraw Hill 1963.

2 Kalman, R.E.: Mathematical description of linear dynamical systems. J. SIAM on Control (1963) vol. 1, 152-192.

3 Gilbert, E.G.: Controllability and observability in multivariable control systems. J. SIAM on Control (1963) vol. 1, 128-151.

4 Jury, E.I., Pavlidis, T.: Stability and aperiodicity constraints for systems design. IEEE Trans. CT (1963) vol. 10, 137-141.

5 Tsypkin, Ya.S.: Fundamentals of the theory of nonlinear pulse control systems. Basel: 2. IFAC Congress 1963.

6 Tsypkin, Ya.S. Die absolute Stabilität nichtlinearer Impuls-Regelsysteme. Regelungstechnik (1963) vol. 11, 145-148.

1964

1 Jury, E.I.: Theory and application of the z-transform method. New York: J. Wiley 1964.

2 Langenhop, C.E.: On the stabilization of linear systems. Proc. American Math. Soc. (1964) vol. 15, 735-742.

3 Oppelt, W.: Kleines Handbuch technischer Regelvorgänge. Weinheim: Verlag Chemie 1964.

4 Parks, P.C.: Ljapunov and the Schur-Cohn stability criterion. IEEE Trans. AC (1964) vol. 9, 121.

5 Mansour, M.: Diskussionsbemerkung zum Aufsatz: Ein Beitrag zur Stabilitätsuntersuchung linearer Abtastsysteme. Regelungstechnik (1964) vol. 12, 267-268.

6 Ackermann, J.: Eine Bemerkung über notwendige Bedingungen für die Stabilität von linearen Abtastsystemen. Regelungstechnik (1964) vol. 12, 308-309.

7 Tsypkin, Ya.S.: Sampling systems theory. New York. Pergamon Press 1964.

8 Luenberger, D.G.: Observing the state of a linear system. IEEE Trans. on Military Electronics (1964) vol. 8, 74-80.

9 Kalman, R.E.: When is a linear control system optimal? Trans. ASME (1964) vol. 86D, 51-60.

10 Šiljak, D.: Generalization of Mitrovic's method. IEEE Trans. pt. II (Applications and Industry), vol. 83 (1964), 314-320.

11 Popov, V.M.: Hyperstability and optimality of automatic systems with several control functions. Rev. Roum. Sci. Techn., Sev. Electrotechn. Energ. (1964), vol. 9, 629-690.

12 Jury, E.I., Lee, B.W.: On the stability of a certain class of nonlinear sampled-data systems. IEEE Trans. AC (1964), vol. 9, 51-61.

1965

1 Wilkinson, J.H.: The algebraic eigenvalue problem. Oxford: Clarendon Press 1965.

2 Bass, R.W., Gura, I.: High order system design via state-space considerations. Preprints Joint Automatic Control Conference (1965), 311-318.

3 Brockett, R.W.: Poles, zeros and feedback: state space interpretation. IEEE Trans. AC (1965) vol. 10, 129-135.

4 Franz, W.: Topologie, vol. 1. Berlin: Sammlung Göschen, 1965.

1966

1 Chidambara, M.R., Johnson, C.D., Rane, D.S., Tuel, W.G.: On the transformation to phase-variable canonical form. IEEE Trans. AC (1966) vol. 11, 607-610.

2 Luenberger, D.G.: Observers for multivariable systems. IEEE Trans. AC (1966) vol. 11, 190-197.

3 Ho, B.L., Kalman, R.E.: Effective construction of linear state-variable models from input/output functions. Regelungstechnik (1966) vol. 14, 545-548.

4 Ackermann, J.: Beschreibungsfunktionen für die Analyse und Synthese von nichtlinearen Abtast-Regelkreisen. Regelungstechnik (1966) vol. 14, 497-544.

1967

1 Wonham, W.M.: On pole assignment in multi-input controllable systems. IEEE Trans. AC (1967), vol. 12, 660-665.

2 Falb, P.L., Wolovich, W.A.: Decoupling in the design and synthesis of multivariable control systems. IEEE Trans. AC (1967) vol. 12, 651-659.

3 Anderson, B.D.O., Luenberger, D.G.: Design of multivariable feedback systems. Proc. IEE (1967) vol. 114, 395-399.

4 Luenberger, D.G.: Canonical forms for multivariable systems. IEEE Trans. AC (1967) vol. 12, 290-293.

5 Jury, E.I.: A note on multirate sampled-data systems. IEEE Trans. AC (1967), vol. 12, 319-320.

1968

1 Ackermann, J.: Anwendung der Wiener-Filtertheorie zum Entwurf von Abtastreglern mit beschränkter Stelleistung. Regelungstechnik (1968) vol. 16, 353-359.

2 Ackermann, J.: Zeitoptimale Mehrfach-Abtastregelsysteme. Preprints IFAC-Symposium Multi-variable Control Systems, Düsseldorf (1968) vol. I.

3 Chen, C.T., Desoer, C.A.: A proof of controllability of Jordan form of state equations. IEEE Trans. AC (1968) vol. 13, 195-196.

4 Bucy, R.S.: Canonical forms for multivariable systems. IEEE Trans. AC (1968) vol. 13, 567-569.

5 Simon, J.D., Mitter, S.K.: A theory of modal control. Information and Control (1968), vol. 13, 316-353.

1969

1 Arbib, M.A., Falb, P.L., Kalman, R.E.: Topics in mathematical system theory. New York: McGraw Hill 1969.

2 Ackermann, J.: Über die Lageregelung von drallstabilisierten Körpern. Zeitschrift für Flugwissenschaften (1969) vol. 17, 199-207.

3 Ackermann, J.: Diskussionsbeitrag zur Arbeit von O. Föllinger "Synthese von Mehrfachregelungen mit endlicher Einstellzeit". Regelungstechnik (1969) vol. 17, 170-173.

4 Chen, C.T.: Design of feedback control systems. Chicago: Proc. Nat. Electronics Conf. (1969), 46-51.

5 Gopinath, B.: On the identification of linear time-invariant systems from input-output data. Bell Syst. Tech. J. (1969) vol. 48.

6 Pearson, J.B., Ding, C.Y.: Compensator design for multivariable linear systems. IEEE Trans. AC (1969) vol. 14, 130-139.

7 Flying qualities of piloted airplanes, MIL-F-8785 B (ASG), Aug. 7, 1969.

8 Šiljak, D.D.: Nonlinear systems, the parameter analysis and design. New York: Wiley 1969.

9 Hautus, M.L.J.: Controllability and observability conditions of linear autonomous systems. Proc. Kon. Ned. Akad. Wetensch., ser. A (1969) vol. 72, 443-448.

10 Duffin, R.J.: Algorithms for classical stability problems. SIAM Review (1969), vol. 11, 196-213.

1970

1 Brunovsky, P.: A classification of linear controllable systems. Kybernetica (1970), Cislo, 173-188.

2 Bucy, R.S., Ackermann, J.: Über die Anzahl der Parameter von Mehrgrößensystemen. Regelungstechnik (1970) vol. 18, 451-452.

3 Chen, C.T.: Introduction to linear system theory. New York: Holt, Rinehart and Winston 1970.

4 Brockett, R.W.: Finite dimensional linear systems. New York: J. Wiley 1970.

5 Rosenbrock, H.H.: State-space and multivariable theory. New-York: J. Wiley 1970, London: Nelson 1970.

6 Athans, M., Levis, A.H., Schlueter, R.A.: On the behavior of optimal linear sampled-data regulators. Preprints, Joint Aut. Contr. Conf. Atlanta, 1970, 659-669.

7 Brasch, F.M., Pearson, J.B.: Pole placement using dynamic compensators. IEEE Trans. AC (1970) vol. 15, 34-43.

8 Åström, K.J., Jury, E.I., Agniel, R.: A numerical method for the evaluation of complex integrals. IEEE Trans. AC (1970), vol. 15, 468-471.

1971

1 Wiberg, D.M.: State space and linear systems. New York: McGraw Hill, Schaum's Outline Series 1971.

2 Kalman, R.E.: Kronecker invariants and feedback. Proc. Conf. on Ordinary Differential Equations. NRL Mathematics Research Center, 14.-23. Juni 1971.

3 Ackermann, J.: Die minimale Ein-Ausgangs-Beschreibung von Mehrgrößensystemen und ihre Bestimmung aus Ein-Ausgangs-Messungen. Regelungstechnik (1971) vol. 19, 203-206.

4 Ackermann, J., Bucy, R.S.: Canonical minimal realization of a matrix of impulse response sequences. Information and Control (1971) vol. 19, 224-231.

5 Jury, E.I.: The inners approach to some problems in system theory. IEEE Trans. AC (1971) vol. 16, 233-239.

6 Wolovich, W.A.: A direct frequency domain approach to state feedback and estimation. IEEE Decision and Control Conf. Miami-Florida 1971.

7 Luenberger, D.G.: An introduction to observers. IEEE Trans. AC (1971) vol. 16, 596-602.

8 Anderson, B.D.O., Moore, J.B.: Linear optimal control. Englewood Cliffs: Prentice Hall 1971.

9 Johnson, C.D.: Accomodation of external disturbances in linear regulator and servomechanism problems. IEEE Trans. AC (1971) vol. 16, 635-644.

1972

1 Ackermann, J.: Der Entwurf linearer Regelungssysteme im Zustandsraum. Regelungstechnik (1972) vol. 20, 297-300.

2 Hautus, M.L.J.: Controllability and stabilizability of sampled systems. IEEE Trans. AC (1972) vol. 17, 528-531.

3 Kwakernaak, H., Sivan, R.: Linear optimal control systems. New York: Wiley 1972.

4 Popov, V.M.: Invariant description of linear time-invariant controllable systems. SIAM J. Control (1972), vol 10, 252-264.

5 Ackermann, J.: Abtastregelung. Berlin: Springer 1972.

1973

1 Anderson, B.D.O., Jury, E.I.: A simplified Schur-Cohn test. IEEE Trans. AC (1973), vol. 18, 157-163.

2 Berger, R.L., Hess, J.R., Anderson, D.C.: Compatibility of maneuver load control and relaxed static stability applied to military aircraft. AFFDL-TR-73-33, 1973.

3 Källström, C.: Computing exp(A) and ∫exp(As)ds. Report 7309, Lund Institute of Technology, Division of Automatic Control, March 1973.

4 Jordan, D., Sridhar, B.: An efficient algorithm for calculation of the Luenberger canonical form, IEEE Trans. AC (1973), vol. 18, 292-295.

1974

1 Wolovich, W.A.: Linear Multivariable Systems. New York: Springer Verlag, 1974.

2 Jury, E.I.: Inners and Stability of Dynamic Systems. New York: Wiley 1974.

3 Lawson, C.L., Hanson, R.J.: Solving least squares problems. Englewood Cliffs: Prentice Hall, 1974.

4 Aplevich, J.D.: Direct computation of canonical forms for linear systems by elementary matrix operations, IEEE Trans. AC (1974), 124-125.

5 Youla, D.C., Bongiorno, J.J., Lu, C.N.: Single-loop feedback stabilization of linear multivariable dynamical plants. Automatica (1974), vol. 10, 159-173.

1975

1 Hirzinger, G., Ackermann, J.: Sampling frequency and controllability region. Computers and Electrical Engineering 1975, vol. 2, 347-351.

2 Forney, G.D.: Minimal bases of rational vector spaces with applications to multivariable linear systems. SIAM J. Control (1975), vol. 13, 493-520.

3 Sirisena, H.R., Choi, S.S.: Pole placement in prescribed regions of the complex plane using output feedback. IEEE Trans. AC (1975), 810-812.

4 Jury, E.I.: A new formulation for inverse z-tranformation. Trans. ASME (1975), 458.

1976

1 Ackermann, J.: Einführung in die Theorie der Beobachter. Regelungstechnik (1976), vol. 24, 217-226.

2 Smith, B.T. et al.: EISPACK guide. Lecture Notes in Computer Science, vol. 6. Berlin: Springer 1976.

3 MacGregor, J.F.: Optimal choice of the sampling interval for discrete process control. Technometrics (1976), vol. 18, 151-160.

1977

1 Schneider, G.: Über die Beschreibung von Abtastsystemen im transformierten Frequenzbereich. Regelungstechnik (1977) vol. 25, Beilage Theorie für den Anwender, Sept., Oct.

2 Hirzinger, G.: Zur Regelung getasteter Totzeitsysteme. DFVLR-Internal Report 552-77/38, Dec. 1977.

3 Grübel, G.: Beobachter zur Reglersynthese. Habilitationsschrift, Ruhr-Universität Bochum, July 1977.

4 Ackermann, J.: Entwurf durch Polvorgabe. Regelungstechnik (1977), vol. 25, 173-179 and 209-215.

5 Ackermann, J.: On the synthesis of linear control systems with specified characteristics. Automatica (1977), vol. 13, 89-94.

6 Nour Eldin, H.A.: Minimalrealisierung der Matrix-Übertragungsfunktion, Regelungstechnik (1977), vol. 25, 82-87.

7 Zeheb, E., Walach, E.: Two-parameter root-loci concepts and some applications. Int. J. Circuit Theory and Applications (1977), vol. 5, 305-315.

8 Müller, P.C., Lückel, J.: Zur Theorie der Störgrößenaufschaltung in linearen Mehrgrößenregelsystemen. Regelungstechnik (1977), vol. 25, 54-59.

1978

1 Fam, A.T., Meditch, J.S.: A canonical parameter space for linear system design. IEEE Trans. AC (1978), vol. 23, 454-458.

2 Whitbeck, R.F., Hofmann, L.G.: Digital control law synthesis in the w'domain. J. Guidance and Control (1978), vol. 1, 319-326.

3 Moler, C.B., Van Loan, C.F.: Nineteen dubious ways to compute the exponential of a matrix. SIAM Rev. (1978) vol. 20, 801-836.

4 Davison, E.J., Gesing, W., Wang, S.H.: An algorithm for obtaining the minimal realization of a linear time-invariant system and determining if a system is stabilizable-detectable. IEEE Trans. AC (1978), vol. 23, 1048-1054.

5 Willems, J., van der Voorde, H.: The return difference for discrete-time optimal feedback systems. Automatica (1978), vol. 14, 511-513.

6 Hautus, M.L.J., Heymann, M.: Linear feedback - an algebraic approach, SIAM J. Control (1978), vol. 16, 83-105.

7 Nour Eldin, H.A.: Berechnung der Matrix-Übertragungsfunktion mittels Hessenbergform. Regelungstechnik (1978), vol. 26, 134-137.

1979

1 Heymann, M.: The pole shifting theorem revisited. IEEE Trans. AC (1979), vol. 24, 479-480.

2 Kučera, V.: Discrete linear control - The polynomial equation approach. Chichester: Wiley, 1979.

3 Ackermann, J.: A robust control system design. Proc. Joint Autom. Contr. Conf. Denver, June 1979, 877-883.

4 Kreisselmeier, G., Steinhauser, R.: Systematische Auslegung von Reglern durch Optimierung eines vektoriellen Gütekriteriums. Regelungstechnik (1979), vol. 27, 76-79.

5 Kalman, R.E.: On partial realization, transfer functions and canonical forms. Acta Polytechnica Scandinavica Ma 31, Helsinki 1979, 9-32.

6 Laub, A.J.: A Schur method for solving algebraic Riccati equations. IEEE Trans. AC (1979), vol. 24, 913-921.

7 Münzner, H.F., Prätzel-Wolters, D.: Minimal bases of polynomial modules, structural indices and Brunovsky transformations, Int. J. Control (1979), vol. 30, 291-318.

1980

1 Franklin, G.F., Powell, J.D.: Digital control of dynamical systems. Reading: Addison-Wesley 1980.

2 Kailath, T.: Linear systems. Englewood Cliffs N.J.: Prentice-Hall 1980.

3 Hickin, J.: Pole assignment in single-input linear systems. IEEE Trans. AC (1980) vol. 25, 282-284.

4 Stoer, J., Bulirsch, R.: Introduction to numerical analysis. New York: Springer 1980.

5 Darenberg, W.: Einsatz eines Lenkwinkel-Beobachters bei der Spurregelung von Kraftfahrzeugen. Regelungstechnik (1980) vol. 28, 323-328.

6 Ackermann, J.: Parameter space design of robust control systems. IEEE Trans. AC. (1980), vol. 25, 1058-1072.

7 Steinhauser, R.: Systematische Auslegung von Reglern durch Optimierung eines vektoriellen Gütekriteriums - Durchführung des Reglerentwurfs mit dem Programm REMVG. DFVLR-Mitteilung 80-18.

8 Föllinger, O.: Regelungstechnik. 3. Auflage, Elitera-Verlag, Berlin 1980.

9 Klema, V.C., Laub, A.J.: The singular value decomposition: Its computation and some applications. IEEE Trans. AC (1980) vol. 25, 164-176.

10 Ackermann, J., Franklin, S.N., Chato, C.B., Looze, D.P.: Parameter space techniques for robust control system design. Report DC-39, Coordinated Science Laboratory, University of Illinois, Urbana, July 1980.

11 Forrest, A.R.: Recent work on geometric algorithms, Chapter 5 in Brodlie, K.W.: Mathematical methods in computer graphics and design. London: Academic Press, 1980.

12 Avriel, M.: Advances in Geometric Programming. New York: Plenum Press, 1980.

1981

1 Wolovich, W.A.: Multipurpose controllers for multi-variable systems. IEEE Trans. AC (1981) vol. 26, 162-170.

2 Patel, R.V.: Computation of matrix fraction descriptions of linear time-invariant systems. IEEE Trans. AC (1981) vol. 26, 148-161.

3 Ackermann, J., Kaesbauer, D.: D-Decomposition in the space of feedback gains for arbitrary pole regions. Preprints VIII IFAC Congress, Kyoto, vol. IV, 12-17.

4 Paige, C.C.: Properties of numerical algorithms related to computing controllability. IEEE Trans. AC (1981) vol. 26, 130-138.

5 O'Reilly, J.: The discrete linear time invariant time-optimal control problem - An overview. Automatica (1981), vol. 17, 363-370.

6 Franklin, S.N., Ackermann, J.: Robust flight control: a design example. AIAA J. Guidance and Control (1981) vol. 4, 597-605.

7 Gutman, S., Jury, E.I.: A general method for matrix root-clustering in subregions of the complex plane. IEEE Trans. AC (1981) vol. 26, 853-863.

8 Jury, E.I., Anderson, B.D.O.: A note on the reduced Schur-Cohn criterion. IEEE Trans. AC (1981), vol. 26, 612-614.

9 Tuschak, R.: Relations between transfer and pulse transfer functions of continuous processes, VIII IFAC Congress, Kyoto, vol. IV, 1-5.

10 Keviczky, L., Kumar, K.S.P.: On the applicability of certain optimal control methods, VIII IFAC Congress. Kyoto, vol. IV, 48-53.

11 Hammer, J., Heymann, M.: Causal factorization and linear feedback. SIAM J. Control and Optimization 19 (1981), 445-468.

12 Nour Eldin, H.A., Heister, M.: Zwei neue Zustandsdarstellungsformen zur Gewinnung von Kroneckerindices, Entkopplungsindices und eines Prim-Matrix-Produktes, Regelungstechnik (1980), 420-425 and (1981), 26-30.

13 Rattan, K.S.: Digital redesign of multiloop continuous control systems. Proc. JACC 1981, U. of Virginia, Charlottesville, June 1981.

14 Van Dooren, P.M.: The generalized eigenstructure problem in linear system theory. IEEE Trans. AC (1981), vol. 26, 111-129.

15 Walker, R.A.: Computing the Jordan form for control of dynamic systems. Stanford University, SUDAAR report 528, March 1981.

16 Varga, A.: A Schur method for pole assignment. IEEE Trans. AC (1981), vol. 26, 517-519.

17 Katz, P.: Digital control using microprocessors. London: Prentice Hall Internat. 1981.

18 Doyle, J.C., Stein, G.: Multivariable feedback design: Concepts for a classical/modern synthesis. IEEE Trans AC (1981), vol. 26, 4-16.

1982

1 Barnett, S., Scraton, R.E.: Location of matrix eigenvalues in the complex plane. IEEE Trans. AC (1982) vol. 27, 966-967.

2 Saeks, R., Murray, J.: Fractional representation, algebraic geometry, and the simultaneous stabilization problem. IEEE Trans. AC (1982), vol. 27, 895-903.

3 Ackermann, J., Türk, S.: A common controller for a family of plant models. 21st IEEE Conference on Decision and Control, Orlando, Dec. 1982, 240-244.

4 Walach, E., Zeheb, E.: Root distribution for the ellipse. IEEE Trans. AC (1982), vol. 27, 960-963. Discussion 1984, vol. 29, 383-384.

5 Vidyasagar, M., Viswanadham, N.: Algebraic design techniques for reliable stabilization. IEEE Trans. AC (1982), vol. 27, 1085-1095.

6 Zeheb, E., Hertz, D.: Complete root distribution with respect to parabolas and some results with respect to hyperbolas and sector. Int. J. Control (1982), vol. 36, 517-530.

1983

1 Sondergeld, K.P.: A generalization of the Routh-Hurwitz stability criteria and an application to a problem in robust controller design. IEEE Trans. AC (1983), vol. 28, 965-970.

2 Hinrichsen, D., Prätzel-Wolters, D.: General Hermite matrices and complete invariants of strict system equivalence, SIAM J. Control (1983), vol. 21, 289-305.

3 Olbrot, A.W.: A simple method of robust stabilization of linear uncertain systems. Proc. Conf. on Measurement and Control, Athens, Sept. 1983.

4 Åström, K.J., Wittenmark, B.: Computer-controlled systems - Theory and design. Englewood Cliffs: Prentice Hall, 1983.

5 Mellichamp, D.A. (Editor): Real-time computing with applications to data acquisition and control. New York: Van Nostrand Reinhold, 1983.

6 Glasson, D.P.: Development and application of multirate digital control. IEEE Control Systems Magazine (1983), vol. 3, 2-8.

7 Barnett, S.: Polynomials and linear control systems. New York: Marcel Dekker 1983.

8 Kreisselmeier, G., Steinhauser, R.: Application of vector performance optimization to a robust control loop design for a fighter aircraft. Int. J. of Control (1983), vol. 37, 251-284.

1984

1 Åström, K.J., Hagander, P., Sternby, J.: Zeros of sampled systems. Automatica (1984), vol. 20, 31-38.

2 Ackermann, J.: Robustness against sensor failures. Automatica (1984), vol. 20, 211-215.

3 Ackermann, J.: Finite-effect sequences - a control oriented system description. Preprints IX IFAC Congress, Budapest, vol. VIII, 59-64.

4 Grübel, G., Joos, D., Kaesbauer, D., Hillgren, R.: Robust back-up stabilization for artificial-stability aircraft. Proc. 14th ICAS Congress, Toulouse/France, 9.-14.9.84.

5 Byrnes, C.I., Anderson, B.D.O.: Output feedback and generic stabilizability. SIAM J. Control and Optimization (1984), vol. 22, 362-380.

6 Desoer, C.A., Lin, C.A.: Simultaneous stabilization of nonlinear systems. IEEE Trans. AC (1984), vol. 29, 455-457.

7 Putz, P., Wozny, M.J.: Parameter space design of control systems using interactive computer graphics. Proc. Aut. Contr. Conf. San Diego, June 1984, 105-114.

8 Jury, E.I.: On the history and progress of sampled-data systems, IEEE Centennial Press Book on History of Control and Systems Engineering, (1984).

9 Ackermann, J.: Simultaneous stabilization of two plant models. Proc. Pre-IFAC Meeting on Current Trends in Control, Dubrovnik-Cavtat, June 1984, Jurema.

10 Phillips, C.L., Nagle, H.: Digital control system analysis and design. Englewood Cliffs: Prentice Hall 1984.

1985

1 Ackermann, J., Åström, K.J., Grübel, G., Kwakernaak, H., Ljung, L., Saridis, G., Tsypkin, Ya.S.: Uncertainty and Control. Berlin: Springer 1985.

2 Putz, P., Wozny, M.J.: An interactive computer graphics environment for CAD of control systems in parameter space. Proc. 3rd IFAC Symposium on CAD in Control and Engineering Systems, Copenhagen, July/Aug. 1985.

Index